The Global Monsoon System

Research and Forecast

3rd Edition

World Scientific Series on Asia-Pacific Weather and Climate

This series will continue the collaboration with the Working Group on Tropical Meteorology Research, a component of the World Weather Research Programme, and other international weather and climate programs in publication projects.

The Editorial Board welcomes book or program/project proposals relevant to weather and climate or related fields, including interdisciplinary topics. Interested scientists please contact the Editor-in-Chief Professor Chih-Pei Chang (cpchang@nps.edu)

Published

Vol. 1: East Asia and Western Pacific Meteorology and Climate: Selected Papers of the Fourth Conference
edited by C.-P. Chang, Guoxiong Wu, Ben Jou and C. Y. Lam

Vol. 2: East Asian Monsoon
edited by C.-P. Chang

Vol. 3: Observation, Theory, and Modeling of Atmospheric Variability: Selected Papers of Nanjing Institute of Meteorology Alumni in Commemoration of Professor Jijia Zhang
edited by Xun Zhu, Xiaofan Li, Ming Cai, Shuntai Zhou, Yuejian Zhu, Fei-Fei Jin, Xiaolei Zou and Minghua Zhang

Vol. 4: Global Perspectives on Tropical Cyclones: From Science to Mitigation
edited by Johnny C. L. Chan and Jeffrey D. Kepert

Vol. 5: The Global Monsoon System: Research and Forecast (2nd Edition)
edited by C.-P. Chang, Yihui Ding, N.-C. Lau, Richard H. Johnson, Bin Wang and Tetsuzo Yasunari

Vol. 6: Climate Change: Multidecadal and Beyond
edited by C.-P. Chang, Michael Ghil, Mojib Latif and John M. Wallace

Vol. 7: Indo-Pacific Climate Variability and Predictability
edited by Swadhin K. Behera and Toshio Yamagata

Vol. 8: Aridity Trend in Northern China
edited by Congbin Fu and Huiting Mao

Vol. 9: The Global Monsoon System: Research and Forecast (3rd Edition)
edited by C.-P. Chang, H.-C. Kuo, N.-C. Lau, R. H. Johnson, B. Wang and M. C. Wheeler

WORLD SCIENTIFIC SERIES ON ASIA-PACIFIC WEATHER AND CLIMATE

Vol. 9

The Global Monsoon System

Research and Forecast

3rd Edition

edited by

Chih-Pei Chang
National Taiwan University & U.S. Naval Postgraduate School

Hung-Chi Kuo
National Taiwan University

Ngar-Cheung Lau
Chinese University of Hong Kong

Richard H. Johnson
Colorado State University

Bin Wang
University of Hawaii

Matthew C Wheeler
Australian Bureau of Meteorology

World Scientific

NEW JERSEY • LONDON • SINGAPORE • BEIJING • SHANGHAI • HONG KONG • TAIPEI • CHENNAI • TOKYO

Published by

World Scientific Publishing Co. Pte. Ltd.

5 Toh Tuck Link, Singapore 596224

USA office: 27 Warren Street, Suite 401-402, Hackensack, NJ 07601

UK office: 57 Shelton Street, Covent Garden, London WC2H 9HE

British Library Cataloguing-in-Publication Data
A catalogue record for this book is available from the British Library.

World Scientific Series on Asia-Pacific Weather and Climate — Vol. 9
THE GLOBAL MONSOON SYSTEM: RESEARCH AND FORECAST (3rd Edition)

ISBN 978-981-3200-90-6

Printed in Singapore by Mainland Press Pte Ltd.

Foreword

Monsoons, the seasonal change in the direction of the prevailing, or strongest winds of a region are known to affect the culture and lifestyle of two-thirds of the world's population. Most prominent of the world's monsoon systems is the Indian monsoon which brings the wet and dry season not only to India but to surrounding areas as well. Recent studies have even linked the fall of China's ruling dynasties to the changes in the strength of monsoon rains. Thus, the importance of monsoons on man's activities can not be underestimated.

This book is the third edition of a book series on the state of the science of monsoon research and forecasting. The series is updated approximately every five years based on the invited papers from the World Meteorological Organization's International Workshop on Monsoons (IWM). The third edition is a collection of peer-reviewed papers presented by leading monsoon experts at IWM-V (Macau, China, 28 October–1 November 2013), with manuscripts revised and updated through 2015 and early 2016. As in previous editions, the book builds on the concept that the monsoon in various parts of the globe can be viewed as components of an integrated global monsoon system, while also emphasizing that significant region-specific characteristics are present in individual monsoon regions. In addition to the regional monsoons, the current volume covers contemporary topics with emphasis on intra-seasonal oscillations, extreme weather, decadal variability, climate change, and summary of recent field experiments including CINDY/DYNAMO and the Asian Monsoon Years.

The Quadrennial International Workshop on Monsoons series has always been well supported by WMO's World Weather Research (WWRP) and World Climate Research Programmes (WCRP). The IWM-V has followed the well-established model of previous IWM's (1995 Bali, Indonesia; 2000 New Delhi, India; 2004 Hangzhou, China; and 2008 Beijing, China). The first edition of the Global Monsoon System was published in 2005 by WMO as a Technical Document (WMO TD 1266). This established the tradition of publishing the invited IWM reviews as peer-reviewed articles. The second edition of the book was published in 2011 (from the IWM in 2008) as a WMO sponsored book by World Scientific Publishing Co. and as Volume 5 of the World Scientific Series on Asia-Pacific Weather and Climate.

On behalf of WMO, I would like to express our appreciation to the Chairperson of WWRP's Monsoon Panel, Professor C.P. Chang, to all authors, reviewers and editors, to the chairpersons and members of the Fifth International Workshop on Monsoon's International and Local Organizing Committees, and to the participants of the workshop for their valuable collaboration.

The Global Monsoon System book series have become important references for researchers and forecasters of National Meteorological and Hydrological Services in the monsoon regions and research institutions. The book has also been adopted as textbooks in leading universities worldwide. I am confident that the third edition in the series will further promote the science of monsoon research and forecasting and bring the science to the attention of a wider community.

Petteri Taalas
Secretary-General, World Meteorological Organization

Contents

Foreword v

Part I Regional Monsoon

Chapter 1. SST Gradients and the East Asian Early-Summer Monsoon 3
Richard H. Johnson and Michael D. Toy

Chapter 2. Synoptic Aspects of the Summertime Australian Monsoon 13
Gareth Berry

Chapter 3. The South American Monsoon System 25
Maria A. F. Silva Dias and Leila M. V. Carvalho

Chapter 4. The West African Monsoon 35
Jean Philippe Lafore, Michela Biasutti, Peter Knippertz and Christopher Taylor

Chapter 5. The East Asian Winter Monsoon 51
Lin Wang and Mong-Ming Lu

Chapter 6. Recent Progress in Research on the Maritime Continent Monsoon 63
Shigeo Yoden, Shigenori Otsuka, Nurjanna Joko Trilaksono and Tri Wahyu Hadi

Chapter 7. Modelling Monsoons 79
Kenneth R. Sperber, Edmee Cusiner, Akio Kitoh, Carlos R. Mechoso, Aurel Moise, Wilfran Moufouma-Okia, Kathleen Schiro and Andrew G. Turner

Part II Intraseasonal Oscillations

Chapter 8. The Planetary-Scale Structure of the Madden-Julian Oscillation 105
Ángel F. Adames and John M. Wallace

Chapter 9. Simulation of the Madden-Julian Oscillation Using General Circulation Models 119
Daehyun Kim and Eric D. Maloney

Chapter 10. Real-Time Forecasting of Modes of Tropical Intraseasonal Variability: The Madden-Julian and Boreal Summer Intraseasonal Oscillations 131
Matthew C. Wheeler, Hae-Jeong Kim, June-Yi Lee and Jon C. Gottschalck

Chapter 11. Propagation Characteristics of Tropical-Subtropical Intraseasonal Variations during Boreal Summer and Their Mechanisms 139
In-Sik Kang and Min-Seop Ahn

Chapter 12. Predictability and Prediction of the Madden-Julian Oscillation: A Review on Progress and Current Status 147
June-Yi Lee, Xiouhua Fu and Bin Wang

Chapter 13. Vertical Structure and Diabatic Processes of the Madden-Julian Oscillation 161
Nicholas P. Klingaman, Xianan Jiang, Prince K. Xavier, Jon Petch, Duane Waliser and Steven J. Woolnough

Chapter 14. Global Cloud/Cloud-System Resolving Models and the Madden–Julian Oscillation 173
Tomoki Miyakawa and Tomonori Sato

Part III Extreme Weather

Chapter 15. Potential Vorticity Structure and Propagation Mechanism of Indian Monsoon Depressions 187
William R. Boos, Brian E. Mapes and Varun S. Murthy

Chapter 16. Recent Progress on Squall Lines in East China 201
Zhiyong Meng

Chapter 17. Advances in Understanding the Early-Summer Heavy Rainfall Over South China 215
Yali Luo

Chapter 18. Recent Development of Heavy Rain Research Associated with Summer Monsoon in East Asia Excited by Advances in Weather Radar 227
Hiroshi Uyeda, Ben Jong-Dao Jou and Dong-In Lee

Chapter 19. Heat Waves over East Asia: Model Simulation of Synoptic Characteristics and Projections for the 21st Century 245
Ngar-Cheung Lau

Chapter 20. Large Increasing Trend of Tropical Cyclone Rainfall in Taiwan and the Roles of Terrain and Southwest Monsoon 255
Hung-Chi Kuo, Chih-Pei Chang, Yi-Ting Yang, Yu-Han Chen, Shih-Hao Su and Lee-Yaw Lin

Chapter 21. Sudden Track Change of Typhoon Megi (2010) and Its Remote Effect on Rainfall over Taiwan — Evaluation of Variability Based on Ensemble Simulations 267
Chun-Chieh Wu and Ting-Chen Chen

Part IV Climate Change and Decadal Variability

Chapter 22. Global Monsoon: Interannual-to-Centennial Variability and Future Changes 279
Bin Wang

Chapter 23. Interdecadal Variability of the Asian Summer Monsoons 289
A. G. Turner and B. Wang

Chapter 24. Monsoon Precipitation in a Future Warmer World 303
Akio Kitoh and Hirokazu Endo

Chapter 25. Review on Studies of Air Pollution and Climate Change Interactions in Monsoon Asia 315
Congbin Fu, Aijun Ding and Jian Wu

Chapter 26. Decadal Change of East Asian Summer Monsoon: Contributions of Internal Variability and External Forcing 327
Tianjun Zhou, Fengfei Song, Kyung-Ja Ha and Xiaolong Chen

Part V Field Experiment

Chapter 27. CINDY/DYNAMO Field Campaign: Advancing Our Understanding of MJO Initiation 339
Chidong Zhang and Kunio Yoneyama

Chapter 28. Intraseasonal Air-Sea Interaction and Convection Observed in DYNAMO/CINDY/AMIE and TOGA COARE 349
Simon P. de Szoeke and James B. Edson

Chapter 29. An Overview of the Asian Monsoon Years 2007–2012 (AMY) and Multi-Scale Interactions in the Extreme Rainfall Events over the Indonesian Maritime Continent 365
Jun Matsumoto, Bin Wang, Guoxiong Wu, Jianping Li, Peiming Wu, Miki Hattori, Shuichi Mori, Manabu D. Yamanaka, Shin-ya Ogino, Hamada Jun-Ichi, Fadli Syamsudin, Toshio Koike, Katsunori Tamagawa, Eiji Ikoma, Hiroko Kinutani, Hirotaka Kamahori, Kenji Kamiguchi and Yayoi Harada

Part I
Regional Monsoon

Chapter 1

SST Gradients and the East Asian Early-Summer Monsoon

Richard H. Johnson* and Michael D. Toy†

*Department of Atmospheric Science,
Colorado State University, Fort Collins, CO 80523, USA

†National Center for Atmospheric Research,
Boulder, CO 80303, USA

johnson@atmos.colostate.edu

www.colostate.edu

The East Asian summer monsoon is characterized by strong, moist southwesterly flow over a broad expanse that includes the seas adjacent to China and the Pacific Ocean south of Japan. Relatively strong sea surface temperature (SST) gradients, particularly at the time of summer monsoon onset, exist over these waters as a result of the previous winter's cold outbreaks. The flow of warm air over cooler water can lead to low-level convergence through the suppression of vertical momentum transport in the boundary layer as a result of increased stability. Given sufficient moisture and instability, deep convection can break out.

Studies have shown that this process is operative off the coast of East Asia during the summer monsoon and frequently contributes to extensive bands of precipitation. Most of the precipitation enhancement occurs over the open ocean; however, an important exception is in the area of Taiwan where the SST gradient intersects the island during May and June. A recent study using data from the 2008 SoWMEX/TiMREX suggests that the SST gradient upstream of Taiwan can increase rainfall over the island by locally enhancing low-level convergence offshore.

1. Introduction

Flow of air across gradients in sea surface temperature (SST) can have a significant impact on the structure of the atmospheric boundary layer. There are several ways in which the impacts can be felt. First, flow across SST gradients can lead to changes in near-surface stability, surface stress and fluxes, such that the wind profile is modified (Sweet *et al.* 1981; Businger and Shaw 1984). A second effect is the modification of turbulent kinetic energy, leading to changes in the vertical transfer of momentum and boundary layer depth (Hayes *et al.* 1989; Wallace *et al.* 1989). In addition to these effects, there can be hydrostatic pressure changes across SST gradients that can drive low-level convergence in the tropics (Lindzen and Nigam 1987). These processes are reviewed in Xie (2004), Small *et al.* (2008), and Chelton and Xie (2010).

To illustrate how SST gradients can modify the vertical transfer of momentum, a schematic diagram is presented in Fig. 1 (from Chelton and Xie 2010). It shows flow of air from cool to warm SSTs across a meandering SST frontal boundary. Winds are stronger on the warm side of the front due to enhanced buoyancy fluxes and downward mixing of higher momentum air from aloft. Where the flow is perpendicular to the front, there is low-level convergence of the flow and the sea surface stress. Where the flow is parallel to the front, positive vorticity is generated as well as a positive curl of the surface stress. Changes in the surface wind stress can

The Global Monsoon System: Research and Forecast (3rd Edition)
Edited by C. P. Chang et al.

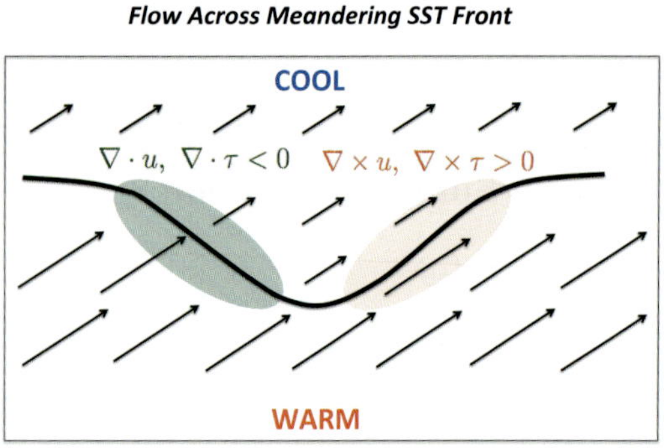

Fig. 1. Idealized depiction of surface flow across a meandering SST front (adapted from Chelton and Xie 2010). Green shaded area indicates convergence of both the surface flow and surface stress τ. Red shaded area indicates positive vorticity and positive curl of the surface stress.

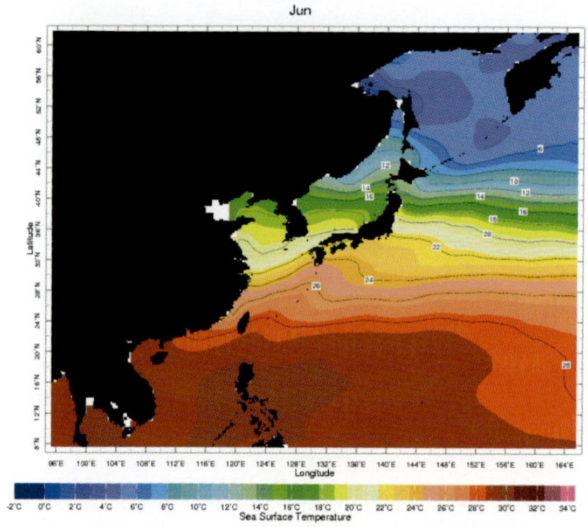

Fig. 2. 1971–2000 climatological mean June SST from Lamont-Doherty Earth Observatory website (http://iridl.ldeo.columbia.edu/) based on 1° Reynolds *et al.* (2002) SST data. White arrows schematically represent typical near-surface monsoonal airflow.

feed back and alter the properties of the upper ocean (Small *et al.* 2008).

The flow configuration depicted in Fig. 1 is representative of that climatologically occurring over East Asian coastal seas and northwest Pacific Ocean during the summer monsoon. Namely, southwesterly monsoon flow passes over warm waters in the south toward cooler waters in the north. This situation for the month of June is illustrated in Fig. 2. Airflow extends from the southern South China Sea and western Pacific Ocean to the Japan/Korea region, where the SST is considerably lower. Although there is not a sharp frontal boundary like that depicted in Fig. 1, there is a flow across an SST gradient such that air mass transformation and modification of the boundary layer occur. The condition of warm-to-cool climatological airflow first begins in mid-May with the onset of the East Asian summer monsoon. The air-sea interaction processes just described will be reviewed for the region of the East Asian summer monsoon.

2. Previous Investigations of Air-Sea Interaction Impacts on Rainfall

The advent of global, satellite-based measurements of surface winds, SST, and precipitation over the past two decades has permitted the documentation of a myriad of fascinating and complex interactions between the ocean and atmosphere (Xie 2004; Small *et al.* 2008; Chelton and Xie 2010). Particularly prominent are situations where flow across relatively strong SST gradients modifies the boundary layer, cloudiness, and precipitation, such as the eastern Pacific cold tongue (de Szoeke and Bretherton 2004), Tropical Instability Waves in the Pacific Ocean (Chelton *et al.* 2001; Liu *et al.* 2000; Hashizume *et al.* 2002), the Gulf Stream (Minobe *et al.* 2008), and the Kuroshio and its extension (Nonaka *et al.* 2003; Xu *et al.* 2011; Miyama *et al.* 2012).

An investigation has been carried out by Tokinaga *et al.* (2009) of air-sea interaction processes associated with warm air flowing over cooler water in the northwest Pacific Ocean, as depicted in Fig. 2. They used ship and satellite observations, and reanalysis data to show that strong SST gradients associated with the Kuroshio Extension impact clouds and precipitation over an extensive area. Results

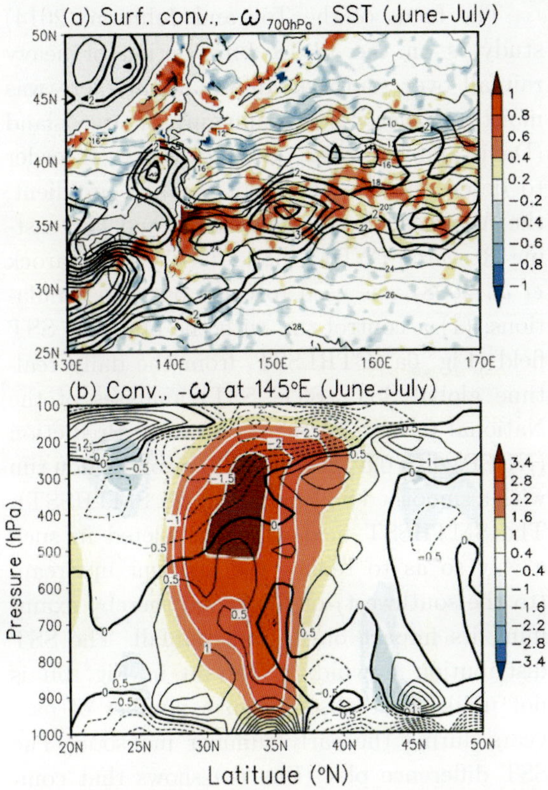

Fig. 3. June–July climatology: (a) QuikSCAT surface convergence (color in 10^{-5} s^{-1}) and AMSR-E SST (thin contours at 1°C intervals) superimposed on negative pressure vertical velocity of JRA-25 at 700 hPa (thick contours greater than 2×10^{-2} Pa s^{-1} at 1×10^{-2} Pa s^{-1} intervals) and (b) vertical section of negative pressure velocity (color in 10^{-2} Pa s^{-1}) and horizontal convergence (contours at 0.5×10^{-6} s^{-1}) along 145°E. From Tokinaga *et al.* (2009).

of their study are shown in Fig. 3. Strong convergence and upward motion at 700 hPa can be seen at the southern edge of the strong SST gradient (Fig. 3a). As evident from Fig. 3b, the surface convergence produces deep upward motion in this convectively unstable region of the western Pacific. In another study, Tanimoto *et al.* (2009) showed that warm southerlies across the Kuroshio Extension can lead to fog formation over the cooler water.

3. SST Gradients Over the Northern South China Sea

The onset of the East Asian summer monsoon over the South China Sea (SCS) typically occurs around mid-to-late May (Wu and Zhang 1998; Ding and Chan 2005). It is marked by a reversal of the winds from north-easterlies to southwesterlies over the northern SCS, where a strong SST gradient exists as a remnant of the winter monsoon (Chu and Chang 1997). Xie *et al.* (2002) showed that for the East China Sea, the cooling is enhanced due to the shallow offshore waters, which should also be a factor for the northern SCS. The mean conditions in this region for May are shown in Fig. 4. A fairly strong SST front exists over the northern SCS with convergence of the surface winds quite noticeable along the boundary. Although the SST moderates somewhat as the monsoon season progresses, an SST gradient still exists in June, as shown Fig. 5.

The mean surface winds for June are generally southerly and a deceleration of the flow is evident as air crosses the strong SST gradient southwest and northeast of Taiwan. The reduction in wind speed, and hence increase in low-level convergence, associated with the passage of the air from warm to cool water

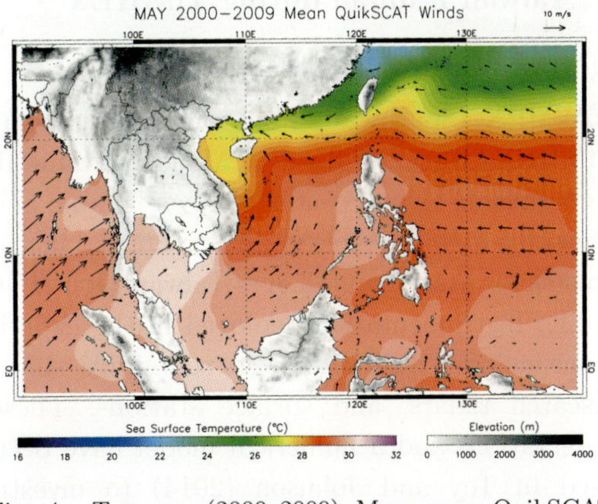

Fig. 4. Ten-year (2000–2009) May mean QuikSCAT winds and SST (1° data; Reynolds *et al.* 2002). Note convergence along SST front over northern SCS.

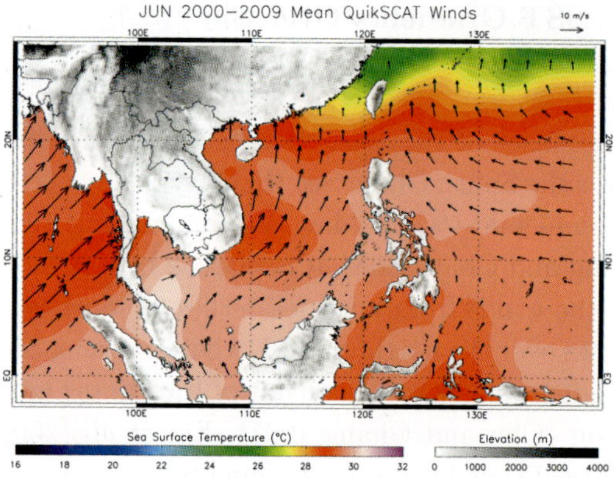

Fig. 5. As in Fig. 4, but for June.

is consistent with the mechanism proposed by Wallace *et al.* (1989). The existence of enhanced low-level convergence upstream of Taiwan during the early summer monsoon or Mei-yu period when the atmosphere is convectively unstable suggests a possible impact of strong SST gradients upwind of Taiwan on island rainfall. The Mei-yu period is a time of very heavy rainfall over Taiwan (Chen *et al.* 2007). The possibility of air-sea interaction affecting Taiwan rainfall has been recently investigated by Toy and Johnson (2014). This work is the subject of the next section.

4. Effects of SST Gradients on Taiwan Rainfall during TiMREX

In May-June 2008, a field campaign — SoWMEX (Southwest Monsoon Experiment)/TiMREX (Terrain-influenced Monsoon Rain-fall Experiment) — was conducted over Taiwan and the surrounding region to investigate the mechanisms for heavy rainfall over Taiwan (Jou *et al.* 2011). Instrumentation deployed in the experiment included aircraft, special soundings, ships, research radars, and surface stations. These observations and a numerical model have been used in Toy and Johnson (2014) to investigate the potential impact of the SST gradient upstream of Taiwan on island rainfall.

The focus of the Toy and Johnson (2014) study is on the 13–18 June period of heavy rainfall over southern Taiwan when there was moist, southerly flow impinging on the island (Davis and Lee 2012; Xu *et al.* 2012). In order to determine the effects of the SST gradient, the Advanced Research Weather and Forecasting Research (WRF) model (ARW; Skamarock *et al.* 2008) was used to carry out two simulations: (1) a control run with the observed SST field (Fig. 6a, CTRLSST, from the daily real-time global SST (RTG SST) analyses of the National Centers for Environmental Prediction (NCEP) on a 0.5° grid) and (2) a companion run with a smoothed SST field (Fig. 6b, SMTHSST). The SMTHSST field was constructed in such a way so as to reduce the gradient upstream (to the southwest) of the island thereby examining its impact on Taiwan rainfall. The SST distribution upwind of Taiwan in Fig. 6b is not unlike that observed in a number of past years during the early summer monsoon. The SST difference plot (Fig. 6c) shows that compared to SMTHSST, the CTRLSST distribution is characterized by cooler water over the northern Taiwan Strait and warmer water off the south tip of Taiwan.

The WRF is used with a convection-permitting horizontal grid spacing of 3 km with a stretched vertical grid having 60 levels from the surface to 50 hPa. The following parameterizations are used: the WRF Single-Moment 6-Class Microphysics scheme (Hong and Lim 2006), Yonsei University boundary layer (Hong *et al.* 2006, Dudhia shortwave radiation, and rapid radiative transfer model (Mlawer *et al.* 1997) for longwave radiation. The control run was initialized with the NCEP Global Operational Analysis fields on the boundaries shown in Fig. 6 starting at 0000 UTC on 12 June with boundary values updated every six hours, and run up until 18 June 2008. An additional six runs were initialized with random perturbations using the WRF Model data assimilation (WRFDA) system (Barker *et al.* 2012). Results

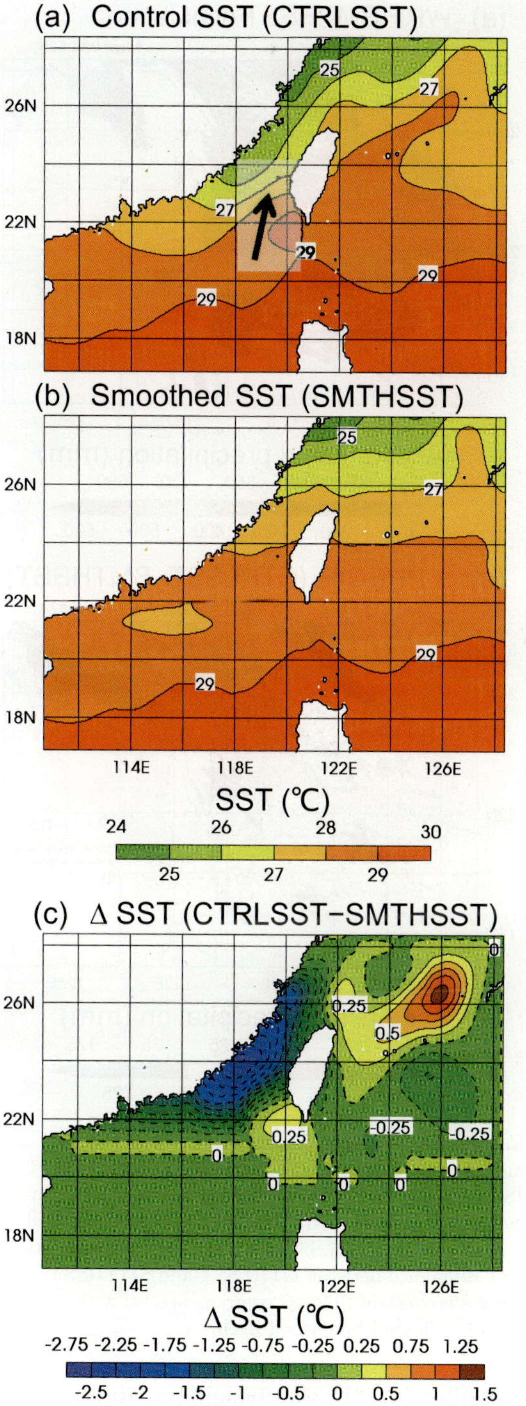

Fig. 6. (a) Control SST for 13–18 Jun and (b) Smoothed-SST used in simulations, and (c) SST difference between the two. Arrow in (a) depicts direction of near-surface flow.

shown are the means from the 7-member ensemble runs. Simulations were carried out with and without the island of Taiwan in the model in order to separate out the effects of the SST gradient from flow blocking by the high terrain of the island. For the sake of brevity, the no-Taiwan results will not be shown here.

The accumulated rainfall over the five-day 13–18 June period from the control simulation is shown along with the Tropical Rainfall Measuring Mission (TRMM) rainfall estimate in Fig. 7. In addition to the large rainfall totals along the south coast of China and over the north-central SCS, the model shows a rainfall maximum along the southwest coast of Taiwan (Fig. 7a). The distribution of the rainfall over the entire domain agrees quite well with TRMM 3B42 estimates (Fig. 7b); however, TRMM places the rainfall maximum offshore whereas the model places it onshore. An operational rainfall product based on composite radar data prepared by Taiwan (not shown) gives better agreement with the model results in having the maximum rainfall just inland of the coastline. The TRMM product, which uses a merger of microwave and outgoing longwave radiation data, likely places the maximum rainfall too far to the southwest as a result of northeasterly flow aloft advecting cirrus from the coastal precipitation offshore. The north end of Taiwan received considerably less precipitation than the south end during this period of the experiment.

The simulated rainfall from the SMTHSST run, along with the rainfall difference (CTRLSST–SMTHSST), is shown in Fig. 8. The general pattern of rainfall over the large domain in the SMTHSST (Fig. 8a) is similar to that for the CTRLSST (Fig. 7a), as might be expected. However, there are important differences between the two simulations in several areas related to the different SST distributions (Fig. 8b). With respect to Taiwan, there is an increase in rainfall up to ∼100 mm along the southwest coast in the control simulation, indicating that the stronger SST gradient

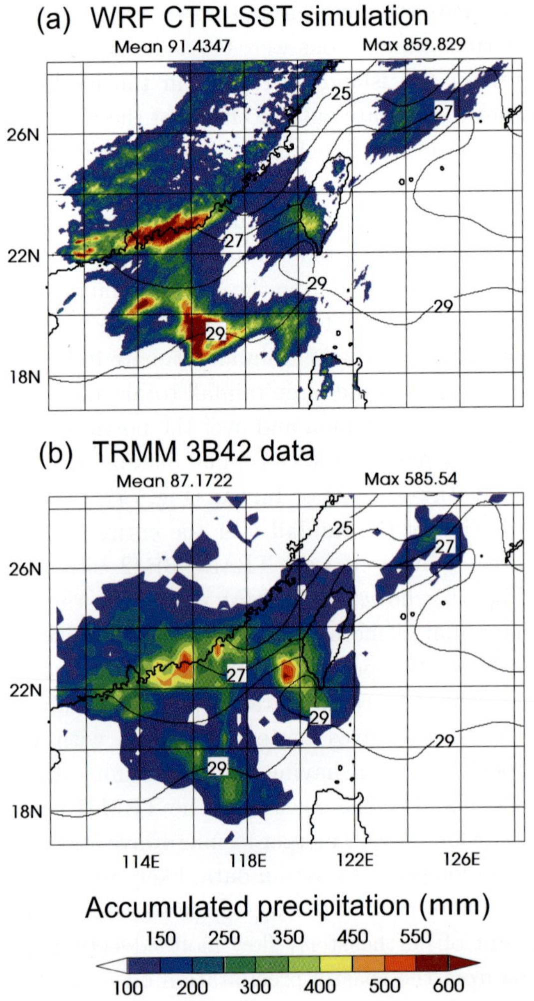

Fig. 7. Accumulated rainfall over model domain from 0000 UTC 13 Jun through 0000 UTC 18 Jun from (a) WRF CTRLSST simulation and (b) TRMM 3B42 data. The SST (°C) is overlaid (contours).

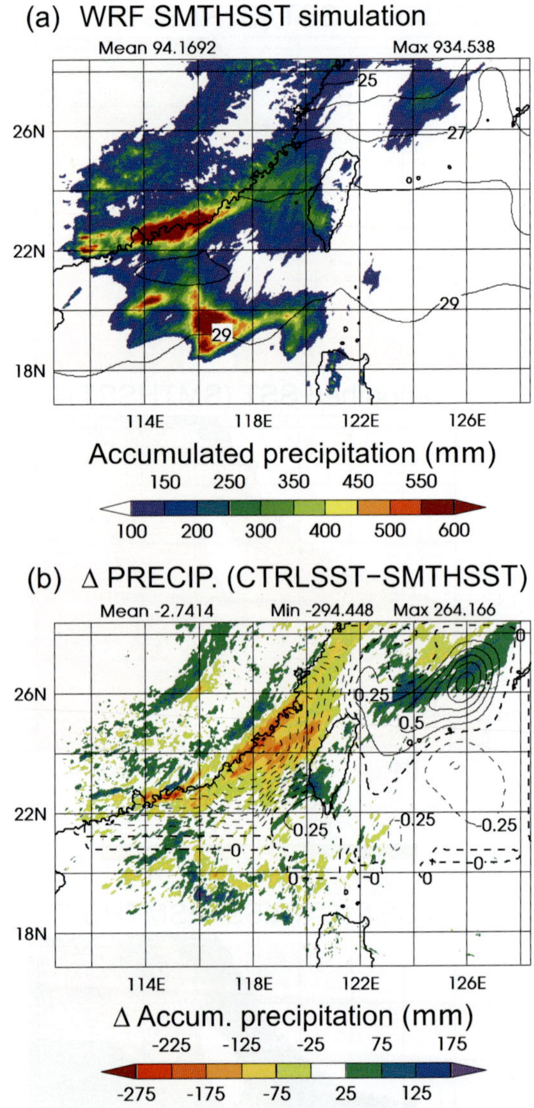

Fig. 8. (a) Accumulated rainfall (color shading) over model domain from 0000 UTC 13 Jun through 0000 UTC 18 Jun from WRF SMTHSST simulation. SST (°C) is overlaid (contours). (b) Rainfall difference (color shading) between CTRLSST and SMTHSST simulations. SST difference between CTRLSST and SMTHSST simulations is overlaid (0.25°C contour interval with negative values indicated by dashed lines).

is enhancing rainfall there. On the other hand, there is a reduction in rainfall over the central and northern Taiwan Strait as a result of the cooler water in the control simulation. The increased rainfall in the CTRLSST simulation over and just offshore the southwest coast of Taiwan is associated with increased convergence in that location (not shown).

In order to understand how the SST gradient increases precipitation in this region, it is necessary to look at the modification of the boundary layer by the changing conditions of the underlying ocean. To do so, plots of the vertical structure of the lower atmosphere are presented for the CTRLSST and SMTHSST simulations (Fig. 9). The longitude of this plot is 120°E,

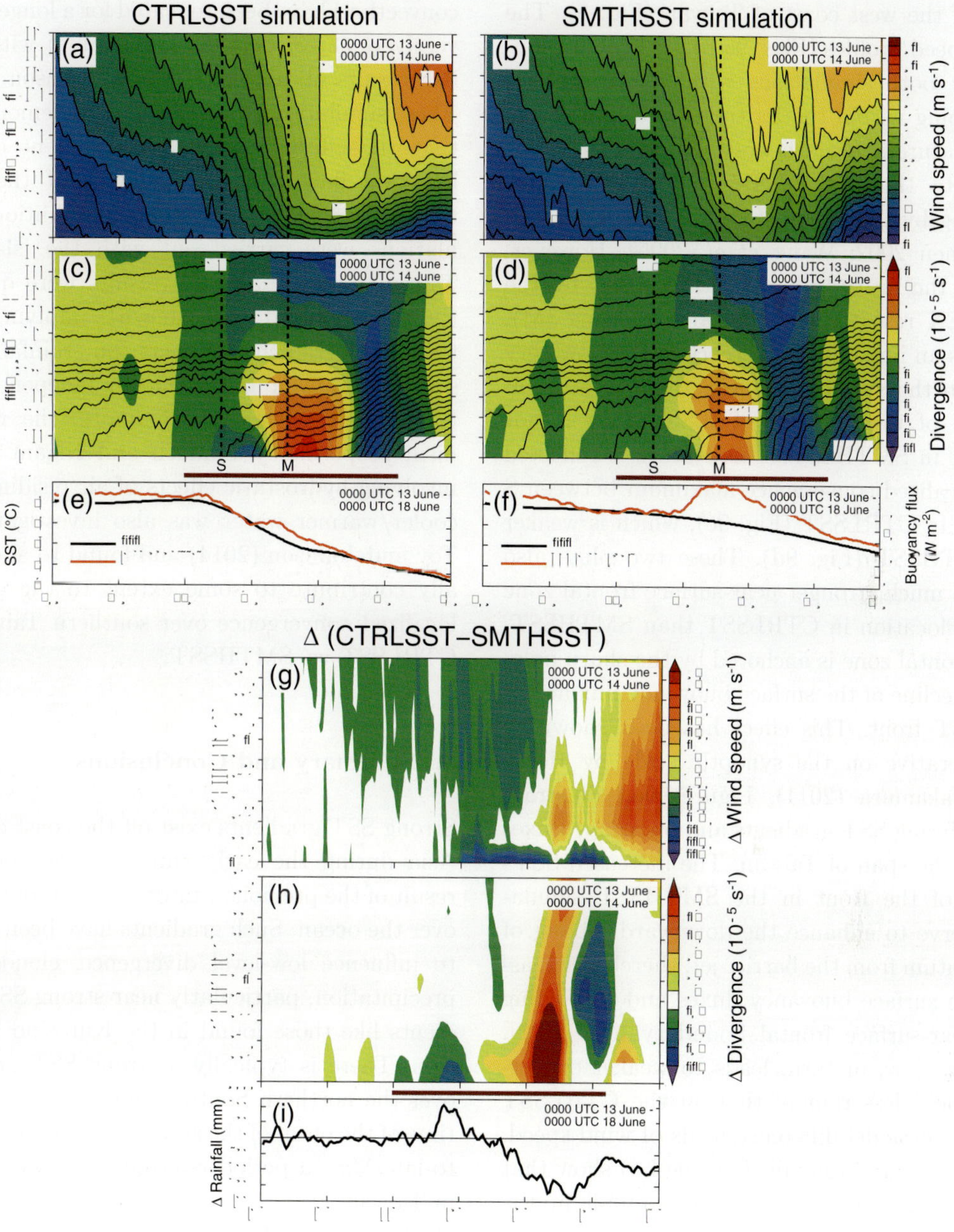

Fig. 9. Vertical cross sections along 120°E of mean 0000 UTC 13 Jun–0000 UTC 14 Jun (a),(b) wind speed, (c),(d) horizontal divergence (color shading) and potential temperature (K) (contours, with 0.2 K interval for $\theta \leq 302\,\text{K}$), and difference (CTRLSST–SMTHSST) in (g) wind speed and (h) horizontal divergence. (e),(f) Mean 0000 UTC 13 Jun–0000 UTC 18 Jun buoyancy flux (red curve) and SST (black curve) along 120°E. (i) Difference (CTRLSST–SMTHSST) in accumulated rainfall along 120°E from 0000 UTC 13 Jun through 0000 UTC 18 Jun. Letters "S" and "M" indicate latitude of ship (99810) and Makung (46734) radiosonde stations referred to in Toy and Johnson (2014). The bold brown line indicates the latitude range of the island of Taiwan for reference.

just off the west coast of Taiwan (Fig. 8). The wind speed profiles (Figs. 9a,b) are broadly similar for both simulations with a low-level jet developing within the latitude band of Taiwan. This feature is a south-westerly barrier jet that forms to the west of Taiwan owing to flow blocking by the island (Li and Chen 1988; Yeh and Chen 2002; Wang *et al.* 2005). However, within the latitude band encompassed by the TiMREX research ship (S) and Makung (M), there is an important difference in the flow very close to the surface. Namely, there is a deceleration of the flow in CTRLSST that is not present in SMTHSST. This feature is reflected in a localized convergence maximum between S and M in CTRLSST (Fig. 9c), which is weaker in SMTHSST (Fig. 9d). These two plots also show a much stronger near-surface frontal zone in this location in CTRLSST than SMTHSST. This frontal zone is anchored by the sharp poleward decline of the surface buoyancy flux across the SST front. This effect has been shown to be operative on the synoptic scale by Hotta and Nakamura (2011). Figures 9e,f illustrate the differing SST gradients and buoyancy fluxes across the span of Taiwan. The increased SSTs north of the front in the SMTHSST simulation serve to enhance the downward mixing of momentum from the barrier jet, thereby increasing the surface buoyancy fluxes and weakening the near-surface frontal boundary near 23°N. This process, in turn, leads to weaker convergence and less rainfall than in the CTRLSST case. The model difference fields of wind speed, divergence, and rainfall (Figs. 9g,h,i) show that the increased convergence and rainfall in the CTRLSST occurs at the leading (southern) edge of the SST front. The greater low-level stability and reduced downward mixing of momentum in the CTRLSST simulation near 23–24°N support the arguments of Hayes *et al.* (1989) and Wallace *et al.* (1989) for SST gradient impacts on boundary layer convergence.

The above findings illustrated for 13–14 June in Fig. 9 (a period prior to the breakout of deep convection) have been repeated for a longer, convectively active period (13–17 June) with similar results obtained (Toy and Johnson 2014). These simulations were designed to demonstrate that the effects of the SST gradient on the low-level flow were independent of the existence of precipitating systems. In addition, simulations were carried out with the island of Taiwan removed and the results were qualitatively unchanged (not shown), implying that the SST gradient impacts on rainfall over southern Taiwan are essentially independent of topographic effects. Furthermore, the mechanism proposed by Lindzen and Nigam (1987) involving hydrostatic effects of air residing over cooler/warmer water was also investigated by Toy and Johnson (2014) and found to additionally contribute to some extent to the greater localized convergence over southern Taiwan in CTRLSST vs. SMTHSST.

5. Summary and Conclusions

Strong SST gradients exist off the coast of East Asia during the early summer monsoon as a result of the previous winter's cold air outbreaks over the ocean. Such gradients have been shown to influence low-level divergence, clouds, and precipitation, particularly near strong SST gradients like those found in the Kuroshio Extension. There is typically a strong SST gradient over the northern South China Sea around the time of the onset of the summer monsoon in mid-to-late May, a period referred to as the Mei-yu in Taiwan.

The effect of this SST gradient on rainfall over Taiwan during the 2008 SoWMEX/TiMREX field campaign has been investigated using the Advanced Research WRF (Toy and Johnson 2014). Through a comparison of rainfall distributions in two experiments, one with the observed SST and one with a smoothed SST distribution, it is found that for the SoWMEX/TiMREX mid-June heavy rainfall period that the stronger SST gradient increases rainfall over southern Taiwan by enhancing low-level convergence along the SST front. A reduction in vertical mixing of higher-momentum air aloft to the surface occurs as the south-to-southwesterly flow crosses from warm to cooler water, consistent with the mechanism proposed by Wallace *et al.* (1989). Since there is year-to-year variability in the strength of the SST gradient over the northern South China Sea as a result varying intensity of previous winter monsoons and other factors, these findings have important implications regarding seasonal prediction of the intensity and location of Mei-yu precipitation over Taiwan.

Acknowledgments

This research has been supported by the National Science Foundation under Grant AGS-0966758. We acknowledge high-performance computing support from Yellowstone (ark:/85065/d7wd3xhc) provided by NCAR's Computational and Information Systems Laboratory, sponsored by the National Science Foundation. We also thank Brian McNoldy for creating several of the figures, Rick Taft for assistance with preparation of the article, and three anonymous reviewers for their helpful comments on the manuscript.

References

Barker, D. and coauthors, 2012: The Weather Research and Forecasting Model's Community Variational/Ensemble Data Assimilation System: WRFDA. *Bull. Amer. Meteor. Soc.*, **93**, 831–843.

Businger, J. A. and W. J. Shaw, 1984: The response of the marine boundary layer to mesoscale variations in sea-surface temperature. *Dyn. Atmos. Oceans*, **8**, 267–281.

Chelton, D. B. and S.-P. Xie, 2010: Coupled ocean-atmosphere interaction at oceanic mesoscales. *Oceanography*, **23**, 52–69.

Chen, C.-S., Y.-L. Chen, C.-L. Liu, P.-L. Lin, and W.-C. Chen, 2007: Statistics of heavy rainfall occurrences in Taiwan. *Wea. Forecasting*, **22**, 981–1002.

Chu, P. C. and C.-P. Chang, 1997: South China Sea warm pool in boreal spring. *Adv. Atmos. Sci.*, **14**, 195–206.

Daisuke Hotta and Hisashi Nakamura, 2011: On the significance of the sensible heat supply from the ocean in the maintenance of the mean baroclinicity along storm tracks. *J. Climate*, **24**, 3377–3401.

Davis, C. A. and W.-C. Lee, 2012: Mesoscale analysis of heavy rainfall episodes from SoWMEX/TiMREX. *J. Atmos. Sci.*, **69**, 521–537.

de Szoeke, S. P. and C. S. Bretherton, 2004: Quasi-Lagrangian large eddy simulations of cross-equatorial flow in the East Pacific atmospheric boundary layer. *J. Atmos. Sci.*, **61**, 1837–1858.

Ding, Y. and J. C. L. Chan, 2005: The East Asian Summer Monsoon: An overview. *Meteor. Atmos. Phys.*, **89**, 117–142.

Hashizume, H., S.-P. Xie, M. Fujiwara, M. Shiotani, T. Watanabe, Y. Tanimoto, W. T. Liu, and K. Takeuchi. 2002: Direct observations of atmospheric boundary layer response to slow SST variations on the Pacific equatorial front. *J. Climate*, **15**, 3379–3393.

Hayes, S. P., M. J. McPhaden, and J. M. Wallace, 1989: The influence of sea-surface temperature on surface wind in the eastern equatorial Pacific: Weekly to monthly variability. *J. Climate*, **2**, 1500–1506.

Hong, S.-Y., Y. Noh, and J. Dudhia, 2006: A new vertical diffusion package with an explicit treatment of entrainment processes. *Mon. Wea. Rev.*, **134**, 2318–2341.

Hong, S.-Y. and J.-O. J. Lim, 2006: The WRF single-moment 6-class microphysics scheme (WSM6). *J. Korean Meteor. Soc.*, **42**, 129–151.

Jou, B. J.-D., W.-C. Lee, and R. H. Johnson, 2011: An overview of SoWMEX/TiMREX. *The Global Monsoon System: Research and Forecasts, 2nd Edition*, C.-P. Chang *et al.*, Eds. World Scientific, pp. 303–318.

Li, J. and Y.-L. Chen, 1998: Barrier jets during TAMEX. *Mon. Wea. Rev.*, **126**, 959–971.

Lindzen, R. S. and S. Nigam, 1987: On the role of sea surface temperature gradients in forcing low-level winds and convergence in the tropics. *J. Atmos. Sci.*, **44**, 2418– 2436.

Liu, W. T., X. Xie, P. S. Polito, S.-P. Xie, and H. Hashizume, 2000: Atmospheric manifestation of tropical instability waves observed by

QuikSCAT and Tropical Rain Measuring Mission. *Geophys. Res. Lett.*, **27**, 2545–2548.

Minobe, S., A. Kuwano-Yoshida, N. Komori, S.-P. Xie, and R. J. Small. 2008: Influence of the Gulf Stream on the troposphere. *Nature*, **452**, 206–209.

Miyama, T., M. Nonaka, H. Nakamura, and A. Kuwano-Yoshida, 2012: A striking early-summer event of a convective rainband persistent along the warm Kuroshio in the East China Sea. *Tellus A*, **64**, 18962.

Mlawer, E. J., S. J. Taubman, P. D. Brown, M. J. Iacono, and S. A. Cough, 1997: Radiative transfer for inhomogeneous atmospheres: RRTM, a validated correlated-k model for the longwave. *J. Geophys. Res.*, **102**(D14), 16663–16682.

Nonaka, M. and S.-P. Xie, 2003: Co-variations? of sea surface temperature and wind over the Kuroshio and its extension: Evidence for ocean-to-atmospheric feedback. *J. Climate*, **16**, 1404–1413.

Reynolds, R. W., N. A. Rayner, T. M. Smith, D. C. Stokes, and W. Wang, 2002: An improved in situ and satellite SST analysis for climate. *J. Climate*, **15**, 1609–1625.

Skamarock, W. C., *et al.*, 2008: A description of the Advanced Research WRF Version 3. NCAR Tech. Note NCAR/TN-475+STR, National Center for Atmospheric Research, 125 pp.

Small, R. J. and Coauthors, 2008: Air–sea interaction over ocean fronts and eddies. *Dyn. Atmos. Oceans*, **45**, 274–319.

Sweet, W., R. Fett, J. Kerling, and P. La Violette, 1981: Air-sea interaction effects in the lower troposphere across the north wall of the Gulf Stream. *Mon. Wea. Rev.*, **109**, 1042–1052.

Tanimoto, Y., S.-P. Xie, K. Kai, H. Okajima, H. Tokinaga, T. Murayama, M. Nonaka, and H. Nakamura, 2009: Observations of marine atmospheric boundary layer transitions across the summer Kuroshio Extension. *J. Climate*, **22**, 1360–1374.

Tokinaga, H., Y. Tanimoto, S.-P. Xie, T. Sampe, H. Tomita, and H. Ichikawa, 2009: Ocean frontal effects on the vertical development of clouds over the western North Pacific: In situ and satellite observations. *J. Climate*, **22**, 4241–4260.

Toy, M. D. and R. H. Johnson, 2014: The influence of an SST front on a heavy rainfall event over coastal Taiwan during TiMREX. *J. Atmos. Sci.*, in press.

Wallace, J. M., T. P. Mitchell, and C. Deser, 1989: The influence of sea-surface temperature on surface wind in the eastern equatorial Pacific: Seasonal and interannual variability. *J. Climate*, **2**, 1492–1499.

Wang, C.-C., G. T.-J. Chen, T.-C. Chen, and K. Tsuboki, 2005: A numerical study on the effects of Taiwan topography on a convective line during the Mei-yu season. *Mon. Wea. Rev.*, **133**, 3217–3242.

Wu, G., Y. Zhang, 1998: Tibetan Plateau forcing and the timing of the monsoon onset over South Asia and the South China Sea. *Mon. Wea. Rev.*, **126**, 913–927.

Xie, S.-P., 2004: Satellite observations of cool ocean–atmosphere interaction. *Bull. Amer. Meteor. Soc.*, **85**, 195–208.

Xie, S.-P., J. Hafner, Y. Tanimoto, W. T. Liu, H. Tokinaga, and H. Xu, 2002: Bathymetric effect on the winter sea surface temperature and climate of the Yellow and East China Seas, *Geophys. Res. Lett.*, **29**(24), 2228, doi:10.1029/2002GL015884.

Xu, H., M. Xu, S.-P. Xie, and Y. Wang, 2011: Deep atmospheric response to the spring Kuroshio Current over the East China Sea. *J. Climate*, **24**, 4959–4972.

Xu, W., E. J. Zipser, Y.-L. Chen, C. Liu, Y.-C. Liou, W.-C. Lee, B. J.-D. Jou, 2012: An orography-associated extreme rainfall event during TiMREX: Initiation, storm evolution, and maintenance. *Mon. Wea. Rev.*, **140**, 2555–2574.

Yeh, H.-C. and Y.-L. Chen, 2002: The role of offshore convergence on coastal rainfall during TAMEX IOP 3. *Mon. Wea. Rev.*, **130**, 2709–2730.

Chapter 2

Synoptic Aspects of the Summertime Australian Monsoon

Gareth Berry
Monash Weather and Climate, Monash University,
Clayton, Victoria, 3800, Australia
Gareth.Berry@monash.edu
www.monash.edu

Much recent research has focused upon the marked intraseasonal variability of the summertime Australian monsoon system. There are significant synoptic scale phenomena embedded within the monsoon system, which have a clear impact on the total monsoon rainfall. In addition to tropical cyclones, there are many monsoon disturbances that resemble synoptic features found over India and Africa during boreal summer. Automated tracking of the Australian disturbances suggests that a significant proportion of these may develop in the subtropics in connection with the equatorward intrusion of extratropical features.

1. Introduction

The northern portion of the Australian continent is subject to well-defined wet and dry seasons. Like many other tropical landmasses that abut oceans there is a seasonal change in the large-scale circulation, which we term the monsoon. The Australian monsoon is driven by the development of persistent inland heat lows, which are created via sensible heating over the interior of the Australian continent (see, e.g., Suppiah (1992), Rácz and Smith (1999)). During Austral summer, the development of the inland heat lows cause the lower tropospheric flow across the Australian continent equatorward of around 15°S to shift from a southeasterly trade regime to a moist northwesterly monsoon flow. The northwesterly monsoon flow is generally associated with an increase in cloudiness and rainfall over a large area of the Australian tropics. On average, the flow transitions to the moist northwesterly flow (monsoon onset) in late December and back to the trade easterlies in mid March (monsoon retreat; Drosdowsky 1996).

The rainfall distribution during anyone particular season is often very different to the mean seasonal cycle, which exhibits a characteristic "build-up" (increase in area-averaged rainfall), monsoon onset and retreat (Fig. 1a). Instead of following a reasonably smooth seasonal cycle, the area-averaged rainfall on any particular year in the Australian monsoon is subject to significant intraseasonal variability (see Wheeler and McBride 2005), where there are periods lasting from several days to several weeks of heavy rain or dry conditions, which are termed monsoon "burst" and "breaks". An example of the area-averaged rainfall from one season is shown in Fig. 1b. The meteorology of burst and break periods, in particular the nature of convective systems is markedly different. In general, the large-scale conditions during break periods are reminiscent of the pre-onset "build-up" conditions, with relatively dry easterly winds throughout the depth of the troposphere. Somewhat counter-intuitively, the most intense convective systems are found during the break period where the individual systems tend to have higher radar reflectivity, have higher

The Global Monsoon System: Research and Forecast (3rd Edition)
Edited by C. P. Chang et al.

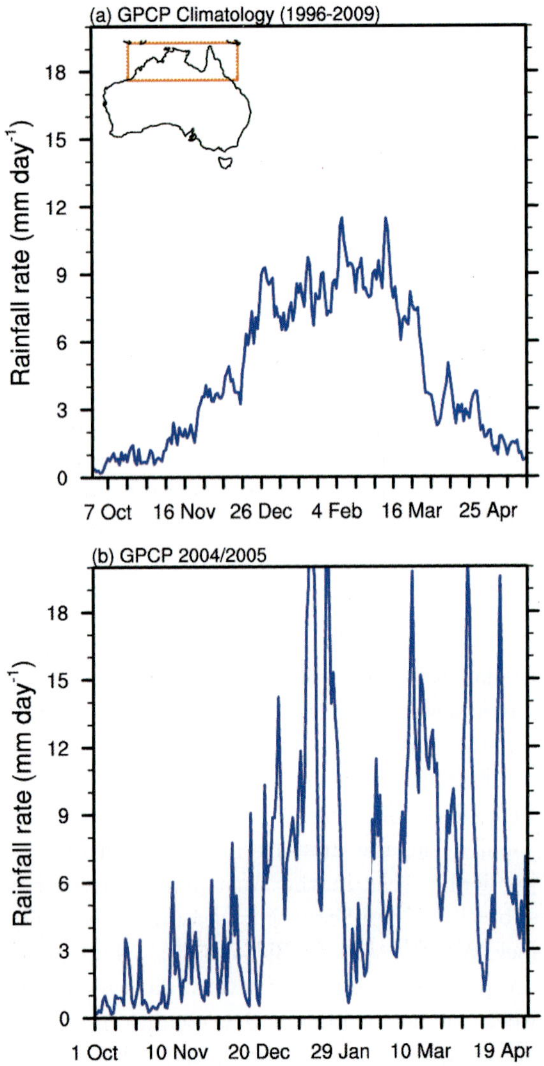

Fig. 1. Seasonal rainfall from the Global Precipitation Climatology Project (GPCP, Huffman *et al.* 1997) shown (a) as a climatology based on the period 1996–2009 and (b) for a single season from October 2004 to May 2005. Data are averaged over the box shown by inset in panel (a).

tops and live longer (e.g. May and Ballinger (2007), Pope *et al.* (2009)). Accordingly, organized mesoscale convective systems are more prevalent during break periods, although they tend to be smaller than those found West Africa during Boreal summer (Frederick and Schumacher 2008), due to differences in vertical wind shear and dust concentration (Zeng *et al.* 2013). During the high precipitation "burst" phases the low level flow becomes westerly and the environment is characterized by increased large scale ascent leading to the convective cells becoming more frequent, but less intense. In general, the convective cells during the monsoon bursts tend to resemble oceanic convection, whereas otherwise the cells are more typical of those found over other continental landmasses (May and Ballinger 2007).

The presence of monsoon bursts and breaks in the Australian monsoon is highlighted as a spectral peak in the 30–60 day frequency range in both the outgoing longwave radiation (OLR; a proxy for convective activity) and the 850 hPa zonal wind (Hendon and Liebmann 1990). This intraseasonal variability is often ascribed to the passage of the Madden Julian Oscillation (MJO), which accounts for approximately half of the observed OLR variability (Wheeler and McBride 2005).

Other studies have suggested that convection in the Australian monsoon is modulated by weather systems in the sub-tropics and mid latitudes. In particular, Keenan and Brody (1988) showed that the intrusion of subtropical troughs and ridges into the tropics modulated the cloudiness in the monsoon region. These authors showed that a mid-latitude trough with an associated upper tropospheric jet streak enhanced convection to its east and suppressed convection to its west. It was speculated that convection is influenced by the secondary circulation associated with the jet streak, which produces large-scale region of ascent and descent. A similar role of the extratropics has been noted during the onset of the Australian monsoon (Davidson *et al.* 2007).

2. Synoptic Weather Systems in the Australian Monsoon

A large body of research has be directed at examining the intraseasonal details of the Australian monsoon (e.g. Hendon and Liebmann 1990), presumably as a result of the well-defined burst and breaks as well as the availability of appropriate diagnostics, such as the MJO defining EOF pair of Wheeler and Hendon (2004). Similarly, convective scale structures have been studied in detail, particularly around Darwin thanks to the wealth of data from comprehensive field experiments, such as the Tropical Warm Pool-International Cloud Experiment (TWP-ICE; May *et al.* 2008). With the exception of tropical cyclone studies, there have been comparatively few studies centred on phenomena that fall within the synoptic and meso-scales in the Australian monsoon.

Synoptic weather systems embedded within large-scale monsoon circulations are frequently found in other regions of the globe. In particular, organized synoptic scale weather systems are frequently encountered in the summertime West African and Indian monsoons. Those that occur over West Africa are known as African Easterly Waves (AEWs) and are perhaps the best documented tropical weather systems, aside from tropical cyclones. AEWs have their origin over the central and eastern parts of tropical North Africa and propagate westwards through the West African monsoon and into the eastern Atlantic basin. These disturbances garner a high level of interest as they strongly modulate the monsoon rainfall and are the primary precursors to North Atlantic tropical cyclones. The typical lengthscale of an AEW is 3000 km, with a period of order 2–5 days and they can be tracked for a week or more as they traverse the African continent, the Atlantic Ocean and on occasion Central America and the Eastern Pacific basin (see, e.g., Carlson 1969). The AEWs develop on a mid-tropospheric easterly jet and it has been demonstrated that the jet is an important source of eddy kinetic energy for the growth of AEWs (e.g. Norquist *et al.* 1977). More recently it has been shown that organised deep convection embedded in the AEW is also important for its growth via the diabatic generation of potential vorticity in the lower troposphere (Berry and Thorncroft 2012). Due to their interaction with the large scale monsoon environment, AEWs exhibit a complex synoptic structure (multiple vorticity centres and variable vertical tilts, etc.) which in many aspects is similar to mid-latitude cyclones (Berry and Thorncroft 2005).

Indian monsoon lows have a less complex structure than AEWs. They develop near the coastline of the Bay of Bengal and tend to propagate inland toward the centre of the broad scale summertime heat low. These cyclonic disturbances are smaller than AEWs (lengthscale of the order 1500 km) and they are shorter lived (2–5 days), which is consistent with a different synoptic environment (Krishnamurti *et al.* 1975). Like tropical cyclones, they are vertically aligned vortices that are characterized by strong cyclonic vorticity from the surface to the upper troposphere with little vertical tilt. These systems are of importance due to their strong surface winds (gusts exceeding 40 kts ($20.6\,\text{ms}^{-1}$)) heavy rainfall (sometimes exceeding 120 mm day^{-1}), which leads to inland flooding (Godbole 1977).

Monsoon depressions are also encountered during the summer in northern Australia, although they have received relatively little attention compared to their counterparts over Africa and India. Synoptic charts issued by the Australian Bureau of Meteorology frequently show long-lived closed lows that form along the inland monsoon trough and are associated with disturbed weather. The most intense of these are colloquially termed "landphoons" due to their apparent similarity to (weak) tropical cyclones. Several monsoon depressions, identified by coherent synoptic scale vortices were encountered during the course of the EMEX experiment, where at least one was a precursor to a tropical cyclone that formed over the Timor sea (Webster and Houze 1991). The relationship between monsoon depressions and tropical cyclone genesis was also noted earlier by

McBride and Keenan (1982), who found that more than half of the tropical cyclones that formed in the Australian sector to the west of 125°E had their origin over land. Tropical cyclogenesis from monsoon disturbances was described by Foster and Lyons (1984), who examined the evolution of two such features after crossing the Australian coast.

The majority of studies examining Australian monsoon disturbances are focused on particular high impact events. For example, Davidson and Holland (1987) investigated two monsoon depressions that developed over the interior of the continent during the summer of 1983, noting that the synoptic structure resembles that of the outer part of a tropical cyclone (i.e. lacking an intense inner core), which is similar to AEWs leaving the West African coast (Berry and Thorncroft 2005). Broadly speaking, the papers that contain case studies of particular Australian monsoon lows find a synoptic structure similar to that of Indian monsoon disturbances, but these papers do not provide a climatological context for their chosen cases (e.g. Hell and Smith (1998), Dengler and Smith (1998), Weber and Smith (2000)).

Dickinson and Molinari (2000) noted the similar geography and lower tropospheric basic state of the Australian and West Africa monsoon systems during their respective summer seasons, in particular, the reversal in the meridional gradient of Potential Vorticity (PV) in the mid-troposphere, which satisfies the necessary condition of instability of the mean flow (Charney and Stern 1962).

Because it has already been shown that AEWs growth via dynamical instabilities Dickinson and Molinari (2000) looked for growing synoptic scale disturbances over northern Australia using the 2–10 day bandpass filtered variance of the meridional wind. Horizontal maps of the 2–10 day basspass filtered meridional wind at 700 hPa computed from the ERA-Interim reanalysis are shown in Fig. 2 for West Africa and Australia during the periods

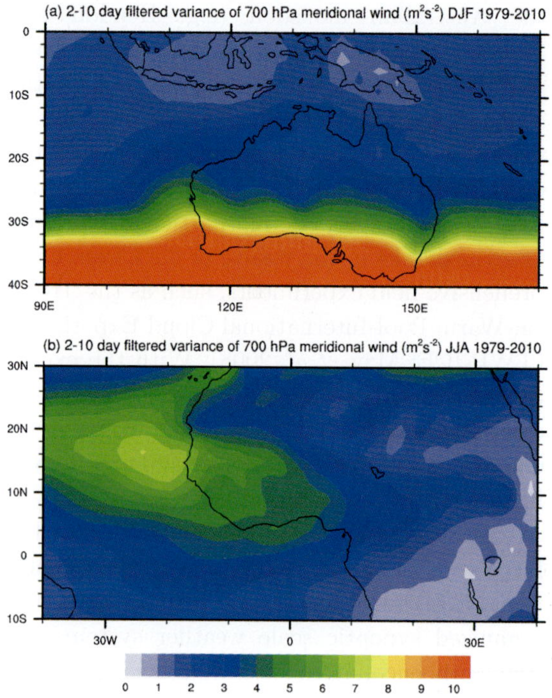

Fig. 2. 2 to 10 day filtered variance of the meridional wind at 700 hPa in the ERA-Interim for the period 1979–2010. Data are shown as (a) for the November–April period over Australia and (b) for the May–October period over Africa.

May–Oct and Nov–Apr, respectively. For the West African region (Fig. 2b), there is well-defined variance maximum over the eastern Atlantic, between 10° and 20°N, corresponding to the location where AEWs are observed to leave the continent. A maximum in bandpass filtered variance extends eastwards across the continent, indicating a synoptic "storm track". The mean flow at 700 hPa along this storm track is easterly, thereby indicating that the synoptic disturbances are growing as they traverse the continent, consistent with what has been show in AEW case studies. In the case of Australia (Fig. 2a), there is much less evidence of a synoptic storm track in tropical latitudes — a small regional maximum exists along the northern coast of Western Australia, but is very weak compared to that over Africa. Dickinson and Molinari (2000) suggest that this is related to tropical cyclones and there is little evidence of growing synoptic systems. The lack of activity based on this synoptic metric that works well with well-known features, such as mid latitude cyclones and AEWs might lead some to conclude that there are very few synoptic weather systems in the Australian monsoon system. However, as will be discussed below this is an artifact of the analysis technique, which assumes that disturbances are periodic.

3. Synoptic Disturbances Diagnosed via Feature Tracking

The dichotomy of monsoon depressions being frequently depicted in synoptic analyses prepared by the Australian Bureau of Meteorology and the apparent lack of activity revealed when using the bandpass filtered metrics provided motivation for a study by Berry *et al.* (2012) to investigate the climatology of Australian monsoon disturbances. Following the case study of Hell and Smith (1998) it was noted that monsoon disturbances shown in the mean sea level pressure analyses are well defined in the isentropic PV in the mid-troposphere. An example of PV on the 315 K isentropic surface (overlaid on satellite and imagery and rainfall estimates) alongside the corresponding mean sea level pressure (MSLP) chart is shown in Fig. 3.

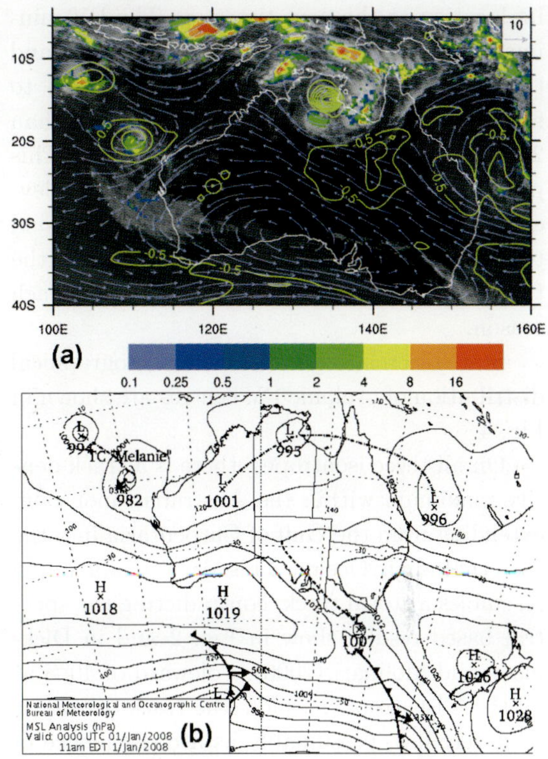

Fig. 3. (a) Example map for 0000 UTC 1 January 2008. Yellow lines show 315-K potential vorticity (multiplied by −1; drawn every 0.1 potential vorticity unit (PVU)) and wind vectors (blue) from the ERA-Interim reanalysis overlaid on TRMM 3B42 rainfall estimate (colored according to legend below panel) and Multifunctional Transport Satellite (MTSAT) infrared imagery (grayscale). (b) BOM surface analysis valid at 0000 UTC 1 January.

This figure shows an intense monsoon low over the Northern Territory, to the south of Darwin (approximately 15°S, 135°E), which is denoted by a closed isobar in the MSLP analysis and a cyclonic PV centre that is co-located with cold cloud tops and heavy rainfall. Two similar features are noted toward the west of the domain over the Indian Ocean (near 105°E and 110°E), one being a named tropical cyclone. Analysis of AEWs using PV on isentropic surfaces was found to be an essential diagnostic by Berry and Thorncroft (2005) as the disturbances were clearly evident and information on the nature of the processes (i.e., diabatic and adiabatic redistribution) could be obtained.

The climatological study of Australian disturbances by Berry *et al.* (2012) used automated tracking of coherent PV centres on isentropic surfaces to derive statistics and generate composites using the 1.5° resolution ERA-Interim dataset for the period 1989–2009. The tracking methodology scanned six-hourly data for all local PV minima less than $-1 \times 10^{-8}\,\mathrm{Km^2\,kgs^{-1}}$ that are separated by at least 3° from adjacent minima (note that these are cyclonic centres as PV is negative by definition in the Southern Hemisphere). The winds at the position of the cyclonic PV centre are then used to predict its location at the next timestep. The PV minima at the next timestep are then found and those within 5° of a prediction are joined to form coherent tracks. Only tracks longer than 24 hours were retained for further analysis. This procedure was conducted on three isentropic levels (315 K, 330 K and 350 K, which correspond to approximately 700, 500 and 200-hPa in the tropics) for the months November to March each season.

Statistics relating to the overall geographical distribution of cyclonic PV centres are shown in Fig. 4.

On all three isentropes, there is a track density maximum within the Australian monsoon, extending from the Gulf of Carpentaria into the eastern Indian Ocean. This density maximum resembles a storm track, contradicting the spectral based results show in Fig. 2 and in Dickinson and Molinari (2000). Poleward of 40°S at all three isentropic surfaces the track density is relatively low, presumably as a consequence of

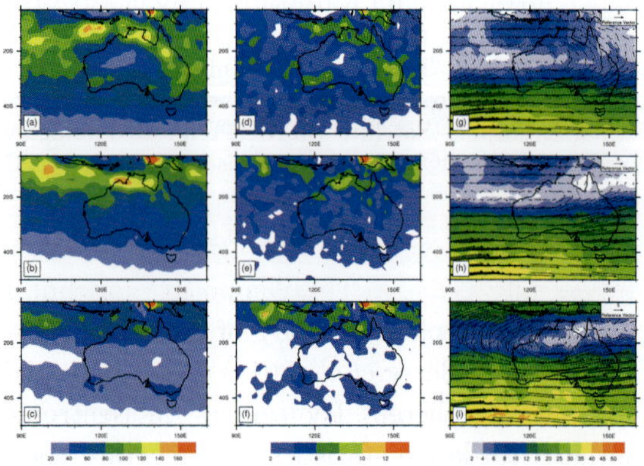

Fig. 4. Statistics of tracking at each of the ERA-Interim grid points at (top) 315, (middle) 330, and (bottom) 350 K. (left) Track density at the ERA-Interim grid points (total number of tracks for the period 1989–2009). (middle) Count of initial locations of PV centres (i.e., first tracked location using all tracks during the period 1989–2009) on the ERA-Interim grid points. (right) Propagation speed (shaded, ms^{-1}) and direction (vector) averaged across all tracks for 1989–2009. Legends for each field are displayed at the bottom of each column.

the PV maxima being defined by closed contours of PV; at these latitudes the cyclonic PV centres are likely to be associated with extended troughs. The first location at which a closed cyclonic PV centre is found is defined as the genesis location. Genesis counts are shown on each of the isentropic surfaces in the central column of Fig. 4. On all three surfaces the high genesis density is co-located with the high track densities. In particular, there are genesis maxima within the tropical storm track, indicating that many cyclonic centres originate within the monsoon itself. The main exception is noted at 315 K, where there is a genesis maximum over southern Queensland (approximately 30°S 150°E), which is away from a track maximum. The mean motion of the cyclonic centres are shown in the right column of Fig. 4. At all levels poleward of about 25°S the motion of coherent PV centers is predominantly from west to east and equatorward of this latitude the motion is generally slower and in the opposite direction. In the tropics on the 315-K level, the motion vectors show that PV centres move equatorward over the eastern half of the Australian continent and curve anticyclonically toward the monsoon region.

PV centres within the monsoon accelerate westward and attain their peak motion (approximately $10\,\mathrm{ms}^{-1}$) over the Timor Sea. This is consistent with the broad anticyclonic flow associated with the heat troughs over northern Australia (see, e.g., Arnup and Reeder 2007) and indicates that PV centres at 315 K frequently move from the subtropics into the tropics.

Some monthly statistics of tracked cyclonic PV centres passing through the region 10°–20°S, 120–147.5°E (approximately the region of high monsoonal rainfall) are displayed in Fig. 5. The number of systems at 315 and 330 K are similar and there are approximately 50% fewer systems tracked at 350 K in all of the months. There is little month-to-month variability in the number of systems tracked at any of the levels; on average there is a coherent PV centre in the monsoon

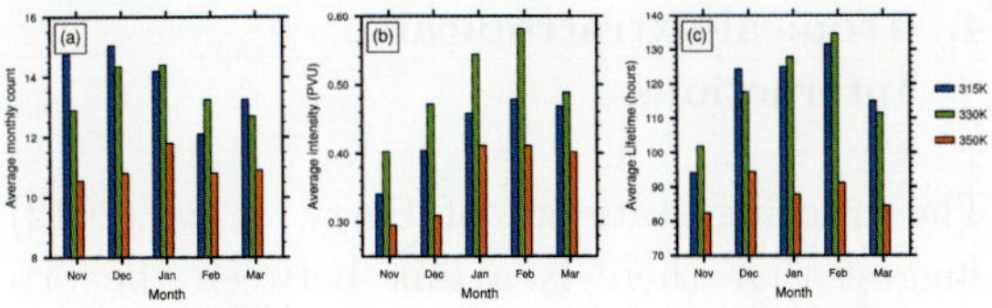

Fig. 5. Monthly mean statistics of coherent PV maxima in the Australian monsoon (defined as the region 10°–20°S, 120°–147.5°E) at each of the three isentropic levels. (a) Average system count, (b) average system intensity (PVU), and (c) average length of track (hours).

region every 2–3 days. In all months the PV centres are most intense at 330 K and least intense at 350 K, consistent with the structure shown for one intense Australian monsoon depression by Hell and Smith (1998). The mean lifetime of PV centres at 315 and 330 K are similar; at both levels the mean lifetime is between 4 and 5 days with peak lifetimes in February when the mean intensities peak. The mean lifetime is shorter and less variable at 350 K with values remaining close to 3.5 days throughout the season. The dichotomy between these tracking statistics and the synoptic metrics used in previous works (e.g. Dickinson and Molinari 2000) was explored by computing the interval between individual PV centres passing through an area near Darwin and comparing it to the same computation carried out in the West African monsoon during Boreal summer. For more than 90% of cases in Africa the time between consecutive events was 1–5 days, whereas over Australia there was a much larger distribution of time intervals. This implies that Australian monsoon disturbances are episodic (consistent with synoptic experience) rather periodic, thus the lack of synoptic activity revealed by spectral analysis is an artifact of the assumptions used in the computation.

With a large number of individual cases a composite cyclonic PV centre structure as constructed using all cases on the 315 K isentropic surface that passed through a box measuring 10 degrees of latitude by 5 degrees of longitude centred to the west of Darwin. This synoptic composite, comprising 454 cases is shown in Fig. 6.

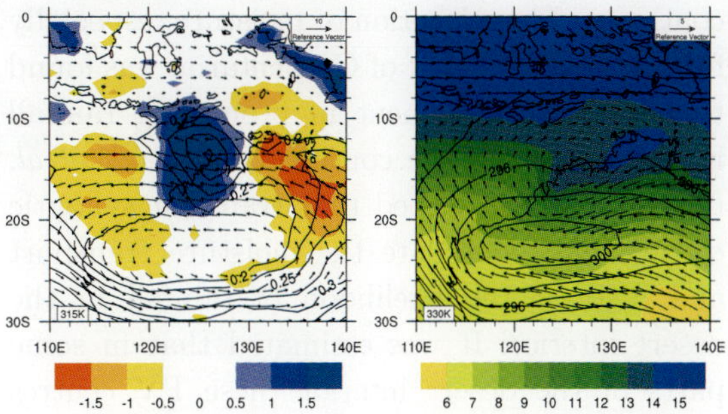

Fig. 6. Composite synoptic structure of all tracked 315-K PV maxima passing through the box 10°–20°S, 125°–130°E (454 cases in the period Nov–Apr): (a) 315-K PV (contoured every 0.1 PVU greater than 0.2 PVU) and wind vectors overlaid on rainfall anomaly from time-varying GPCP climatology (mm day^{-1}); (b) 900-hPa mixing ratio (shaded, gkg^{-1}) wind vectors and potential temperature (black contours, K).

The composite disturbance is comprised of a 315 K PV centre (Fig. 6a) that is of order 1000 km, similar to Indian monsoon disturbances and the embedded subsynoptic scale structures in AEWs. The rainfall anomalies (also shown in Fig. 6a) associated with the PV centre show a significant modulation of the mean rainfall. In the western (leading) part of the 315 K PV centre there is an area of enhanced rainfall, which is similar in size to the composite PV centre. To the east and west of this rainfall enhancement there are two regions of rainfall suppression (near 120° and 140°E), giving an approximate lengthscale of the disturbance of around 2000 km. The composite horizontal wind vectors with specific humidity and temperature fields at 900 hPa are shown in Fig. 6b. The cyclonic circulation of the PV centre is evident along the northern coast, although it is relatively weak with wind vectors on the order of 1 ms^{-1}.The highest moisture content is located at low latitudes (equatorward of 10°S), but there is a local maximum on the eastern side of the cyclonic circulation near Darwin. Although the magnitude of the wind vectors is small, the flow pattern around the disturbance acts across the gradients of specific humidity and advects moisture into the interior of the Australian continent. The advection of moisture, especially from around the Gulf of Carpentaria, was found to be a key component in promoting rainfall in the interior of the continent by Berry *et al.* (2011), who speculated that transient synoptic disturbances modulate the moisture transport and increase the likelihood of rainfall in the desert interior. It was estimated that in some parts of the desert interior these PV centres could be associated with up to 50% of the annual rainfall.

The relationship between some synoptic systems and rainfall was evaluated by Lavender and Abbs (2013) who were motivated by an observed trend of increasing rainfall in the western portion of the Australian tropics. These authors used an automated tracking system to find all minima in mean sea level pressure (MSLP) and geopotential height on pressure levels that were associated with cyclonic circulation in the ERA-Interim reanalysis between 1989–2009. It was then assumed that any precipitation in the Australian water availability project (AWAP) dataset that falls within 5 degrees of a tracked MSLP or geopotential height minimum at any level is directly attributed to it. Using this method, Lavender and Abbs (2013) find that approximately 30–50% of the rainfall over Australia can be attributed to closed lows and that the majority of the extreme precipitation over much of the tropics falls in association with closed lows. They note that the recent trends cannot be explained by changes in tropical cyclones, but up to 60% of the observed trend in total rainfall in northwestern Australia is explained by changes in overall closed low systems. Because of the simple way that the tracking of the closed lows is conducted many different types of phenomena are combined in this study. This means that the contributions due to different phenomena, such as closed centres within the climatological heat low, upper level troughs and monsoon depressions cannot be separated and remains an area of active research.

4. Tropical-extratropical Interactions

The tracking statistics of Berry *et al.* (2012) suggest that there is a link between the origin of synoptic disturbances that are present in the monsoon region and those from higher latitude. The mean motion field shown in Fig. 4 indicates that PV centres move anticyclonically around the continent, with a substantial meridional component along the eastern and western coasts of Australia. The tracking statistics on the 315 K isentrope (Figs. 4d and 4g) suggest that a significant proportion of PV centres on this surface are generated in the subtropics, over southern Queensland, and move into the tropics. Furthermore, the precipitation field from a lag composite created by Berry *et al.* (2012), based on the passage of a 315 K PV to a box west of Darwin shows a coherent synoptic scale eastward propagating signal in the subtropics and extratropics. Given the large sample size (in excess of 400 individual events) it was concluded that there is a phase relationship between synoptic systems in the monsoon and those at higher latitudes.

A coherent relationship between the activity in the Australian tropics and extratropics was also noted more than a quarter of a century earlier by Davidson *et al.* (1983), who examined synoptic aspects surrounding the onset of the Australian monsoon during seven years. These authors noted that monsoon onset was coincident with a characteristic evolution of synoptic systems in which there was anticyclogenesis over southwestern Australia and in four of the six cases a distinct band of midlatitude frontal cloud over the eastern part of the continent. It was hypothesized that the seasonal build up of land-sea temperature contrasts needed to wait for a favorable phase of the midlatitude high and low pressure systems before monsoon onset could occur.

The impact of the extratropics upon monsoon onset has be examined more recently by

Davidson *et al.* (2007) using composites constructed from reanalysis data and idealized dry simulations. From the synoptic composite it was found that the onset of the Australian monsoon was preceded by cyclogenesis at high latitudes over the Indian Ocean (near 60°S, 90°E). It was concluded that the propagation of Rossby waves in the upper troposphere from this cyclogenesis event was likely responsible for the monsoon onset. The location and intensity of the upper tropospheric subtropical jet was identified as a key component of the interaction as the presence of this jet displaces the critical latitude and permits extratropical trough to intrude further equatorward (see discussion of "westerly ducts" in Webster and Holton (1982)).

In addition to the link to the onset of the Australian monsoon extratropical troughs and breaking Rossby waves have been linked to a modulation of convective activity. Allen *et al.* (2008) examined the circumstances surrounding two "tropical drought periods" (a 3 and 4 day period of reduced rainfall and cloud cover) with a wet period in between during the "build-up" phase of the monsoon during November 2005. These authors found that the tropical rainfall was modulated on the synoptic timescale by the penetration of upper tropospheric air from mid-latitudes as a result of Rossby wave breaking along the subtropical jet. The breaking Rossby waves are manifested as a distinct lowering of the tropopause (or even a fold) that may extend as low as 500 hPa. The high static stability of the stratospheric air inhibits convection and the low humidity supresses convection due to the entrainment into rising convective plumes. Convection was found to be enhanced ahead of high PV air, giving rise to a distinct northwest-southeast orientated cloud band.

5. Conclusions

Up until the recent few years, there has been a paucity of studies relating to the synoptic scale aspects of the Australian monsoon. Perhaps this was due to an interest in intraseasonal variability, or because traditional metrics of synoptic activity find little activity in the region. Monsoon depressions are frequently found in northern Australia during summer and are identifiable in the mid-tropospheric potential vorticity field as coherent cyclonic centres. On average, these features track slowly toward the west and there is one such system somewhere within the monsoon region every 2–3 days, although the time interval between consecutive events at a particular point is highly variable. It is this variability that is responsible for the previously perceived lack of synoptic activity in the monsoon region. The monsoon depressions consist of an almost upright "column" of cyclonic PV that enhances rainfall on its leading (western) flank, similar to the structure of Indian monsoon depressions. Some work suggests that monsoon depressions in this region are connected with a significant proportion (30–50%) of the rainfall in Northern Australia.

Recent research has highlighted some important pathways in which the extratropics can influence rainfall in the monsoon region. The Australian sector is noted as a region where the breaking of mid-latitude upper tropospheric Rossby waves is favoured. Automated tracking of coherent PV centres suggests there is an indirect effect of the extratropics whereby "debris" from breaking Rossby waves, characterized by cyclonic PV, may be advected into the tropics by the large scale flow around the continent and act as the seed for the development of monsoon depressions. In other cases, the location of the subtropical jet can provide a "westerly duct" whereby breaking extratropical Rossby waves can directly impact the monsoon. It is hypothesized that this direct influence can trigger monsoon onset and modulate the monsoon convection. This tropical-extratropical interaction is an area of active research and may be able to provide an additional tool for predicting the behaviour of the Australian monsoon.

Acknowledgments

The author wishes to thank Prof. Todd Lane of the University of Melbourne and Prof. Michael Reeder of Monash University for discussion of this topic. Thanks also to Aurel Moise and one anonymous reviewer for their helpful comments on the manuscript.

References

Allen, G., G. Vaughan, D. Brunner, P. T. May, W. Heyes, P. Minnis, and J. K. Ayers, 2009: Modulation of tropical convection by breaking Rossby waves. *Quart. J. Roy. Meteor. Soc.*, **135**, 125–137.

Arnup, S. J. and M. J. Reeder, 2007: The diurnal and seasonal variation of the northern Australian dryline. *Mon. Wea. Rev.*, **135**, 2995–3008.

Berry, G., M. J. Reeder, and C. Jakob. 2011: Physical mechanisms regulating summertime rainfall over northwestern Australia. *J. Climate*, **24**, 3705–3717.

Berry, G. J., M. J. Reeder, and C. Jakob, 2012: Coherent synoptic disturbances in the Australian monsoon. *J. Climate*, **25**, 8409–8421.

Berry, G. J. and C. D. Thorncroft, 2005: Case study of an intense African easterly wave. *Mon. Wea. Rev.*, **133**, 752–766.

Berry, G. J. and C. D. Thorncroft, 2013: African easterly wave dynamics in a mesoscale numerical model: The upscale role of convection. *J. Atmos. Sci.*, **69**, 1267–1283.

Carlson, T. N., 1969: Synoptic histories of three African disturbances that developed into Atlantic hurricanes. *Mon. Wea. Rev.*, **97**, 256–276.

Charney, J. G. and M. E. Stern, 1962: On the stability of internal baroclinic jets in a rotating atmosphere. *J. Atmos. Sci.*, **19**, 159–172.

Davidson, N., K. J. Tory, M. J. Reeder, and W. L. Drosdowsky, 2007: Extratropical–tropical interaction during onset of the Australian monsoon: Reanalysis diagnostics and idealized dry simulations. *J. Atmos. Sci.*, **64**, 3475–3498.

Davidson, N. E., J. L. McBride, and B. J. McAvaney, 1983: The onset of the Australian monsoon during winter MONEX: Synoptic aspects. *Mon. Wea. Rev.*, **111**, 496–516.

Davidson, N. E. and G. J. Holland, 1987: A diagnostic analysis of two intense monsoon depressions over Australia. *Mon. Wea. Rev.*, **115**, 380–392.

Dengler, K. and R. K. Smith, 1998: A monsoon depression over northwestern Australia. Part II: A numerical model study. *Aust. Met. Mag.*, **47**, 131–144.

Dickinson, M. and J. Molinari, 2000: Climatology of sign reversals of the meridional potential vorticity gradient over Africa and Australia. *Mon. Wea. Rev.*, **128**, 3890–3900.

Drosdowsky, W., 1996: Variability of the Australian summer monsoon at Darwin: 1957–1992. *J. Climate*, **9**, 85–96.

Foster, I. J. and T. J. Lyons, 1988: The development of tropical cyclones in the north-west of Australia. *Quart. J. Roy. Meteor. Soc.*, **114**, 1187–1199.

Frederick, K. and C. Schumacher, 2008: Anvil Characteristics as Seen by C-POL during the Tropical Warm Pool International Cloud Experiment (TWP-ICE). *Mon. Wea. Rev.*, **136**, 206–222.

Godbole, R. V., 1977: The composite structure of the monsoon depression. *Tellus*, **29**, 25–40.

Hell, R. M. and R. K. Smith, 1988: A monsoon depression over northwestern Australia. Part I: A case study. *Aust. Met. Mag.*, **47**, 21–40.

Hendon, H. H. and B. Liebmann, 1990: A composite study of onset of the Australian summer monsoon. *J. Atmos. Sci.*, **47**, 2227–2240.

Huffman, G. J., R. F. Adler, P. Arkin, A. Chang, R. Ferraro, A. Gruber, and U. Schneider, 1997: The global precipitation climatology project (GPCP) combined precipitation dataset. *Bull. Amer. Meteor. Soc.*, **78**, 5–20.

Keenan, T. D. and L. R. Brody, 1988: Synoptic-scale modulation of convection during the Australian summer monsoon. *Mon. Wea. Rev.*, **116**, 71–85.

Krishnamurti, T. N., M. Kanamitsu, R. Godbole, C. B. Chang, F. Carr, and J. H. Chow, 1975: Study of a monsoon depression (I): Synoptic structure. *J. Meteor. Soc. Japan*, **53**, 227–240.

Lavender, S. L. and D. J. Abbs, 2013: Trends in Australian rainfall: Contribution of tropical cyclones and closed lows. *Clim. Dyn.*, **40**, 317–326.

May, P. T. and A. Ballinger, 2007: The statistical characteristics of convective cells in a monsoon regime (Darwin, Northern Australia). *Mon. Wea. Rev.*, **135**, 82–92.

May, P. T., J. H. Mather, G. Vaughan, K. N. Bower, C. Jakob, G. M. McFarquhar, and G. G. Mace, 2008: The tropical warm pool international cloud experiment. *Bull. Amer. Meteor. Soc.*, **89**, 629–645.

McBride, J. L. and T. D. Keenan, 1982: Climatology of tropical cyclone genesis in the Australian region. *J. Climatology*, **2**, 13–33.

Norquist, D. C., E. E. Recker, and R. J. Reed, 1977: The energetics of African wave disturbances as observed during phase III of GATE. *Mon. Wea. Rev.*, **105**, 334–342.

Pope, M., C. Jakob, and M. J. Reeder, 2008: Convective systems of the north Australian monsoon. *J. Clim.*, **21**, 5091–5112.

Rácz, Z. and R. K. Smith, 1999: The dynamics of heat lows. *Quart. J. Roy. Meteor. Soc.*, **125**, 225–252.

Suppiah, R., 1992: The Australian summer monsoon: A review. *Progress in Physical Geography*, **16**, 283–318.

Weber, H. C. and R. K. Smith, 2000: A monsoon depression over northwestern Australia. Part III: Motion. *Aust. Met. Mag.*, **49**, 1–21.

Webster, P. J. and J. R. Holton, 1982: Cross-equatorial response to middle-latitude forcing in a zonally varying basic state. *J. Atmos. Sci.*, **39**, 722–733.

Webster, P. J. and R. A. Houze Jr., 1991: The Equatorial Mesoscale Experiment (EMEX): An overview. *Bull. Amer. Meteor. Soc.*, **72**, 1481–1506.

Wheeler, M. C. and H. H. Hendon, 2004: An all-season real-time multivariate MJO index: Development of an index for monitoring and prediction. *Mon. Wea. Rev.*, **132**, 1917–1932.

Wheeler, M. and G. N. Kiladis, 1999: Convectively coupled equatorial waves: Analysis of clouds and temperature in the wavenumber-frequency domain. *J. Atmos. Sci.*, **56**, 374–399.

Wheeler, M. C. and J. L. McBride, 2005: Australian-Indonesian monsoon. In *Intraseasonal Variability in the Atmosphere-Ocean Climate System* (pp. 125–173). Springer, Berlin, Heidelberg.

Zeng, X., W. K. Tao, S. W. Powell, R. A. Houze Jr., P. Ciesielski, N. Guy, and T. Matsui, 2013: A Comparison of the Water Budgets between Clouds from AMMA and TWP-ICE. *J. Atmos. Sci.*, 487–503.

Chapter 3

The South American Monsoon System

Maria A. F. Silva Dias* and Leila M. V. Carvalho

Department of Atmospheric Sciences University of Sao Paulo
05508-090 Sao Paulo, Brazil
Earth Research Institute and Department of Geography
University of California, Santa Barbara, USA
**maria.silvadias@iag.usp.br*
www.usp.br

The multiscale nature of the South American Monsoon System is presented here in a review of the interannual to decadal variability, intraseasonal variability and regional and mesoscale features of the clouds systems associated with the break and active phases. Although the lack of long records in the Amazon and surrounding areas limits multidecadal analyses of the monsoon, associations of monsoon features with the Atlantic Multidecadal Oscillation and the Pacific Decadal Oscillation have been found. There is evidence that El Niño/La Niña modulate the South Atlantic Convergence Zone and extreme daily rainfall events in the region. The onset/demise of the SAMS is variable throughout the region and associated to different features of convective activity, and the convective/stratiform partition. Aerosols emitted by biomass burning have been shown to act as Could Condensation Nuclei whereby they may affect clouds although during the premonsoon phase not much sensitivity has been found to variations in aerosol concentration. During the wet season, however, aerosol concentration appears as a possible cause to explain convective intensity. Predictability of onset, amplitude, duration and demise of SAMS have shown increased skill through empirical relationships between bimonthly average precipitation and modes of interannual variability of upper level winds in South America.

1. Introduction

The dominant climatic feature of tropical South America (SA) is the seasonal change in precipitation and moisture that accompanies variations in the trade winds, diabatic heating, surface pressure, thermodynamic instability, low level convergence and upper level divergence of mass in the atmosphere. These features characterize the South American Monsoon System (SAMS). Unlike other monsoon systems, easterly winds dominate year-round over northern SA and tropical Atlantic. However, when the annual mean is removed from the winter and summer mean circulation, a clear reverse in the low-level circulation monthly anomalies becomes evident and resembles the seasonal change in the wind direction observed in other monsoon systems (Zhou and Lau 1998). Figure 1 illustrates these features by showing monthly variations of 850hPa wind anomalies (long term annual mean removed) and precipitation. The SAMS seasonal cycle is essentially driven by the differential heating between South America and the Atlantic Ocean. In the pre-monsoon season, the incoming solar radiation increases the diabatic heating over the central western tropical SA and land-ocean contrasts initiate the monsoonal circulations. The differential heating strengthens the cross equatorial moisture transport by the trade winds toward the center of the continent. Thermodynamic instability increases as the onset of the rainy season approaches and remains large throughout the wet season

The Global Monsoon System: Research and Forecast (3rd Edition)
Edited by C. P. Chang et al.

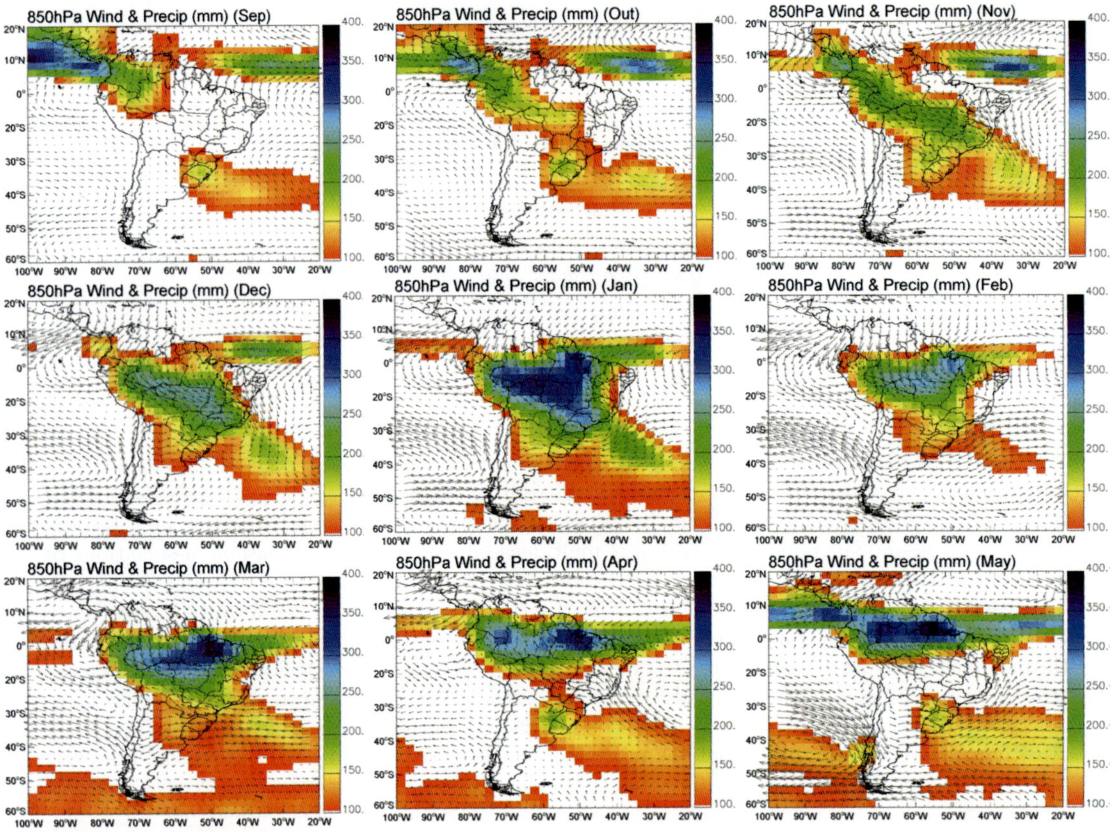

Fig. 1. Seasonal cycle of 850 hPa wind anomalies (long term mean removed) and Precipitation. Winds are from the National Center for Environmental Prediction/National Center for Atmospheric Research (NCEP/NCAR) Reanalysis-2 (Kanamitsu *et al.* 2002). Precipitation from Global Precipitation Climatology Project (GPCP) (Xie *et al.* 2003).

(Fu *et al.* 1999; Fu and Li 2004, Fisch *et al.* 2004). Convective activity intensifies on broad ranges of spatial and temporal scales to redistribute the excess of moist static energy accumulated near the surface (Rickenbach *et al.* 2011). The poleward migration of the upper-level mid-latitude jet creates an anticyclonic shear over South America that alters the characteristics of frontal systems and favors the organization of the SACZ and the transition from winter to summer monsoon regime (Nieto-Ferreira and Rickenbach 2011). Land-atmosphere processes control evapotranspiration and play a significant role in the onset and maintenance of the SAMS (Fu and Li 2004). It is generally accepted that the mean onset of the wet season over the Brazilian High plains in the present climate is between mid-October to mid-November (Kousky *et al.* 1988; Gan *et al.* 2006; Silva and Carvalho 2007; Raia and Cavalcanti 2008), with some regional differences (Nieto-Ferreira *et al.* 2010). The peak of the monsoonal precipitation is between December–February and its demise between March–April.

The South Atlantic Convergence Zone (SACZ) is a component of the SAMS and the most significant feature of cloudiness and precipitation in the wet season over eastern South America and subtropical Atlantic Ocean. It is characterized by a northwest–southeast persistent cloud band associated with low level convergence of winds and moisture (Kodama 1992;

Carvalho *et al.* 2002a; 2004) that affects densely populated areas over tropical and subtropical SA. Convection in the oceanic portion of the SACZ is often observed in association with transient atmospheric wave trains that propagate from midlatitudes of the Southern Hemisphere during the summer (Grimm and Silva Dias 1995; Liebmann *et al.* 1999; Nogués-Paegle and Mo 1997; Carvalho *et al.* 2004; Muza *et al.* 2009; Carvalho *et al.* 2010a, Quadro *et al.* 2013). These systems alter low-level winds and moisture advection that produce important modification along the convective margins of the SACZ (Ma *et al.* 2011; Robertson and Mechoso 2000).

2. SAMS Interannual-to-Decadal Variability

Most of the explained interannual variability of the SAMS precipitation comes from El Niño Southern Oscillation (ENSO; Paegle and Mo 2002; Grimm 2003; Marengo *et al.* 2010 and references therein). Nonetheless, ENSO impacts vary throughout the monsoon season and reverse between spring (September–November) and summer (December–February), especially in central-eastern South America (Grimm and Zilli 2009). These variations suggest that regional surface-atmosphere interactions compete with remote influences during the wet season (Grimm *et al.* 2007). Land-surface and land-cover processes have also impacts on interannual variations of the SAMS (Collini *et al.* 2008; Ma *et al.* 2010; Saulo *et al.* 2010). Moreover, there is evidence that ENSO modulates the intensity and position of the SACZ and the occurrence of extreme precipitation at the southeast coast of Brazil (Carvalho *et al.* 2002a; and Liebmann *et al.* 2001). Interannual variations in sea surface temperatures (SST) of the tropical Atlantic also play a significant role in variations of the SAMS and the SACZ (Venegas *et al.* 1997; Robertson and Mechoso 2000; Robertson *et al.* 2003; Taschetto and Wainer 2008; Bombardi *et al.* 2013). Cold (warm) SST anomalies over the tropical South Atlantic along with warm (cold) SST anomalies over the extratropical South Atlantic have been associated with early (late) onsets (Bombardi and Carvalho 2011) and increased (decreased) total daily precipitation over southeastern Brazil (Bombardi *et al.* 2013). Modeling studies have suggested significant ocean-atmospheric feedbacks over southwest tropical Atlantic upon the organization of the SACZ (Misra 2008; Chaves and Nobre 2004).

The SAMS variability on decadal to multidecadal timescales is less known largely due to short, sparse and intermittent observations in the Amazon basin and surrounding areas. The majority of these studies examined rain and river-gauge data, reanalysis and satellite based rainfall estimates that extended no earlier than the middle of the 20th century. Few studies have investigated SAMS variations using proxies from paleoclimatic records collected in sites under influence of the monsoonal rainfall. Krishnamurthy and Misra (2010) found two modes of SAMS variability on decadal time scales that suggest associations with the Atlantic Multidecadal Oscillation and the Pacific Decadal Oscillation. Carvalho *et al.* (2010b) showed a significant shift in the mean onset, demise and duration of the SAMS in the early 1970s. Additional evidence for large-scale climate shift in SA in the early 1970s has been reported in Jacques-Coper and Garreaud (2014) using multiple data sets and in Piovano *et al.* (2002) based on sediment records in Laguna Mar Chiquita.

Hsu *et al.* (2011) indicated that the global monsoon total precipitation has strengthened in the past 30 years along with an increase in the global mean surface temperature. They argue that the expansion of the monsoon regions occur primarily at the poleward edges of the tropical monsoon regions at 20°–30° north and south. Teixeira and Satyamurty (2011) and

Silva Dias *et al.* (2013) found positive trends in the frequency of heavy and extreme rainfall events in southern and southeastern Brazil during 1960–2010 that is consistent with this hypothesis.

3. SAMS Intraseasonal Variability

SAMS exhibits persistent active and inactive ("break") periods in the intensity of rainfall as may be seen in the examples in Fig. 2. Tropical intraseasonal oscillations (i.e. oscillations roughly between 20–100 days) modulate these alternating periods and impact circulation (Casarin and Kousky 1986; Silva Dias and Marengo 1999; Jones and Carvalho 2002), mesoscale activity (Carvalho *et al.* 2002b; Petersen *et al.* 2002), extreme precipitation (Carvalho *et al.* 2004) and the seesaw of wet and dry periods over eastern South America (Nogues-Paegle and Mo 1997; Muza *et al.* 2009). The Madden-Julian Oscillation (MJO-Madden and Julian 1994) has been identified among the most important tropical modes modulating convection and circulation over tropical and subtropical SA on intraseasonal time-scales (Carvalho *et al.* 2004; Jones and Carvalho 2002; Liebmann *et al.* 2004).

Changes in circulation caused by intraseasonal oscillations, associated or not with the MJO, impact moisture transport and water balance in the monsoon region (Herdies *et al.* 2002; Carvalho *et al.* 2010b). Active phases of SAMS are associated with westerly and break phases with easterly wind anomalies over central Brazil (Jones and Carvalho 2002). The anomalous circulation accompany changes in the profile of moist static stability, modify the diurnal cycle of convection and the characteristics of the mesoscale convective systems (Petersen *et al.* 2002; Cifelli *et al.* 2002; Rickenbach *et al.* 2002; Carvalho *et al.* 2002b). Active and break phases of the SAMS' regime and related variations in circulation and thermodynamic properties are associated with synoptic scale activity. The propagation of frontal systems organize convection in the SACZ when westerly wind regimes dominate (active phase) over central

Fig. 2. Examples of Active (left) and Break (right) phases of the SAMS. Arrows indicate the direction of the low-level circulation. Active phases (left) are associated with westerly winds over central South America and cross equatorial northerly winds; break phases (right) are associated with intensified low-level jet transporting moisture to subtropical areas and easterly winds over central Brazil.

and eastern South America (Garreaud 2000; Rickenbach 2002). These systems are modulated by the envelope of intraseasonal disturbances and are associated with patterns of intense precipitation over tropical South America and western South Atlantic (Vera *et al.* 2006; Marengo *et al.* 2010; Cunningham and Cavalcanti 2006).

4. SAMS Regional and Mesoscale Features

The SAMS shows regional aspects and specific mesoscale features across South America. Nieto_Ferreira and Rickenbach (2011) indicate that the onset of SAMS starts around October 18–22 when precipitation begins in the northwestern part of the continent and gradually progresses towards the south and southeast. The second stage is marked by the abrupt establishment of the South Atlantic Convergence Zone (SACZ). This stage occurs on average around October 28–November 1. The third stage of monsoon onset involves the late arrival of the monsoon to the mouth of the Amazon River, associated with the slow migration of the Atlantic Intertropical Convergence Zone (ITCZ) observed around December 27–31. Mesoscale Convective Systems are also regionally specific and contrasting between the before and after onset. Rickenbach *et al.* (2011) show that the west-central Amazon and the mouth of the Amazon river had the greatest pre- to post-monsoon onset change in rainfall. Metrics related to the vertical intensity of MCSs showed distinct maxima 2 months before monsoon onset in these regions. The timing of the maximum in MCS intensity was different in each region, with the west-central region maximum in August and a delayed maximum the mouth of the Amazon during October. In both regions, MCS intensity subsequently decreased leading up to and following onset. For the SACZ region, Rickenbach *et al.* (2011) show that the timing of the onset is similar to the west-central Amazon region although with less average rainfall and more variable intensity due to rain variations associated with the influence of transient midlatitude baroclinic systems (Siqueira and Machado, 2004; Siqueira *et al.* 2005). The lightning flash rate and convective intensity indicators was greater between August and October (premonsoon) than during the monsoon phase. Southeastern South America, although with a weak annual cycle, has the monsoon precipitation dominated by Mesoscale Convective Complexes (MCC) documented by several authors (Velasco and Fritsch, 1987, Durkee and Mote 2009) with peak lightning flash rate following onset more than twice the value observed in the other regions.

For the Amazon region Artaxo *et al.* (2002) show that in the premonsoon phase, or trasition season from dry to wet seasons, local farmers prepare the land by setting fire to pasture areas. These fires outbreaks release large amounts of aerosols into the atmosphere, contributing considerably to the increase in the number of cloud condensation nuclei (CCN). Higher CCN concentrations have been suggested to modulate the convection development in various parts of the world, particularly in the SAMS (Williams *et al.* 2002). Albrecht *et al.* (2011) show that during the premonsoon phase Southwestern Amazon has more intense thunderstorms, with more lightning and more positive cloud-to-ground lightning, than during the monsoon season. They analyze the differences in thunderstorm intensity as a function of aerosol loading and find equal amounts in days with thunderstorms days and nonthunderstorm days during the pre-monsoon season. They argue that the atmosphere has an overwhelming number of aerosol particles so that thunderstorm intensity will be defined mainly by the cloud base updraft. Regions with deforestation are thus prone to have more intense thunderstorms during the dry season due to higher surface temperatures (Fisch *et al.* 2004) than cooler regions over forest.

Albrecht *et al.* (2011) also show that as the wet season progresses and active and break phases are observed, the aerosol appears associated to higher convective intensity. However, causality in the case of aerosol and thunderstorm features is very hard to ascertain and is still a question for further research.

Herdies *et al.* (2002) associated active and break phases, respectively, to period where the SACZ was active and when it was not (NSACZ). Rickenbach *et al.* (2002) showed that low level winds were from the west in active phases and from the east in break phases and these systems tended to be more convective in the break periods. Anagnostou and Morales (2002) indicated that the diurnal cycle of convection was more pronounced during the break phases and Carvalho *et al.* (2002b) associated active and break phases to different features of convective activity such as larger and shallower systems during active phases smaller but deeper systems during the break phases. Most of the results on convective activity during the SAMS in the Amazon region are a result of field campaigns that addressed the biosphere-atmosphere interactions and climate reviewed by Betts and Silva Dias (2009).

Recent work by Romatschke and Houze (2013) addresses the summer rainfall in South America using 10 years of TRMM radar data. They show that most of the Amazon region and Central part of Brazil, or Brazilian Plateau have monsoon rainfall dominated by convective systems with large stratiform components. In the La Plata basin the rainfall is dominated by systems that are highly convective and severe (Zipser *et al.* 2006, Durkee *et al.* 2009, Romatschke and Houze 2010, Silva Dias 2011). Over the SACZ, Romatschke and Houze (2013) show that most of the rainfall is associated with medium to very large systems with a diurnal cycle that shows a broad maximum during the day, contrasting with Amazon systems that are smaller and have a pronounced afternoon peak. As indicated by Rasmussen and Houze (2011) the central and northern foothills of the Andes, located at the western edge of the SAMS, have storms triggered by the local upslope circulation that propagate eastward while evolving into storms with wide convective and broad stratiform areas (cf. Angelis *et al.* 2004, Negri *et al.* 2000).

5. SAMS Predictability

Forecasting monsoon features such as onset, duration, amplitude and demise date represents a grand challenge due to the societal applications associated with it, especially in agriculture and hydropower generation. Jones *et al.* (2012) indicate that an improved skill over climatology may be obtained using the NCEP Climate Forecast System reforecast version 2, by the use of empirical orthogonal functions applied to precipitation anomalies for SAMS. Zamboni *et al.* (2009) also indicate potential for reliable predictions based on correlations between bimonthly mean precipitation anomalies and modes of interannual variability of upper level winds over South America. Regional features within SAMS are not addressed in neither of this articles thus pointing to needed research in the area.

Acknowledgments

MAFSD acknowledges support of FAPESP (2009/15235-8, 2013/05014-0) and of NAP/USP-INCLINE. LMVC thanks the support of NOAA Climate Program Office (NA10OAR-4310170) and CGIAR Research Program on Climate Change, Agriculture and Food Security (CCAFS); sub-contract with the International Potato Center (SB120184)

References

Albrecht, R. I., C. A. Morales, and M. A. F. Silva Dias, 2011: Electrification of precipitating systems over the Amazon: Physical processes of thunderstorm development. *J. Geophys. Res.*, **116**, D08209.

Anagnostou, E. N. and C. A. Morales, 2002: Rainfall estimation from TOGA radar observations during LBA field campaign. *J. Geophys. Res.*, **107**, D20, 8068. doi:10.1029/2001JD000377.

Angelis, C. F., G. R. McGregor, and C. Kidd, 2004: Diurnal cycle of rainfall over the Brazilian Amazon. *Climate Res.*, **26**, 139–149.

Betts, A. K. and M. A. F. Silva Dias, 2010: Progress in understanding land-surface-atmosphere coupling from LBA research. *Journal of Advances in Modeling Earth Systems*, **2**, 1–20.

Bombardi, R., L. M. V. Carvalho, C. Jones, and M. S. Reboita, 2013: Precipitation over eastern South America and the south Atlantic sea surface temperature during neutral ENSO periods. *Climate Dynamics*. DOI 10.1007/s00382-013-1832-7.

Bombardi, R. J. and L. M. V. Carvalho, 2011: The South Atlantic dipole and variations in the characteristics of the South American Monsoon in the WCRP-CMIP3 multimodel simulations. *Climate Dynamics*, **36**, 2091–2102.

Carvalho, L. M. V., C. Jones, and B. Liebmann, 2002a: Extreme precipitation events in Southeastern South America and large-scale convective patterns in the South Atlantic convergence zone. *J. Climate*, **15**, 2377–2394.

Carvalho L. M. V., C. Jones, and M. A. F. Silva Dias, 2002b: Intraseasonal large-scale circulations and mesoscale convective activity in tropical South America during the TRMM-LBA campaign. *J. Geophys. Res. Atmos.*, **107**, doi:10.1029/2001JD000745.

Carvalho, L. M. V., C. Jones, and B. Liebmann, 2004: The South Atlantic convergence zone: Intensity, form, persistence, and relationships with intraseasonal to interannual activity and extreme rainfall. *J. Climate*, **17**, 88–108.

Carvalho, L. M. V., A. E. Silva, C. Jones, B. Liebmann, P. L. Silva Dias, and H. R. Rocha, 2010a: Moisture transport and Intraseasonal Variability in the South America Monsoon System. *Climate Dynamics*, DOI 10.1007/s00382-010-0806-2.

Carvalho, L. M. V., C. Jones, A. E. Silva, B. Liebmann, and P. L. Silva Dias, 2010b: The South American Monsoon System and the 1970s climate transition. *Int. J. Climatol.*, DOI: 10.1002/joc.2147.

Casarin D. P. and V. E. Kousky, 1986: Anomalias de precipitac¸a˜o no sul do Brasil e variacoes na circulacao atmosferica. *Rev Bras Meteorol.* **1**, 83–90.

Cunningham, C. C. A. and I. F. A. Cavalcanti, 2006: Intraseasonal modes of variability affecting the South Atlantic Convergence Zone. *Int. J. Climatol.*, **26**, 1165–1180.

Chaves R. R. and P. Nobre, 2004: Interactions between sea surface temperature over the South Atlantic Ocean and the South Atlantic Convergence Zone. *Geophys. Res. Lett.*, **31**(3), L0320410.1029/2003GL018647.

Cifelli, R., W. A. Petersen, L. D. Carey, S. A. Rutledge, and M. A. F. da Silva Dias, Radar observations of the kinematic, microphysical, and precipitation characteristics of two MCSs in TRMM LBA. *J. Geophys. Res.*, **107**(D20), 8077.

Collini E. A., E. H. Berbery, V. R. Barros, and M. E. Pyle, 2008: How does soil moisture influence the early stages of the South American monsoon? *J. Climate*, **21**, 195–213.

Durkee, J. D. and T. L. Mote, 2010: A climatology of warm-season mesoscale convective complexes in subtropical South America. *Int. J. Climatol.*, **30**, 418–431.

Durkee, J. D., T. L. Mote, and J. M. Shepherd, 2009: The contribution of mesoscale convective complexes to rainfall across subtropical South America. *J. Climate*, **22**, 4590–4605.

Fisch, G., J. Tota, L. A. T. Machado, M. A. F. Silva Dias, R. F. da F. Lyra, C. Nobre, A. Dolman, and J. Gash, 2004: The convective boundary layer over pasture and forest in Amazonia. *Theor. Appl. Climatol.*, **78**, 47–59.

Fu, R. and W. Li, 2004: The influence of the land surface on the transition from dry to wet season in Amazonia. *Theor. Appl. Climatol.*, **78**, 97–110, DOI 10.1007/s00704-004-0043-x.

Fu, R., B. Zhu, and R. E. Dickinson, 1999: How do atmosphere and land surface influence seasonal changes of convection in the tropical Amazon? *J. Climate*, **12**, 1306–1321.

Gan M. A., V. B. Rao, and M. C. L. Moscati, 2006: South American monsoon indices. *Atmos. Sci. Lett.*, **6**, 219–223.

Garreaud, R. D., 2000: Cold air incursions over subtropical South America: Mean structure and dynamics. *Mon. Wea. Rev.*, **128**, 2544–2559.

Grimm, A. M. and P. L. Silva Dias, 1995: Analysis of tropical-extratropical interactions with influence functions of a barotropic model. *J. Atmos. Sci.*, **52**, 20, 3538–3555.

Grimm A. M., 2003: The El Nino impact on the summer monsoon in Brazil: Regional processes versus remote influences. *J. Climate*, **16**, 263–280.

Grimm A. M., J. Pal, and F. Giorgi, 2007: Connection between spring conditions and peak summer monsoon rainfall in South America: Role of soil moisture, surface temperature, and topography in eastern Brazil. *J. Climate*, **20**, 5929–5945.

Grimm, A. M. and M. T. Zilli, 2009: Interannual variability and seasonal evolution of summer monsoon rainfall in South America. *J. Climate*, **22**, 2257–2275.

Herdies D. L., A. da Silva, M. A. F. Silva Dias, and R. N. Ferreira, 2002: Moisture budget of the bimodal pattern of the summer circulation over South America. *J. Geophys. Res. Atmos.*, **107**, 8079.

Hsu, P., T. Li, and B. Wang, 2011: Trends in global monsoon area and precipitation over the past 30 years. *Geophys. Res. Lett.*, **38**, L08701, doi:10.1029/2011GL046893.

Jacques-Coper, M. and R. D. Garreaud, 2014: Characterization of the 1970s climate shift in South America. *Int. J. Climatol.*, doi: 10.1002/joc.412.

Jones, C. B. and L. M. V. Carvalho, 2002: Active and break periods in the South America monsoon system. *J. Climate*, **15**, 905–914.

Jones, C. B., L. M. V. Carvalho, and B. Liebmann, 2012: Forecast skill of the South American Monsoon System. *J. Climate*, **25**, 1883–1889.

Kanamitsu, M., W. Ebisuzaki, J. Woollen, S.-K. Yang, J. J. Hnilo, M. Fiorino, and G. L. Potter, 2002. NCEP-DOE AMIP-II Reanalysis (R-2). *Bulletin of the American Meteorological Society*, **83**, 1631–1643.

Kodama, Y.-M., 1992: Large-scale common features of subtropical precipitation zones (the Baiu Frontal Zone, the SPCZ, and the SACZ). Part I: characteristics of subtropical frontal zones. *J. Meteor. Soc. Japan*, **70**, 813–835.

Kousky, V. E., 1988: Pentad outgoing longwave radiation climatology for the South American sector. *Rev. Bras. de Meteorol.*, **3**, 217–231.

Krishnamurthy, V. and V. Misra, 2010: Daily atmospheric variability in the South American Monsoon System, *Clim. Dyn.*, doi:10.1007/s00382-010-0881-4.

Madden, R. A. and P. R. Julian, 1994: Observations of the 40–50 day tropical oscillation: A review. *Mon. Wea. Rev.*, **122**, 814–837.

Negri, A. J., E. N. Anagnostou, and R. F. Adler, 2000: A 10-yr climatology of Amazonian rainfall derived from passive microwave satellite observations. *J. Appl. Meteor.*, **39**, 42–56.

Piovano, E. L., D. Ariztegui, and S. D. Moreira, 2002: Recent environmental changes in Laguna Mar Chiquita (central Argentina): A sedimentary model for a highly variable saline lake. *Sedimentology* **49**, 1371–1384.

Petersen, Walter A., Stephen W. Nesbitt, Richard J. Blakeslee, Robert Cifelli, Paul Hein, and Stephen A. Rutledge, 2002: TRMM Observations of Intraseasonal Variability in Convective Regimes over the Amazon. *J. Climate*, **15**, 1278–1294.

Liebmann, B., G. N. Kiladis, J. A. Marengo, T. Ambrizzi, and J. D. Glick, 1999: Submonthly convective variability over South America and the South Atlantic Convergence Zone. *J. Climate*, **12**, 1877–1891.

Liebmann, B., G. N. Kiladis, C. S. Vera, A. C. Saulo, and L. M. V. Carvalho, 2004: Subseasonal variations of rainfall in South America in the vicinity of the low-level jet east of the Andes and comparison to those in the South Atlantic Convergence Zone. *J. Climate*, **17**, 3829–3842.

Ma, H.-Y., X. Ji, J. D. Neelin, and C. R. Mechoso, 2011: Mechanisms for precipitation variability of the Eastern Brazil/SACZ convective margin. *J. Climate*, **34**, 3445–3456. doi: 10.1175/2011JCLI4070.1.

Marengo, J. A., B. Liebmann, A. M. Grimm, V. Misra, P. L. Silva Dias, I. F. A. Cavalcanti, L. M. V. Carvalho, E. H. Berbery, T. Ambrizzi, C. S. Vera, A. C. Saulo, J. Nogués-Paegle, E. Zipser, A. Seth, and L. M. Alves, 2010: Review recent developments on the South American Monsoon System. *Int. J. Climatol.*, DOI.1002/joc.2254.

Misra, V., 2008: Coupled air, sea, and land interactions of the South American Monsoon. *J. Climate*, **21**, 6389–6403.

Muza, M. N., L. M. V. Carvalho, C. Jones, and B. Liebmann, 2009: Intraseasonal and interannual variability of extreme dry and wet events over Southeastern South America and subtropical Atlantic during the Austral summer. *J. Climate*, **22**, 1682–1699.

Nieto-Ferreira, R. and T. M. Rickenbach, 2010: Regionality of monsoon onset in South America: A three-stage conceptual model. *Int. J. Climatol.*, doi:10.1002/joc.2161.

Nieto-Ferreira, R., T. M. Rickenbach, and E. A. Wright, 2011: The role of cold fronts in the onset of the monsoon season in the South Atlantic convergence zone. *Q. J. R. Meteorol. Soc.*, **137**, 908–922. doi: 10.1002/qj.810.

Nogues-Paegle, J. N. and K. C. Mo, 1997: Alternating wet and dry conditions over South America during summer, *Mon. Wea. Rev.*, **125**, 279–291.

Quadro, M. F. L., E. H. Berbery, M. A. F. Silva Dias, D. L. Herdies, and L. G. G. Goncalves, 2013: The atmospheric water cycle over South America as seen in the new generation of global reanalyses. In *Radiation Processes in the Atmosphere and Ocean* (*IRS2012*): *Proceedings of the International Radiation Symposium (IRC/IAMAS) Dahlem Cube 732.*

Raia, A. and I. F. A. Cavalcanti, 2008: The life cycle of the South American monsoon system. *J. Climate*, **21**, 6227–6246.

Rasmussen, K. L. and R. A. Houze Jr., 2011: Orogenic convection in South America as seen by the TRMM satellite. *Mon. Wea. Rev.*, **139**, 2399–2420.

Rickenbach, T. M., R. Nieto-Ferreira, R. P. Barnhill, and S. W. Nesbitt, 2011: Regional contrasts of mesoscale convective system structure prior to and during monsoon onset across South America. *J. Climate*, **24**, 3753–3763.

Rickenbach, T. M., R. Nieto Ferreira, J. Halverson, D. L. Herdies, and M. A. F. Silva Dias, 2002: Modulation of convection in the southwestern Amazon basin by extratropical stationary fronts, *J. Geophys. Res.*, **107**, doi:10.1029/2000JD000263.

Robertson, A. W. and C. R. Mechoso, 2000: Interannual and interdecadal variability of the South Atlantic convergence zone. *Mon. Wea. Rev.*, **128**, 2947–2957.

Robertson, A. W., J. D. Farrara, and C. R. Mechoso, 2003: Simulations of the atmospheric response to South Atlantic sea surface temperature anomalies. *J. Climate*, **16**, 2540–2551.

Romatschke, U. and R. A. Houze Jr., 2010: Extreme summer convection in South America. *J. Climate*, **23**, 3761–3791.

Romatschke, U. and R. A. Houze Jr., 2013: Characteristics of Precipitating Convective Systems Accounting for the Summer Rainfall of Tropical and Subtropical South America. *J. Hydromet.*, **14**, 25–46.

Saulo, A. C., L. Ferreira, J. Nogués-Paegle, M. E. Seluchi, and J. Ruiz, 2010: Land-atmosphere interactions during the Northwestern Argentina Low Event. *Mon. Wea. Rev.*, **138**, 2481–2498.

Silva A. E. and L. M. V. Carvalho, 2007: Large-scale index for South America monsoon (LISAM). *Atmos. Sci. Lett.*, **8**, 51–57.

Silva Dias, M. A. F., 2011: An increase in the number of tornado reports in Brazil. *Wea. Climate Soc.*, **3**, 209–217.

Silva Dias, M. A. F., J. Dias, L. M. V. Carvalho, E. D. Freitas, and P. L. Silva Dias, 2013: Changes in extreme daily rainfall for São Paulo, Brazil, *Climatic Change*, **116**, 705–722.

Siqueira, J. R. and L. A. T. Machado, 2004: Influence of the frontal systems on the day-today convection variability over South America. *J. Climate*, **17**, 1754–1766.

Siqueira, J. R., W. B. Rossow, L. A. T. Machado, and C. Pearl, 2005: Structural characteristics of convective systems over South America related to cold-frontal incursions. *Mon. Wea. Rev.*, **133**, 1045–1064.

Taschetto, A. S. and I. Wainer, 2008: The impact of the subtropical South Atlantic SST on South American precipitation. *Ann. Geophys.*, **26**, 3457–3476.

Teixeira, M. S. and P. Satyamurty, 2011: Trends in the Frequency of Intense Precipitation Events in Southern and Southeastern Brazil during 1960–2004. *J. Climate*, **24**, 1913–1921.

Velasco, I. and J. M. Fritsch, 1987: Mesoscale convective complexes in the Americas. *J. Geophys. Res.*, **92**(D8), 9591–9613.

Venegas, S. A., L. A. Mysak, and D. N. Straub, 1997: Atmosphere–Ocean Coupled Variability in the South Atlantic. *J. Climate*, **10**, 2904–2920.

Vera, C., *et al.*, 2006: Toward a unified view of the American monsoon systems. *J. Climate*, **19**(20), 4977–5000.

Williams, E. and coauthors, 2002: Contrasting convective regimes over the Amazon: Implications for cloud electrification. *J. Geophys. Res.*, **107**, 10.1029/2001JD000380.

Xie, P. and coauthors, 2003: GPCP pentad precipitation analyses: An experimental dataset based on gauge observations and satellite estimates. *J. Climate*, **16**, 2197–2214.

Zamboni, L., C. R. Mechoso, and F. Kucharski, 2010: Relationships between upper-level circulation over South America and rainfall over southeastern South America: A physical base for seasonal predictions. *J. Climate*, **23**, 3300–3315.

Zhou, J. and K.-M. Lau, 1998: Does a Monsoon Climate Exist over South America? *J. Climate*, **11**, 1020–1040.

Zipser, E. J., D. J. Cecil, C. Liu, S. W. Nesbitt, and D. P. Yorty, 2006: Where are the most intense thunderstorms on Earth? *Bull. Amer. Meteor. Soc.*, **87**, 1057–1071.

Chapter 4

The West African Monsoon

Jean Philippe Lafore*,†, Michela Biasutti‡, Peter Knippertz§ and Christopher Taylor¶

†*CNRM-GAME, Toulouse, France*

‡*Columbia U., NY, USA*

§*KIT, Germany*

¶*CEH, Wallingford, UK*

This chapter provides a summary of the recent advances made on the understanding and simulation of the coupled atmosphere-ocean-land system of the West African Monsoon (WAM) and of its variability on a wide range of space and timescales. More specifically we highlight the interannual to climate scales, the intraseasonal variability, weather systems and their predictability, the surface-atmosphere coupling and the coupling with dust.

1. Introduction

The first phase of the international AMMA[a] project (Redelsperger *et al.* 2006) contributed to major progress in the documentation and understanding of the coupled atmosphere-ocean-land West African Monsoon (WAM) system as summarised in the 3rd edition of the WMO book *The Global Monsoon System: Research and Forecast.* Since 2010, the second phase of AMMA hinged on three key interacting research themes: (1) interactions between society, environment and climate, which necessitates the second theme: (2) study of predictability and improvement of meteorological, seasonal and climate forecasting, which itself requires the third theme: (3) continued effort to enrich our knowledge of the monsoon system. For this second phase, AMMA increased its collaboration with other programs, integrated some of them, and benefited from new field experiments such as FENNEC for dust (2010–2012) and DACCIWA[b] for clouds (planned for 2016) to reach the above ambitious goals.

In the present 5th edition of this book, we summarize efforts and major results on themes (2) and (3) in the last four years. More specifically we highlight the following points: the interannual to climate scales, the intraseasonal variability, weather systems and their predictability, the surface-atmosphere coupling and the coupling with dust. For a more complete presentation of the WAM system and its variability, we refer the reader to the previous edition.

2. WAM from the Interannual to Climate Scales

The instrumental records of the 20th century (Nicholson, 1979; New *et al.* 2002) and longer reconstructions (Nicholson 1978; Shanahan *et al.* 2009) have shown that seasonal rainfall in West Africa is highly variable at timescales that range from interannual to decadal and multi-decadal.

At interannual time scales, anomalies in sea surface temperature (SST) in the Gulf of Guinea

*Corresponding address: Jean Philippe Lafore CNRM-GAME Météo-France and CNRS, 42 av. G. Coriolis 31 057 Toulouse Cedex, France. jean-philippe.lafore@meteo.fr

[a]African Monsoon Multidisciplinary Analysis, http://www.amma-international.org

[b]Dynamics-aerosol-chemistry-cloud interactions in West Africa, http://www.dacciwa.eu

The Global Monsoon System: Research and Forecast (3rd Edition)
Edited by C. P. Chang et al.

and the eastern Pacific are a dominant driver, as indicated by numerous observational studies, idealized experiments with atmospheric atmospheric global circulation models (AGCMs), and hindcast ensemble integrations (Ward 1998; Janicot *et al.* 2001; Giannini *et al.* 2005). In recent years, the recognition that West African rainfall is potentially predictable, due to its sensitivity to SST anomalies, has spurred the development of several seasonal prediction systems. Unfortuately, predictive skill is hindered by widespread model biases in the teleconnections to SST (see for example Rowell 2013) and in the mean and seasonal cycle of the equatorial Atlantic SST (which mostly affect the Gulf of Guinea, see for example Tompkins and Feudale, 2010). For example, Philippon *et al.* (2010) show that the coupled models of the ENSEMBLES European project (initialized on May 1st) have no skill in forecasting July–August–September variability in Sahel rainfall, but do much better in the Gulf of Guinea. Arribas *et al.* (2011) show low Briar scores in West Africa for the GloSea4 system, but at the same time indicate an encouraging ability of the model to forecast the correct teleconnection to ENSO. In many instances, the skill of the prediction is better for circulation variables (Philippon *et al.* 2010; Ndiaye *et al.* 2011), suggesting that a Model Output Statistics (MOS) correction can lead to skillful rainfall prediction. Another encouraging development is the ability of the coupled CFS forecast system to produce skillful forecast of the West African monsoon with long lead times and without the help of a MOS correction (Ndiaye *et al.* 2011). It is possible that more skill will follow from incorporating a fuller range of external forcings (King *et al.* 2010; Joseph and Zeng 2011), including soil moisture (Koster *et al.* 2004; Agustí-Panareda *et al.* 2010).

The large ensemble simulations with coupled ocean-atmosphere global circulation models (OAGCMs) and Earth System Models (EaSMs) that have been performed in support of the fourth (Meehl *et al.* 2007) and fifth (Taylor *et al.* 2012) IPCC assessment reports have spurred substantial progress in understanding both past variations at multi-decadal scales and future trends.

As was the case at interannual time scales, a large fraction of rainfall variability in West Africa at decadal and longer time scales is also paced by variability in the SST (Folland *et al.* 1986; Giannini *et al.* 2003). The SST pattern related to the long-term trend has been related to the Atlantic Multidecadal Oscillation (Zhang and Delworth, 2006; Ting *et al.* 2011), but anthropogenic forcings have started to play a substantial role (Rotstayn and Lohmann, 2002; Held *et al.* 2005). In particular, there is a growing consensus on the partial attribution of centennial trends in Sahel rainfall to anthropogenic emissions of sulfate aerosols in the Atlantic sector, which cause an inter-hemispheric pattern of SST anomalies (Biasutti and Giannini 2006; Ackerley *et al.* 2011; Hwang *et al.* 2013; see also Fig. 1).

At the same time, a firm attribution of the multi-decadal swings and extreme drought remains elusive, due to the inability of the models to capture the full range of observed variability (Biasutti 2013). This is a common bias (notwithstanding the remarkably good performance of a few models in both CMIP3 and CMIP5, e.g., Held *et al.* 2005) that has not been fully explained. Possible causes are wrong teleconnections (Rowell 2013), missing dust (Mahowald *et al.* 2010), or missing vegetation processes (Kucharski *et al.* 2012). Alternatively, the cause might be an underestimation of the external forcing, especially the strength of the indirect aerosol forcing (Booth *et al.* 2013, but see also Zhang *et al.* 2013).

Climate projections for the 21st century annual-mean Sahel rainfall are uncertain even in sign in both CMIP3 (Cook and Vizy 2006; Biasutti and Giannini 2006) and CMIP5 (Roehrig *et al.* 2013; Biasutti 2013), but some other aspects of the projections, namely the

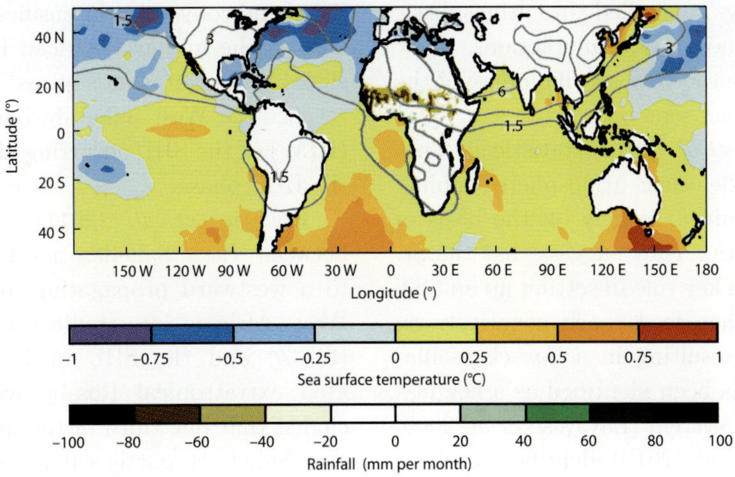

Fig. 1. An increase in the concentration of sulphate aerosols over the Northern Hemisphere between 1940 and 1985 (grey contour lines in mg of sulphate) cooled sea surface temperatures in the Northern Hemisphere, contributing to an inter-hemispheric thermal gradient (blue and orange hues). Simulations indicate that this gradient forced in part the decline in Sahel rainfall observed over the same period (brown and green hues). The change in rainfall is calculated from the linear trend in average monthly rainfall over July to September between 1940 and 1985; the change in sea surface temperatures is calculated from the linear trend in annual average data over the same period; and the change in aerosol burden is calculated as the difference in four-year averages centred on 1940 and 1985. Rainfall data are from the University of East Anglia Climate Research Unit TS3p0 dataset; aerosol data was provided by Ben Booth at the Hadley Centre, Exeter, UK; sea surface temperature data are from the National Oceanic and Atmospheric Administration (NOAA) National Climatic Data Center (NCDC) ERSST version3b dataset. (from Biasutti, 2011).

seasonal evolution of the anomalies and their spatial pattern, are less uncertain (Monerie *et al.* 2012a,b; Biasutti and Sobel 2009; Biasutti 2013). Dry anomalies in the first half of the year indicate a later beginning of the rainy season, while wet anomalies in September and October suggest a strengthening and concentration of the rainy season. These changes are best described as an amplification and delay of the rainy season and are coherent with global patterns of change (Seth *et al.* 2013; Chou *et al.* 2013; Dwyer *et al.* 2013). The CMIP5 models reproduce the CMIP3 projection of a growing zonal gradient in rainfall anomalies across the Sahel with stronger agreement, with more conspicuous drying in West Africa at the beginning of the rainy season and more conspicuous wetting in the central and eastern Sahel at the end of the rainy season (Monerie *et al.* 2012a,b). This is consistent with circulation anomalies forced by a strengthening of the Saharan heat low (SHL; Monerie *et al.* 2012b; Lavaysse *et al.* 2010).

3. The Intraseasonal Variability

The intraseasonal time scale is critical in West Africa where resources are highly rainfall dependent. Nevertheless, contrary to considerable efforts dedicated to seasonal-to-decadal timescales, the intraseasonal variability (ISV) has been explored only recently. Three main modes of variability have been identified, two with a mean periodicity of 15 days called the "quasi-biweekly zonal dipole" (QBZD; Mounier *et al.* 2008) and the "Sahelian mode" (Sultan *et al.* 2003) and one with a mean periodicity around 40 days called "African MJO" mode (Matthews, 2004; Janicot *et al.* 2009). These modes have a regional scale and can strongly influence precipitation and convective activity.

They are mainly controlled by atmospheric dynamics and land-surface interactions. They can also modulate the very specific phase of the African summer monsoon onset.

De Coëtlogon *et al.* (2010) statistically analyzed the ISV of the ocean–atmosphere coupling in the Gulf of Guinea, largely on the basis of satellite observation. They suggest that the St. Helena High has a key role in setting up an SST anomaly which then feeds back negatively on the atmosphere, resulting in a 2-week oscillation. The SHL has been identified as a key feature of the WAM system (Lavaysse *et al.* 2009, 2010). Chauvin *et al.* (2010) identified a robust propagative mode of variability of the SHL over North Africa and the Mediterranean in the 10–25 days range. This mode is forced by midlatitude Rossby waves over Europe. When penetrating into Africa these waves modulate the low-level cold advection (ventilation hereafter) from the eastern North Atlantic and Mediterranean, thus modifying the location and intensity of the SHL as illustrated by Fig. 2 for a specific case study. When the ventilation is increased along the Moroccan coast, it is weakened over the Lybian coast, and the SHL is shifted and reinforced eastward corresponding to the East phase (HLE) of the SHL. The monsoon flux is thus reinforced over the eastern Sahel and convection intensifies a few days later. Due to the midlevel African Easterly Jet, this signal propagates at $\sim 6\,\mathrm{ms}^{-1}$ westward and reaches the West Sahel during the West phase (HLW) of the SHL occurring about a week after the HLE phase.

Roehrig *et al.* (2011) studied the link between the "Sahelian mode" corresponding to a westward propagating modulation of the West African intertropical convergence zone (ITCZ) and the SHL mode and the associated extratropical Rossby waves. They concluded that one third of the dry and wet spells over Sahel are partly explained by the midlatitude ISV, through a major role played by the SHL. They also showed that the WAM onset is favoured by the HLE phase of the SHL, when the Mediterranean ventilation is weakened and the SHL is able to strengthen. For instance, the onset of the 2011 monsoon occurred during the transition between a strong HLE event and a strong HLW one as illustrated by Fig. 2. This extratropical origin of Sahelian dry and wet spells is complementary to that proposed by Janicot *et al.* (2010), who invoked equatorial Rossby wave dynamics.

Poan *et al.* (2013) assessed the Sahelian intraseasonal variability from the perspective of the distribution of precipitable water (PW).

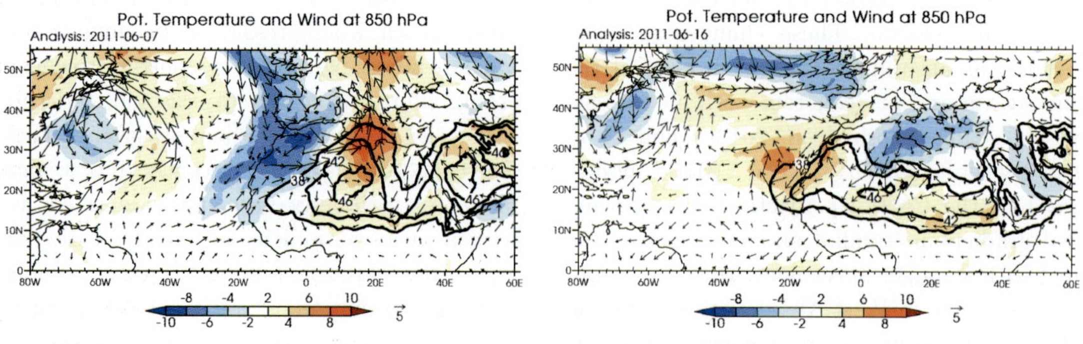

Fig. 2. Illustration of a HLE phase on 7 June (left panel) and of a HLW phase 10 days later on 16 June 2011 (right panel). Horizontal wind anomalies at 850 hPa (in ms^{-1}) are shown in vector. 850-hPa potential temperature (in K) anomalies are indicated in colors. Contours indicates the raw 850-hPa potential temperature (in °C), with one contour every 4°C, from 38°C to outline the HL. All anomalies are determined against a climatological daily seasonal cycle, computed with ERA-I. 6-hourly ECMWF operational data were averaged to obtain daily data.

Whereas eastern Sahel PW variability is dominated by time scales longer than 10 days, synoptic scales dominate in the western Sahel, especially because of the influence of African easterly waves (AEWs). A composite study of AEWs, as seen from their moisture footprints, provides an interesting documentation of their structure and of the role of diabatic processes on their propagation and growth that the dynamics alone cannot explain (Poan *et al.* 2014). Corresponding dry and wet PW anomalies significantly modulate convective activity and can be tracked back to the eastern Sahel and over a long period. This implies a potential skill for short- to medium-range forecasts of wet and dry events over the Sahel.

To explore the potential forecast skill brought about by the advanced understanding of the ISV, several initiatives have been launched to monitor the tropical ISV in general[c] and more recently with a focus on Africa (CPC[d], SUNY[e]). Among them, the MISVA[f] collaborative project between Senegal and France aims at monitoring and forecasting the ISV over Africa in real-time. This project has provided a series of diagnostic indices during the WAM season since 2011 and the experience has showed that dry and moist events can be detected several days in advance, even as Numerical Weather Prediction (NWP) models have almost no skill at these scales, at least for precipitation. In general such events are the result of the combination of several synoptic-to-intraseasonal modes. The physical understanding of the WAM discussed above helped to improve forecasting methods over West Africa and led to the current writing of the "Forecaster's Handbook for West Africa" in close collaboration between forecasters and scientific international communities, under the World Meteorological Organization (WMO) umbrella.

4. Weather Systems and Their Prediction

Compared with midlatitudes, the improvement of NWP in the Tropics is slow, even when using metrics better adapted to low latitudes characterized by a large Rossby radius. The skill of an AEW forecast within different operational NWP models is less than 48 h (Torn, 2010), and the synoptic variability and the coupling between AEWs and precipitation differ considerably among NWP systems.

There are many reasons for such biases and poor skills at short forecast ranges; they depend on the quality of observations and data assimilation, as well as on models, but also on the theoretical limit to predictability in West Africa. The AMMA project contributed to the reactivation and modernization of the radiosonde network (Parker *et al.* 2008) over West Africa and provided complementary observations to document all components of the WAM system. The assimilation of such improved observation networks (Agustí-Panareda *et al.* 2010; Faccani *et al.* 2009) and of alternative data such as satellite microwave channels over land (Karbou *et al.* 2010) significantly improves initial fields as compared with independent observations provided by AMMA such as driftsondes or GPS precipitable water. Nevertheless forecasts initialized from those analyses lost this advantage within the first 24 h. The ensemble-based sensitivity analysis of Torn (2010) confirms this rapid decay of the positive impact of the initial fields for the dynamics. Nevertheless the impact of the initial moisture increases with the forecast range in agreement with an increasing sensitivity to the model representation of diabatic convective terms.

The main difficulty to forecast the synoptic activity over West Africa has its roots in the

[c]http://www.cawcr.gov.au/staff/mwheeler/maproom/OLR_modes/index.htm
[d]http://www.cpc.ncep.noaa.gov/products/Global_Monsoons/African_Monsoons/precip_monitoring.shtml
[e]http://mikeventrice.weebly.com/tropical-waves.html
[f]http://misva.sedoo.fr/

strong interactions between the AEWs and convection, which are not properly represented by current NWP models, especially for the triggering stage. Convection is very sensitive to the humidity field due to numerous non-linear and small-scale processes (source of energy, surface fluxes, detrainment of convection, evaporation of precipitation generating convective wakes...) that are not well represented in models. The difficulties are magnified over West Africa due to the dryer mean conditions and to the large variability of the humidity field characteristic of that semi-aride region. Cloud resolving models (CRMs) are now a powerfull tool to study this coupling between AEWs, convection and the surface, and to improve the representation of the key processes involved in NWP and climate models. During the last decade, a new generation of operational CRMs has been developed (WRF in the US, MetUM LAM in the UK, AROME in France, and COSMO in Germany, among many others) and the models are now able to properly resolve the life-cycles of mesoscale convective systems (MCSs) bringing new perpectives to forecast weather system over Africa (Lean *et al.* 2008; Pearson *et al.* 2010; Marsham *et al.* 2011; Beucher *et al.* 2013).

5. Surface-atmosphere Coupling

The Sahel drought of the early 1970s initiated a long and continuing series of land-atmosphere studies, beginning with the Charney hypothesis (1975) of a feedback between a desertifying continent and the large-scale circulation. In the decades since then, the vegetation of the region has largely recovered, in tandem with rainfall (Hickler *et al.* 2005, Govaerts and Lattanzio 2009). Various modelling studies (e.g. Xue and Shukla 1993, Zeng *et al.* 1999) have emphasised the role of vegetation feedback in intensifying and prolonging the drought, though its oceanic origins are now recognized (e.g. Giannini *et al.* 2003). On much shorter timescales, the region was identified as a hotspot of soil moisture — precipitation coupling in a multi-model study (Koster *et al.* 2004). A recent global analysis of satellite data (Taylor *et al.* 2012) highlighted Africa, and the Sahel in particular, as a region where soil moisture impact on precipitation is especially strong.

That study benefited directly from knowledge gained during the AMMA programme on short time scale processes (hours to days). Rainfall variations on these scales controls soil moisture, in turn affecting fluxes of sensible and latent heat, and the properties of the planetary boundary layer (Schwendike *et al.* 2010). Driven by strong spatial variability in rainfall, soil moisture can thus feedback on the atmosphere at scales from a few to 1000s km. Observations during the AMMA campaign identified low level convergence induced by mesoscale gradients in soil moisture (Taylor *et al.* 2007; Dixon *et al.* 2012) and vegetation (Garcia-Carreras 2010). Various atmospheric modelling studies have shown the importance of soil moisture patchiness on these circulations and their role in the initiation of deep convection (e.g. Gaertner *et al.* 2010, Gantner and Kalthoff 2010, Birch *et al.* 2012). Analysis of initiation of over 3000 MCSs using satellite data provided a firm observational basis for this effect (Taylor *et al.* 2011). This study found a clear preference for the initiation of the first deep convective cells on strong soil moisture gradients induced by rain in previous days (see example in Fig. 3). The effect is clearest for soil moisture patches on scales of 10–40 km, and was estimated to account for 1 in 8 MCS initiations in the western Sahel. As a result of this feedback, rain from MCSs in their early stages is more likely over drier soil (see Fig. 3), but may shift to heavier rain over wetter soil in some circumstances (e.g. Gantner and Kalthoff 2010, Wolters *et al.* 2010).

Larger-scale circulations are also sensitive to soil moisture variability created by dry spells. These can increase sensible heat fluxes over many hundreds of kilometres, inducing

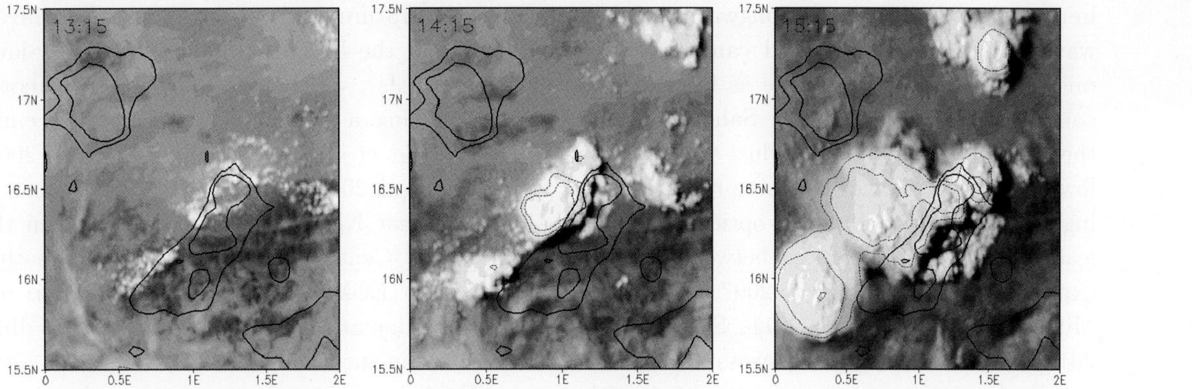

Fig. 3. Example of the early afternoon development of an MCS in Mali (Taylor *et al.* 2010) from visible channel satellite data. Solid contours indicate wet patches created by rain the preceding day, and cloud top temperatures below −40 and −60°C are enclosed by dashed contours.

nocturnal heat low circulation anomalies (Taylor 2008, Parker 2008), and interacting with larger-scale variability, notably AEWs and the Sahelian mode (see previous Sec. 3) of intra-seasonal variability (Lavender *et al.* 2010) to favour westward propagation of coupled surface and atmospheric features. On longer time-scales, the seasonal evolution of land surface properties (soil moisture, vegetation) also influences characteristics of the WAM. For example, Yamada *et al.* (2012) found that prescribing sub-surface soil moisture in a set of global seasonal simulations allowed the model to capture key circulation features associated with the onset period of the WAM, leading to substantial improvements in the simulation of the abrupt latitudinal shift in rainfall associated with the ITCZ.

Many challenges remain for modelling the land surface influence on the atmosphere in West Africa. Surface processes are often poorly described (e.g. rapid vegetation phenology) or ignored (interaction with groundwater). Current convection schemes in large-scale models tend to be over-sensitive to surface evaporation, erroneously favouring daytime rain over wet soils (Taylor *et al.* 2012), in contrast to cloud-resolving models (Hohenegger *et al.* 2009, Taylor *et al.* 2013). This is of particular importance in the Sahel, where soil moisture controls on convection are especially strong, and convective variability plays a key role in the overall monsoon (Marsham *et al.* 2013a). The direct effect of poorly simulated seasonal evolution of precipitation on land fluxes was highlighted by Boone *et al.* (2010), who illustrated the diversity of meridional gradients of surface sensible and latent heating in an ensemble of AGCMs. As discussed above, these biases can in turn feedback on circulation. In terms of future climate, changes in land use may play an important role (Paeth *et al.* 2009) though these are highly uncertain (Xue *et al.* 2012). Even without changes in land use, Sahelian vegetation is likely to change in the coming decades in response to rising temperature and CO_2 levels, and in particular, to potentially large fluctuations in precipitation. These land feedbacks, combined with remote climatic forcings, make the provision of future climatic projections in the Sahel particularly challenging.

6. Coupling with Aerosols

The WAM interacts closely with the largest dust source on Earth, the Sahara and Sahel. Multi-scale circulations and convection associated with the monsoon can lead to dust emission and transport (Knippertz and Todd 2012);

in turn, dust modulates the longwave and shortwave radiation budgets, and can affect cloud microphysics through its role as ice and cloud condensation nuclei. Over the Sahel, the sign of the precipitation response to dust appears to follow the local top of atmosphere radiative forcing, which is sensitive to dust optical properties, leading to gross differences between modeling experiments (Yoshioka *et al.* 2007, Perlwitz and Miller 2010, Yue *et al.* 2011, Solmon *et al.* 2012). Positive forcing increases the export of energy from the convecting region, resulting in anomalous low-level moisture convergence, which increases conditional instability (Chou *et al.* 2005).

During the boreal winter dry season, most of West Africa is dominated by cool northeasterly winds. Periods of enhanced pressure gradients at the southern edge of the subtropical anticyclones or behind long trailing cold fronts of Mediterranean/North African cyclones can lead to prolonged dust periods in the Sahel and dust advection to the tropical Atlantic Ocean and into the Sudan and Guinea zones (Klose *et al.* 2010, Knippertz *et al.* 2011). Here, the dust often mixes with biomass burning aerosol (Heinold *et al.* 2011). The large diurnal cycle in temperature leads to strong surface inversions and the formation of nocturnal low-level jets (NLLJs), which break down in the morning creating a distinctive peak in dust emission (Todd *et al.* 2008, Schepanski *et al.* 2009, Fiedler *et al.* 2013). Fiedler *et al.* (2013) recently showed that NLLJs are underestimated in the European Centre for Medium-Range Weather Forecasts (ECMWF) analysis data due to too much mixing at night, leading to a too flat diurnal cycle in surface wind.

During summer, dust storms can be caused by NLLJs associated with flow acceleration into the SHL. Strong dust-lifting winds also occur at the leading edge of cold pools created by evaporation of precipitation from convective storms ("haboobs"), mostly over the Sahel and southern Sahara (Fig. 4). Haboobs are difficult to detect from space due to masking by convective clouds and hard to model due to the difficulties with representing cold pools and MCSs in models with parameterised convection (Marsham *et al.* 2011, Heinold *et al.* 2013). Dust emission and transport and haboob activity are strongly modulated by AEWs and their interactions with extratropical disturbances (Knippertz and Todd 2010). Recent high-resolution modelling

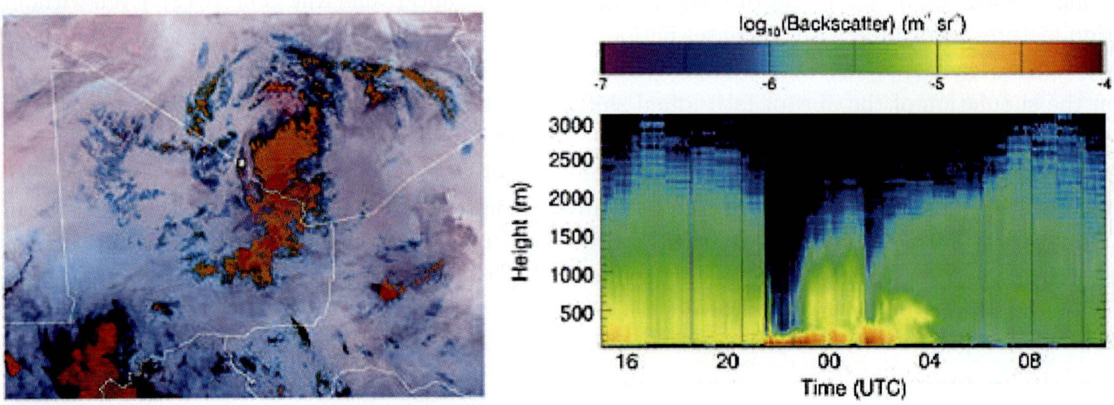

Fig. 4. Passage of a major haboob duststorm over the Fennec supersite at Bordj Badji Mokhtar (Algeria) on 29–30 June 2011. *Left*: Meteosat dust product for 2130 UTC showing a deep convective system stretching from eastern Mali into southern Algeria with a dusty cold pool emerging on the western side. The supersite is marked with a white square. *Right*: Time series of ground-based lidar backscatter. The leading part of the cold pool is optically thick enough to totally attenuate the lidar signal above about 200 m. (Images taken from Marsham *et al.* 2013b.)

studies have shown that NLLJs can form within aged cold pools, thereby creating a strong link between convective activity, flow acceleration into the SHL and dust emission (Heinold *et al.* 2013). CRM simulations of Kocha *et al.* (2013) over a full June month demonstrate the importance of the diurnal cycle of aerosol optical thickness (AOT) associated with dust, which is mainly driven by the breakdown of the NLLJ and by haboobs in the afternoon. The undersampling of the diurnal cycle by current satellites plus the impact of cloud masks on AOT retrievals induce an underestimation of AOT by ~40% over the convective regions and an overestimation of AOT by 17% over morning source areas like the Bodélé Depression. An additional factor relevant for dust emission is dry convective mixing in the daytime planetary boundary layer, which can lead to the formation of dust devils or dust plumes, which are not resolved in current dust models.

A major limitation to our understanding of the mechanisms of dust production in the Sahara is the lack of observations from this vast and uninhabited region. Long-term station records are mostly restricted to the Sahel (Klose *et al.* 2010). The AMMA Sahelian dust transect with three stations in Senegal, Mali and Niger generated an extremely valuable record over several years (Marticorena *et al.* 2010). More recently, the UK Fennec project has for the first time provided very detailed observations from the central Sahara through the deployment of automatic weather stations in remote parts of Mauritania and Algeria (Hobby *et al.* 2013), the set-up of two supersites in Zouerate (Mauritania, Todd *et al.* 2013) and Bordj Baji Mokhtar (Algeria, Marsham *et al.* 2013b) as well as research flights with the UK BAe146 and French Falcon.

Figure 4 shows the passage of a haboob dust storm over the Bordj Badji Mokhtar supersite, emerging form a large MCS over eastern Mali and adjacent Algeria. Ground-based lidar observations show the sharp leading edge of the shallow cold pool and dust concentrations high enough to cause a complete attenuation of the lidar signal above about 200 m. Cold pools are a frequent phenomenon in this region, emerging from afternoon convection in the Sahel and arriving in southern Algeria during the night (Allen *et al.* 2013; Marsham *et al.* 2013b).

7. Final Comments

Although recent advances have been made in terms of improved understanding and model simulation of the WAM system as summarized in the present chapter, its representation in numerical models still presents major bias and short-comings at all scales. Results from the Transpose Atmospheric Model Intercomparison Project II suggest that systematic errors seen in long-term climate runs are already present by day 5 in ensemble average hindcasts (Ma *et al.* 2014), allowing to conclude that it is not an initial value problem. Therefore, these bias are likely initiated by model parameterizations so that improvements in forecast models could lead to a better mean state in climate models and potentially future projections. For the WAM, key issues are the improvement of the convection parameterization, with specific attention on its triggering, diurnal cycle, propagation, coupling with the surface and the representation of cold pools, for example as proposed by Grandpeix and Lafore (2010) and implemented in the LMDz for certain CMIP5 runs (Rio *et al.* 2012; Hourdin *et al.* 2013). The better understanding of the ISV and of the role played by diabatic sources could provide a way to test and improve the coupling between convection, clouds and waves. EaSMs remain very simplified in terms of representing the real Earth system, so two challenges are being able to model the strong scale interactions and Earth-system feedbacks with the inclusion of some missing processes such as dust and vegetation for Africa. Given that we are not going to see an EaSM that resolves the mesoscale any time soon, we should advocate enhanced observations and increased interactions across the various scientific disciplines that are tackling the different aspects of the WAM dynamics.

References

Ackerley, D., B. B. B. Booth, S. H. E. Knight, E. J. Highwood, D. J. Frame, M. R. Allen, and D. P. Rowell, 2011: Sensitivity of twentieth-century Sahel rainfall to sulfate aerosol and CO_2 forcing. *J. Climate*, **24**(19), 4999–5014. doi:10.1175/JCLI-D-11-00019.1.

Agustí-Panareda A., A. Beljaars, C. Cardinali, I. Genkova, and C. Thorncroft, 2010: Impact of assimilating AMMA soundings on ECMWF analyses and forecasts. *Wea. Forecasting* **25**, 1142–1160. doi: 10.1175/2010WAF2222370.1.

Agustí-Panareda, A., G. Balsamo, and A. Beljaars, 2010: Impact of improved soil moisture on the ECMWF precipitation forecast in West Africa. *Geophys. Res. Lett.*, **37**(20), n/a–n/a. doi:10.1029/2010GL044748.

Allen, C. J. T., R. Washington, and S. Engelstaedter, 2013: Dust emission and transport mechanisms in the central Sahara: Fennec ground-based observations from Bordj Badji Mokhtar, June 2011, *J. Geophys. Res. Atmos.*, **118**, doi:10.1002/jgrd.50534.

Arribas, A., M. Glover, A. Maidens, K. Peterson, M. Gordon, C. MacLachlan, *et al.* 2011: The GloSea4 ensemble prediction system for seasonal forecasting. *Mon. Wea. Rev.*, **139**(6), 1891–1910. doi:10.1175/2010MWR3615.1.

Beucher F., J.-P. Lafore, F. Karbou, and R. Roca, 2013: High resolution prediction of a major convective period over West Africa. *Q. J. R. Meteorol. Soc.*, **139**. DOI:10.1002/qj.2225.

Biasutti, M., 2011: Atmospheric science: A man-made drought. *Nature Climate Change*, **1**(4), 197–198. doi:10.1038/nclimate1151.

Biasutti, M., 2013: Forced Sahel rainfall trends in the CMIP5 archive. *J. Geophys. Res.*, **118**(4), 1613–1623. doi:10.1002/jgrd.50206.

Biasutti, M. and Giannini, A. 2006: Robust Sahel drying in response to late 20th century forcings. *Geophys. Res. Lett.*, **33**(11), L11706. doi:10.1029/2006GL026067.

Biasutti, M. and A. H. Sobel, 2009: Delayed seasonal cycle and African monsoon in a warmer climate. *Geophys. Res. Lett.*, **36**, L23707. doi:10.1029/2009GL041303.

Birch, C. E., D. J. Parker, A. O'Leary, J. H. Marsham, C. M. Taylor, P. P. Harris, and G. M. S. Lister, 2012: Impact of soil moisture and convectively generated waves on the initiation of a West African mesoscale convective system. *Q. J. R. Meteorol. Soc.*, **139**(676), 1712–1730. doi:10.1002/qj.2062.

Boone, A., I. Poccard-Leclercq, Y. Xue, J. Feng, and P. Rosnay, 2010: Evaluation of the WAMME model surface fluxes using results from the AMMA land-surface model intercomparison project. *Clim. Dyn.*, **35**(1), 127–142.

Booth, B. B. B., N. J. Dunstone, P. R. Halloran, T. Andrews, and N. Bellouin, 2012: Aerosols implicated as a prime driver of twentieth-century North Atlantic climate variability. *Nature*, **484**(7393), 228–232. doi:10.1038/nature10946.

Chang, C. Y., J. C. H. Chiang, M. F. Wehner, A. R. Friedman, and R. Ruedy, 2011: Sulfate aerosol control of tropical atlantic climate over the twentieth century. *J. Climate*, **24**(10), 2540–2555. doi:10.1175/2010JCLI4065.1.

Charney, J. G., 1975: Dynamics of deserts and drought in the Sahel, *Q. J. R. Meteorol. Soc.*, **101**, 193–202.

Chauvin, F., R. Roehrig, and J.-P. Lafore, 2010: Intraseasonal variability of the Saharan Heat Low and its link with mid-latitudes. *J. Climate*, **23**, 2544–2561.

Chou, C., J. D. Neelin, U. Lohmann, and J. Feichter, 2005: Local and remote impacts of aerosol climate forcing on tropical precipitation, *J. Climate*, **18**, 4621–4636.

Chou, C., J. Chiang, C. W. Lan, C. H. Chung, and Y. C. Liao, 2013: Increase in the range between wet and dry season precipitation. *Nature*. doi:published online: 3 March 2013. DOI: 10.1038/NGEO1744.

Coëtlogon G., S. Janicot, and A. Lazar, 2010: Intraseasonal variability of the ocean-atmosphere coupling in the gulf of Guinea during the northern spring and summer. *QJRMS* **136**, 426–441. DOI:10.1002/qj.554.

Cook, K. H. and E. K. Vizy, 2006: Coupled model simulations of the West African monsoon system: 20th and 21st century simulations. *J. Climate*, **19**(15), 3681–3703. doi:10.1175/JCLI3814.1.

Dixon, N. S., D. J. Parker, C. M. Taylor, L. Garcia-Carreras, P. P. Harris, J. H. Marsham, J. Polcher, and A. Woolley 2012: The effect of background wind on mesoscale circulations above variable soil moisture in the Sahel. *Q. J. R. Meteorol. Soc.*, doi: 10.1002/qj.2012.

Dwyer, J. G., M. Biasutti, and A. H. Sobel, 2013: The effect of greenhouse-gas-induced changes in SST on the seasonality of tropical precipitation. *J. Climate*, Accepted.

Faccani C., F. Rabier, N. Fourrié, A. Agustí-Panareda, F. Karbou, P. Moll, J. P. Lafore, M. Nuret, F. Z. Hdidou, and O. Bock, 2009: The impact of the AMMA radiosonde data on the French global assimilation and forecast system. *Weather and Forecasting*, **24**, 1268–1286.

Fiedler S., K. Schepanski, B. Heinold *et al.* 2013: Climatology of nocturnal low-level jets over North Africa and implications for modeling mineral dust emission. *J. Geophys. Res.*, **118**(12), 6100–6121. doi:10.1002/jgrd.50394.

Folland, C. K., T. N. Palmer, and D. E. Parker, 1986: Sahel rainfall and worldwide sea temperature, 1901–85. *Nature*, **320**, 602–687.

Gaertner, M. A., M. Dominguez, and M. Garvert, 2010: A modelling case-study of soil moisture-atmosphere coupling, *Q. J. R. Meteorol. Soc.*, **136**, 483–495.

Gantner, L. and N. Kalthoff, 2010: Sensitivity of a modelled life cycle of a mesoscale convective system to soil conditions over West Africa. *Q. J. R. Meteorol. Soc.*, **136**, 471–482.

Garcia-Carreras, L., D. J. Parker, C. M. Taylor, C. E. Reeves, and J. G. Murphy, 2010: Impact of mesoscale vegetation heterogeneities on the dynamical and thermodynamic properties of the planetary boundary layer. *J. Geophys. Res. — Atmospheres*, **115**.

Giannini, A., R. Saravanan, and P. Chang, 2003: Oceanic forcing of Sahel rainfall on interannual to interdecadal time scales. *Science*, **302**(5647), 1027–1030.

Govaerts, Y. and A. Lattanzio, 2008: Estimation of surface albedo increase during the eighties Sahel drought from Meteosat observations. *Global and Planetary Change*, **64**(3–4), 139–145.

Grandpeix, J.-Y. and J. P. Lafore, 2010: A density current parametrization coupled to Emanuel's convection scheme. Part I: The models. *J. Atmospheric Sciences*, **67**(4), 881–897. DOI: 10.1175/2009JAS3044.1.

Heinold, B., I. Tegen, K. Schepanski, M. Tesche, M. Esselborn, V. Freudenthaler, S. Gross, K. Kandler, P. Knippertz, D. Müller, A. Schladitz, C. Toledano, B. Weinzierl, A. Ansmann, D. Althausen, T. Müller, A. Petzold, and A. Wiedensohler, 2011: Regional modelling of Saharan dust and biomass burning smoke — Part 1: Model description and evaluation. *Tellus*, **63B**, 781–799. doi:10.1111/j.1600-0889.2011.00570.x.

Heinold, B., P. Knippertz, and J. H. Marsham *et al.*, 2013: The role of deep convection and low-level jets for dust emissions in West Africa. *J. Geophys. Res.*, **118**(10), 4385–4400, doi:10.1002/jgrd.50402.

Held, I. M., T. L. Delworth, J. Lu, K. L. Findell, and T. R. Knutson, 2005: Simulation of Sahel drought in the 20th and 21st centuries. *Proc. Natl. Acad. Sci.*, **102**(50), 17891–17896. doi:10.1073/pnas.0509057102.

Hickler, T., L. Eklundh, J. W. Seaquist, B. Smith, J. Ardo, L. Olsson, M. T. Sykes, and M. Sjostrom, 2005: Precipitation controls Sahel greening trend. *Geophys. Res. Lett.*, **32**(21).

Hobby, M., M. Gascoyne, J. Marsham, M. Bart, C. Allen, S. Engelstaedter, D. Fadel, A. Gandega, R. Lane, J. McQuaid, B. Ouchene, A. Ouladichir, D. Parker, P. Rosenberg, M. Ferroudj, A. Saci, F. Seddik, M. Todd, D. Walker, and R. Washington, 2013: The Fennec Automatic Weather Station (AWS) Network: Monitoring the Saharan Climate System. *J. Atmos. Oceanic Technol.*, **30**, 709–724, doi: 10.1175/JTECH-D-12-00037.1.

Hohenegger, C., P. Brockhaus, C. S. Bretherton, and C. Schar 2009: The soil moisture-precipitation feedback in simulations with explicit and parameterized convection. *J. Climate*, **22**(19), 5003–5020.

Hourdin, F., *et al.*, 2012: LMDZ5B: The atmospheric component of the IPSL climate model with revisited parameterizations for clouds and convection. *Clim. Dyn.*, doi: 10.1007/s00382-012-1343-y.

Hwang, Y.-T., D. M. W. Frierson, and S. M. Kang, 2013: Anthropogenic sulfate aerosol and the southward shift of tropical precipitation in the late 20th century. *Geophys. Res. Lett.*, n/a–n/a. doi:10.1002/grl.50502.

Janicot, S., S. Trzaska, and I. Poccard, 2001: Summer Sahel-ENSO teleconnection and decadal time scale SST variations. *Clim. Dyn.*, **18**, 303–320.

Janicot, S., F. Mounier, N. Hall, S. Leroux, B. Sultan, and G. Kiladis, 2009: The West African monsoon dynamics. Part IV: Analysis of 25-90-day variability of convection and the role of the Indian monsoon. *J. Climate*, **22**, 1541–1565.

Janicot, S., F. Mounier, S. Gervois, B. Sultan, and G. Kiladis, 2010: The dynamics of the

West African monsoon. Part V: The detection and role of the dominant modes of convectively coupled equatorial Rossby waves. *J. Climate*, doi:10.1175/2010JCLI3221.1,

Joseph, R. and N. Zeng, 2011: Seasonally modulated tropical drought induced by volcanic aerosol. *J. Climate*, **24**(8), 2045–2060. doi:10.1175/2009JCLI3170.1.

Karbou F., F. Rabier, J. P. Lafore, J. L. Redelsperger, and O. Bock, 2010: Global 4D-Var assimilation and forecast experiments using AMSU observations over land. Part II: Impact of assimilating surface sensitive channels on the African Monsoon during AMMA. *Weather and Forecasting*, **25**, 20–36.

King, M. P., Kucharski, F., and F. Molteni, 2010: The roles of external forcings and internal variabilities in the northern hemisphere atmospheric circulation change from the 1960s to the 1990s. *J. Climate*, **23**(23), 6200–6220. doi:10.1175/2010JCLI3239.1.

Klose M., Y. Shao, M. K. Karremann, and A. H. Fink, 2010: Sahel dust zone and synoptic background. *Geophys. Res. Lett.* **37**, L09802, doi:10.1029/2010GL042816.

Knippertz, P. and M. C. Todd, 2012: Mineral dust aerosols over the Sahara: Processes of emission and transport, and implications for modeling. *Rev. Geophys.*, **50**, RG1007. doi:10.1029/2011RG000362.

Knippertz P., M. Tesche, and B. Heinold, 2011: Dust mobilization and aerosol transport from West Africa to Cape Verde — A meteorological overview of SAMUM-2. *Tellus*, **63B**, 430–447.

Kocha C., P. Tulet, J.-P. Lafore, and C. Flamant, 2013: The importance of the diurnal cycle of Aerosol optical depth in West Africa. *Geophys. Res. Lett.*, **40**, 785–790. doi:10.1002/grl.50143.

Koster, R. D., P. A. Dirmeyer, Z. Guo, G. Bonan, E. Chan, P. Cox, *et al.*, 2004: Regions of strong coupling between soil moisture and precipitation. *Science*, **305**(5687), 1138–1140. doi:10.1126/science.1100217.

Kucharski, F., N. Zeng, and E. Kalnay, 2013: A further assessment of vegetation feedback on decadal Sahel rainfall variability. *Clim. Dyn.*, **40**(5), 1453–1466. doi:10.1007/s00382-012-1397-x.

Lavaysse, C., C. Flamant, S. Janicot, D. Parker, J.-P. Lafore, B. Sultan, and J. Pelon, 2009: Seasonal evolution of the West African heat low: A climatological perspective. *Clim. Dyn.*, **33**(2), 313–330. doi:10.1007/s00382-009-0553-4.

Lavaysse, C., C. Flamant, and S. Janicot, 2010: Regional-scale convection patterns during strong and weak phases of the Saharan heat low. *Atmospheric Science Letters*, **11**(4), 255–264. doi:10.1002/asl.284.

Lavender, S. L., C. M. Taylor, and A. J. Matthews, 2010: Coupled land–atmosphere intraseasonal variability of the West African monsoon in a GCM. *J. Climate*, **23**(21), 5557–5571.

Lean, H. W., P. A. Clark, M. Dixon, N. M. Roberts, A. Fitch, R. Forbes, and C. Halliwell, 2008. Characteristics of high-resolution versions of the Met Office Unified Model for forecasting convection over the United Kingdom. *Mon. Weat. Rev.*, **136**, 3408–3424.

Ma, H.-Y., S. Xie, S. A. Klein, K. D. Williams, J. S. Boyle, S. Bony, H. Douville, S. Fermepin, B. Medeiros, S. Tyteca, M. Watanabe, and D. Williamson, 2014: On the correspondence between mean forecast errors and climate errors in CMIP5 models. *J. Climate*, **27**, 1781–1798. doi: 10.1175/JCLIM-D13-00474.1.

Mahowald, N. M., Kloster, S., S. Engelstaedter, J. K. Moore, S. Mukhopadhyay, J. R. McConnell, *et al.*, 2010: Observed 20th century desert dust variability: Impact on climate and biogeochemistry. *Atmospheric Chemistry and Physics*, **10**(22), 10875–10893. doi:10.5194/acp-10-10875-2010.

Marsham, J. H., P. Knippertz, N. S. Dixon, D. J. Parker, and G. M. S. Lister, 2011. The importance of the representation of deep convection for modeled dust-generating winds over West Africa during summer. *Geophys. Res. Lett.*, **38**, L16803. doi: 10.1029/2011GL048368.

Marsham J., N. Dixon, L. Garcia-Carreras, *et al.*, 2013a: The role of moist convection in the West African monsoon system — insights from continental-scale convection-permitting simulations. *Geophys. Res. Lett.*, **40**(9), 1843–1849, doi:10.1002/grl.50347.

Marsham, J. H., M. Hobby, C. J. T. Allen, *et al.*, 2013b: Meteorology and dust in the central Sahara: Observations from Fennec supersite-1 during the June 2011 Intensive Observation Period. *J. Geophys. Res.*, **118**(10), 4069–4089. doi: 10.1002/jgrd.50211.

Marticorena, B., B. Chatenet, J. L. Rajot, S. Traoré, M. Coulibaly, A. Diallo, I. Koné, A. Maman, T. NDiaye, and A. Zakou, 2010: Temporal variability of mineral dust concentrations over West Africa: Analyses of a pluriannual monitoring from the AMMA Sahelian Dust Transect. *Atmos. Chem. Phys.*, **10**, 8899–8915.

Matthews A. J., 2004. Intraseasonal variability over tropical Africa during northern summer. *J. Climate*, **17**, 2427–2440.

Meehl, G. A., C. Covey, K. E. Taylor, T. Delworth, R. J. Stouffer, M. Latif, *et al.*, 2007: THE WCRP CMIP3 multimodel dataset: A new era in climate change research. *Bulletin of the American Meteorological Society*, **88**(9), 1383–1394. doi:10.1175/BAMS-88-9-1383.

Monerie, P.-A., B. Fontaine, and P. Roucou, 2012a: Expected future changes in the African monsoon between 2030 and 2070 using some CMIP3 and CMIP5 models under a medium-low RCP scenario. *J. Geophys. Res.*, **117**(D16), D16111. doi:10.1029/2012JD017510.

Monerie, P.-A., P. Roucou, and B. Fontaine, 2012b: Mid-century effects of Climate Change on African monsoon dynamics using the A1B emission scenario. *Int. J. Climatology*, **33**(4), 881–896. doi:10.1002/joc.3476.

Mounier, F., S. Janicot, and G. N. Kiladis, 2008. The West African monsoon dynamics. Part III: The quasi-biweekly zonal dipole. *J. Climate*, **21**, 1911–1928.

Ndiaye, O., M. N. Ward, and W. M. Thiaw, 2011: Predictability of seasonal Sahel rainfall using GCMs and lead-time improvements through the use of a coupled model. *J. Climate*, **24**(7), 1931–1949. doi:10.1175/2010JCLI3557.1.

New, M., M. Todd, M. Hulme, and P. Jones, 2002: Precipitation measurements and trends in the twentieth century. *Int. J. Climatology*, **21**(15), 1889–1922. doi:10.1002/joc.680.

Nicholson, S. E., 1978: Climatic variations in the Sahel and other African regions during the past five centuries. *J. Arid Environments*, **1**(1), 3–24.

Nicholson, S. E., 1979: Revised rainfall series for the West African subtropics. *Mon. Weat. Rev.*, **107**(5), 620–623.

Paeth, H., K. Born, R. Girmes, R. Podzun, and D. Jacob 2009: Regional climate change in tropical and Northern Africa due to greenhouse forcing and land use changes. *J. Climate*, **22**(1), 114–132.

Parker, D. J. 2008: A simple model of coupled synoptic waves in the land surface and atmosphere of the northern Sahel. *Q. J. R. Meteorol. Soc.*, **134**(637), 2173–2184.

Parker, D. J., A. H. Fink, S. Janicot, J. B. Ngamini, M. Douglas, E. Afiesimama, A. Agustí-Panareda, A. Beljaars, F. Dide, A. Diedhiou, T. Lebel, J. Polcher, J. L. Redelsperger, C. Thorncroft, and G. A. Wilson, 2008. The AMMA radiosonde program and its implications for the future of atmospheric monitoring over Africa. *Bull. Am. Meteorol. Soc.*, **89**, 1015–1027.

Pearson, K. J., R. J. Hogan, R. P. Allan, G. M. S. Lister, and C. E. Holloway, 2010: Evaluation of the model representation of the evolution of convective systems using satellite observations of outgoing longwave radiation. *J. Geophys. Res.*, **115**, D20206. doi:10.1029/2010JD014265.

Perlwitz, J. and R. Miller, 2010: Cloud cover increase with increasing aerosol absorptivity — a counterexample to the conventional semi-direct aerosol effect. *J. Geophys. Res.*, **115**, D08203. doi:10.1029/2009JD012637.

Philippon, N., F. J. Doblas-Reyes, and P. M. Ruti, 2010: Skill, reproducibility and potential predictability of the West African monsoon in coupled GCMs. *Clim. Dyn.*, **35**(1), 53–74. doi:10.1007/s00382-010-0856-5.

Poan, D. E., R. Roehrig, Fleur Couvreux, and Jean-Philippe Lafore, 2013: West African monsoon intraseasonal variability: A precipitable water perspective. *J. Atmos. Sc.*, **70**(4), 1035–1052. doi: http://dx.doi.org/10.1175/JAS-D-12-087.1.

Poan, E., J.-P. Lafore, R. Roehrig and F. Couvreux, 2014: Internal processes within African Easterly Waves system. *Q. J. R. Meteorol. Soc.*, in revision.

Redelsperger, J.-L., C. D. Thorncroft, A. Diedhiou, T. Lebel, D. J. Parker, and J. Polcher, 2006: African monsoon multidisciplinary analysis: An international research project and field campaign. *Bull. Am. Meteorol. Soc.*, **87**, 1739–1746.

Rio, C. *et al.* 2012: Control of deep convection by sub-cloud lifting processes: The ALP closure in the LMDZ5B general circulation model. *Cli. Dyn.* DOI: 10.1007/s00382-012-1506-x.

Roehrig, R., F. Chauvin, and J.-P. Lafore, 2011: 10-25-day intraseasonal variability of convection over the Sahel: A role of the Saharan heat low and midlatitudes. *J. Climate*, **24**, 5863–5878. doi:10.1175/2011JCLI3960.1.

Roehrig, R., D. Bouniol, F. Guichard, F. Hourdin, and J. L. Redelsperger, 2013: The present and future of the West African monsoon: A process-oriented assessment of CMIP5 simulations along the AMMA transect. *J. Climate*, 1–88. doi:10.1175/JCLI-D-12-00505.1.

Rotstayn, L. D. and U. Lohmann, 2002: Tropical rainfall trends and the indirect aerosol effect. *J. Climate*, **15**, 2103–2116.

Rowell, D. P., 2013: Simulating SST teleconnections to Africa: What is the state of the art? *J. Climate*. doi:10.1175/JCLI-D-12-00761.1.

Seth, A., S. A. Rauscher, M. Biasutti, A. Giannini, S. J. Camargo, and M. Rojas, 2013: CMIP5 projected changes in the annual cycle of precipitation in monsoon regions. *J. Climate*, 1–48.

Shanahan, T. M., J. T. Overpeck, K. J. Anchukaitis, J. W. Beck, J. E. Cole, D. L. Dettman, *et al.*, 2009: Atlantic forcing of persistent drought in West Africa. *Science*, **324**(5925), 377–380. doi:10.1126/science.1166352.

Schepanski, K., I. Tegen, M. C. Todd, *et al.*, 2009: Meteorological processes forcing Saharan dust emission inferred from MSG-SEVIRI observations of subdaily dust source activation and numerical models. *J. Geophys. Res.*, **114**, D10201. doi:10.1029/2008JD010325.

Schwendike, J., N. Kalthoff, and M. Kohler, 2010: The impact of mesoscale convective systems on the surface and boundary-layer structure in West Africa: Case-studies from the AMMA campaign 2006. *Q. J. R. Meteorol. Soc.*, **136**(648), 566–582.

Solmon, F., N. Elguindi, and M. Mallet, 2012: Radiative and climatic effects of dust over West Africa, as simulated by a regional climate model. *Clim. Res.*, doi:10.3354/cr01039.

Sultan, B., S. Janicot, and A. Diedhiou, 2003: The West African monsoon dynamics. Part I: Documentation of intraseasonal variability. *J. Climate*, **16**, 3407–3427.

Taylor, K. E., R. J. Stouffer, and G. A. Meehl, 2012: An overview of CMIP5 and the experiment design. *Bull. Am. Meteorol. Soc.*, **93**(4), 485–498. doi:10.1175/BAMS-D-11-00094.1.

Taylor, C. M., D. J. Parker, and P. P. Harris, 2007: An observational case study of mesoscale atmospheric circulations induced by soil moisture. *Geophys. Res. Lett.*, **34**, L15801. doi:10.1029/2007GL030572.

Taylor, C. M., 2008: Intraseasonal land-atmosphere coupling in the West African monsoon. *J. Climate*, **21**(24), 6636–6648.

Taylor, C. M., P. P. Harris, and D. J. Parker, 2010: Impact of soil moisture on the development of a Sahelian mesoscale convective system: A case study from the AMMA special observing period. *Quart. J. Roy. Meteorol. Soc.*, **136**, 456–470.

Taylor, C. M., A. Gounou, F. Guichard, P. P. Harris, R. J. Ellis, F. Couvreux, and M. De Kauwe, 2011: Frequency of Sahelian storm initiation enhanced over mesoscale soil-moisture patterns. *Nature Geosci.*, **4**(7), 430–433.

Taylor, C. M., R. A. M. de Jeu, F. Guichard, P. P. Harris, and W. A. Dorigo, 2012: Afternoon rain more likely over drier soils. *Nature*, **489**(7416), 423–426.

Taylor, C. M., C. E. Birch, D. J. Parker, N. Dixon, F. Guichard, G. Nikulin, and G. M. S. Lister, 2013: Modeling soil moisture-precipitation feedback in the Sahel: Importance of spatial scale versus convective parameterization. *Geophys. Res. Lett.*, **40**(23), 2013GL058511.

Ting, M., Y. Kushnir, and R. Seager, 2011: Robust features of Atlantic multi-decadal variability and its climate impacts. *Geophys. Res. Lett.*

Todd, M. C., R. Washington, S. Raghavan, *et al.*, 2008: Regional model simulations of the Bodélé low-level jet of northern Chad during the Bodélé Dust Experiment (BoDEx 2005). *J. Climate*, **21**, 995–1012.

Todd, M. C., C. J. T. Allen, M. Bart, *et al.*, 2013: Meteorological and dust aerosol conditions over the Western Saharan region observed at Fennec supersite-2 during the Intensive Observation Period in June 2011. *J. Geophys. Res.*, doi: 10.1002/jgrd.50470.

Tompkins, A. M. and L. Feudale, 2010: Seasonal ensemble predictions of West African monsoon precipitation in the ECMWF system 3 with a focus on the AMMA special observing period in 2006. *Weat. Forecasting*, **25**(2), 768–788. doi:10.1175/2009WAF2222236.1.

Torn, R. D., 2010: Ensemble-based sensitivity analysis applied to African easterly waves. *Weat. Forecasting*, **25**, 20–61-78. doi: 10.1175/2009WAF2222255.1.

Ward, M. N., 1998: Diagnosis and short-lead time prediction of summer rainfall in Tropical North Africa at interannual and multidecadal timescales. *J. Climate*, **11**, 3167–3191.

Wolters, D., C. C. van Heerwaarden, J. V.-G. de Arellano, B. Cappelaere, and D. Ramier, 2010: Effects of soil moisture gradients on the path and the intensity of a West African squall line. *Q. J. R. Meteorol. Soc.*, **136**(653), 2162–2175.

Xue, Y. and J. Shukla, 1993: The influence of land-surface properties on Sahel climate. Part 1. Desertification. *J. Climate*, **6**, 2232–2245.

Xue, Y., A. Boone, and C. M. Taylor, 2012: Review of recent developments and the future prospective in West African atmosphere/land interaction studies. *Int. J. Geophys.*, **2012**, 12.

Yamada, T. J., S. Kanae, T. Oki, and Y. Hirabayashi, 2011: The onset of the West African monsoon simulated in a high-resolution atmospheric general circulation model with reanalyzed soil moisture fields. *ASL*, doi: 10.1002/asl.367.

Yoshioka, M., N. M. Mahowald, A. J. Conley, W. D. Collins, D. W. Fillmore, C. S. Zender, and D. B. Coleman 2007: Impact of desert dust radiative forcing on Sahel precipitation: Relative importance of dust compared to sea surface temperature variations, vegetation changes, and greenhouse gas warming. *J. Climate*, **20**, 1445–1467.

Yue, X., H. Liao, H. J. Wang, S. L. Li, and J. P. Tang, 2011: Role of sea surface temperature responses in simulation of the climatic effect of mineral dust aerosol. *Atmos. Chem. Phys.*, 6049–6062. doi:10.5194/acp-11-6049-2011.

Zeng, N., J. D. Neelin, K. M. Lau, and C. J. Tucker 1999: Enhancement of interdecadal climate variability in the Sahel by vegetation interaction. *Science*, **286**, 1537–1540.

Zhang, R., T. L. Delworth, Sutton, R., D. L. R. Hodson, K. W. Dixon, I. M. Held, *et al.*, 2013: Have aerosols caused the observed Atlantic multidecadal variability? *J. Atmos. Sci.*, 130104154622004. doi:10.1175/JAS-D-12-0331.1.

Chapter 5

The East Asian Winter Monsoon

Lin Wang*,‡ and Mong-Ming Lu†,§

*Center for Monsoon System Research, Institute of Atmospheric Physics, Chinese Academy of Sciences, Beijing

†APEC Research Center for Typhoon and Society, Taipei

‡wanglin@mail.iap.ac.cn,

§lu@rdc.cwb.gov.tw

The East Asian winter monsoon (EAWM) research of the past several years is briefly reviewed. On the interannual time scale, the EAWM variability features two major modes: the strength and pathway of the EAWM, which are linked to the changes in the amplitude and phase of atmospheric planetary waves, respectively. The blocking activity over Ural and Pacific regions exert significant impacts on the strength of the EAWM in the form of Eurasian and Western Pacific patterns. The former also contributes largely to the recent decadal amplification of the EAWM. El Niño-Southern Oscillation (ENSO) and Arctic sea ice are important external forcing for the EAWM variability, while the impacts of ENSO on the EAWM vary with decades. On the intraseasonal time scale, the Siberian high shows clear out-of-phase relationship between early and peak winter in recent thirty years. Both the stratospheric variability and the Madden-Julian Oscillation could influence the characteristics of cold surges over East Asia. All these provide intraseasonal prediction potentials for the EAWM.

1. Introduction

The East Asian winter monsoon (EAWM) is a distinct meteorological phenomenon whose fluctuations influence the weather and climate in Eurasian continent and Asia-Pacific region from sub-polar zone to the equator (Chang *et al.* 2011; Huang *et al.* 2012). Near the surface, it is characterized by an intense Siberian high (SH) with the Aleutian low (AL) to its east and the Maritime Continent low-pressure to its south. The resultant strong northwesterlies and northeasterlies are observed to the north and south of about 30°N, respectively. These surface features are dynamically coupled to the mid-tropospheric East Asian trough and the upper-tropospheric East Asian jet stream. Primarily driven by the large-scale thermal contrast between the cold Asian continent and the adjacent warm oceans, the EAWM varies on various time scales. These variations of the EAWM are closely linked to the atmospheric planetary waves (PWs) excited by topography and land-sea thermal contrast. Therefore, the fluctuations of external factors such as tropical sea surface temperature (SST) and Arctic sea ice extent can influence the EAWM via exciting atmospheric wave-like responses. In this chapter, we will briefly review the progress in the EAWM research reported during the past few years with a particular focus on its variability in a variety of time scales ranging from intraseasonal to decadal.

The Global Monsoon System: Research and Forecast (3rd Edition)
Edited by C. P. Chang et al.

2. Interannual Variations

2.1. *Major modes of the EAWM and the associated PW activities*

The interannual variability of the EAWM can be described by the major modes of the circulation (Wu *et al.* 2006; Wang *et al.* 2009a; Chen *et al.* 2014) or the near-surface temperature (Kang *et al.* 2006; 2009) fields. The leading temperature mode is characterized by a homogeneous pattern south of 50°N over East Asia (Kang *et al.* 2006; 2009). It is related to the strength of the EAWM that reflects the overall coldness of East Asia with below (above) normal temperature in strong (weak) EAWM winters (Figs. 1a, 1b; Wang *et al.* 2009a). The second mode exhibits the out-of-phase temperature oscillation straddling at 40°N over East Asia. It is related to the pathway of the EAWM, which reflects the preferred route of the cold air intruding East Asia in the monthly or seasonal mean sense (Wang *et al.* 2009a). For example, in strong EAWM winters with a strong southern (northern) pathway, the cold air tends to intrude (be confined) to the southern (northern) part of East Asia; hence, significant low temperature anomalies are observed centering at about 25°N (45°N) (Figs. 1c, 1d), which is about 10° south (north) of the position without clarifying the pathway (Fig. 1a).

The variability of the two major modes of the EAWM and the atmospheric PWs are inherently related because both EAWM and the stationary PWs can be influenced by the large-scale thermal contrast between land and oceans (Chen *et al.* 2005; Wang *et al.* 2009b; Takaya and Nakamura 2013; Wang and Chen 2014a). This explains, to some extent, why the variability of the EAWM is closely linked to the middle and upper-tropospheric wave activities and teleconnections although the prevailing northerly wind of the EAWM is shallow and confined mainly below 850 hPa level. The excitation of the PWs from atmospheric lower boundary can be

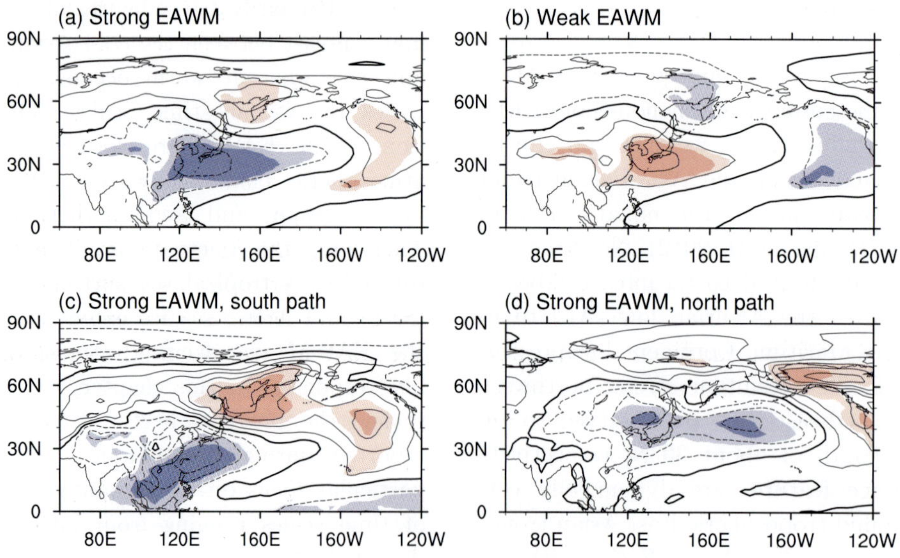

Fig. 1. Composite of DJF mean 850-hPa air temperature anomalies in (a) strong EAWM winters without clarifying the monsoon pathway, (b) weak EAWM winters without clarifying the monsoon pathway, (c) strong EAWM winters with strong southern pathway, and (d) strong EAWM winters with strong eastern pathway. Zero contour lines are bolded and negative values are dashed. Contour intervals (CIs) are 0.5°C. Dark and light shading indicate the 99% and 95% confidence levels based on Student's *t* test, respectively. Adapted from Wang *et al.* (2009a).

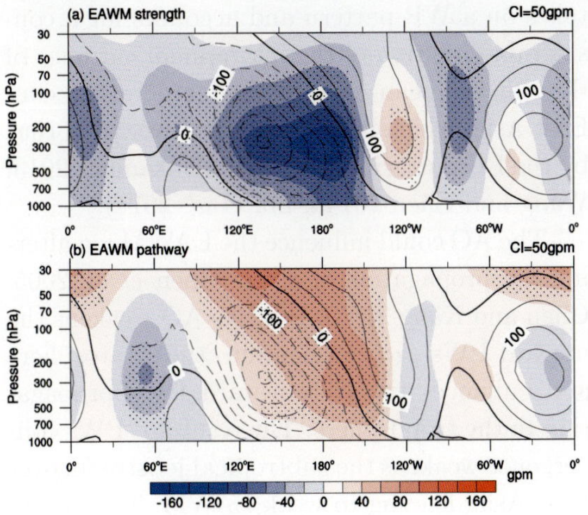

Fig. 2. The climatology of geopotential height departures from zonal means (contours) along 50°N overlaid with composite anomalies (shading) of DJF mean geopotential height (a) between strong and weak EAWM winters, and (b) between winters with southern and northern EAWM pathway. Contour and shading intervals are 50 gpm and 20 gpm, respectively. Zero contour lines are bold. Stippling indicates the 95% confidence level based on Student's t test. Adapted from Wang (2014).

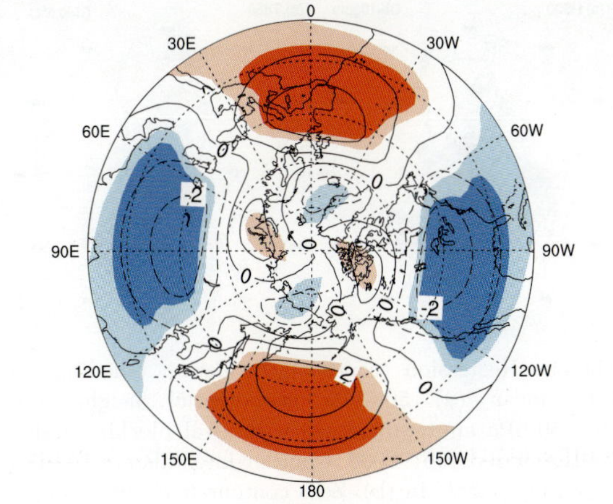

Fig. 3. Composite difference of the zonal wavenumber-2 of DJF mean SLP between weak and strong PW activity winters. Contour intervals are 1 hPa. Dark and light shading indicate the 99% and 95% confidence levels based on Student's t test, respectively. Adapted from Chen *et al.* (2005).

enhanced by the intensified land-sea thermal contrast over East Asia and western North Pacific (Chen *et al.* 2005; Wang *et al.* 2009b), and it consequently enhances the PW amplitude throughout the troposphere especially over East Asia-North Pacific region (Fig. 2a). This process is dominated by the zonal wavenumber-2 waves that can deepen the mid-tropospheric East Asian trough and enhance the surface SH and AL (Figs. 2a, 3), leading to a strong EAWM with intensified northerly wind and cold condition in East Asia. When the phase of PWs shifts westward over the mid-latitude North Pacific, it facilitates a more north-south-oriented mid-tropospheric East Asian trough without changing its amplitude too much (Fig. 2b). The resultant meridional flow eventually produces a strong southern pathway and steers the cold air moving to the southern part of East Asia (Wang 2014).

By analyzing the surface air temperature over East Asia, Wang, B. *et al.* (2010a) also identified two leading modes of the EAWM that are potentially predictable in climate models (Wu, Z. *et al.* 2011; Lee, J. Y. *et al.* 2013a), known as the northern and southern modes. The center of the northern temperature mode is located in western and central Siberia, and the center of the southern temperature mode is located in temperate East Asia including China, Korea, Japan and their coastal waters. The patterns of the two modes are different from those related to the aforementioned strength and the pathway of the EAWM, so their relations need to be examined in the future. The cold northern mode is preceded by excessive autumn snow cover over southern SH, whereas the cold southern mode is preceded by the development of a La Niña episode and reduced snow cover over northeast Siberia.

2.2. *Influences of blocking and teleconnection patterns*

A key circulation feature for the interannual variability of the EAWM is the Eurasian (EU)-like pattern (Wallace and Gutzler, 1981) (Fig. 4a) modulated by PWs (Takaya and

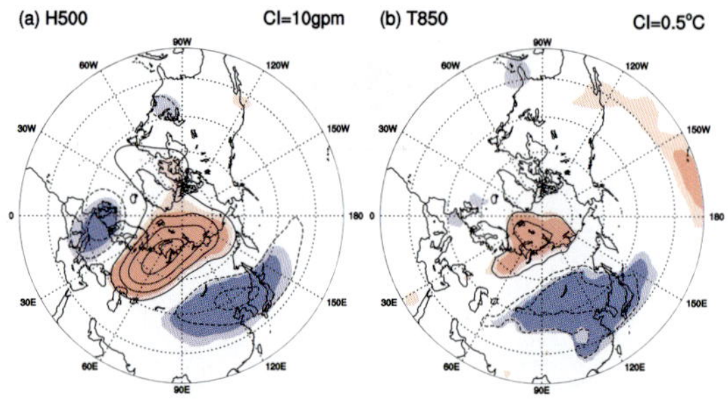

Fig. 4. Regression (contour)/correlation (shading) of DJF mean (a) 500-hPa geopotential height and (b) 850-hPa air temperature with Ural blocking index for the period 1957–2000. Contour intervals are 10 gpm in (a) and 0.5°C in (b). Zero contour lines are omitted and negative values are dashed. Dark and light shading indicate the 99% and 95% confidence levels based on Student's *t* test, respectively. Adapted from Wang *et al.* (2010b).

Nakamura 2013). The EU pattern is often associated with the enhanced blockings over Europe and Ural regions accompanied by a barotropic upper-tropospheric Rossby wave train on intraseasonal (Takaya and Nakamura 2005a) or interannual (Wang, L. *et al.* 2010b; Liu *et al.* 2014) time scales. The EU-like pattern is tightly linked to the amplifications of the SH and the mid-tropospheric East Asian trough and cold condition over East Asia (Fig. 4b). It explains approximately 30% of the interannual variance of the winter mean surface air temperature over East Asia (Liu *et al.* 2014). The relationship between the EU-like blocking pattern and the EAWM was enhanced after the mid-1970s (Wang, L. *et al.* 2010b) with some modulations by the El Niño-Southern Oscillation (ENSO) and the Arctic Oscillation (AO) (Cheung *et al.* 2012; Chen *et al.* 2013). The Western Pacific (WP) pattern (Wallace and Gutzler, 1981) is another key circulation feature for the interannual variability of the EAWM (Takaya and Nakamura 2005b). The blockings over North Pacific, driven by the vorticity feedback forcing of transient eddies, could retrogress from the Aleutian region into eastern Siberia on the intraseasonal or interannual time scales. It takes on a WP pattern and accounts for a considerable fraction of the interannual variance of the winter mean surface air temperature over East Asia, which is only slightly lower than that by the EU pattern (Takaya and Nakamura 2013; Wang and Chen 2014a; Liu *et al.* 2014),

The AO could influence the EAWM via altering the propagation of PWs (Chen *et al.* 2005; Chen and Kang 2006). Positive AO weakens the upward PWs propagation into the stratosphere and enhances the equatorward PWs propagation in the troposphere. The resultant PWs convergence weakens the subtropical jet stream over East Asia, leading to weak EAWM. This mechanism is especially significant in the easterly phase of the stratospheric tropical quasi-biennial oscillation (Chen and Li 2007).

2.3. *Influences of SST and sea ice*

Warm (cold) ENSO events often lead to anomalous lower-tropospheric anticyclone (cyclone) around Philippine via Gill-type Rossby wave response in their peak winters and facilitate weak (strong) EAWM (Zhang *et al.* 1996; Wang *et al.* 2000). Such relationship is valid only during the negative phase of the Pacific Decadal Oscillation (PDO) due to the modulation effect of PDO on the anomalous anticyclone (cyclone) (Figs. 5a, 5b, Wang *et al.* 2008). The PDO is in its positive phase from mid-1970s to early-2000s; hence, a weakened ENSO-EAWM linkage is observed in this epoch (Fig. 5c, Wang and He 2012). In coupled climate prediction models, ENSO is an important source of predictability of the EAWM with efficient prediction leading up to seven months (Jiang *et al.* 2013; Lee, J. Y. *et al.* 2013). The successful reproduce of the ENSO-EAWM relationship in coupled climate models mainly depends on the reasonable simulation of both the ENSO's spatial structure and the ENSO's strength of interannual variability (Sohn *et al.* 2011; Gong *et al.* 2014; 2015). The ocean-atmosphere coupling, especially that over the western Maritime

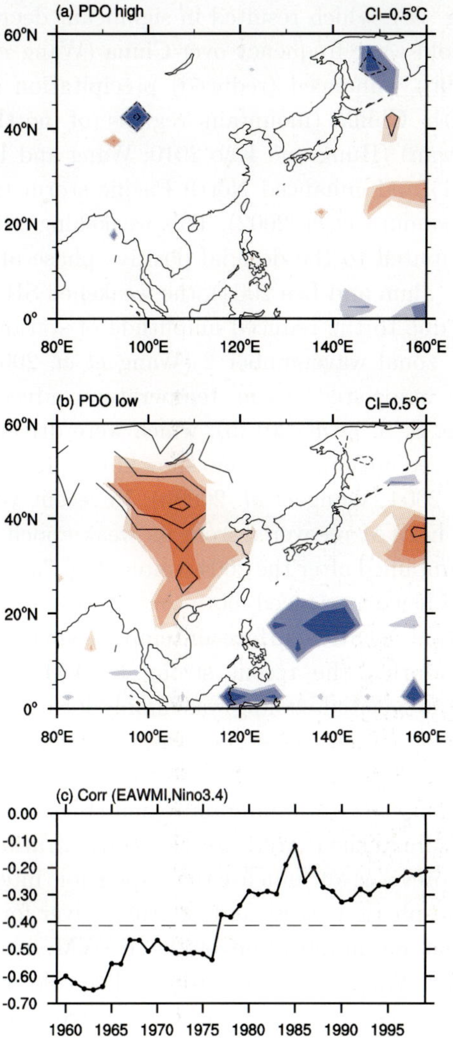

Fig. 5. Regression (contour)/correlation (shading) of the observed DJF mean surface air temperature from CRUTEM3 on the normalized Niño3.4 index in (a) high and (b) low phase of the PDO. (c) 23-year sliding correlation coefficient between DJF mean Niño3.4 index and EAWM index. Contour intervals are 0.5°C in (a) and (b). Zero contour lines are omitted and negative values are dashed. Dark, medium and light shading indicate the 99%, 95%, and 90% confidence levels based on Student's *t* test, respectively. The dashed line in (c) indicates the 95% confidence levels based on Student's *t* test. Adapted from Wang *et al.* (2008) and Wang and He (2012).

Continent is important for the EAWM prediction (Jiang *et al.* 2013). The tropical Indian Ocean may provide additional information for the EAWM predictions besides ENSO (Wang and Chen 2014a). The associated mechanism is not fully understood, but it is quite likely related to an upper-tropospheric Rossby wave train excited by the anomalous tropical precipitation over the Indo-Western Pacific (Zheng *et al.* 2013).

Reduced autumn Arctic sea ice over the Atlantic sector of the Arctic often favor strong EAWM with cold condition and increased snowfall (e.g., Honda *et al.* 2009; Wu, B. *et al.* 2011; Liu *et al.* 2012). On one hand, the enhanced heating due to reduced sea ice over the Barents-Kara-Laptev Seas could serve as a source of a Rossby wave train that propagates towards East Asia with positive SLP and geopotential height anomalies around the Ural Mountain (Honda *et al.* 2009; Wu, B. *et al.* 2011). On the other hand, the diminished sea ice in this region could shift the surface baroclinic zone northward due to the reduced meridional temperature gradient (Inoue *et al.* 2012). The tracks of wintertime cyclone then shifts northward to the polar region, resulting in positive SLP and geopotential height anomalies around the Kara Sea (Fig. 6). Both mechanisms cause anomalous cold advection to the eastern Siberia. The resultant surface cold anomalies can serve as surface anticyclonic potential vorticity anomalies

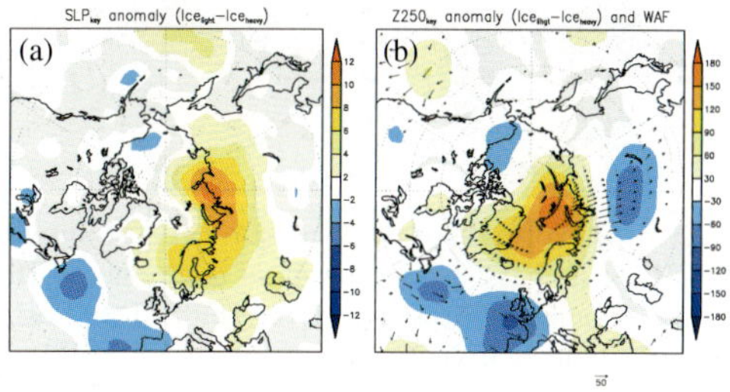

Fig. 6. Difference maps of (a) SLP (hPa) and (b) geopotential height (m) at 250 hPa and the wave activity flux (vector, $m^2\ s^{-2}$, Takaya and Nakamura 2001) between light and heavy Arctic sea ice years. These values were obtained by subtracting the composite response for averaged heavy ice case events from those for averaged light ice case events. Gray shading indicates areas with less than a 99% confidence level based on Student's *t* test. Adapted from Inoue *et al.* (2012).

and reinforce the upper-tropospheric wave train, thereby inducing an amplification of the SH and strong EAWM (Honda *et al.* 2009). These mechanisms are often difficult to reproduce in numerical experiments owing to energetic and large atmospheric internal variability, while ensemble simulations with enough members could partially overcome the difficulty and detect a robust atmospheric response to the diminished sea ice (Mori *et al.* 2014).

3. Decadal and Long-term Variations

A most significant decadal variation of the EAWM is its weakening after the 1986 winter (Fig. 7a), which resulted in significant decrease of cold wave frequency over China (Wang *et al.* 2009b), enhanced (reduced) precipitation over South China (mountain regions of northern Taiwan) (Hung and Kao 2010; Wang and Feng 2011) and enhanced North Pacific storm track (Nakamura *et al.* 2002). This weakening can be attributed to the decadal positive phase of the AO (Jhun and Lee 2004), the weakened SH and AL due to the reduced amplitude of stationary PW zonal wavenumber 2 (Wang *et al.* 2009b), and associated warm temperature advection (Lee, S. S. *et al.* 2013b), which were driven by reduced autumn Eurasian snow cover (Jhun and Lee 2004, Wang *et al.* 2009b). In recent years, the EAWM recovered from its weak epoch and re-amplified after the 2004 winter (Fig. 7a). East Asia has experienced more cold winters and significant negative surface air temperature anomalies during the recent strong EAWM epoch. The associated cooling was mainly located over inland northern East Asia with a west-east orientation. Enhanced wintertime blocking activity around the Ural Mountain region (Figs. 7b, 7c) and diminished Arctic sea ice concentration in the previous autumn are the responsible internal atmospheric process and external driver for the recent re-amplification of the EAWM, respectively (Wang and Chen 2014b). The cold waves in this strong EAWM epoch were stronger and lasted longer compared to those that occurred in the 1990s (Woo *et al.* 2013), which shared similar features with those cold waves in the strong EAWM epoch before 1986.

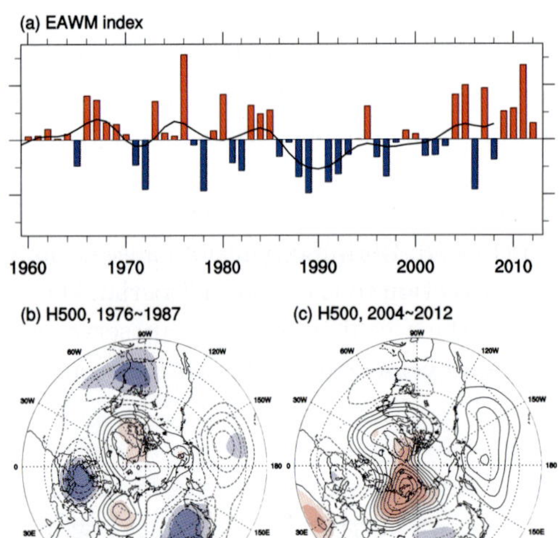

Fig. 7. (a) Normalized DJF mean EAWM intensity index (bar) and its 9-year low-pass components (line) using the Lanczos filter based on HadSLP2. Positive (negative) indices indicate strong (weak) EAWM winters. Composite anomalies of DJF mean 500 hPa geopotential height in two strong EAWM epochs (a) 1976–1987 and (b) 2004–2012 based on the NCEP/NCAR reanalysis dataset. The composite is performed with respect to the weak EAWM epoch 1988-2003. Contour intervals are 10 gpm in (c) and (d). Zero contour lines are omitted and negative values are dashed. Dark and light shading indicate the 99% and 95% confidence levels based on Student's t test, respectively. Adapted from Wang and Chen (2014b).

4. Intraseasonal Variations

4.1. *The Siberian high*

From the perspective of the intraseasonal evolution, the EAWM can be divided into initial, developing, and final stages and show clear interannual variations (Kim *et al.* 2013; 2014). The intraseasonal variations of the EAWM is closely linked to that of the SH, which is influenced by blocking originated from the regions to its west and east in the form of wave train and westward retrogression, respectively (Takaya and Nakamura 2005b; Park *et al.* 2014). Lu and Chang (2009) found that only in a few extremely strong cases over the last 50 years has there been a relationship between the Atlantic blocking and the intensity of the SH. Meanwhile, Wang, L. *et al.* (2010b) reported the emergence of a closer relationship between the Ural blocking and the SH after mid-1970s.

Chang and Lu (2012) showed the SH had an intraseasonal variation that tended to be seasonally-synchronized in the recent 30 years, which produced an out-of-phase relationship between November and December-January. This implies a special intraseasonal predictability that did not exist in the two previous decades. If this relationship continues, the EAWM will be the only known major circulation system whose intensity can be predicted to reverse from the previous month. The spatial pattern of this seasonally-synchronized intraseasonal out-of-phase relationship is not related with ENSO but related with the intraseasonal reversal of blocking frequency over Ural and Pacific regions (Fig. 8). This out-of-phase relationship was not observed before the climate shift in 1970s. It is suggested that the predictability is enhanced by the reduced frequency of blocking events during the positive phase of the AO. While it suggests the predictability may diminish if the AO phase is reversed, it may become more prevalent in the future if the prediction of more frequent positive AO-like patterns in a warming world forced by greenhouse gases is borne out.

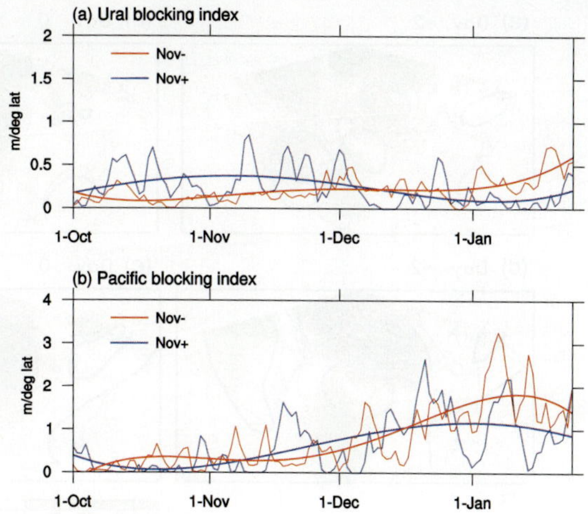

Fig. 8. Composite time series of (a) Ural blocking index and (b) Pacific blocking index with respect to the sign of the November Siberian high anomaly for 1979–2008. Thick lines are the 4th order polynomial fit of the time series. Adapted from Chang and Lu (2012).

4.2. *Influences of the AO and the stratosphere*

The AO exerts distinct impacts on the cold surges of East Asia. In the positive AO months, the tropospheric background winds are more zonally oriented. Accordingly, cold surges mainly feature an upper-tropospheric wave train that propagates eastward to the North Pacific (Figs. 9a–9c). Durations of the cold surges are relatively short and the climatic impacts are mainly confined to inland China due to southward shift of cold surge tracks. In the negative AO months, the background winds are more meandering. The cold surges feature a well-organized anticyclone-cyclone couplet over East Asia (Figs. 9d–9f). The resultant climatic impacts are long lasting and strong with eastward shift of cold surge tracks (Park *et al.* 2011). On the other hand, the dynamics of the AO involves apparent coupling between the troposphere and stratosphere. Hence, the stratospheric information may provide prediction potentials for the cold surges. Increased cold days can be observed over Eurasian continent especially over East Asia after the stratospheric polar vortex weakens (Fig. 10, Kolstad *et al.* 2010). Some case studies also support the importance of the stratosphere in the cold surges (Jeong *et al.* 2006; Ding and Ma 2007; Wang and Chen 2010).

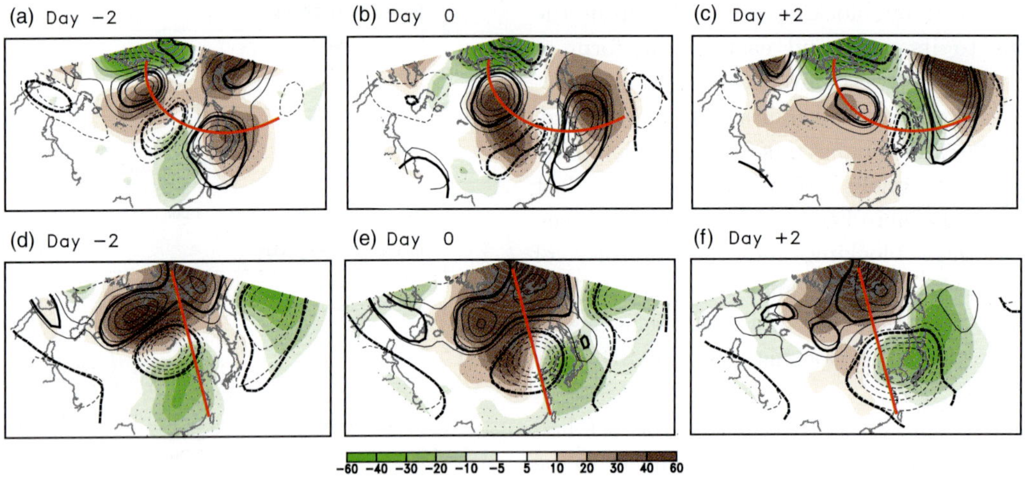

Fig. 9. Composite anomalies of geopotential height at 300 hPa (contours; in intervals of 20 m, negative values are dashed, significant values at the 95% confidence level are represented by thick lines) and 850 hPa (shadings; significant values at the 95% confidence level are represented by gray dots) for cold surge occurrences during (a)–(c) positive AO months and (d)–(f) negative AO months. Adapted from Park *et al.* (2011).

4.3. *Influences of the Madden-Julian Oscillation*

The Madden-Julian Oscillation (MJO), which is a major predictability source on the intraseasonal time scale, could exert significant impacts on the EAWM. In phases 2–3 of the MJO (enhanced convection is over the Indian Ocean), the EAWM is featured with cold condition, more frequent strong cold surges (Jeong *et al.* 2005), and enhanced precipitation (Jeong *et al.* 2008; Jia *et al.* 2011). In phases 6–7 (enhanced convection is located over western Pacific), the climate anomalies over East Asia are roughly opposite to those in phases 2–3. Such influences are realized through changes of local Hadley circulation and the Northern Hemisphere branch of the equatorially trapped Rossby wave gyres forced by the MJO heating (He *et al.* 2011) and the associated quasi-geostrophic forcing of the vertical motion (Jeong *et al.* 2008). The ENSO-induced seasonal anomalies can significantly modify the pattern of MJO teleconnection, leading to different influences of the MJO on the EAWM between El Niño and La Niña winters (Moon *et al.* 2011).

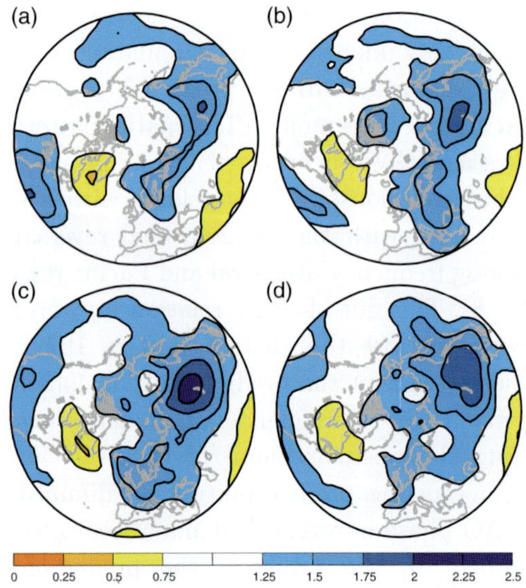

Fig. 10. The composite number of cold days divided by the date-wise climatological mean number of cold days for each grid point during the (a) peak, (b) mature, (c) decline, and (d) decay stage of weak vortex events. Adapted from Kolstad *et al.* (2010).

5. Discussion

The EAWM is an essential part of the global monsoon system. Understanding its variability has been a challenging task. Despite the progresses in recent years, many issues remain open. For example, what are the major characteristics and mechanism of the EAWM variability on the intraseasonal time scales? What are the atmospheric external factors that dominate the EAWM variability on interannual and decadal time scales and what are the involved physical processes? How important is the atmospheric internal variability compared with the atmospheric external forcing for the EAWM variability? How will the EAWM response to the increasing anthropogenic forcing? How reproducible and predicable is the EAWM in numerical models? All these issues deserve further research in the future.

Acknowledgments

We thank the anonymous reviewer for the constructive comments. LW was supported by the National Natural Science Foundation of China (41422501, 41230527) and the Excellent Young Scientists Project of the Chinese Academy of Sciences.

References

Chang, C. P., M. M. Lu, and B. Wang, 2011: The East Asian winter monsoon. *The Global Monsoon System: Research and Forecast (2nd Ed)*, C. P. Chang, *et al.*, Eds., World Scientific, pp. 99–109.

Chang, C. P. and M. M. Lu, 2012: Intraseasonal predictability of Siberian high and East Asian winter monsoon in recent decades. *J. Climate*, **25**, 1773–1778.

Chen, W., S. Yang, and R. H. Huang, 2005: Relationship between stationary planetary wave activity and the East Asian winter monsoon. *J. Geophys. Res.*, **110**, doi: 10.1029/2004JD005669.

Chen, W. and L. Kang, 2006: Linkage between the Arctic Oscillation and winter climate over East Asia on the interannual timescale: Roles of quasi-stationary planetary waves. *Chinese J. Atmos. Sci.*, **30**, 863–870.

Chen, W. and T. Li, 2007: Modulation of Northern Hemisphere wintertime stationary planetary wave activity: East Asian climate relationships by the quasi-biennial oscillation. *J. Geophys. Res.*, **112**, doi: 10.1029/2007JD008611.

Chen, W., X. Q. Lan, L. Wang, and Y. Ma, 2013: The combined effects of the ENSO and the Arctic Oscillation on the winter climate anomalies in East Asia. *Chin. Sci. Bull.*, **58**, 1355–1362.

Chen, Z., R. Wu, and W. Chen, 2014: Distinguishing interannual variations of the northern and southern modes of the East Asian winter monsoon. *J. Climate*, **27**, 835–851.

Cheung, H. N., W. Zhou, H. Y. Mok, and M. C. Wu, 2012: Relationship between Ural-Siberian blocking and the East Asian winter monsoon in relation to the Arctic Oscillation and the El Niño-Southern Oscillation. *J. Climate*, **25**, 4242–4257.

Ding, Y. H. and X. Q. Ma, 2007: Analysis of isentropic potential vorticity for a strong cold wave in 2004/2005 winter. *Acta Meteorol. Sin.*, **65**, 695–707.

Gong, H. N., L. Wang, W. Chen, R. Wu, K. Wei, and X. Cui, 2014: The climatology and interannual variability of the East Asian winter monsoon in CMIP5 models. *J. Climate*, **27**, 1659–1678.

Gong, H. N., L. Wang, W. Chen, D. Nath, G. Huang, and W. C. Tao, 2015: Diverse influences of ENSO on the East Asian–Western Pacific winter climate tied to different ENSO properties in CMIP5 models. *J. Climate*, **28**, 2187–2202.

He, J. H., H. Lin, and Z. W. Wu, 2011: Another look at influences of the Madden-Julian Oscillation on the wintertime East Asian weather. *J. Geophys. Res.*, **116**, doi: 10.1029/2010JD014787.

Honda, M., J. Inoue, and S. Yamane, 2009: Influence of low Arctic sea-ice minima on anomalously cold Eurasian winters. *Geophys. Res. Lett.*, **36**, doi: 10.1029/2008GL037079.

Huang, R., J. Chen, L. Wang, and Z. Lin, 2012: Characteristics, processes, and causes of the spatio-temporal variabilities of the East Asian monsoon system. *Adv. Atmos. Sci.*, **29**, 910–942.

Hung, C. W. and P. K. Kao, 2010: Weakening of the winter monsoon and abrupt increase of winter rainfalls over northern Taiwan and southern China in the early 1980s. *J. Climate*, **23**, 2357–2367.

Inoue, J., M. E. Hori, and K. Takaya, 2012: The role of Barents sea ice in the wintertime cyclone track and emergence of a warm-Arctic cold-Siberian anomaly. *J. Climate*, **25**, 2561–2568.

Jeong, J. H., C. H. Ho, B. M. Kim, and W. T. Kwon, 2005: Influence of the Madden-Julian Oscillation on wintertime surface air temperature and cold surges in East Asia. *J. Geophys. Res.*, **110**, doi: 10.1029/2004JD005408.

Jeong, J. H., B. M. Kim, C. H. Ho, D. L. Chen, and G. H. Lim, 2006: Stratospheric origin of cold surge occurrence in East Asia. *Geophys. Res. Lett.*, **33**, doi: 10.1029/2006GL026607.

Jeong, J. H., B. M. Kim, C. H. Ho, and Y. H. Noh, 2008: Systematic variation in wintertime precipitation in East Asia by MJO-induced extratropical vertical motion. *J. Climate*, **21**, 788–801.

Jhun, J. G. and E. J. Lee, 2004: A new East Asian winter monsoon index and associated characteristics of the winter monsoon. *J. Climate*, **17**, 711–726.

Jia, X. L., L. J. Chen, F. M. Ren, and C. Y. Li, 2011: Impacts of the MJO on winter rainfall and circulation in China. *Adv. Atmos. Sci.*, **28**, 521–533.

Jiang, X., S. Yang, Y. Li, A. Kumar, W. Wang, and Z. Gao, 2013: Dynamical prediction of the East Asian winter monsoon by the NCEP Climate Forecast System. *J. Geophys. Res.*, **118**, 1312–1328.

Kang, L., W. Chen, and K. Wei, 2006: The interdecadal variation of winter temperature in China and its relation to the anomalies in atmospheric general circulation. *Clim. Environ. Res.*, **11**, 330–339.

Kang, L., W. Chen, L. Wang, and L. Chen, 2009: Interannual variations of winter temperature in China and their relationship with the atmospheric circulation and sea surface temperature. *Clim. Environ. Res.*, **14**, 45–53.

Kim, Y., K. Y. Kim, and J. G. Jhun, 2013: Seasonal evolution mechanism of the East Asian winter monsoon and its interannual variability. *Clim. Dyn.*, **41**, 1213–1228.

Kim, Y., K. Y. Kim, and S. Park, 2014: Seasonal scale variability of the East Asian winter monsoon and the development of a two-dimensional monsoon index. *Clim. Dyn.*, **42** 2159–2172.

Kolstad, E. W., T. Breiteig, and A. A. Scaife, 2010: The association between stratospheric weak polar vortex events and cold air outbreaks in the Northern Hemisphere. *Q. J. Roy. Meteor. Soc.*, **136**, 886–893.

Lee, J. Y., S. S. Lee, B. Wang, K. J. Ha, and J. G. Jhun, 2013: Seasonal prediction and predictability of the Asian winter temperature variability. *Clim. Dyns.*, **41**, 573–587.

Lee, S. S., S. H. Kim, J. G. Jhun, K. J. Ha, and Y. W. Seo, 2013: Robust warming over East Asia during the boreal winter monsoon and its possible causes. *Environ. Res. Lett.*, **8**, 034001.

Liu, J. P., J. A. Curry, H. J. Wang, M. Song, and R. M. Horton, 2012: Impact of declining Arctic sea ice on winter snowfall. *Proc. Natl. Acad. Sci. USA*, **109**, 4074–4079.

Liu, Y., L. Wang, W. Zhou, and W. Chen, 2014: Three Eurasian teleconnection patterns: Spatial structures, temporal variability, and associated winter climate anomalies. *Clim. Dyn.*, **42**, 2817–2839.

Lu, M. M. and C. P. Chang, 2009: Unusual late-season cold surges during the 2005 Asian winter monsoon: Roles of Atlantic blocking and the Central Asian anticyclone. *J. Climate*, **22**, 5205–5217.

Mori, M., M. Watanabe, H. Shiogama, J. Inoue, and M. Kimoto, 2014: Robust Arctic sea-ice influence on the frequent Eurasian cold winters in the recent past. *Nat. Geosci.*, **7**, 869–873.

Moon, J. Y., B. Wang, and K. J. Ha, 2011: ENSO regulation of MJO teleconnection. *Clim. Dyn.* **37**, 1133–1149.

Nakamura, H., T. Izumi, and T. Sampe, 2002: Interannual and decadal modulations recently observed in the Pacific storm track activity and East Asian winter monsoon. *J. Climate*, **15**, 1855–1874.

Park, T. W., C. H. Ho, and S. Yang, 2011: Relationship between the Arctic Oscillation and cold surges over East Asia. *J. Climate*, **24**, 68–83.

Park, T. W., C. H. Ho, and Y. Deng, 2014: A synoptic and dynamical characterization of wave-train and blocking cold surge over East Asia. *Clim. Dyns.*, **43**, 753–770.

Sohn S. J., C. Y. Tam, and C. K. Park, 2011: Leading modes of East Asian winter climate variability and their predictability: An assessment of the APCC multi-model ensemble. *J. Meteor. Soc. Japan*, **89**, 455–474.

Takaya, K. and H. Nakamura, 2005a: Mechanisms of intraseasonal amplification of the cold Siberian high. *J. Atmos. Sci.*, **62**, 4423–4440.

Takaya, K. and H. Nakamura, 2005b: Geographical dependence of upper-level blocking formation associated with intraseasonal amplification of the Siberian high. *J. Atmos. Sci.*, **62**, 4441–4449.

Takaya, K. and H. Nakamura, 2013: Interannual variability of the East Asian winter monsoon and associated modulations of the planetary waves. *J. Climate*, **26**, 9445–9461.

Wallace, J. M. and D. S. Gutzler, 1981: Teleconnections in the geopotential height field during the Northern Hemisphere winter. *Mon. Wea. Rev.*, **109**, 784–812.

Wang, B., R. Wu, and X. Fu, 2000: Pacific–East Asian teleconnection: How does ENSO affect East Asian climate? *J. Climate*, **13**, 1517–1536.

Wang, B., Z. W. Wu, C. P. Chang, J. Liu, J. P. Li, and T. J. Zhou, 2010: Another look at interannual-to-interdecadal variations of the East Asian winter monsoon: The northern and southern temperature modes. *J. Climate*, **23**, 1495–1512.

Wang, H. J. and S. P. He, 2012: Weakening relationship between East Asian winter monsoon and ENSO after mid-1970s. *Chin. Sci. Bull.*, **57**, 3535–3540.

Wang, L., W. Chen, and R. H. Huang, 2008: Interdecadal modulation of PDO on the impact of ENSO on the East Asian winter monsoon. *Geophys. Res. Lett.*, **35**, L20702. doi:10.1029/2008GL035287.

Wang, L., W. Chen, W. Zhou, and R. H. Huang, 2009a: Interannual variations of East Asian trough axis at 500 hPa and its association with the East Asian winter monsoon pathway. *J. Climate*, **22**, 600–614.

Wang, L., R. H. Huang, L. Gu, W. Chen, and L. H. Kang, 2009b: Interdecadal variations of the East Asian winter monsoon and their association with quasi-stationary planetary wave activity. *J. Climate*, **22**, 4860–4872.

Wang, L. and W. Chen, 2010: Downward Arctic Oscillation signal associated with moderate weak stratospheric polar vortex and the cold December 2009. *Geophys. Res. Lett.*, **37**, L09707. doi:10.1029/2010GL042659.

Wang, L., W. Chen, W. Zhou, J. C. L. Chan, D. Barriopedro, and R. Huang, 2010: Effect of the climate shift around mid 1970s on the relationship between wintertime Ural blocking circulation and East Asian climate. *Int. J. Climatol.*, **30**, 153–158.

Wang, L. and J. Feng, 2011: Two major modes of the wintertime precipitation over China. *Chinese J. Atmos. Sci.*, **35**, 1105–1116.

Wang, L., 2014: Stationary wave activity associated with the East Asian winter monsoon pathway, *Atmos. Oceanic Sci. Lett.*, **7**, 7–10.

Wang, L. and W. Chen, 2014a: An intensity index for the East Asian winter monsoon. *J. Climate*, **27**, 2361–2374.

Wang, L. and W. Chen, 2014b: The East Asian winter monsoon: Re-amplification in the mid-2000s. *Chin. Sci. Bull.*, **59**, 430–436.

Woo, S. H., B. M. Kim, J. H. Jeong, S. J. Kim, and G. H. Lim, 2012: Decadal changes in surface air temperature variability and cold surge characteristics over northeast Asia and their relation with the Arctic Oscillation for the past three decades (1979-2011). *J. Geophys. Res.*, **117**, doi:10.1029/2011JD016929.

Wu, B., R. Zhang, and R. D'Arrigo, 2006: Distinct modes of the East Asian winter monsoon. *Mon. Wea. Rev.*, **134**, 2165–2179.

Wu, B., J. Su, and R. Zhang, 2011: Effects of autumn-winter Arctic sea ice on winter Siberian high, *Chin. Sci. Bull.*, **56**, 3220–3228.

Wu, Z., J. Li, Z. Jiang, and J. H. He, 2011: Predictable climate dynamics of abnormal East Asian winter monsoon: Once-in-a-century snowstorms in 2007/2008 winter. *Clim. Dyns.*, **37**, 1661–1669.

Zhang, R., A. Sumi, and M. Kimoto, 1996: Impact of El Niño on the East Asian monsoon: A diagnostic study of the '86/87 and '91/92 events. *J. Meteor. Soc. Japan*, **74**, 49–62.

Zheng, J., Q. Liu, C. Wang, and X. T. Zheng, 2013: Impact of heating anomalies associated with rainfall variations over the Indo-Western pacific on Asian atmospheric circulation in winter. *Clim. Dyn.*, **40**, 2023–2033.

Chapter 6

Recent Progress in Research on the Maritime Continent Monsoon

Shigeo Yoden

Division of Earth and Planetary Sciences, Kyoto University, Japan
yoden@kugi.kyoto-u.ac.jp

Shigenori Otsuka

RIKEN Advanced Institute for Computational Science, Japan

Nurjanna Joko Trilaksono

Faculty of Earth Sciences and Technology
Institut Teknologi Bandung, Indonesia

Tri Wahyu Hadi

Faculty of Earth Sciences and Technology
Institut Teknologi Bandung, Indonesia

Recent progress and achievements in the research on the Maritime Continent monsoon are reviewed by focusing on the studies reported in the last five years or so. The progress has been promoted by the advancement of computing facilities for data analysis and numerical simulations, technical innovations in observations, and well-designed numerical experimentations for better understanding. This review is divided into five subsections based on their time scales as follows: (a) annual cycle and seasonal transitions, (b) diurnal cycle of rainfall and its multiscale interactions, (c) subseasonal variability, (d) interannual variations and seasonal predictability, and (e) interdecadal variability, trends, and geological time-scale climate changes.

1. Introduction

In the previous edition of *The Global Monsoon System* (Chang *et al.* 2011), Robertson *et al.* (2011) reviewed Maritime Continent (MC) monsoon research and related topics based on selected publications mainly in the period of about five or ten years before that time. The MC monsoon is related to both Asian and Australian monsoons with seasonal asymmetries under the geographical asymmetry with respect to the equator in continental scales (Fig. 1). Complex topography in the MC region introduces wind-terrain interactions that cause multiscale variations in time and space, from local diurnal variations to global-scale interannual variations. Madden-Julian Oscillation (MJO) and El Niño-Southern Oscillation (ENSO) are the most prominent and important features in tropical intraseasonal and interannual variations, respectively, which could modulate and interact with the MC monsoon with their own time scales. Climate change in the MC over geological time scales has also been reported by analyzing new proxy data. This article reviews recent progress and achievements in the research on the MC monsoon since the previous review by Robertson *et al.* (2011).

The Global Monsoon System: Research and Forecast (3rd Edition)
Edited by C. P. Chang et al.

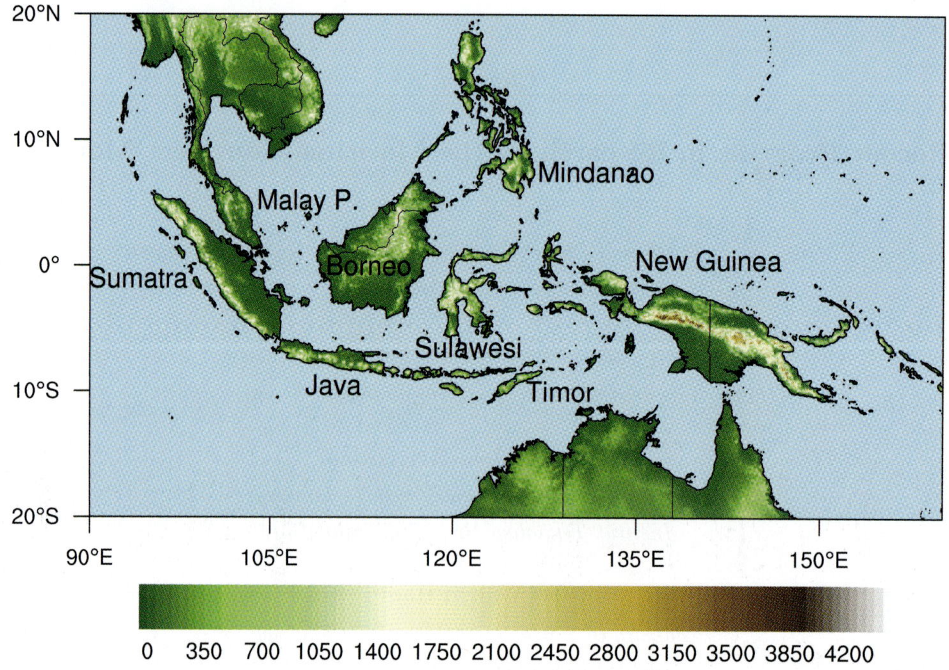

Fig. 1. Topography of the Maritime Continent.

The atmosphere has diurnal and seasonal responses to the periodic variations of the solar energy input onto the Earth due to its rotation and revolution, respectively, and the latter is the dynamical driver of monsoon. The periodic response to the annual insolation variation contains higher harmonics in addition to the basic sinusoidal variation with a 12-month period because of the nonlinear nature of the response of the atmospheric circulation. The higher harmonics play an important role in producing rather quick onset of monsoon circulations and thus asymmetry between spring and fall. For longer time scales, hydrosphere (oceans), cryosphere, and biosphere play more important roles in climate variations and modulations of monsoon circulations, because of their large heat capacity (i.e. thermal inertia). These are some fundamental and general background for common feature of global monsoon systems, among which the MC monsoon is a unique subset because of the complex coast lines and topography of MC and its location close to the equator.

2. Review of the Recent Progress and Achievements

2.1. *Annual cycle and seasonal transitions*

The MC monsoon has asymmetry between boreal summer and austral summer in each hemisphere due to the geographical distribution of Asian and Australian continents as well as the spring-fall asymmetry in each location as described by Chang *et al.* (2005). Figure 2 shows the 30-year climatology of seasonal 925 hPa horizontal winds and divergence over the MC and the 3-year (2003–2005) average of seasonal rainfall (Qian 2008). The monsoonal migration of rainfall is clearer over the southern MC; more rainfall in the austral summer months (December–February, or DJF) than the

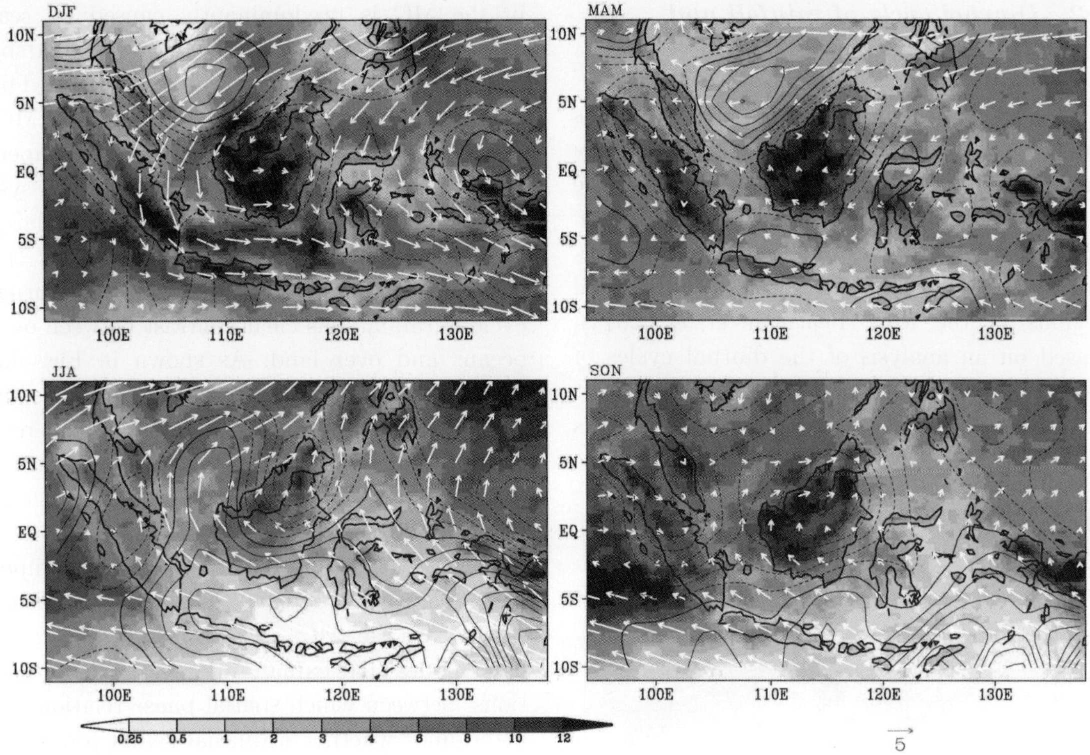

Fig. 2. Climatology (1971–2000) of horizontal winds (vectors) and divergence (contours) at 925 hPa over the Maritime Continent obtained from the NCEP-NCAR Reanalysis Project (NNRP) dataset, and 2003–2005 averaged seasonal precipitation (mm day^{-1}, shaded) obtained from the NOAA/Climate Prediction Center's Morphing Technique (CMORPH) high-resolution precipitation dataset (Qian 2008).

austral winter months (June–August, or JJA) with intermediate values during the austral fall (March–May, or MAM) and spring (September–November, or SON).

Some detailed descriptive studies have been added to the previous studies of monsoon rainfall in and around the MC and disturbances that bring the rainfall, by using archived observational data sources such as surface rainfall measurements, satellite observations, and reanalysis datasets of meteorological fields. For example, as for the monsoon rainfall in Peninsular Malaysia, Suhaila *et al.* (2010) analyzed daily rainfall data from 50 rain gauge stations for the period of 1975–2004 to show the spatial patterns and trends of five rainfall indices, and Varikoden *et al.* (2011) used the rain rate from Tropical Rainfall Measuring Mission (TRMM) 3-hourly data, available for a 10-year period from 1999 to 2008 at a high spatial resolution of 25 km to explore the diurnal variation of rainfall occurrences in the area. Chen *et al.* (2013) also analyzed rainfall datasets together with reanalysis covering the past three decades from 1979 to 2009 to describe the major cause of heavy rainfall events in the peninsular with synoptic and dynamic analyses. Similar analyses of the monsoon rainfall and disturbances for the other regions in the MC have been done intensively with such new datasets to describe the details of annual cycle and seasonal transitions of the MC monsoon, though publication in international journals might be limited.

2.2. *Diurnal cycle of rainfall and its multiscale interactions*

As shown in Fig. 2, rainfall over the MC is mostly concentrated over islands, and the diurnal cycle of rainfall plays an important role to form the climatological spatial distribution. Figure 3a shows large amplitudes of the diurnal harmonic of rainfall over islands and coastal regions in the MC (Peatman *et al.* 2014). Based on an analysis of the diurnal cycles of rainfall and winds, Qian (2008) pointed out that the concentration of rainfall over islands

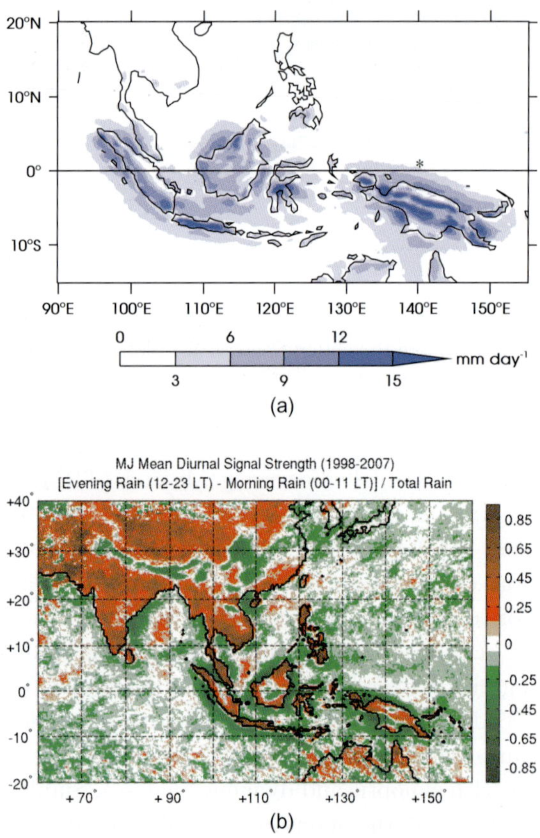

Fig. 3. (a) Climatological amplitude of the diurnal harmonic of precipitation rate calculated from TRMM 3B42HQ data (Peatman *et al.* 2014). (b) Average May–June evening (1200–2300 LT) minus morning (0000–1100 LT) rainfall normalized by May–June mean rainfall for ten years (1998–2007) using TRMM 3B42 data (Johnson 2011).

in the MC is predominantly caused by sea-breeze convergence over islands, reinforced by mountain-valley winds, and further amplified by the cumulus merger processes. Over the MC, therefore, the diurnal cycle of local winds superposed on large-scale slowly-evolving wind systems is an important process in forming large amounts of rainfall and latent heating of the atmosphere. Furthermore, phase of the diurnal cycle of rainfall has clear contrast between over oceans and over land. As shown in Fig. 3b, morning rainfall maxima can be seen in coastal regions throughout most of the MC, whereas evening rainfall maxima over land except for the central mountain area of Borneo Island (Johnson 2011).

Teo *et al.* (2011) made a principal component analysis of observed and modeled diurnal rainfall in the MC and obtained two dominant modes that explain most of diurnal variations, between which spatial phase relationship determines whether a diurnal signal propagating or stationary. Their space-time analysis clarified the impact of coastlines and mountains as sources of propagating signals.

The diurnal cycle of rainfall over the MC has also been studied with high-resolution numerical models. Trilaksono *et al.* (2011) demonstrated that daily propagation of enhanced rainfall signals was simulated well in an explicit convection model with horizontal resolution of 2 km (Fig. 4) and that during the Jakarta flood event in 2007 the heavy rainfall in the northern coast region of western Java occurred at the time when the propagating signals from north and south meet there. They pointed out that finer resolution results show better performance on the rainfall over complex terrain of the mountain region. Based on the numerical result of a 4-km resolution explicit convection model, Love *et al.* (2011) also showed an important role of diurnally forced gravity waves that propagate away from their point of generation in the diurnal cycle of convection over the MC. They also noted the necessity of the inclusion of such a small-scale effect in

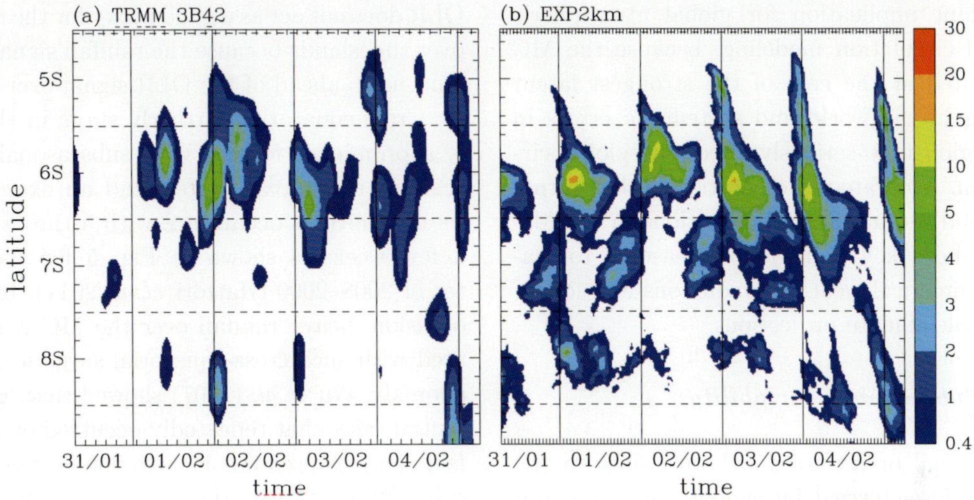

Fig. 4. Time-latitude cross sections of the precipitation rate (mm/hr) of (a) the TRMM 3B42 estimation and (b) the ensemble mean of a 2-km regional model experiment from 0700 LT 31 January to 0600 LT 5 February 2007 averaged between 106.5°E–108.5°E (West Java). Modified from Fig. 2 of Trilaksono *et al.* (2011). The vertical lines denote 0000 LT. Note that time and horizontal resolutions for the TRMM estimation are 3 hours and 25 km, respectively, and time resolution for the plot of the 2-km model result is 1 hour.

climate model simulation from the comparison with the result obtained in a 40-km resolution model with convective parametrization.

Multiscale interactions of the diurnal cycle of rainfall over the MC and impact of the MJO have been studied with TRMM and outgoing long-wave radiation (OLR) satellite observations and reanalysis datasets. Rauniyar and Walsh (2011) showed both of the amplitude and phase of the diurnal cycle of rainfall are modulated by the MJO categories (active, suppressed, and weak) over land and ocean, and Oh *et al.* (2012) showed an enhancement of the diurnal cycle during the active phases of the MJO and the difference of the changes in the precipitation rate due to the MJO over land and ocean. Peatman *et al.* (2014) showed 80% of the MJO rainfall signal in the MC is accounted for by changes in the amplitude of the diurnal cycle over land where the mean diurnal cycle of rainfall is strong as shown in Fig. 3a.

More studies pertaining to the MJO and diurnal cycle interactions around Sumatra Island have been reported. Combined analyses of observational data and numerical experiment data by Fujita *et al.* (2011) showed that, under active phases of the MJO, the evening convection over the land is mainly induced by solar radiative heating and the midnight convection over the ocean is triggered by convergence of the low-level westerly wind and the land breeze. The effects of topography of small islands off the western coast of Sumatra also strongly characterize the diurnal convection (Kamimera *et al.* 2012).

Qian *et al.* (2013) found that rainfall over Borneo is strongly affected by the diurnal cycle of land-sea breeze and that the spatial distribution of rainfall, which depends on the direction of monsoonal winds, is also associated with ENSO. Their weather typing analysis shows that the spatial structure of rainfall anomalies associated with ENSO is caused by the variability in the frequency of occurrence of different weather types.

As pointed out by Qian (2008) and others, the multiscale interaction in the MC between the diurnal cycle of local rainfall and winds and the longer-time variations of large-scale weather and climate systems has an important implication for global atmospheric general circulation modeling, because the MC is situated at the core of the strongest latent heating in the world and systematic errors in this region may seriously affect the global circulation. Accurate representation of the diurnal cycle and its multiscale interactions in the MC should be a key point for subseasonal to seasonal numerical weather predictions and longer time-scale climate projections.

2.3. *Subseasonal variability*

Subseasonal or intraseasonal variability in the MC is characterized by synoptic features and events such as MJO, cold surge, and Borneo vortex, and their interactions (Robertson *et al.* 2011). Convectively coupled equatorial waves, such as Kelvin waves and equatorial Rossby waves, have been well observed to produce rather regular and coherent intraseasonal variations in near equatorial monsoons, as reviewed by Wheeler and McBride (2011). Complex relationship among those synoptic features and the effects of the topography contribute to the variability in convection patterns over a variety of space and time scales in the MC.

As for the subseasonal variations associated with the MJO, Wu and Hsu (2009) pointed out the influence of topography on the MJO in the MC such that the eastward movement of the MJO features tends to skirt around mountainous islands, such as New Guinea. Rauniyar and Walsh (2011) showed that the average daily rainfall rate over the islands of the MC is higher during convectively suppressed phase of MJO, whereas for surrounding oceans, more rainfall occurs during convectively active phase of MJO. These are associated with distinct variations in large-scale atmospheric conditions in each phase of MJO; during the active phase, more humid but more thermodynamically stable conditions result in less rainfall over the islands of the MC, with much less upward motion. Peatman *et al.* (2014) also pointed out that in the MJO cycle OLR does not act as a good proxy for the rainfall over the islands because the rainfall signals over land jump ahead of the OLR signal over land.

Cross-equatorial northerly surge in the MC is a prominent feature with subseasonal time-scale in the boreal winter, and an example of its intermittent occurrence with a time scale of a few weeks is shown in Fig. 5 for the winter of 2008–2009 (Hattori *et al.* 2011). In some occasion, heavy rainfall over the MC is associated with such cross-equatorial surge with cold anomaly. Wu *et al.* (2007) showed that the torrential rains that repeatedly occurred over Java Island and caused the Jakarta flood event in the end of January and the beginning of February in 2007 coincided with a strong and persistent cross-equatorial cold surge event. Hattori *et al.* (2011) analyzed the 10 winters from December 2000 to March 2009 and identified that about a half (27 events) of the cross-equatorial northerly surge over the MC (62 events) were cold surges. The cold surges associated with active phase of the MJO were shown to have significant relationship to the increased rainfall to the north of Java Island.

Trilaksono *et al.* (2012) performed a series of time-lagged ensemble hindcast experiments for January and February in 2007 using a regional nonhydrostatic model to investigate the evolutions of synoptic field and the probability of precipitation rate and to show the specific conditions related to the heavy rainfall that caused the Jakarta flood event in 2007. Their probability analysis for pentad time scale showed the possibility of heavy precipitation was highest for the pentad of the cold surge event in the analyzed two months.

Takahashi *et al.* (2011) investigated the impact of northerly surges of the east Asian winter monsoon on tropical cyclogenesis over the eastern Indian Ocean and the MC, by using Japanese 25-year Reanalysis (JRA-25) and OLR datasets for 28 winters in 1979/1980–2006/2007. They described the seasonally dependent impacts of long-lasting surges associated with

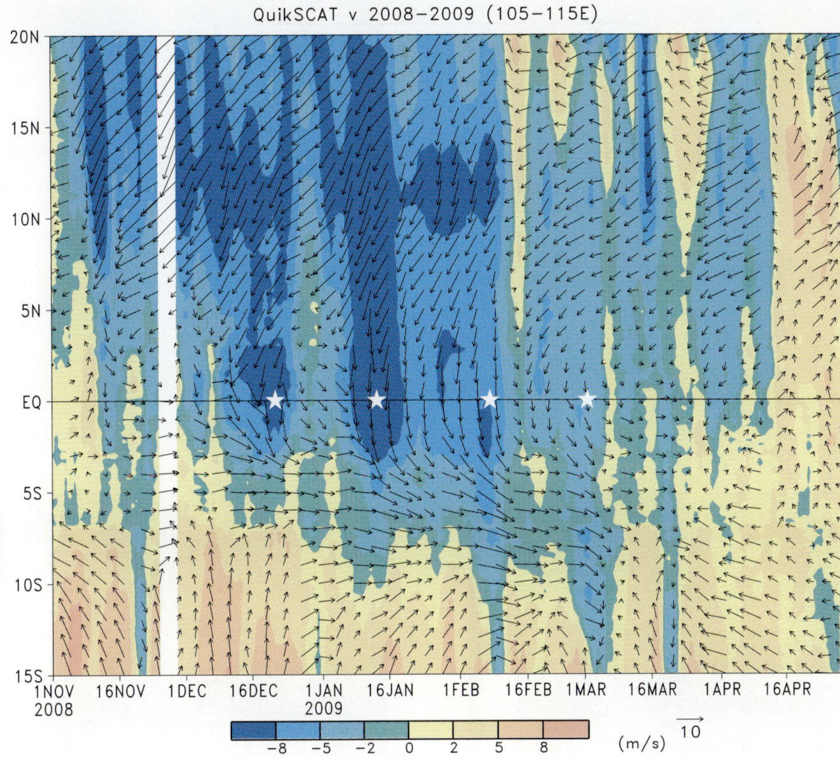

Fig. 5. Time-latitude cross-sections of the meridional component of sea surface wind (shading) and the wind vectors from November in 2008 to April in 2009 over the South China Sea between 105E and 115E. Stars represent some events of southward extension of the northerly wind (Hattori *et al.* 2011).

the background conditions of atmospheric circulation. In the MC, they noticed the development of so-called Borneo vortex in December through February, and pointed out the associated intensification of low-level wind along a channel between the islands of Borneo, Sumatra, and Java.

The Borneo vortex is a geographically fixed, closed circulation with a shallow structure near the surface, causing enhanced convection and heavy rainfall through the interaction with the northerly cold surge (Robertson *et al.* 2011). Ooi *et al.* (2011) made a case study of a Borneo vortex event induced by the cold surge during January 2010, and Braesicke *et al.* (2012) described climatological properties of strong offshore Borneo vortex by a composite analysis of flow pattern and ozone and water vapor distributions. By analyzing reanalysis and satellite datasets for the month of December from 1981 to 2008, Koseki *et al.* (2014) showed that Borneo vortices during strong cold surge events (133 days) occur mostly around the west of Borneo near the equator where tropical cyclones usually do not develop (Fig. 6a). The composite maps show axi-asymmetric structure around the center of Borneo vortex in horizontal wind and absolute vorticity at 850 hPa (Fig. 6b), indicating the influence of the strong cold surges. The pattern of heavy rainfall is approximately consistent with the pattern of strong convergence with a slight northeastward displacement of the former from the latter (Fig. 6c).

Based on the hindcast experiments data for January and February in 2007, Trilaksono *et al.* (2012) described three-dimensional

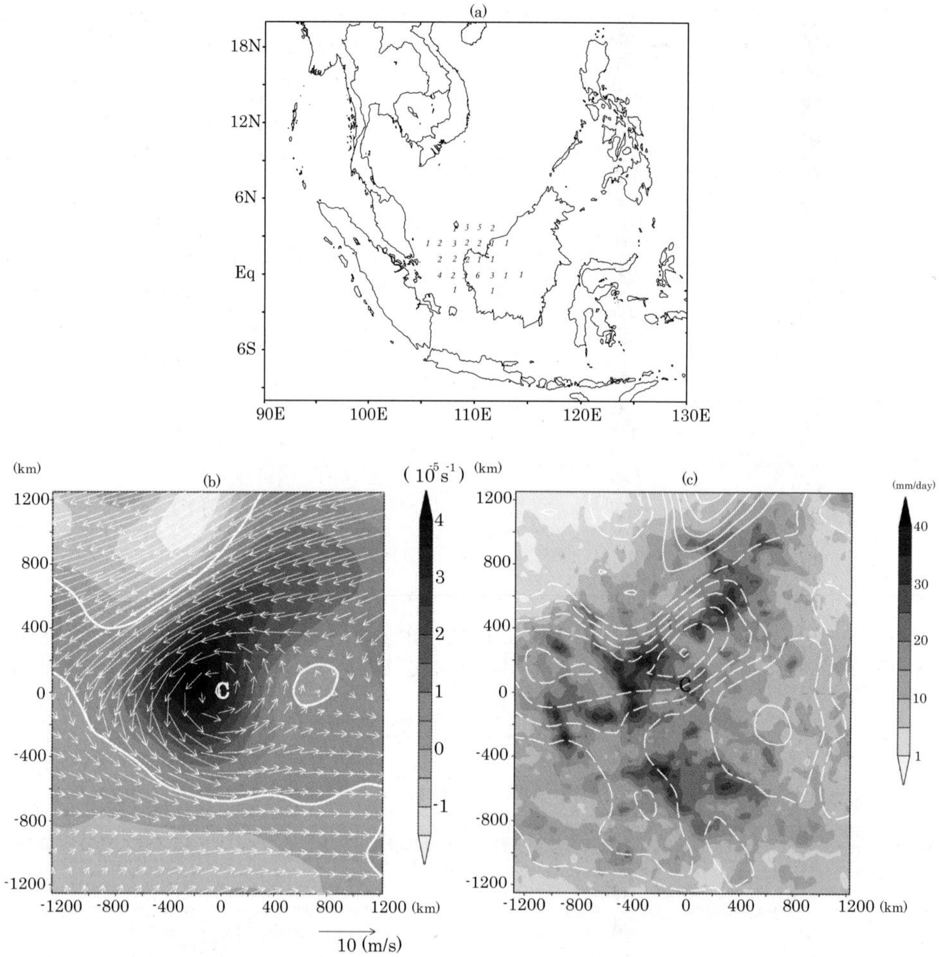

Fig. 6. Composites of Borneo vortex obtained from JRA25/JCDAS and TRMM data (Koseki *et al.* 2014). (a) The location of Borneo vortex center. The grid numbers give the number of days that the center is located at each grid. (b) The composite of absolute vorticity (shaded) and horizontal wind (vector) at 850 hPa around the center of Borneo vortex. The white thick contour denotes the value of zero absolute vorticity. (c) The composite of horizontal divergence (contour interval is $10^{-6}\,s^{-1}$) at 850 hPa and hourly rainfall (shaded) around the center of the Borneo vortex. Solid (dashed) line means positive (negative) value. Symbol C denotes the center of the composite Borneo vortex.

synoptic feature of the Borneo vortex prior to a cross-equatorial cold surge; it is characterized by shallow structure with a vertical scale of 3 km from the surface and a horizontal scale of 1000 km. Koseki *et al.* (2014) performed a semi-idealized numerical experiment on the Borneo vortex during a "perpetual" cold surge. They revealed new detailed features of the meso-β-scale structure of the Borneo vortex in the organization of the comma-shaped rainband and analyzed the vorticity budget for the growth and maintenance of the meso-α-scale vortex.

During the peer review process of this article, a review paper on monsoon convection in the MC by Chang *et al.* (2016) became available. It is a thorough review and summary of published studies on the interactions between the MC terrain and large-scale transient systems of the Borneo vortex and cold surges during the boreal winter monsoon; it describes two dynamic processes, i.e., the terrain and cold surges, associated with northeasterly monsoon winds that can lead to the initiation and strengthening of the Borneo Vortex. A section is devoted to review the formation of Typhoon Vamei near the equator in 2001 as an extreme case of the interaction between a Borneo vortex and a persistent strong cold surge over open water. It includes a classical theory by Lim and Chang (1981) on simple dynamics of surge-forced equatorial cyclogenesis in the absence of terrain.

2.4. *Interannual variations and seasonal predictability*

Interannual rainfall variability in the MC is strongly related to ENSO; anomalously low rainfall during El Niño events is associated with anomalous large-scale subsidence, a weakened Walker circulation with anomalous surface westerly, and anomalously low sea surface temperature (SST) (Robertson *et al.* 2011). Kubota *et al.* (2011) further investigated ENSO-related interannual rainfall variability over the eastern MC with station rainfall data in the Republic of Palau and eastern Indonesia. They showed fine spatial structure of a delay of the start of the dry condition during the El Niño developing year. Hamada *et al.* (2012) pointed out that interannual rainfall variability over northwestern Java is influenced not only by ENSO but also by the Indian Ocean Dipole events.

Feng *et al.* (2010) studied different impacts of two types of Pacific Ocean warming, El Niño and El Niño Modoki (Ashok *et al.* 2007), on Southeast Asian rainfall during boreal winter. As shown in Fig. 7, dry conditions are observed over the MC and the Philippines except for the vicinity of Sumatra and the western part of Borneo for the conventional El Niño (a), whereas the negative rainfall anomaly signals are much weaker for the El Niño Modoki cases (b), suggesting a relatively weaker influence of El Niño Modoki on the MC rainfall compared with the conventional El Niño. Interannual relationships between the tropical SST and summertime subtropical anticyclone over the western North Pacific were investigated by Chung *et al.* (2011), and it was shown that the 2–3 year oscillation of the western North Pacific subtropical high is driven by the SST anomaly forcing over the MC triggered by enhanced convection.

Chowdary *et al.* (2010) investigated the skills of 11 coupled atmosphere-oceans general

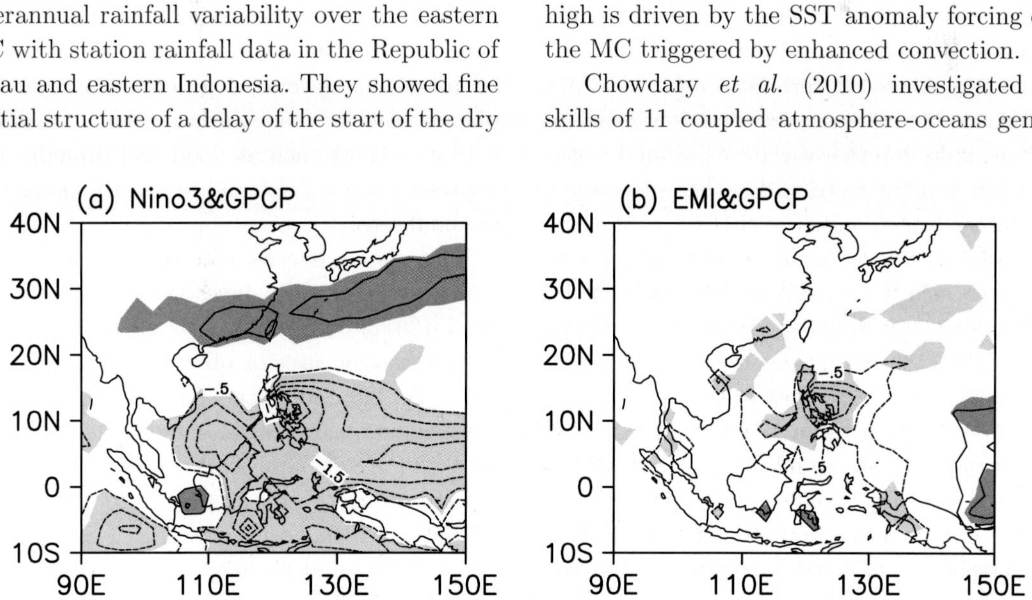

Fig. 7. Partial regressions of winter mean (DJF) rainfall for the time period from 1979–1980 to 2007–2008 with (a) the Niño3 index and (b) the El Niño Modoki index (EMI) based on the Global Precipitation Climatology Project (GPCP), covering the period from 1979 to the present on a 2.5° latitude–longitude grid precipitation data (Feng *et al.* 2010). The contour interval is 0.5 mm day^{-1}. Zero contours are omitted. Shading denotes the 95% confidence level from a two-tailed Student's *t* test.

circulation models in the prediction of seasonal rainfall and circulation anomalies of northwest Pacific with a focus on the summer following the mature phase of El Niño, and showed coupled models are capable of predicting suppressed rainfall over the focused region and season. It was demonstrated that remote forcing by tropical Indian Ocean SST variability and teleconnections are important for northwest Pacific rainfall prediction in those models. As for seasonal forecasts of winter monsoon rainfall in Malaysia, Juneng *et al.* (2010) compared the skills of four different forecasting approaches with a one-month lead time and showed statistical downscaling forecasts using multimodel output variables could result in some skill enhancement.

2.5. *Interdecadal variability, trends, and geological time-scale climate changes*

Sub-systems with large thermal inertia of the Earth's climate system, such as hydrosphere (oceans), cryosphere, and biosphere, are the main components to produce and maintain interdecadal and longer time variations that are fundamentally internal variations around a stable mean climate state. The other groups for the climate variations of such time scales are responses to the time variations of external forcings, such as rather periodic 11-year cycle of solar activity and irregular occurrence of major volcanic eruptions with time interval of decades, whereas monotonic changes of the climate system, approximately estimated as a linear trend for several decades, could be attributable to responses of the climate system to anthropogenic effects, including the increase of carbon dioxide, chlorofluorocarbons (during the last half century), and so on. In geological time scales, periodic variations of the Earth's orbital parameters, such as eccentricity, axial tilt, and precession, regulate solar energy input to the Earth in the order of 10^4 years (Milankovitch theory). Movements of continents and evolution of the atmospheric constituents introduce irregular or monotonic change of the Earth's climate, including global monsoon systems, in the longer time scales.

As an extension of the studies reviewed in Robertson *et al.* (2011), Juneng and Tangang (2010) examined long-term trends of winter monsoon synoptic circulations over the MC for 1962–2007 with a reanalysis dataset. They showed a significant increase in the frequency of Borneo vortex within the study period and the offshore shift of its center and discussed possible relationship to the strengthened easterly component of the cold surge winds. Deni *et al.* (2010) studied trends of dry spells during monsoon seasons over Peninsular Malaysia with the daily rainfall data from 36 stations for 1975–2004 and showed significant decrease in several dry spell indices. A scientific overview of climate change and variability over Malaysia and the MC of the Southeast Asian region was given by Tangang *et al.* (2012). They concluded that knowledge gap in the science of climate change over the region remains wide, with regards to influence on monsoon and other regional phenomena by the anthropogenic warming, extreme events such as flood and drought, and regional sea level rise and long-term changes in regional seas.

Regional patterns of tropical Indo-Pacific climate change were investigated by Tokinaga *et al.* (2012) over the last six decades based on a synthesis of *in situ* observations and ocean model simulations, with a focus on physical consistency among sea surface wind, sea level pressure (SLP), SST, and some other quantities and showed an evidence of slowdown of the Walker circulation with positive trends of SLP change over the MC and negative trends over the central equatorial Pacific. The observed decreases in land rainfall over the MC (Fig. 8) and marine cloudiness are also consistent with the surface wind changes which should suppress moisture convergence over the MC.

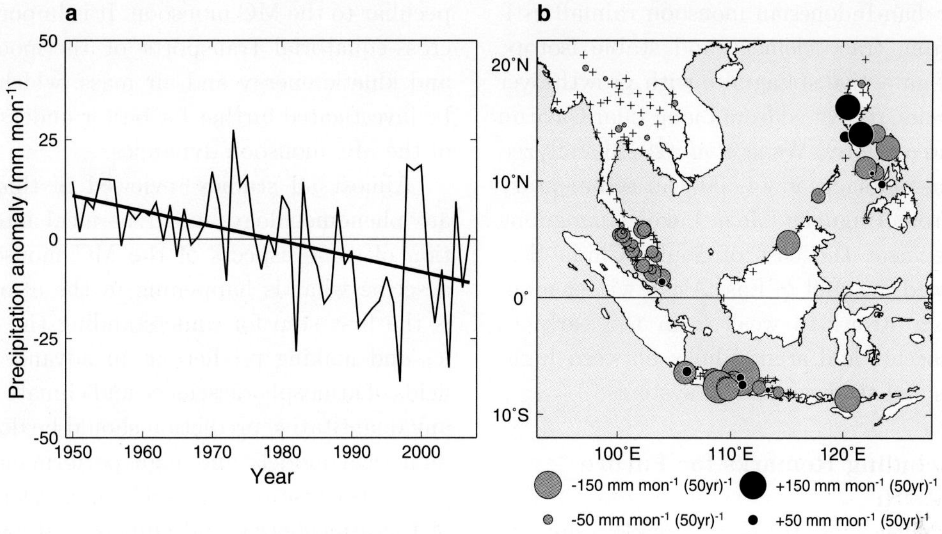

Fig. 8. Trends in rainfall over the MC (Tokinaga *et al.*, 2012). (a) Time series of land precipitation anomaly (thin line; mm month^{-1}) averaged over the MC (80E–150E, 20S–20N) and their linear regression (thick line). The precipitation data is based on the gridded monthly product provided by University of Delaware. (b) Land precipitation changes obtained from the Association of Southeast Asian Nations (ASEAN) Climatic Atlas Project (ACAP) data on rain gauge observations from 1950 to 1999 [mm month^{-1} (50 yr)$^{-1}$]. The decreasing (increasing) trends exceeding the 95% confidence level are indicated with gray (black) circles, and their magnitudes are shown by the size of circles below the panel.

Climate variations and change in geological time scales are studied intensively for these years, largely influenced by the increase of scientific motivation and the advancement of technology to analyze trace quantities of minor components. For decadal time-scale variations, for example, Watanabe *et al.* (2010) compared stable isotope time series of stalagmite from Ciawitali Cave of western Java and rainfall data near the site to confirm their usefulness as proxy data, and the analysis of the annual bands over millennium time scale is under way. For centennial time scales, D'Arrigo *et al.* (2011) reconstructed stream flow for Citarum River of western Java with teak tree-ring records for 1759–2006 and argued linkages to tropical climate dynamics dominated by the ENSO-Indian Ocean dipole systems. On the other hand, Rodysill *et al.* (2012) presented a new paleolimnologic record of rainfall and drought of eastern Java with a multi-proxy dataset from Lake Logung for the last 1,400 years. They showed that eastern Java became wetter over the last millennium and discussed the consistency with evidence for the southward migration of the Intertropical Convergence Zone during that period. Intensification of rainfall over the past millennium in eastern Java was also found by Konecky *et al.* (2013) with high-resolution proxy data of rainfall from Lake Lading of eastern Java based on the δD (hydrogen isotopic ratio) of terrestrial plant waxes from 850 Common Era to present.

For millennium and longer period records, Tierney *et al.* (2010) analyzed δD of terrestrial higher plant waxes in marine sediments from southwest of Sulawesi, and showed centennial-scale oscillations in the data following well-known variations in climate, such as the Little Ice Age and Medieval Warm Period, implying a dynamical link between the northern hemisphere temperatures and Indo-Pacific Warm Pool hydrology. Griffiths *et al.* (2010) showed an evidence for Holocene changes in Australian-Indonesian monsoon rainfall estimated from trace element and stable isotope ratios of an active stalagmite with growth over 12,640 years, recovered from Liang Luar Cave on the island of Flores. Wang *et al.* (2012) analyzed diatom assemblages in 14,500-year sedimentary record from Huguang Maar Lake, Guangdong Province, near the cost of South China Sea, and showed the shift of East Asian winter monsoon from strong to weak from the early to late Holocene, and argued links between high-latitudes and tropical climate systems.

3. Concluding Remarks for Future Research

Recent progress and achievements in the research on the MC monsoon were reviewed by focusing on the studies published after the previous review by Robertson *et al.* (2011). Unique features of the MC monsoon in the global monsoon systems are the complex coast lines and topography of the MC and its location close to the equator (Fig. 1). As shown in Figs. 3a and b in Sec. 2.2, the diurnal harmonic of precipitation rate has large amplitudes over islands in the MC with evening rainfall maxima, in contrast to that with morning rainfall maxima in coastal regions. Amplitude and phase of the diurnal cycle of rainfall have complicated dependence on the topography because of the multiscale interactions among the diurnally varying local winds and larger-scale wind systems with subseasonal (e.g., MJO), seasonal (monsoon), and interannual (e.g., ENSO) time scales.

Borneo vortex is a unique meso-scale vortex near the equator and our recognition and understanding of the vortex has been improved by recent works in data analyses and numerical experiments. Its spatial structure has some clear difference from that of tropical cyclones, with a shallow structure confined to the lower troposphere, which might be attributable to its location close to the equator. Cross-equatorial cold surge is another interesting dynamical feature peculiar to the MC monsoon. It is important for cross-equatorial transports of thermodynamic and kinetic energy and air mass, which should be investigated further for better understanding of the MC monsoon dynamics.

Almost all studies reviewed in this article are phenomenology and dynamical interpretation of some aspects of the MC monsoon. To describe what is happening in the atmosphere is the first step for understanding the dynamics and making predictions in advance. In the fields of atmospheric science and climate science, any quantitative prediction should be done with numerical models and high performance computers because of the highly nonlinear nature of the atmospheric and climate systems. However, current numerical weather prediction models and climate projection models are far from the perfect. As reviewed in this article, even the state-of-the-art models have biases and uncertainties in reproducing diurnal cycles of rainfall and winds or parametrizing their roles in resolved scales in global models. Reproduction of convective systems and their organizations in the tropics, particularly over the MC with complex topography, is one of the major challenging subjects in the modeling for numerical weather predictions and future climate projections.

Development of new diagnosis metrics to characterize moist convection and its organization as a system is another challenge. Otsuka *et al.* (2013) analyzed the statistical nature of size distributions of tropical convective systems in satellite observation data and the numerical output data by Trilaksono *et al.* (2011), and found a scaling law depending on the model resolution. The spatial scale of convective systems is one of the parameters to measure the reproducibility of a numerical model, and its statistical property depends on model resolution, cumulus parametrization, cloud microphysics, and so on. To understand difference depending on the models, it would be a first step to know the dependence of the size distribution of convective systems on the model horizontal resolutions (2, 4, 5, and 20 km, in their study). Such process-oriented evaluations of moist convective systems in the tropics are important for improving numerical models of weather prediction and climate projection.

Acknowledgments

We thank Chih-Pei Chang, the chief editor of this volume, for his very helpful comments and suggestions on an early version of the manuscript. Thanks also go to Edvin Aldrian, Huang-Hsiung Hsu, Matthew Wheeler, and two anonymous reviewers for their helpful comments. This research has been supported under JSPS Core-to-Core Program (Asia-Africa Science Platforms) for FY2015–2017 "International research collaborations and networking on extreme weather in changing climate in the Maritime Continent", and Kyoto Univ. Supporting Program for Interaction-based Initiative Team Studies (SPIRITS) for its preparation period in FY2013–2014.

References

Ashok, K., S. K. Behera, S. A. Rao, H. Weng, and T. Yamagata, 2007: El Niño Modoki and its possible teleconnection. *J. Geophys. Res.*, **112**, C11007. doi:10.1029/2006JC003798.

Braesicke, P., O. S. Hai, and A. A. Samah, 2012: Properties of strong off-shore Borneo vortices: A composite analysis of flow pattern and composition as captured by ERA-Interim. *Atmos. Sci. Lett.*, **13**, 128–132.

Chang, C.-P., Z. Wang, J. McBride, and C.-H. Liu, 2005: Annual cycle of Southeast Asia — Maritime Continent rainfall and the asymmetric monsoon transition. *J. Climate*, **18**, 287–301. doi: http://dx.doi.org/10.1175/JCLI-3257.1.

Chang, C.-P., Y. Ding, N.-C. Lau, R. H. Johnson, B. Wang, and T. Yasunari (Eds.), 2011: *The Global Monsoon System: Research and Forecast* (2nd Edition). World Scientific Series on Asia-Pacific Weather and Climate: Volume 5, World Scientific, 594 pp.

Chang, C.-P., M.-M. Lu, and H. Lim, 2016: Monsoon convection in the Maritime Continent: Interaction of large-scale motion and complex terrain. *Multiscale Convection-Coupled Systems in the Tropics*, Eds. R. Fovell and W. Tung, *Meteoro. Monogr.*, **56**. doi: 10.1175/AMSMONOGRAPHS-D-15-0011.1

Chen, T. C., J. D. Tsay, M. C. Yen, and J. Matsumoto, 2013: The winter rainfall of Malaysia. *J. Climate*, **26**, 936–958.

Chowdary, J. S., S.-P. Xie, J.-Y. Lee, Y. Kosaka, and B. Wang, 2010: Predictability of summer northwest Pacific climate in 11 coupled model hindcasts: Local and remote forcing. *J. Geophys. Res.*, **115**, D22121. doi:10.1029/2010JD014595.

Chung, P.-H., C.-H. Sui, and T. Li, 2011: Interannual relationships between the tropical sea surface temperature and summertime subtropical anticyclone over the western North Pacific. *J. Geophys. Res.*, **116**, D13111. doi:10.1029/2010JD015554.

D'Arrigo, R., N. Abram, C. Ummenhofer, J. Palmer, and M. Mudelsee, 2011: Reconstructed streamflow for Citarum River, Java, Indonesia: Linkages to tropical climate dynamics. *Clim. Dyn.*, **36**, 451–462.

Deni, S. M., J. Suhaila, W. Z. W. Zin, and A. A. Jemain, 2010: Spatial trends of dry spells over Peninsular Malaysia during monsoon seasons. *Theor. Appl. Climatol.*, **99**, 357–371.

Feng, J. A., L. Wang, W. Chen, S. K. Fong, and K. C. Leong, 2010: Different impacts of two types of Pacific Ocean warming on Southeast Asian rainfall during boreal winter. *J. Geophys. Res.*, **115**, D24122. doi:10.1029/2010JD014761.

Fujita, M., K. Yoneyama, S. Mori, T. Nasuno, and M. Satoh, 2011: Diurnal convection peaks over the eastern Indian Ocean off Sumatra during different MJO phases. *J. Meteor. Soc. Japan*, **89A**, 317–330. doi:10.2151/jmsj.2011-A22.

Griffiths, M. L., R. N. Drysdale, M. K. Gagan, S. Frisia, J. Zhao, L. K. Ayliffe, W. S. Hantoro, J. C. Hellstrom, M. J. Fischer, Y.-X. Feng, and B. W. Suwargadi, 2010: Evidence for Holocene changes in Australian-Indonesian monsoon rainfall from stalagmite trace element and stable isotope ratios. *Earth and Planetary Science Letters*, **292**, 27–38.

Hamada, J.-I., S. Mori, H. Kubota, M. D. Yamanaka, U. Haryoko, S. Lestari, R. Sulistyowati, and F. Syamsudin, 2012: Interannual rainfall variability over northwestern Jawa and its relation to the Indian Ocean Dipole and El Niño-Southern Oscillation events. *SOLA*, **8**, 69–72.

Hattori, M., S. Mori, and J. Matsumoto, 2011: The cross-equatorial northerly surge over the Maritime Continent and its relationship to precipitation patterns. *J. Meteor. Soc. Japan*, **89A**, 27–47.

Johnson, R. H., 2011: Diurnal cycle of monsoon convection. *The Global Monsoon System: Research and Forecast* (2nd Edition). Eds. C.-P. Chang *et al.*, World Scientific Series on Asia-Pacific Weather and Climate Vol. 5, World Scientific, pp. 257–276.

Juneng, L. and F. T. Tangang, 2010: Long-term trends of winter monsoon synoptic circulations over the Maritime Continent: 1962–2007. *Atmos. Sci. Lett.*, **11**, 199–203.

Juneng, L., F. T. Tangang, H. Kang, W.-J. Lee, and Y. K. Seng, 2010: Statistical downscaling forecasts for winter monsoon precipitation in Malaysia using multimodel output variables. *J. Climate*, **23**, 17–27.

Kamimera, H., S. Mori, M. D. Yamanaka, and F. Syamsudin, 2012: Modulation of diurnal rainfall cycle by the Madden-Julian oscillation based on one-year continuous observations with a meteorological radar in west Sumatera. *SOLA*, **8**, 111–114. doi:10.2151/sola.2012-028.

Konecky, B. L., J. M. Russell, J. R. Rodysill, M. Vuille, S. Bijaksana, and Y. Huang, 2013: Intensification of southwestern Indonesian rainfall over the past millennium. *Geophys. Res. Lett.*, **40**, 386–391. doi:10.1029/2012GL054331.

Koseki, S., T.-Y. Koh, and C.-K. Teo, 2014: Borneo vortex and mesoscale convective rainfall. *Atmos. Chem. Phys.*, **14**, 4539–4562. doi:10.5194/acp-14-4539-2014.

Kubota, H., R. Shirooka, J.-I. Hamada, and F. Syamsudin, 2011: Interannual rainfall variability over the eastern Maritime Continent. *J. Meteor. Soc. Japan*, **89A**, 111–122.

Lim, H. and C.-P. Chang, 1981: A theory for midlatitude forcing of tropical motions during winter monsoons. *J. Atmos. Sci.*, **38**, 2377–2392.

Love, B. S., A. J. Matthews, and G. M. S. Lister, 2011: The diurnal cycle of precipitation over the Maritime Continent in a high-resolution atmospheric model. *Quart. J. Royal Meteor. Soc.*, **137**, 934–947.

Oh, J.-H., K.-Y. Kim, and G.-H. Lim, 2012: Impact of MJO on the diurnal cycle of rainfall over the western Maritime Continent in the austral summer. *Clim. Dyn.*, **38**, 1167–1180.

Ooi, S. H., A. A. Samah, and P. Braesicke, 2011: A case study of the Borneo Vortex genesis and its interactions with the global circulation. *J. Geophys. Res.*, **116**, D21116. doi:10.1029/2011JD015991.

Otsuka, S., N. J. Trilaksono, and S. Yoden, 2013: A statistical analysis on the size distributions of tropical convective systems during the Jakarta flood event in 2007 simulated by JMA-NHM. Abstract of a paper presented at DACA-13, Davos, Switzerland.

Peatman S. C., A. J. Matthews, and D. P. Stevens, 2014: Propagation of the Madden-Julian Oscillation through the Maritime Continent and scale interaction with the diurnal cycle of precipitation. *Quart. J. Royal Meteor. Soc.*, **140**, 814–825. DOI:10.1002/qj.2161.

Qian, J. H, 2008: Why precipitation is mostly concentrated over islands in the Maritime Continent. *J. Atmos. Sci.*, **65**, 1428–1441.

Qian, J.-H., A. W. Robertson, and V. Moron, 2013: Diurnal cycle in different weather regimes and rainfall variability over Borneo associated with ENSO. *J. Climate*, **26**, 1772–1790.

Rauniyar, S. P. and K. J. E. Walsh, 2011: Scale interaction of the diurnal cycle of rainfall over the Maritime Continent and Australia: Influence of the MJO. *J. Climate*, **24**, 325–348.

Robertson, A. W., V. Moron, J.-H. Qian, C.-P. Chang, F. Tangang, E. Aldrian, T.-Y. Koh, and L. Juneng, 2011: The maritime continent monsoon. *The Global Monsoon System: Research and Forecast* (2nd Edition), Eds. C.-P. Chang *et al.*, World Scientific Series on Asia-Pacific Weather and Climate Vol. 5, World Scientific, pp. 85–98. DOI: 10.1142/9789814343411_0006

Rodysill, J. R., J. M. Russell, S. Bijaksana, E. T. Brown, L. O. Safiuddin, and H. Eggermont, 2012: A paleolimnological record of rainfall and drought from East Java, Indonesia during the last 1,400 years. *J. Paleolimnol.*, **47**, 125–139.

Suhaila, J., S. M. Deni, W. Z. W. Zin, and A. A. Jemain, 2010: Spatial patterns and trends of daily rainfall regime in Peninsular Malaysia during the southwest and northeast monsoons: 1975–2004. *Meteor. Atmos. Phys.*, **110**, 1–18.

Takahashi, H. G., Y. Fukutomi, and J. Matsumoto, 2011: The impact of long-lasting northerly surges of the East Asian winter monsoon on tropical cyclogenesis and its seasonal march. *J. Meteor. Soc. Japan*, **89A**, 181–200.

Tangang, F. T., L. Juneng, E. Salimun, K. M. Sei, L. J. Le, and H. Muhamad, 2012: Climate change and variability over Malaysia: Gaps in science and research information. *Sains Malaysiana*, **41**, 1355–1366.

Teo, C.-K., T.-Y. Koh, J. C.-F. Lo, and B. C. Bhatt, 2011: Principal component analysis of observed and modeled diurnal rainfall in the Maritime Continent. *J. Climate*, **24**, 4662–4675.

Tierney, J. E., D. W. Oppo, Y. Rosenthal, J. M. Russell, and B. K. Linsley, 2010: Coordinated hydrological regimes in the Indo-Pacific region during the past two millennia. *Paleoceanography*, **25**, PA1102. doi:10.1029/ 2009PA001871.

Tokinaga, H., S.-P. Xie, A. Timmermann, S. McGregor, T. Ogata, H. Kubota, and Y. M. Okumura, 2012: Regional patterns of tropical Indo-Pacific climate change: Evidence of the Walker circulation weakening. *J. Climate*, **25**, 1689–1710.

Trilaksono, N. J., S. Otsuka, S. Yoden, K. Saito, and S. Hayashi, 2011: Dependence of model-simulated heavy rainfall on the horizontal resolution during the Jakarta flood event in January–February 2007. *SOLA*, **7**, 193–196.

Trilaksono, N. J., S. Otsuka, and S. Yoden, 2012: A time-lagged ensemble simulation on the modulation of precipitation over West Java in January–February 2007. *Mon. Wea. Rev.*, **140**, 601–616.

Varikoden, H., B. Preethi, A. A. Samah, and C. A. Babu, 2011: Seasonal variation of rainfall characteristics in different intensity classes over Peninsular Malaysia. *J. Hydrology*, **404**, 99–108.

Wang, L., J. Li, H. Lu, Z. Gu, P. Rioual, Q. Hao, A. W. Mackay, W. Jiang, B. Cai, B. Xu, J. Han, and G. Chu, 2012: The East Asian winter monsoon over the last 15,000 years: Its links to high-latitudes and tropical climate systems and complex correlation to the summer monsoon. *Quaternary Science Reviews*, **32**, 131–142.

Watanabe Y., H. Matsuoka, S. Sakai, J. Ueda, M. Yamada, S. Ohsawa, M. Kiguchi, T. Satomura, S. Nakai, B. Brahmantyo, K. A. Maryunani, T. Tagami, K. Takemura, and S. Yoden, 2010: Comparison of stable isotope time series of stalagmite and meteorological data from West Java, Indonesia. *Palaeogeography, Palaeoclimatology, Palaeoecology*, **293**, 90–97.

Wheeler, M. C. and J. L. McBride, 2011: Australasian monsoon. *Intraseasonal Variability in the Atmosphere-Ocean Climate System* (2nd edition). Eds. W. K. M. Lau, and D. E. Waliser, Springer, pp. 147–198.

Wu, P., M. Hara, H. Fudeyasu, M. D. Yamanaka, J. Matsumoto, F. Syamsudin, R. Sulistyowati, and Y. S. Djajadihardja, 2007: The impact of trans-equatorial monsoon flow on the formation of repeated torrential rains over Java Island. *SOLA*, **3**, 93–96.

Wu, C.-H. and H.-H. Hsu, 2009: Topographic influence on the MJO in the Maritime Continent. *J. Climate*, **22**, 5433–5448.

Chapter 7

Modelling Monsoons

Kenneth R. Sperber

PCMDI, Lawrence Livermore National Laboratory,
Livermore, CA 94551 USA
sperber1@llnl.gov
www-pcmdi.llnl.gov

Edmee Cusiner

Department of Atmospheric and Oceanic Sciences, UCLA,
Los Angeles, CA 90005, USA
ecuisinier@atmos.ucla.edu
www.atmos.ucla.edu

Akio Kitoh

Faculty of Life and Environmental Sciences, University of Tsukuba,
Tsukuba, Ibaraki 305-8572, Japan
kito.akio.ff@u.tsukuba.ac.jp
www.life.tsukuba.ac.jp/en/index.html

Carlos R. Mechoso

Department of Atmospheric and Oceanic Sciences, UCLA,
Los Angeles, CA 90005, USA
mechoso@atmos.ucla.edu
www.atmos.ucla.edu

Aurel Moise

CAWCR Australian Bureau of Meteorology,
Melbourne, VIC 3001, Australia
A.Moise@bom.gov.au
www.cawcr.gov.au

Wilfran Moufouma-Okia

Hadley Centre, The Met Office,
Exeter Devon, EX1 3PB, UK
wilfran.moufouma-okia@metoffice.gov.uk
www.metoffice.gov.uk

The Global Monsoon System: Research and Forecast (3rd Edition)
Edited by C. P. Chang et al.

Kathleen Schiro
Department of Atmospheric and Oceanic Sciences, UCLA,
Los Angeles, CA 90005, USA
kshiro@atmos.ucla.edu
www.atmos.ucla.edu

Andrew G. Turner
NCAS-Climate, University of Reading
Reading, RG6 6BB, UK
a.g.turner@reading.ac.uk
www.ncas.ac.uk/index.php/en/home-page

The simulation of monsoons remains a challenging problem given the complexity of the multi-scale interactions and the modulating influences that operate on a broad range of time scales. Recent studies have demonstrated an improvement in the simulation of monsoons in CMIP5 relative to CMIP3, though many outstanding issues remain. For example, although mean state biases have been reduced in amplitude, the spatial error pattern remains virtually unchanged between the two vintages of experimentation. Systematic errors in the onset time of monsoons indicate that the simulated atmosphere-land-ocean interactions are not responding properly to the annual cycle of solar forcing. As such, the CMIP5 models have early monsoon onset over the Sahel and the North American domain, and late monsoon onset over India, the South American domain, and the Gulf of Guinea. This indicates that a regional process study approach is warranted for improving our understanding of atmosphere-land-ocean interactions inherent to monsoon development and onset. The simulation of intraseasonal variability remains a grand challenge problem, especially given its importance for initiating monsoon onset and being associated with precipitation extremes. Experimental prediction of intraseasonal variability suggests forecast skill to 3 weeks, with the potential for increased skill with the implementation of different projection basis function for boreal summer vs. boreal winter. Despite improvements in the simulation of the El Nino/Southern Oscillation, the interannual monsoon teleconnections are sensitive to biases in the regional rainfall that can compromise the response to the remote forcing. At best, interdecadal variations of Sahel rainfall are qualitatively captured in CMIP5, with the amplitude of the 1970's-1980's drought strongly underestimated. Conversely, mechanisms for observed interdecadal trends in rainfall over East Asia and northern Australia are yet to be understood, and these trends remain to be simulated. Higher horizontal resolution has been beneficial in the representation of orographic rainfall, as well as larger scale aspects of the circulation (e.g. Baiu front) though improvements to model physics are most essential for improving the simulation and prediction of monsoons.

1. Introduction

Understanding and forecasting monsoon variations remains a "Grand Challenge" problem. This undertaking, comprising diurnal through centennial time scales (including the effects of climate change), requires numerical weather prediction models, climate models, and regional models. From a physics standpoint, the fidelity in the representation of the land-ocean-atmosphere-cryosphere system is critical. This attests to the variety of forcings and time scales of interaction that influence monsoons (Webster *et al.* 1998). Sensible heating and soil moisture are processes through which monsoons can be affected on short time scales. Snow cover was recognized a potential predictor of Asian monsoon (Blanford 1884, Becker *et al.* 2000) since it can affect the partitioning of sensible vs. latent heating, as does soil moisture. Remote large-scale teleconnections were also noted, especially the sea-level pressure variation over the Pacific (Walker 1924). Slowly varying boundary conditions, such as sea-surface temperature

(Charney and Shukla 1981), impart memory to the system from subseasonal through centennial time scales, with the El Niño/Southern Oscillation (ENSO) being one of the major factors modulating monsoon variability across the globe. The multi-scale interactions of these factors, under a background of evolving greenhouse gas (GHG) forcing and aerosol impacts, conspire to make monsoon forecasting a difficult problem.

Traditionally, the term "monsoon" refers to the seasonal reversal of prevailing winds that is accompanied by excessive rainfall. Using the "global monsoon" definition of Wang and Ding (2006), Fig. 1 shows the observed and simulated CMIP5 multi-model ensemble mean representation of the global monsoon. With the exception of East Asia, and to a lesser extent the poleward extension of the North and South American monsoons, the mean model qualitatively represents the extent of the global monsoon domain. Typically, the multi-model mean shows better agreement with observations than any individual model, indicating that monsoon fidelity in individual models needs to be evaluated and improved.

While investigating the monsoon using a global approach is useful for summarizing model performance and isolating large-scale coherent climate change signals (Lee and Wang 2014), there are unique local aspects to each monsoon

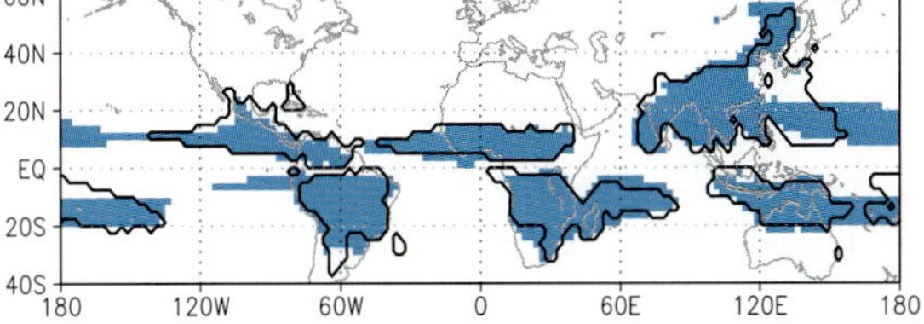

Fig. 1. Observed (thick contour) and simulated (shading) global monsoon domain, defined where the annual range of rainfall (difference between May–September and November–March) is greater than 2.5 mm/day. The observed domain is from GPCP data, and the multi-model mean precipitation is from CMIP5 historical simulations for 1986–2005. After Kitoh *et al.* (2013).

domain that require critical evaluation. The goal of this chapter is to highlight the model errors that are most critical for simulation and prediction of monsoons, and where possible highlight commonalities to errors across the domains that may benefit from coordinated process/diagnostic studies.

2. The Asian Monsoon

2.1. *Time-mean state*

In terms of rainfall, the Asian summer monsoon extends from Pakistan to China, Korea, and Japan, including the western North Pacific (Fig. 1). Though the monsoon over these regions is mutually interactive (Annamalai and Sperber 2005), there are regional influences that distinguish the Indian component from the East Asian component. Though the meridional temperature gradient and the subsequent latent heat release after monsoon onset are the main drivers of the monsoon, the monsoon over East Asia is also strongly affected by the subtropical high over the western Pacific and the East Asian Jet. The pressure gradient that develops in response to the temperature gradient drives the monsoon circulation. The low-level cross-equatorial flow and the southwesterlies that develop over the western Indian Ocean and the Arabian Sea contribute to the moisture transport that is the source of the monsoon rainfall, especially over India. Additional moisture from the Bay of Bengal and the South China Sea also contribute to rainfall over these regions and further downstream due to the influence of the west Pacific subtropical high.

Intercomparisons of the Asian summer monsoon in CMIP3 (Meehl *et al.* 2007a) and CMIP5 (Taylor *et al.* 2012) simulations have been performed by numerous authors. Sperber *et al.* (2013) find that, in terms of the multi-model means, the CMIP5 models slightly improve upon the CMIP3 models for all of the diagnostics performed. Wang *et al.* (2014) also showed

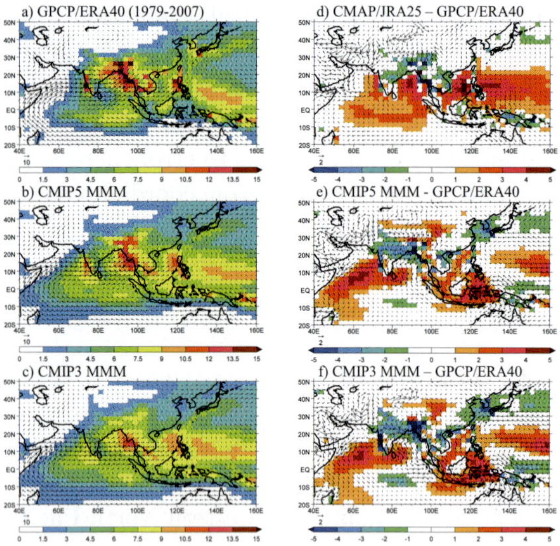

Fig. 2. June–September climatologies of precipitation (mm/day) and 850 hPa wind vectors. (a) GPCP and ERA-40 (1979–2004), (b) CMIP5 MMM (1961–1999), and (c) CMIP3 MMM (1961–1999). Precipitation and wind error. (d) CMAP/JRA25 minus GPCP/ERA-40, (e) CMIP5 minus GPCP/ERA-40, and (f) CMIP3 MMM minus GPCP/ERA-40. After Sperber *et al.* (2013).

that the CMIP5 models are more skillful than the CMIP3 models in terms of additional Asian-Australian monsoon metrics. However, there are systematic errors that are consistent between the two vintages of experimentation (Figs. 2e and 2f). For example, the time mean rainfall error has a consistent pattern between CMIP5 and CMIP3, though the amplitude of the error is smaller in CMIP5 relative to CMIP3. Part of the error reduction results from the better simulation of precipitation maxima in the vicinity of steep orography, implying an advantage of using higher horizontal resolution in the CMIP5 models. The models systematically under-estimate the rainfall over India, China, Korea, and Japan, consistent with the underestimation of the monsoon domain in Fig. 1.

2.2. *Variability*

Over the Asian monsoon domain, the 30–60 day boreal summer intraseasonal variability (BSISV) causes active and break phases of monsoon rainfall (e.g. Yasunari 1979, 1981, Annamalai and Sperber 2005). Simulation of the BSISV has been problematic in climate and NWP models, in which eastward propagation of near-equatorial convection into the western Pacific and the northward propagation of convection near India are not well represented (e.g. Sperber *et al.* 2001; Waliser *et al.* 2003; Sperber and Annamalai 2008). Though most climate models still lack an adequate simulation of the BSISV, relative to CMIP3 there are an increasing number of CMIP5 models that have an improved capability to simulate the BSISV (Sperber *et al.* 2013; Sabeerali *et al.* 2013). From examination of the CMIP5 multi-model mean, improvement is still needed to realistically represent the full extent of the northward propagation of convection over India, which may be related to the underestimate of mean state rainfall over northern India (Sperber *et al.* 2013). As a consequence the northwest to southeast rainband is not as tilted as observed. Also, improvement is need in representing the equatorial eastward propagation through the Maritime Continent into the western Pacific, and the subsequent north-northwestward propagation over the western tropical Pacific. Additionally, predictability of the BSISV is limited since the leading boreal summer intraseasonal mode is not related to slowly varying boundary conditions (Sperber *et al.* 2000).

On interannual time scales, the ability to predict regionally averaged ENSO SST anomalies (e.g. Niño3.4) with a 6–12 month lead-time led to the anticipation of skillful monsoon forecasts. However, the pattern of ENSO SST anomalies, which is less well forecasted, is critical to the characteristics of the teleconnection, as are the conditions over the Indian Ocean (Slingo and Annamalai 2000). Thus, even during years of strong ENSO forcing, seasonal monsoon forecasting is a challenge. Additionally, the strength of the ENSO-monsoon teleconnection exhibits pronounced decadal scale variability (Annamalai *et al.* 2007, Turner *et al.* 2007).

Over East Asia, a northward migrating zonal-oriented frontal zone of precipitation is a distinct feature in early summer. The CMIP3 models have difficulty in reproducing the latitude of the observed rainfall zones (Ninomiya 2009, 2011). Moreover the majority of the models do not reproduce realistic year-to-year variations in the Meiyu/Changma/Baiu frontal zones (Ninomiya 2012). Large spread also exists in CMIP5 models over East Asia (Huang *et al.* 2013). The ability of models to reproduce these year-to-year variations is closely related to their ability to reproduce the modulating influences, including the East Asian summer monsoon, the North Pacific subtropical anticyclone, and the SST over the Pacific Ocean.

2.3. *Climate change*

Difficulty in simulating the regional characteristics of rainfall and it variability in present-day climate model simulations poses a challenge for estimating the impact of climate change on monsoon variability (e.g. Turner and Annamalai 2012). At least two competing sets of processes associated with GHG changes impact climate change over South Asia, (1) thermodynamic changes whereby increased tropical SST gives rise to enhanced evaporation, increased moisture transport, and increased rainfall, and (2) dynamic effects whereby the monsoon circulation weakens due to a reduction in the meridional temperature gradient (e.g. Ueda *et al.* 2006).

A broad assessment of the prospective future change of the global monsoon has been performed by Lee and Wang (2014). This analysis evaluated 20 CMIP5 models, with the best 4 models used to generate a multi-model ensemble to assess climate change under RCP4.5. Their results suggest that (i) the global extent of the monsoon is not expected to change appreciably, although the monsoon over western Asia expands by ~11%; (ii) the annual mean, the annual cycle range, and the percentage of rainfall that falls during the boreal summer are expected to increase; (iii) the monsoon onset will be advanced, and the monsoon withdrawal delayed, and compared to the Southern Hemisphere (SH), the Northern Hemisphere (NH) monsoon will increase significantly due to a change in the interhemispheric temperature gradient that enhances the Hadley circulation.

Under RCP4.5, Wang *et al.* (2014) indicate that the extent of the Asian-Australian monsoon domain will not change appreciably, though it is suggested that the domain may extend westward. Over Australia and the subtropical Asian region, the summer rainfall will make up a larger percentage of the annual mean. By the end of the 21st century, an Asian-Australian Monsoon (AAM) rainfall increase of 4.5% is evident, primarily due to ISM and EASM increases. Due to hemispheric asymmetry in the temperature change, the rainfall increases over the ASM are larger than over the Australian monsoon domain. The ENSO-AAM teleconnection is expected to strengthen.

Increases in global monsoon area and global monsoon intensity are further amplified for RCP8.5 (Hsu *et al.* 2012, 2013; Kitoh *et al.* (2013). As discussed in Kitoh *et al.* (2013), over East Asia, (China, Korea, and Japan) average rainfall is projected to increase by ~7%, while over the remainder of the Asian summer monsoon domain the projected increase is ~13%. Extremes in precipitation during the Asian monsoon are also projected to amplify, with the seasonal maximum 5-day total rainfall increasing by ~20%, and the number of consecutive dry days increasing by ~5%. Different circulation changes are reported between CMIP3 and CMIP5 models in the Asian summer monsoon, i.e. westerly acceleration over the northern Indian Ocean in CMIP5 contrasted with slowdown of monsoon circulation in CMIP3 (Ogata *et al.*, 2014).

Liu *et al.* (2013) performed a regional modeling study using RegCM3 with multiple large-scale driving datasets to downscale present-day and future scenarios A1B and A1FI over China. For present-day simulations the RegCM3 model is typically in better agreement with observations than the Community Climate System Model (CCSM) in terms of the pattern correlation of surface temperature and precipitation, likely reflecting the increased resolution. Under climate change, the RegCM3 produces larger regional precipitation trends than CCSM, with both models showing greater warming under scenario A1FI than A1B. However, such results must be treated with caution owing to the lack of ocean-atmosphere coupling in the regional modelling framework. The largest temperature increases occur over the Tibetan Plateau and northern China. However, there is substantial uncertainty in the projections as reflected by RegCM3 simulations forced by CCSM projections as compared to MIROC projections. Furthermore, it was found that present-day model biases affect future climate projections, indicating the need to reduce systematic model error (see also Levine *et al.*, 2013 for the case of the Indian monsoon).

3. The Australian Monsoon

3.1. *Time-mean state*

Tropical Australia is strongly influenced by the surrounding oceans, and is subject to a strong seasonal cycle of precipitation and surface pressure, dominated by the Australian component of the Asian-Australian monsoon (e.g. McBride 1998). During the November to April period typically more than 90% of annual rainfall occurs across northern Australia (Nicholls *et al.* 1982) which also is traditionally referred to as Australia's wet season. The onset of the monsoon typically occurs in late December (Drosdowsky, 1996) but can vary considerably between regions.

On the large scale, the Australian monsoon is mainly driven by the differential heating of the continental land mass compared to the surrounding ocean in the build-up period of October–November resulting in the reversal of the meridional temperature gradient. This gradient produces a meridional circulation that brings moist air through low level westerlies across the maritime continent to tropical Australia (Hung *et al.* 2004). One of the defining characteristics of the large scale circulation is the reversal of the lower tropospheric winds from dry (wintertime) easterlies to moist (summertime) westerlies and a corresponding change to upper tropospheric easterlies (e.g. Troup 1961, Ramage 1971, Hung *et al.* 2004).

Coupled atmosphere-ocean climate models have traditionally shown difficulties in simulating the rainfall across tropical Australia during the monsoon season (Fig. 3). This has been shown for CMIP2 models (Moise *et al.* 2005), and CMIP3 models (Christensen *et al.* 2007, Colman *et al.* 2011, Moise *et al.* 2012, 2015a, 2015b, Brown *et al.* 2016) with many GCMs simulating considerable biases (often too wet across the northwest and too dry bias across northeast Australia). Compensating biases amongst the ensemble lead to a surprisingly good representation of the multi-model mean (Colman *et al.* 2011; also see Figs. 3f and 3g).

Pattern correlation skill scores of climatological precipitation and 850 hPa wind (not shown) for the Australian monsoon show encouraging improvement in model skill from CMIP3 to CMIP5 (Moise *et al.* 2012, 2015a, 2015b; Brown *et al.* 2016), though the gross structure of the precipitation error is similar between the two vintages of models (Figs. 3m and 3n). Furthermore, there is an overall improvement in the seasonality of the Australian monsoon from CMIP3 to CMIP5, with most CMIP5 models better simulating the monsoon timing and very low rainfall rates outside of the monsoon season (Li *et al.* 2012). Some of the reasons for

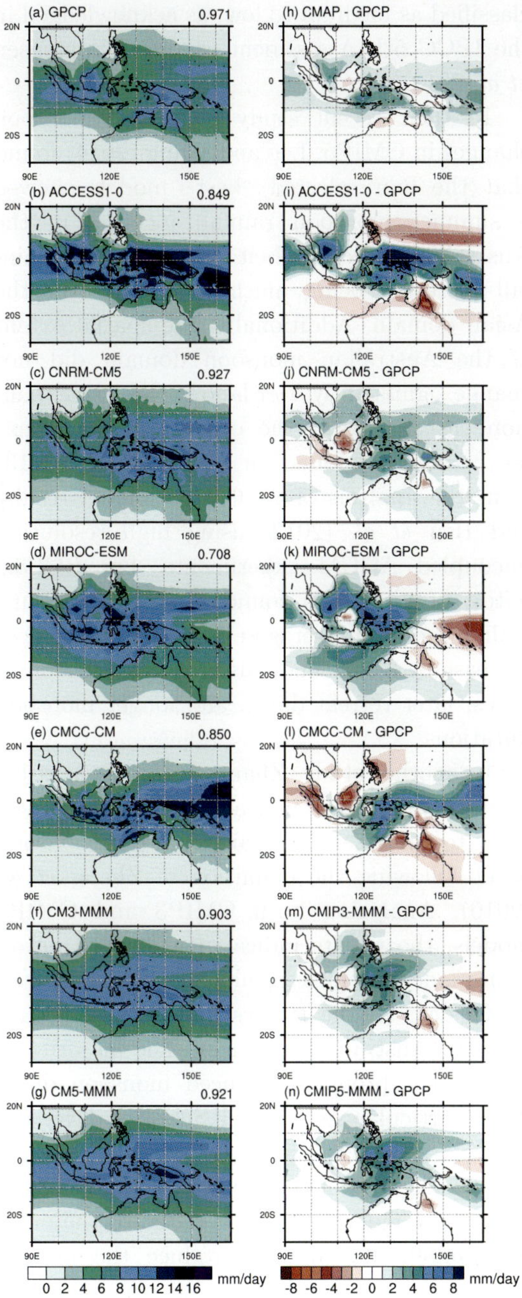

Fig. 3. a–g DJFM precipitation rate climatology from (a) GPCP, (b) ACCESS, (c) CNRM-CM5, (d) MIROC-ESM, (e) CMCC-CM, (f) CMIP3 MMM, and (g) CMIP5 MMM. Also given in (a) is the pattern correlation of GPCP with CMAP, and in b–g are the model pattern correlations with GPCP over the region 90°E–200°E, 30°S–20°N. (h) CMAP minus GPCP, i–n as b–g but for model minus GPCP. The units are (mm day^{-1}). GPCP and CMAP data is from 1979 to 1999 and the model data is from 1980 to 1999.

the lack of skill in the simulation of the monsoon have been attributed to SST pattern errors and also an unrealistic representation of the Indian Ocean dipole (Achuthavarier *et al.* 2012, Boschat *et al.* 2012). Meehl *et al.* (2012) reports on improvements in the pattern of rainfall in the Australian monsoon in CCSM4 compared to CCSM3 because of an improved representation of the surrounding SST's.

However, some significant geographical biases in the CMIP5 models are present (Moise *et al.* 2015a, 2015b; Brown *et al.* 2016), with precipitation extending too far south into the continental interior. An inverse relationship is also found between precipitation amounts and co-located continental surface temperature. Precipitation biases in models are related to differences in occurrence of convection and suppressed vertical motion.

3.2. *Variability*

The annual cycle of precipitation over tropical Australia shows strong variability on higher frequencies related to synoptic-scale variability (e.g. Pope *et al.* 2009), with the Madden-Julian Oscillation (MJO) affecting the onset/bursts/breaks/retreat of the monsoon. Meehl *et al.* (2012) report intraseasonal variability of the Asian–Australian monsoon is also much improved, though it is still somewhat weaker in CCSM4 compared to observations. In CMIP5 errors in the phase speed and eastward propagation of MJO rainfall anomalies from the Indian Ocean into the western Pacific still persist (Hung *et al.* 2013).

Total precipitation in northern and eastern Australia shows marked interannual variability, being strongly influenced by ENSO. Interannual variability in precipitation is too weak in the model ensemble (although the model range in variability is considerable). Improvement can be partially attributed to the enhanced ENSO skill in CMIP5 compared to CMIP3. Most models reproduce the observed Australian monsoon-ENSO teleconnection (e.g. Cai *et al.* 2009, 2011), with its strength dependent on the ENSO variance. However, over the maritime continent, the simulated monsoon-ENSO connection is generally weaker than observed (Jourdain *et al.* 2013). Similar results have been reported by Catto *et al.* (2012a, 2012b) with respect to the evolution of North Australian SST's during composite El Niño and La Niña events.

Despite significant wind anomalies in the Indian Ocean related to Indian Ocean Dipole (IOD) events, the Australian monsoon-IOD relationship remains relatively weak both in observations and in CMIP3 and CMIP5 models (e.g. Jourdain *et al.* 2013).

3.3. *Climate change*

The projected weakening of the large-scale tropical atmospheric overturning circulation has been detected in both CMIP3 (e.g. Vecchi and Soden, 2007; IPCC 2007) and CMIP5 models (Chadwick *et al.* 2013; IPCC 2013), and may have a direct impact on changes to the Australian monsoon. Similarly, the Walker circulation is projected to weaken under global warming (Power and Kociuba, 2011). Translating large scale changes of the tropical climate onto regional scales can incur large uncertainties, particular for rainfall (IPCC, 2007, 2013), and the Australian Monsoon region is especially affected. The factors contributing to this large uncertainty at different temporal and spatial scales are due to the complexity of the Australian monsoon system; the multitude of climate drivers impacting on the monsoon system; the model skill in representing these climate drivers, and finally the uncertainty surrounding possible changes to these drivers under enhanced greenhouse warming. Particularly over land, this can lead to inter-model differences in projected local rainfall changes becoming larger than differences between scenarios (Frieler *et al.* 2012; Moise *et al.* 2012), emphasizing the importance of process-based model evaluation. As a consequence, confidence in the projections is classified as medium to low, as acknowledged in the IPCC 5th Assessment report (Christensen *et al.* 2013).

In their recent study on global monsoon changes in CMIP5, Lee and Wang (2014) found that the four selected "best" models showed a summer monsoon rainfall increase in the Australian region, albeit with summer rainfall increases being much less than over the Asian domain. Additionally, the spatial extent of the Australian monsoon domain did not change significantly over land with some expansion eastward over the oceanic region. Similar results are found in Kitoh *et al.* (2013) from the analysis of 29 CMIP5 model results and Hsu *et al.* (2012) using high resolution uncoupled AGCM experiments. For RCP8.5 Kitoh *et al.* (2013) found that for the Australian monsoon average rainfall is projected to increase by $\sim$7%. They also found earlier onset dates, later retreat dates, and longer monsoon durations.

In another study, Zhang *et al.* (2013, 2015) used daily 850 hPa winds and precipitable water data to analyze the change in the monsoon onset following the definition of Zhang *et al.* (2010). Analyzing both CMIP3 and CMIP5 models, the results indicated a delayed onset, with model-ensemble results showing a delay of 10–20 days in the tropics and around 7 days over the Australian continent. The majority of the models produced reduced monsoon duration, particular in tropical Sumatra-Java and nearby waters.

The significant uncertainties that accompany these model results is related to the models' limited ability to reproduce the present monsoon climate and the large scatter among the model projections that limit the confidence in the projected changes. While these studies highlight the large uncertainty in future changes to the mean state of the Australian monsoon, there is much less uncertainty in projected changes to both rainfall variability (at different time scales) and extreme rainfall events

(IPCC 2013; Kitoh *et al.* 2013; Zhang and Moise 2016). For example, Kitoh *et al.* (2013) found that extremes in Australian precipitation are projected to amplify, with the seasonal maximum 5-day total rainfall increasing by ~15%, and the number of consecutive dry days increasing by ~7%. At the same time, projections for the formation of tropical cyclones (TCs) in the vicinity of tropical Australia suggest a decrease in TC genesis frequency (Tory *et al.* 2013) along with an intensification of TCs (Christensen *et al.* 2014).

Australian monsoon rainfall variability is set to increase on inter-annual time scales (following its strong relationship to ENSO) and in particular on intra-seasonal time scales (Zhang and Moise 2016). Following the analysis from Menon *et al.* (2013), Zhang and Moise (2016) investigated the change in daily summer rainfall variability for the Australian monsoon region (land only) from 25 CMIP5 models (Fig. 4). On average they found a 15% increase with all but two models agreeing on the direction of change.

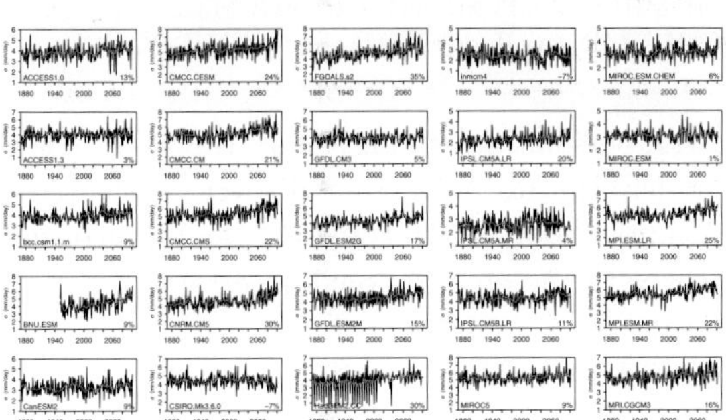

Fig. 4. Time series of sub-seasonal standard deviation of tropical Australia summer (December–March) monsoon rainfall for the 25 CMIP5 climate models for RCP-8.5 scenario. The variability (σ) is computed as the root-mean-square of the daily deviation from the seasonal mean of each individual year. Please note that the y-axis range differs for some models. The trend in seasonal standard deviation is shown (red line) and the percentages show the relative change in standard deviation in the future with respect to the past, i.e. (σ 2071—2100 − σ 1871—1900)/σ 1871—1900.

4. The African Monsoon

The African monsoon system is composed of the Western African Monsoon (WAM) and the Eastern African Monsoon. Prominent droughts and flooding events associated with the monsoon variability have often struck across different parts of sub-Saharan Africa and led to substantial economic and human losses (Peterson *et al.* 2012, Druyan 2011). African monsoon systems usually interact with monsoon systems from remote regions worldwide (Biasutti *et al.* 2003), suggesting that both regional and global perspectives are needed to improve our understanding of monsoon variability.

4.1. *Time-mean state*

Improved understanding of the WAM has been gained through the African Multidisciplinary Monsoon Analysis project (AMMA, Redelsperger *et al.* 2006). The WAM system is a thermally direct land-ocean-atmosphere coupled circulation driven by the meridional gradients of dry and moist static energy in the boundary layer between the Gulf of Guinea and the continent from April through November (Lafore *et al.* 2011). The system is complicated, with the Saharan heat low pressure minimum driving the convergence of two opposing low-level flows along the inter-tropical discontinuity. The resulting baroclinicity is responsible for the African easterly jet (AEJ), with the convection located within the inter-tropical convergence zone (ITCZ). African easterly waves are the primary weather systems, but understanding their interactions with mesoscale convective systems remains a scientific challenge. Lafore *et al.* (2011) exposed the critical role of the diurnal cycle of the atmospheric boundary layer (ABL), as well as the exchanges between different atmospheric layers and the large mesoscale fluctuations which feedback to the atmospheric circulations.

The WAM Modelling and Evaluation project (WAMME, Xue *et al.* 2010) and the AMMA-Model Intercomparison project (AMMA-MIP, Hourdin *et al.* 2010) have provided insights on the ability of GCMs to simulate WAM climate. They showed that AGCMs could reasonably simulate the spatial patterns and intensity of mean seasonal rainfall in West Africa. The main error is the failure to capture the regional circulations, including the AEJ and the upper-troposphere Tropical Easterly Jet.

For a range of time scales, RCM simulations from the COordinated Regional climate downscaling EXperiment (CORDEX; Giorgi *et al.* 2009, Jones *et al.* 2011) outperform the driving GCMs (Nikulin *et al.* 2012). Kim *et al.* (2014) further examined the monthly mean precipitation and surface air temperature produced by these 10 CORDEX RCMs and exposed the contrasting ability of models between the western and eastern parts of Africa, although significant biases are found in individual models depending on the region and seasons. Moufouma-Okia and Rowell (2009) showed that the HadRM3P regional model was more sensitive to lateral boundary conditions as compared to the initialization of soil moisture, while other studies of RCMs forced with the same reanalysis as lateral boundary conditions produce a wide-range of results. Thus, caution is warranted in drawing conclusions based on simulation with a single RCM.

Coupled GCMs, such as those in CMIP5, have difficulties with capturing both the climatology and prominent modes of variability over Africa (Fig. 5). Models tend to misplace the latitudinal location of the tropical rainbelt, which is a major factor of rainfall variability in the tropics, and produce excessive summer rainfall over the Gulf of Guinea, leading to a substantial dry bias over the Sahel region (Bielli *et al.* 2011, Cook and Vizy 2006). This error structure, prevalent in the CMIP3 and CMIP5 multi-model means, is similar to the difference between CMAP and GPCP rainfall (Sperber 2014, personal communication). However, in the models the excessive rainfall over the tropical Atlantic Ocean extends to 15°S, and is inconsistent with the observational uncertainty.

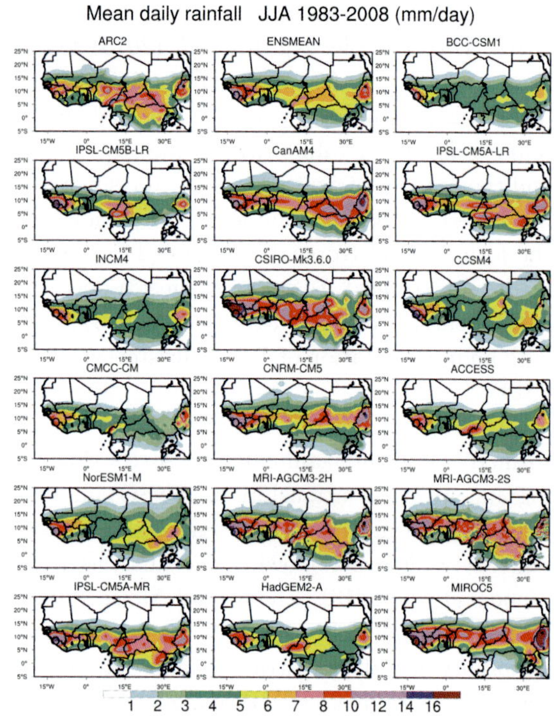

Fig. 5. June–August rainfall climatologies for 1983–2008 from (a) Africa Rainfall Climatology-V2, (b) the CMIP5 ensemble mean, (c) BCC-CSM1, (d) IPSL-CM5B-LR, (e) CanAM4, (f) IPSL-CM5A-LR, (g) INMCM4, (h) CSIRO-Mk3.6.0, (i) CCSM4, (j) CMCC-CM, (k) CNRM-CM5, and (l) ACCESS, (m) NorESM1-M, (n) MRI-AGCM3-2H, (o) MRI-AGCM3-2S, (p) IPSL-CM5A-MR, (q) HadGEM2-A, and (r) MIROC5.

4.2. *Variability*

The WAM intraseasonal variability and timing of the rainfall onset are critical issues for communities in West Africa, since most of the annual rainfall occurs in only a few months and even small variations can have a dramatic consequence. The meridional migration of the ITCZ with the Sun is not smooth but rather marked by a succession of active phases and pauses (Sultan and Janicot 2003). The first rainy season along the Guinean coasts occurs between April and June. It is followed by an abrupt northward shift of the tropical rainbelt from 5°N to 11°N, associated with the ITCZ, bringing rainfall to the Sahel. This abrupt northward shift of the WAM rainfall contrasts with the smooth retreat of the ITCZ, which gives rise to a second rainy season over the Guinean coasts in October–November. The mean onset date occurs between 24 and 30 June with a standard deviation of 8 days, depending on the variable used for the estimate.

Janicot *et al.* (2011) exposed three main modes of intraseasonal variability. Two modes, corresponding to the "quasi-biweekly zonal dipole" and the "Sahel" mode, have a mean periodicity of 15 days. The third mode is the "African MJO" mode, with a mean periodicity of ~40 days. All these modes depict a regional scale structure and can strongly influence rainfall and convective activity over West Africa.

Other important features of the WAM are the interannual and decadal variability. The ENSO phenomenon, Pacific Ocean, Indian Ocean, Mediterranean Sea, and the Atlantic Ocean have been identified as important drivers of the WAM rainfall at interannual time scales. The current consensus is that the Sahelian drought that developed during the second half of the 20th century is a result of remote effects of oceanic anomalies (Giannini *et al.* 2008). At decadal time scales and beyond, the variability of Sahelian rainfall coincides more closely with marked low frequency changes in the background state such as the Pacific decadal oscillation (PDO), the Atlantic multi-decadal oscillation (AMO) and the low frequency component of the Atlantic-Niño. In CMIP5, historical simulations reasonably capture the magnitude of the centennial Sahel drying over the span of the 20th century and confirm that anthropogenic forcings may have contributed to it (Biasutti 2013). Yet, the models do not reproduce the amplitude of observed oscillations at multidecadal time scales, suggesting that either oscillations in the forcing or the strength of natural variability are underestimated.

Unfortunately, little attention has been devoted to understanding the underlying physical mechanisms of East African monsoon (EAM) system. The EAM system possesses a number of unique features. The most important of these is the relative dryness, attributable to low-level divergent flow over East Africa in boreal summer. Most rainfall in the region occurs during the boreal spring and autumn season, when the flow pattern is more or less convergent with the monsoon rainfall total being relatively low. The interannual variations of EAM rainfall, primarily associated with surface and atmospheric perturbations over the Indian Ocean basin, are also paired with ENSO, which can alter the local Walker circulation. In coastal areas of East Africa for example, moist conditions are usually associated with cool sea surface temperatures (SSTs) and related to vertical motions in the eastern Indian Ocean (Tierney *et al.* 2013, Webster *et al.* 1999).

Simulated features of the East African monsoon have been studied by Cook et al. (2012), using CCSM4. Comparisons have also been made to the atmospheric component of CCSM4 (CAM4) forced with observed SSTs, in order to isolate the contribution of land-atmosphere processes in the model errors. The results indicate that CCSM4 produces an improved simulation of the EAM rainfall compared to the CAM4. While CCSM4 and CAM4 both reasonably capture the spatial distribution of seasonal rainfall totals, but with slightly higher values, the intraseasonal variability of the EAM rainfall is more realistic in the coupled model simulations.

4.3. *Climate change*

Accurately predicting the diversity of climate regimes across Africa poses major challenges for coupled general circulation models (GCMs). This is due partly to our limited understanding of the complex interactions across weather and climate time scales, the interactions between local and remote processes, and the coarse horizontal grid resolutions used in coupled GCMs. The CMIP5 coupled GCMs strengthen the previous assessment for a substantial role of anthropogenic emissions in driving rainfall changes in the Sahel.

Regarding future projections, the IPCC 4th Assessment Report indicates that the CMIP3 coupled GCMs (Meehl *et al.* 2007b) disagree on both the magnitude and sign of anthropogenically forced rainfall changes over Africa (Druyan 2011, Caminade and Terray 2010). Patricola and Cook (2010) used the Weather Research and Forecasting (WRF) RCM to downscale nine CMIP3 simulations and found that the precipitation change signal is very mixed over West Africa in the second half of the twenty-first century.

Compared to the CMIP3 results, in CMIP5 there has been little evolution of model performance in terms of both the present-day model errors and uncertainty in future projections (Roehrig *et al.* 2013). In CMIP5 80% of the simulations suggest ~20% drying of western Sahel and 75% of the models project increased rainfall over the eastern Sahel, with large spread in the amplitude of the projected change in rainfall. Thus, over the whole of the Sahel, the CMIP5 RCP8.5 average precipitation response is rather muted, with a projected increase of only 2.4%. The projected change in rainfall extremes are rather larger, with the seasonal maximum 5-day total rainfall increasing by ~14%, and the number of consecutive dry days increasing by ~9% (Kitoh *et al.* 2013). However, given difficulties in simulating the phase of the diurnal cycle of precipitation and the propagating mesoscale systems that contribute the bulk of rainfall to the Sahel, and the errors in the simulation of SST in the tropical Atlantic, Roehrig *et al.* (2013) indicate "that over West Africa, CMIP5 models have not reached yet a degree of maturity that makes it possible to directly rely on them to anticipate climate changes and their impacts, especially with regard to rainfall."

5. The South American Monsoon System (SAMS)

5.1. *Time-mean state*

The CMIP5 models underestimate rainfall in Amazonia during both wet and dry seasons (Yin *et al.* 2013). Most models overestimate convective rainfall and underestimate large-scale precipitation during the wet season. This is a consequence of complex feedbacks between the land surface heat flux, rainfall, radiation, and the large-scale circulation. Jones and Carvalho (2013) reported that only a few CMIP5 models were able to simulate the maximum precipitation over the core of the monsoon region, as seen in Fig. 6, while others had serious deficiencies, such as "excessive precipitation over Northeast Brazil, displaced ITCZ, double ITCZ, and too little precipitation over eastern Amazon." All ten models analyzed in their paper show a trend of delayed SAMS retreat dates over the later

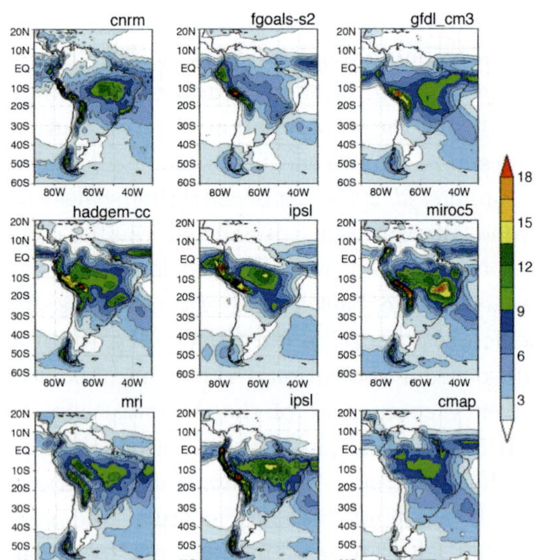

Fig. 6. Rainfall climatology (mm/day) for December–February 1979–2010 for (a) CNRM, (b) FGOALS-s2, (c) GFDL-CM3, (d) HadGEM2-cc, (e) IPSL-CM5A-LR, (f) MIROC5, (g) MPI-ESM-LR, (h) MRI-CGCM3, and (i) CMAP.

portion of the historical runs, consistent with observations, with several models showing average retreat dates consistent with the observed climatology.

5.2. *Variability*

The interannual variability of SAMS has been investigated using EOF analysis of the January–March (JFM) 200 hPa mean winds over 50°S–10°S, 70°W–0°W (Robertson and Mechoso 2000). This mode consists of an equivalent barotropic vortex centered on the coast of southeast Brazil (around 30°S, 45°W). As seen in Fig. 7, for the CMIP5 models there are intermodel differences in the location and intensity of the vortex, as well as in explained variance. The cyclonic (anticyclonic) phase of the vortex corresponds to an enhanced (weakened) South Atlantic Convergence Zone (SACZ), and to anomalously low (high) precipitation over most of La Plata basin (Ma *et al.* 2011). During the anticyclonic (cyclonic) phase the low-level winds bring air with lower (higher) moist static energy to the SACZ from the Atlantic (Amazon) resulting in decreased (enhanced) convection.

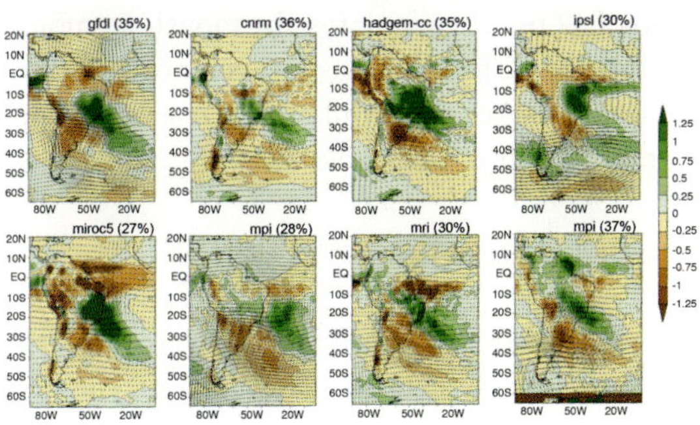

Fig. 7. Regression maps of January–March (JFM) 200 hPa wind on PC1 from the EOF analysis of the 200 hPa wind (arrows, m s^{-1}) over 50°S-10°S, 70°W-0°W domain for seven CMIP5 models and the NCEP-NCAR reanalysis dataset, for the cyclonic phase. Also shown are the regression maps of the JFM precipitation on PC1 (shading, mm day^{-1}), multiplied by the mean value of PC1 during years of strong cyclonic phase (when PC1 $> \sigma$(PC1) or $< -\sigma$(PC1)).

During years with a strong cyclonic (anticyclonic) phase, this mode of variability accounts for a large part of the precipitation anomalies in both the simulation and the observation. The relationship between wind and precipitation anomalies is generally consistent with the theory of convective margins (Ma *et al.* 2011) and with the ventilation mechanism (Lintner and Neelin 2010).

5.3. *Climate change*

Precipitation over South America during the wet season is projected to increase in the CMIP5 models (Seth *et al.* 2013; Kitoh *et al.* 2013). A similar increase was obtained by the CMIP3 models (Seth *et al.* 2010) and is present in the historical record (Skansi *et al.* 2013). The percentage change in median precipitation for the SAMS is a 1.5% increase within the RCP4.5 emissions scenario and a 2.4% increase within the RCP8.5 scenario by the end of the 21st century (Kitoh *et al.* 2013). The precipitation increase seems mostly related to an intensification of heavy rainfall, particularly over the Amazon and southeast South America, rather than to increases in the duration or frequency of consecutive wet days (Skansi *et al.* 2013). During the monsoon months there is a projected increase of ~18% for the model median five-day seasonal maximum in precipitation and the median number of consecutive dry days by 2100 in the RCP8.5 scenario (Kitoh *et al.* 2013). The seasonal amplitude of the monsoon, inferred from precipitation, moisture, temperature and monsoonal circulation, is also projected to increase by 30% from the current amplitude by mid-century (Jones and Carvalho 2013). Lastly, the length of the monsoon season is generally projected to increase. By mid-century, the ensemble mean of selected CMIP5 models projects an earlier onset by 14 days and a later demise by 17 days in the RCP8.5 scenario (Jones and Carvalho 2013). However, the different precipitation indices used to quantify this change in different studies result in somewhat inconsistent results (Kitoh *et al.* 2013).

6. The North American Monsoon System (NAMS)

6.1. *Time-mean state*

Compared to the CMIP3 results, in CMIP5 the mean annual cycle of monthly precipitation over a core NAMS domain shows little improvement in magnitude, but clear improvement in phasing (Geil *et al.* 2013). For most models, NAMS onset is early and monsoon retreat is poorly captured. This difficulty has been associated with an unrealistic flux of low-level moisture from the tropics into the NAMS region that persists after the monsoon season. There is large intermodal variability in the simulation of NAMS.

The North American Monsoon System (NAMS) has been evaluated using the CAM4 with three different dynamical cores at multiple horizontal resolutions. The three dynamical cores are Spectral Eulerian, HOMME, and MPAS. As seen in Fig. 8, at T85 the ITCZ is too weak and there is too much precipitation over the Sierra Madre Occidental (SMO). The T341 simulation is more successful in representing the ITCZ and precipitation in the USA states bordering the Gulf of Mexico. HOMME, at 110 km resolution, is more successful than T85 in the precipitation over the SMO, but there is still too little precipitation over the Gulf States, Panama, and Costa Rica. HOMME at 27 km resolution can better simulate the ITCZ, but precipitation is too high over the SMO. It is hypothesized that higher resolution simulations can improve the overall spatial characteristics of the ITCZ and reduce dry biases in the southeastern U.S. and northern Mexico by better resolving the mesoscale dynamics contributing to the NAMS.

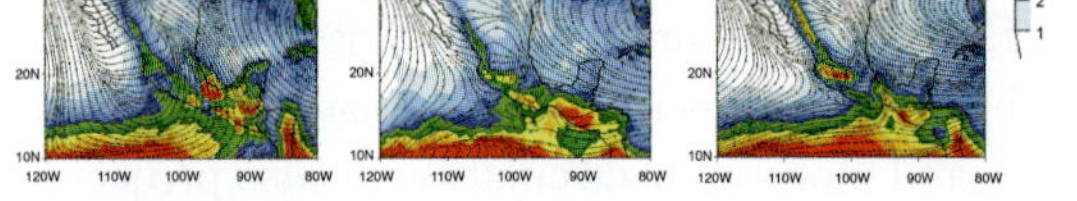

Fig. 8. Precipitation climatologies from: (a) CMAP data at 2.5° global resolution; (b) TRMM data at 0.25° global resolution; (c) CAM4 with the spectral Eulerian dynamical core at ~1° global resolution (T85); (d) CAM4 with the spectral Eulerian dynamical core at ~0.25° global resolution (T341); (e) CAM4 with the HOMME dynamical core at 110 km global resolution (HOMME 110 km); and (f) CAM4 with the HOMME dynamical core at 27 km global resolution (HOMME 27 km) in the NH summer (JJA). Climatologies are for 2000–2006.

Previous results, performed with the NCEP Global Forecasting System (GFS; Mo *et al.* 2005) and CAM3 (Collier and Zhang 2007) also indicate that increased model horizontal resolution has a positive impact on the simulation of NAMS. In GFS, precipitation over the southwest U.S. and northwest Mexico was improved when the resolution was increased from T62 to T126. Relative to CAM3 T42, T85 rainfall over the SMO region is better simulated, and a phase bias in the diurnal cycle is somewhat reduced. Also, T42 poorly simulates the month-to-month evolution of the wet season over northwestern Mexico and Arizona, while these biases are considerably reduced with T85.

6.2. *Climate change*

For the NAMS, projected climate changes to the end of the 21st century show a weakening of the monsoon through a weakening and poleward extension of the Hadley cell (Lu *et al.* 2007; Vera *et al.* 2013). The CMIP5 models project an overall decrease in annual mean precipitation (Cook and Seager 2013) and a decrease in total monsoon rainfall by ~7% in the RCP8.5 scenario in the core NAMS region by 2100 (Kitoh *et al.* 2013). The CMIP5 models show that in RCP8.5

there are significant decreases in early monsoon (June–July) precipitation and increases in late monsoon (September–October) precipitation; this is consistent with CMIP3 model results (Cook and Seager 2013; Maloney *et al.* 2014). CMIP5 models consistently project an increase in summer mean temperatures, which may be associated with higher extreme temperatures (Duffy and Tebaldi 2012) and may also impact extremes of precipitation. The median number of consecutive dry days is projected to increase by ∼11% in RCP4.5 and by ∼23% in RCP8.5 by 2100 (Kitoh *et al.* 2013). Additionally the median seasonal maximum in five-day precipitation is projected to increase by ∼3% in the RCP4.5 scenario and by ∼6% in the RCP8.5 scenario by 2100 (Kitoh *et al.* 2013).

7. Challenges

A variety of mean state errors can be found over the various monsoon domains. For the Asian monsoon, improvement is needed in the simulation of the mean state and the annual cycle phasing and amplitude (Sperber *et al.* 2013). While coupled ocean-atmosphere simulations are critical for establishing monsoon teleconnections (Wang *et al.* 2005), SST errors can also lead to biases in monsoon. For example, Levine *et al.* (2013) find that CMIP5 models that have a substantial cold bias over the Arabian Sea have a delayed onset due to a weakened moisture flux. Other possible contributions include inadequate simulation of land surface processes and sensitivity to convection scheme parameters (Chen *et al.* 2010, Zou *et al.* 2014).

Over Africa present-day simulations have difficulties with capturing the climatology and prominent modes of variability. Additionally, the simulation of reliable atmospheric water cycles associated with the African monsoon systems requires major improvements in the physical parameterizations of climate and weather models. Importantly, more attention needs to be devoted to understanding the physical mechanisms and variability of the East African monsoon system. Over Australia the CMIP5 mean precipitation extends too far southward into the continental interior. For SAMS, most models overestimate convective rainfall and underestimate large-scale precipitation during the wet season, and underestimate rainfall in Amazonia during both the wet and dry seasons. For NAM, extension of rainfall over northern Mexico and the southern U.S, is problematic, as is the representation of the coastal vs. ocean rainfall.

The biases in the annual cycle of monsoon precipitation are particularly acute, and regionally diverse. As seen in Fig. 9, Sperber and Annamalai (2014) analyzed climatologies of pentad rainfall from GPCP, CMAP, and 21 CMIP5 historical simulations. From an evaluation of the fractional accumulation of rainfall they found that the models were systematically late in monsoon onset for All-India rainfall, the South American monsoon, and the Gulf of Guinea. However, over the Sahel and the NAM domain the models were systematically early in the onset of monsoon rainfall. For WAM, the models poorly represented the sudden transition

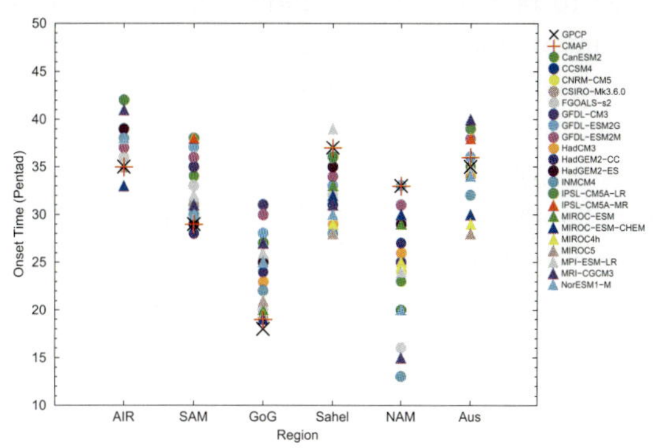

Fig. 9. The pentad of climatological summer monsoon rainfall onset from GPCP and CMAP (1979–2004) and from 21 CMIP5 historical simulations (1961–1999). See Sperber and Annamalai (2014) for a discussion of the technique of fractional accumulation for model evaluation, and for the spatial extent of the various domains. Note for Aus and SAM the pentads were reordered to July–June prior to analysis.

of rainfall from the Gulf of Guinea to the Sahel. Similarly, the models did not represent well the ocean vs. land behavior for NAM.

Pronounced interdecadal variations in rainfall are observed over all monsoon land domains (Zhang and Zhou 2011). Adequate simulation and understanding of processes that drive secular trends in rainfall variability are basic challenges. Over the Sahel, simulation of the observed decadal drying trend is a particularly daunting task for most GCMs, due to the wide range of physical processes competing to drive the rainfall trend. Similarly, over Australia one of the major challenges for current climate models is the failure to simulate the recent upward trend in northwest Australian rainfall. Neither the CMIP3 nor CMIP5 models are able to capture this trend, with suggestions that it is related to either aerosols (Rotstayn *et al.* 2012) or changes in diurnal/synoptic patterns of rainfall (Catto *et al.* 2012c; Berry *et al.* 2012). Similarly, for the past 60 years, the East Asian summer monsoon has been weakening, resulting in a tendency toward increased droughts in northern China and floods in the Yangtze River Valley. The long time scales suggest that teleconnections remote from the monsoon domains need to be better understood. This is also true on interannual time scales, where for example, excessive snow cover over Eurasia has been linked to dry biases of the Indian summer monsoon (Saha *et al.* 2013). Also remaining are systematic biases that are particularly important at regional scales. These biases include adequate simulation of SST and its variability, and the representation of the multi-scale interactions that modulate monsoons.

The impact of aerosols on the Asian monsoon rainfall remains a subject of much debate, and adequate modelling of both their direct and indirect effects is still a source of uncertainty for attribution studies (Song *et al.* 2014), present-day climate reproducibility and future climate change scenarios. The elevated heat pump hypothesis is a potential, but debated, mechanism by which aerosols influence the Asian monsoon (Lau and Kim 2006, Lau *et al.* 2006). Over Africa the role of aerosols, including wind-blown dust, need further investigation to assess their impact on the radiation budget and hydrological cycle.

While improved representation of monsoon variability is intimately tied to realistic physical processes, higher horizontal resolution has a demonstrated benefit for the simulation of monsoons. Over Asia there is improvement in the simulation of the southwest monsoon circulation, orographic rainfall, and the monsoon trough over the Bay of Bengal. Along the monsoon trough, synoptic scale disturbances and mesoscale convective systems are noticeably improved (Sabin *et al.* 2013), as is the Baiu frontal structure, and orographic rainfall over East Asia (Kitoh and Kusunoki 2008). Over North America the ITCZ and precipitation in the states bordering the Gulf of Mexico improve at higher resolution, and over South America circulations are improved with better representation of the Andes Mountains. Additionally, increased vertical resolution in the upper ocean has been shown to be important for capturing diurnal variations of sea surface temperature. These diurnal variations project onto intraseasonal time scales, giving rise to the SST perturbations that improve the organization of intraseasonal convection (Bernie *et al.* 2005, Klingaman *et al.* 2011).

Estimating the potential impact of climate change is a daunting problem, especially in light of model bias. Until recently, the typical approach was "one-model, one-vote" when performing a climate change study. Recently, a broader range of model evaluations has been facilitated due to open access to the CMIP database (Taylor *et al.* 2012). As such, a wider variety of process evaluations and skill metrics have been applied to the historical simulations. This has promoted the selective use of models in climate change assessments, by critically assessing the quality of the historical simulations, with the goal of reducing uncertainty in projections. For example, in the case of Asian monsoon, the spatial pattern of the mean state precipitation, the quality of ENSO, and the ENSO-monsoon teleconnection have been used as filters for model selection (e.g. Annamalai *et al.* 2007). Alternatively, for global monsoon assessments, the ability of models to represent the leading modes of the annual variation of precipitation has been used (e.g. Lee and Wang 2014). Thus, for relevant processes and interactions the sub-selected models exhibit better overall skill in their simulation of the present-day interactions as compared to the full suite of models. An important byproduct of the sub-selection process is that it provides modelers with an indication of the relevant processes that need to be improved in the next generation of models.

A hallmark test of any model is its ability to make predictions. From a practical standpoint, intraseasonal prediction is a highly coveted, given the impact of the ISO on monsoon onset, extreme rain events, flood and drought conditions, and the associated practicalities for farming and hydrometeorological services. The modal structures of the MJO (e.g. Wheeler and Hendon 2004) and the boreal summer intraseasonal variability (BSISV; e.g. Lee *et al.* 2013) are well known from observations. As such experimental real-time forecasts of the MJO and BSISV are ongoing through the effort of the WGNE MJO Task Force and the contributions of most of the world's numerical weather prediction centers (Gottschalck 2010). From projecting the relevant NWP forecast variables onto the observed basis functions there is useful skill out to about 3 weeks for predicting the large-scale intraseasonal structure. While the life-cycle of the MJO and the BSISV has been characterized in detail (Sperber 2003, Annamalai and Sperber 2005), a theory of intraseasonal variability still eludes us (Hendon *et al.* 2011). As such, progress in modelling the intraseasonal variability has been limited (Hung *et al.* 2013, Sperber and Annamalai 2008, Sperber *et al.* 2013), with sensitivities to the details of convective parameterization and their interaction with environmental moisture identified as potential limiting factors in our ability to model intraseasonal variability (Kim *et al.* 2014). Modeling intraseasonal variability has been recognized as a grand challenge problem, with a coordinated international effort underway to improve our understanding of the essential processes that give rise to this mode of variability (Petch *et al.* 2011).

The challenges of predicting and simulating monsoon variability from diurnal through centennial time scales attests to the broad scope of processes and multi-scale interactions that need to be represented with fidelity in order to produce more realistic simulations of monsoons.

Acknowledgments

K. R. Sperber was supported by the Office of Science (BER), U.S. Department of Energy through Lawrence Livermore National Laboratory contract DE-AC52-07NA27344. A. Kitoh was supported by the SOUSEI program. A. F. Moise was supported by the Bureau of Meteorology, CSIRO and the Department of Industry, Innovation, Climate Change, Science, Research and Tertiary Education through the Australian Climate Change Science Program (ACCSP) and the Regional Natural Resource Management Planning for Climate Change Fund — A Clean Energy Future Initiative. A. G. Turner was funded under a NERC Fellowship (NE/H015655/1).

References

Achuthavarier, D., V. Krishnamurthy, B. P. Kirtman, and B. H. Huang, 2012: Role of the Indian Ocean in the ENSO-Indian Summer Monsoon Teleconnection in the NCEP Climate Forecast System. *J. Climate*, **25**, 2490–2508.

Annamalai, H. and K. R. Sperber, 2005: Regional heat sources and the active and break phases of boreal summer intraseasonal (30–50 day) variability. *J. Atmos. Sci.*, **62**, 2726–2748.

Annamalai, H., K. Hamilton, and K. R. Sperber, 2007: The South Asian summer monsoon and its relationship with ENSO in the IPCC AR4 simulations. *J. Climate*, **20**, 1017–1092.

Becker, B. D., J. M. Slingo, L. Ferranti, and F. Molteni, 2000: Seasonal predictability of the Indian summer monsoon: What role do land surface processes play? *Mausam*, **52**, 47–62.

Bernie, D. J., S. J. Woolnough, and J. M. Slingo, 2005: Modelling diurnal and intraseasonal variability of the ocean mixed layer. *J. Climate*, **18**, 1190–1202.

Berry, G., M. Reeder, and C. Jakob 2012: Coherent synoptic disturbances in the Australian monsoon. *J. Climate*, **25**, 8409–8421.

Biasutti, M., D. S. Battisti, and E. S. Sarachnik, 2003: The annual cycle over the tropical Atlantic, South America and Africa. *J. Climate*, **16**, 2491–2508.

Biasutti, M., 2013: Forced Sahel rainfall trends in the CMIP5 archive. *J. Geophys. Res.*, **118**, 1613–1623. doi:10.1002/jgrd.50206.

Blanford, H. F., 1884: On the connection of the Himalaya snowfall with dry winds and seasons of drought in India. *Proc. Roy. Soc. London*, **37**, 3–22.

Bielli, S., H. Douville, and B. Pohl 2011: Understanding the West African monsoon variability and its remote effects: An illustration of the grid point nudging methodology. *Clim. Dynam.*, **35**, 159–174. doi: 10.1007/s00382-009-0667-8.

Boschat, G., P. Terray, and S. Masson, 2012: Robustness of SST teleconnections and precursory patterns associated with the Indian summer monsoon. *Clim. Dynam.*, **38**, 2143–2165.

Brown, J. R., A. F. Moise, R. Colman, and H. Zhang, 2016: Will a warmer world mean a wetter or drier Australian monsoon? *J. Climate*, **29**, 4577–4596, doi:10.1175/JCLI-D-15-0695.1.

Cai, W., A. Sullivan, and T. Cowan, 2011: Interactions of ENSO, the IOD, and the SAM in CMIP3 models. *J. Climate*, **24**, 1688–1704.

Cai, W. J., A. Sullivan, and T. Cowan, 2009: Rainfall teleconnections with Indo-Pacific variability in the WCRP CMIP3 models. *J. Climate*, **22**, 5046–5071.

Caminade, C. and L. Terray, 2010: Twentieth century Sahel rainfall variability as simulated by the ARPEGE AGCM, and future changes. *Clim. Dynam.*, **35**, 75–94. doi:10.1007/s00382-009-0545-4.

Catto, J. L., N. Nicholls, and C. Jakob, 2012a: North Australian sea surface temperatures and the El Niño Southern Oscillation in observations and models. *J. Climate*, **25**, 5011–5029.

Catto, J. L., N. Nicholls, and C. Jakob, 2012b: North Australian sea surface temperatures and the El Niño Southern Oscillation in the CMIP5 models. *J. Climate*, **25**, 6375–6382.

Catto, J. L., C. Jakob, and N. Nicholls, 2012c: The influence of changes in synoptic regimes on North Australian wet season rainfall trends. *J. Geophys. Res.*, **117**, D10102. doi:10.1029/2012JD017472.

Charney, J. and J. Shukla, 1981: Predictability of monsoons. In *Monsoon Dynamics*, Eds. J. Lighthill and R. P. Pearce, Cambridge Univ. Press, pp. 99–109.

Chen, H., T. Zhou, R. B. Neale, X. Wu, and G. J. Zhang, 2010: Performance of the new NCAR CAM3.5 in East Asian summer monsoon simulations: Sensitivity to modifications of the convection scheme. *J. Climate*, **23**, 3657–3675.

Christensen, J. H. *et al.*, 2007: Regional climate projections. In: Climate Change 2007: The Physical Science Basis. Contribution of Working Group I to the Fourth Assessment Report of the Intergovernmental Panel on Climate Change, edited by Solomon, S. *et al.*, Cambridge Univ. Press, Cambridge and New York.

Collier, J. C. and G. J. Zhang, 2007: Effects of increased horizontal resolution on simulation of the North American monsoon in the NCAR CAM3: An evaluation based on surface, satellite, and reanalysis data. *J. Climate*, **20**, 1843–1861.

Colman, R. A., A. F. Moise, and L. I. Hanson 2011: Tropical Australian climate and the Australian monsoon as simulated by 23 CMIP3 models. *J. Geophys. Res.*, **116**. D10116. doi:10.1029/2010JD015149.

Cook, K. H. and E. K. Vizy, 2006: Coupled model simulations of the West African monsoon system: Twentieth- and twenty-first century simulations. *J. Climate*, **19**, 3681–3703.

Cook, B. I. and R. Seager, 2013: The response of the North American monsoon to increased greenhouse gas forcing. *J. Geophys. Res.*, **118**, 1690–1699. doi:10.1002/jgrd.50111.

Cook, K. H., G. A. Meehl, and J. M. Arblaster, 2012: Monsoon regimes and processes in CCSM4. Part II: African and American monsoon systems. *J. Climate*, **25**, 2609–2621. doi:10.1175/JCLI-D-11-00185.1.

Drosdowsky, W., 1996: Variability of the Australian summer monsoon at Darwin: 1957–1992. *J. Climate*, **9**, 85–96.

Druyan, L., 2011: Review studies of 21st-century precipitation trends over West Africa. *Int. J. Climatol.*, **31**, 1415–1424. doi:10.1002/joc.2180.

Duffy, P. B. and C. Tebaldi, 2012: Increasing prevalence of extreme summer temperatures in the U.S. *Climatic Change*, **111**, 487–495.

Frieler, K., M. Meinshausen, M. Mengel, N. Braun, and W. Hare, 2012: A scaling approach to probabilistic assessment of regional climate. *J. Climate*, **25**, 3117– 3144.

Geil, K. L. Y. L. Serra, and X. Zeng, 2013: Assessment of CMIP5 model simulations of the North American monsoon system. *J. Climate*, **26**, 8787–8801. doi:10.1175/JCLI-D-13-00044.1.

Giahnini, A., M. Biasutti, I. Held, and A. Sobel, 2008: A global perspective on African climate. *Climatic Change*, **90**, 359–383. doi:10.1007/s10584-008-9396-y.

Giorgi, F., C. Jones, and G. R. Asrar, 2009: Addressing climate information needs at the regional level: The CORDEX framework. *WMO Bulletin*, **58**, 175–183.

Gottschalck, J., M. Wheeler, K. Weickmann, F. Vitart, N. Savage, H. Lin, H. Hendon, D. Waliser, K. Sperber, M. Nakagawa, C. Prestrelo, M. Flatau, and W. Higgins, 2010: A framework for assessing operational Madden-Julian Oscillation forecasts. *Bull. Amer. Meteorol. Soc.*, **91**, 1247–1258. doi:10.1175/2010 BAMS2816.1.

Hendon, H. H., K. R. Sperber, D. E. Waliser, and M. C. Wheeler, 2011: Modeling monsoon intraseasonal variability. *Bull. Amer. Meteor. Soc.*, **92**, 32–35. doi:10.1175/2011BAMS3164.1.

Hourdin, F. and coauthors, 2010: AMMA model intercomparison project. *Bull. Amer. Meteor. Soc.*, **91**, 95–104. doi:10.1175/2009BAMS2791.1.

Hsu, P.-C., T. Li, J.-J. Luo, H. Murakami, A. Kitoh, and M. Zhao, 2012: Increase of global monsoon area and precipitation under global warming: A robust signal? *Geophys. Res. Lett.*, **39**, L06701.

Hsu, P.-C., T. Li, H. Murakami, and A. Kitoh, 2013: Future change of the global monsoon revealed from 19 CMIP5 models. *J. Geophys. Res.*, **118**, 1247–1260. doi:10.1002/jgrd.50145.

Huang, D.-Q., J. Zhu, Y.-C. Zhang, and A.-N. Huang, 2013: Uncertainties on the simulated summer precipitation over Eastern China from the CMIP5 models. *J. Geophys. Res. Atmos.*, **118**, 9035–9047. doi:10.1002/jgrd.50695.

Hung, C-W., X. Liu, and M. Yanai, 2004: Symmetry and asymmetry of the Asian and Australian summer monsoons. *J. Climate*, **17**, 2413–2426.

Hung M.-P., J.-L. Lin, W. Wang, D. Kim, T. Shinoda, and S. J. Weaver, 2013: MJO and convectively coupled equatorial waves simulated by CMIP5 climate models. *J. Climate*, **26**, 6185–6214. doi: 10.1175/JCLI-D-12-00541.1.

Intergovernmental Panel on Climate Change (IPCC) 2007: Climate Change 2007: The Physical Science Basis. Contribution of Working Group I to the Fourth Assessment Report of the IPCC, edited by S. Solomon *et al.*, Cambridge Univ. Press, Cambridge, UK.

Intergovernmental Panel on Climate Change (IPCC) 2013: Climate Change 2013: The Physical Science Basis. Contribution of Working Group I to the Fifth Assessment Report of the IPCC, edited by Stocker *et al.*, Cambridge Univ. Press, Cambridge, UK.

Janicot S. *et al.*, 2011: Intraseasonal variability of the West African monsoon. *Atmos. Sci. Lett.*, **12**, 66–74. doi:10.1002/asl.280.

Jones, C. and L. M. Carvalho, 2013: Climate change in the South American Monsoon System: Present climate and CMIP5 projections. *J. Climate*, **26**, 6660–6678. doi:10.1175/JCLI-D-12-00412.1.

Jones, C., F. Giorgi, and G. Asrar, 2011: The COordinated Regional Downscaling Experiment CORDEX: An international downscaling link to CMIP5, *International CLIVAR Project Office*, 56, Southampton, UK, 34–40.

Jourdain, N. C., A. Sen Gupta, A. S. Taschetto, C. C. Ummenhofer, A. F. Moise, and K. Ashok, 2013: The Indo-Australian monsoon and its relationship to ENSO and IOD in reanalysis data and the CMIP3/CMIP5 simulations. *Clim. Dynam.*, **41**, 3073–3102. doi:10.1007/s00382-013-1676-1.

Kang, I.-S., P.-H. Jang, and M. Almazroui, 2013: Examination of multi-perturbation methods for ensemble prediction of the MJO during boreal summer. *Clim. Dynam.* doi:10.1007/s00382-013-1819-4.

Kim, D., P. Xavier, E. Maloney, M. Wheeler, D. Waliser, K. Sperber, H. Hendon, C. Zhang, R. Neale, Y.-T. Hwang, and H. Liu, 2014: Process-oriented MJO simulation diagnostics: Moisture sensitivity of simulated convection. *J. Climate*, **27**, 5379–5395. doi:10.1175/JCLI-D-13-00497.1.

Kim, J. and coauthors, 2014: Evaluation of the CORDEX-Africa multi-RCM hindcasts: Systematic errors. *Clim. Dynam.*, **42**, 1189–1202. doi:10.1007/s00382-013-1751-7.

Kitoh, A. and S. Kusunoki, 2008: East Asian summer monsoon simulation by a 20-km mesh AGCM. *Clim. Dynam.*, **31**, 389–401.

Kitoh, A., H. Endo, K. Krishna Kumar, I. F. A. Cavalcanti, P. Goswami, and T. Zhou, 2013: Monsoons in a changing world: A regional perspective in a global context. *J. Geophys. Res.*, **118**, 3053–3065. doi:10.1002/jgrd.50258.

Klingaman, N. P., S. J. Woolnough, H. Weller, and J. M. Slingo, 2011: The impact of finer-resolution air-sea coupling on the intraseasonal oscillation of the Indian monsoon. *J. Climate*, **24**, 2413–2430. doi: 10.1175/2010JCLI3868.1.

Lafore, J.-P., *et al.* 2011: Progress in understanding of weather systems in West Africa. *Atmos. Sci. Lett.*, **12**, 7–12. doi:10.1002/asl.335.

Laprise R., L. Hernandez-Diaz, K. Tete, L. Sushama, L. Separovic, A. Martynov, K. Winger, and M. Valin, 2012: Climate projections over CORDEX Africa domain using the fifth-generation Canadian Regional Climate Model (CRCM5). *Clim. Dynam.*, **41**, 3219–3246. doi:10.1007/s00382-012-1651-2.

Lau, K. M. and K. M. Kim, 2006: Observational relationships between aerosol and Asian monsoon rainfall, and circulation. *Geophys. Res. Lett.*, **33**, L21810.

Lau, K. M., M. K. Kim, and K. M. Kim, 2006: Asian summer monsoon anomalies induced by aerosol direct forcing: The role of the Tibetan Plateau. *Clim. Dynam.*, **26**, 855–864.

Lee, J.-Y. and B. Wang, 2014: Future change of global monsoon in the CMIP5. *Clim. Dynam.*, **42**, 101–119. doi:10.1007/s00382-012-1564-0.

Lee, J.-Y., B. Wang, M. C. Wheeler, X. Fu, D. E. Waliser, and I.-S. Kang, 2013: Real-time multivariate indicies for the boreal summer intraseasonal oscillation over the Asian summer monsoon region. *Clim. Dynam.*, **40**, 493–509. doi:10.1007/s00382-012-1544-4.

Levine, R. C., A. G. Turner, D. Marathayil, and G. M. Martin, 2013: The role of northern Arabian Sea temperature biases in CMIP5 model simulations and future projections of Indian summer monsoon rainfall. *Clim. Dynam.*, **41**, 155–172. doi: 10.1007/s00382-012-1656-x.

Li, Y., N. C. Jourdain, A. S. Taschetto, C. C. Ummenhofer, K. Ashok, and A. Sen Gupta, 2012: Evaluation of monsoon seasonality and the tropospheric biennial oscillation transitions in the CMIP models. *Geophys. Res. Lett.*, **39**, L20713. doi:10.1029/2012GL053322, 2012.s.

Lintner, B. R. and J. D. Neelin, 2010: Tropical South America-Atlantic sector convective margins and their relationship to low-level inflow. *J. Climate*, **23**, 2671–2685.

Liu, S., W. Gao, and X.-H. Liang, 2013: A regional climate model downscaling projection of China future climate change. *Clim. Dynam.*, **41**, 1871–1884. doi: 10.1007/s00382-012-1632-5.

Lu, J., G. Vecchi and T. Reichler, 2007: Expansion of the Hadley cell under global warming. *Geophys. Res. Lett.*, **34**, LO6805. doi:10.1029/2006GL028443.

Ma, H.-Y., X. Ji, J. D. Neelin, and C. R. Mechoso, 2011: Mechanisms for precipitation variability of the Eastern Brazil/SACZ convective margin. *J. Climate*, **24**, 3445–3456.

Maloney, E. and coauthors, 2014: North American Climate in CMIP5 Experiments: Part III: Assessment of 21st Century Projections. *J. Climate*, **27**, 2230–2270. doi:10.1175/JCLI-D-13-00273.1.

McBride, J. L., 1998: Indonesia, Papua New Guinea, and Tropical Australia, Chapter 3A of Meteorology of the Southern Hemisphere, Meteorological Monograph No. 49, American Meteorological Society.

Meehl, G. A., C. Covey, T. Delworth, M. Latif, B. McAvaney, J. F. B. Mitchell, R. J. Stouffer, and K. E. Taylor, 2007a: The WCRP CMIP3 multimodel dataset: A new era in climate change research. *Bull. Amer. Meteor. Soc.*, **88**, 1383–1394.

Meehl, G. A. and coauthors, 2007b: Global climate projection. Climate change 2007: The physical science basis. Contribution of Working Group I to the fourth assessment report of the Intergovernmental Panel on Climate Change, 2S, edited by S. Solomon *et al.*, Cambridge Univ. Press, Cambridge and New York, pp. 747–846.

Meehl, G. A., J. M. Arblaster, J. M. Caron, H. Annamalai, M. Jochum, A. Chakraborty, and R. Murtugudde, 2012: Monsoon regimes and processes in CCSM4. Part I: The Asian-Australian monsoon. *J. Climate*, **25**, 2583–2608.

Menon, A., A. Levermann, and J. Schewe 2013: Enhanced future variability during India's rainy season. *Geophys. Res. Lett.*, **40**, 3242–3247. doi:10.1002/grl.50583.

Mo, K. C., J.-K. Schemm, H. M. H. Juang, R. W. Higgins, and Y. Song, 2005: Impact of model resolution on the prediction of summer precipitation over the United States and Mexico. *J. Climate*, **18**, 3910–3927.

Moise, A. F, R. A. Colman, and H. Zhang, 2005: Coupled model simulations of current Australian surface climate and its changes under greenhouse warming: An analysis of 18 CMIP2 models. *Aust. Met. Mag.*, **54**, 291–307.

Moise, A. F., R. A. Colman, and J. R. Brown, 2012: Behind uncertainties in projections of Australian tropical climate: Analysis of 19 CMIP3 models. *J. Geophys. Res.*, **117**, D10103. doi:10.1029/2011JD017365.

Moise, A. F., L. Wilson, M. Grose, P. Whetton, I. Watterson, J. Bhend, J. Bathols, L. Hanson, T. Erwin, T. Bedin, C. Heady, and T. Rafter, 2015a: Evaluation of CMIP3 and CMIP5 models over the Australian region to inform confidence in projections.*Australian Meteorological and Oceanographic Journal*, **65**, 19–53.

Moise, A. F., J. Bhend, I. Watterson, and L. Wilson, 2015b: Evaluation of climate models, Chapter 5 in: Climate Change in Australia — Information for Australia's Natural Resource Management Regions. Technical Report, CSIRO and Bureau of Meteorology, Australia.

Moufouma-Okia, W. and D. P. Rowell, 2009: Impact of soil moisture initialisation and lateral boundary conditions on regional climate model simulation of the West African monsoon. *Clim. Dynam.*, **35**, 213–229. doi:10.1007/s00382-009-0638.

Nicholls, N., J. L. McBride, and R. J. Ormerod, 1982: On predicting the onset of the Australian wet season at Darwin. *Mon. Wea. Rev.*, **110**, 14–17.

Nikulin, G., *et al.* 2012: Precipitation climatology in an ensemble of CORDEX-Africa regional climate simulations. *J. Climate*, **25**, 6057–6078. doi:10.1175/JCLI-D-11-00375-1.

Ninomiya, K., 2009: Characteristics of precipitation in the Meiyu-Baiu season in the CMIP3 20th century simulation. *J. Meteor. Soc. Japan*, **87**, 829–843.

Ninomiya, K., 2011: Characteristics of the Meiyu and Baiu frontal precipitation zone in the CMIP3 20th century simulation and 21st century projection. *J. Meteor. Soc. Japan*, **89**, 151–159.

Ninomiya, K., 2012: Year-to-year variations in the Meiyu and Baiu frontal precipitation zones in the CMIP3 20th century simulation. *J. Meteor. Soc. Japan*, **90A**, 317–325.

Ogata, T., H. Ueda, T. Inoue, M. Hayasaki, A. Yoshida, S. Watanabe, M. Kira, M. Ooshiro, and A. Kumai, 2014: Projected future changes in the Asian monsoon: A comparison of CMIP3 and CMIP5 model results. *J. Meteor. Soc. Japan*, **92**, 207–225.

Patricola, C. M. and K. H. Cook, 2010: Northern African climate at the end of the twenty-first century: An integrated application of regional and global climate models. *Clim. Dynam.*, **35**, 193–212. doi:10.1007/s00382-009-0623-7.

Petch, J. D. Waliser, X. Jiang, P. Xavier, and S. Woolnough, 2011: A global model intercomparison of the physical processes associated with the Madden-Julian Oscillation. *GEWEX News*, **21**, 3–5.

Peterson, C. T., P. A. Stott, and S. Herring, 2012: Explaining extreme events of 2011 from a climate perspective. *Bull. Amer. Meteor. Soc.*, **93**, 1041–1067.

Pope, M., C. Jakob, and M. Reeder, 2009: Variability of the North Australian monsoon. *J. Climate*, **22**, 6699–6715.

Power, S. B. and G. Kociuba, 2011: What caused the observed twentieth-century weakening of the walker circulation? *J. Climate*, **24**, 6501–6514.

Ramage, C., 1971: *Monsoon Meteorology.* International Geophysics Series, Vol. 15, Academic Press, 256 pp.

Redelsperger, J.-L., C. D. Thorncroft, A. Diedhiou, T. Lebel, D. J. Parker, and J. Polcher, 2006: African monsoon multidisciplinary analysis: An international research project and field campaign. *Bull. Amer. Meteor. Soc.*, **87**, 1739–1746.

Robertson, A. W. and C. R. Mechoso, 2000: Interannual and interdecadal variability of the South Atlantic convergence zone. *Mon. Wea. Rev.*, **128**, 2947–2957.

Roehrig, R., D. Bouniol, F. Guichard, F. Hourdin, and J. Redelsperger, 2013: The present and future of the West African monsoon: A process-oriented assessment of CMIP5 simulations along the AMMA transect. *J. Climate*, **26**, 6471–6505, doi:10.1175/JCLI-D-12-00505.1.

Rotstayn, L. D., S. J. Jeffrey, M. A. Collier, S. M. Dravitzki, A. C. Hirst, J. I. Syktus, and K. K. Wong, 2012: Aerosol- and greenhouse gas-induced changes in summer rainfall and circulation in the Australasian region: A study using single-forcing climate simulations. *Atmos. Chem. Phys.*, **12**, 6377–6404. doi:10.5194/acp-12-6377-2012.

Sabeerali, C. T., A. Ramu Dandi, A. Dhakate, K. Salunke, S. Mahapatra, and S. A. Rao, 2013: Simulation of boreal summer intraseasonal oscillations in the latest CMIP5 coupled GCMs. *J. Geophys. Res.*, **118**, 4401–4420. doi:10.1002/jgrd.50403.

Sabin, T. P., R. Krishnan, J. Ghattas, S. Denvil, J.-L. Dufresne, F. Hourdin, and T. Pascal, 2013:

Clim. Dynam., **41**, 173–194. doi: 10.1007/s00382-012-1658-8.

Saha, S. K., S. Pokhrel, and H. S. Chaudhari, 2013: Influence of Eurasian snow on Indian summer monsoon in NCEP CFSv2 freerun. *Clim. Dynam.*, **41**, 1801–1815. doi: 10.1007/s00382-012-1617-4.

Seth, A., S. A. Rauscher, M. Biasutti, A. Giannini, S. J. Camargo, and M. Rojas, 2013: CMIP5 Projected changes in the annual cycle of precipitation in monsoon regions. *J. Climate*, **26**, 7328–7351.

Seth, A., M. Rojas, and S. A. Raucher, 2010: CMIP3 projected changes in the annual cycle of the South American monsoon. *Climatic Change*, **98**, 331–357. doi: 10.1007/s10584-009-9736-6.

Skansi, M. D. L. M., *et al.* 2013: Warming and wetting signals emerging from analysis of changes in climate extreme indices over South America. *Global and Planetary Change*, **100**, 295–307.

Slingo, J. M. and H. Annamalai, 2000: The El Niño of the century and the response of the Indian summer monsoon. *Mon. Wea. Rev.*, **128**, 1778–1797.

Song, F., T. Zhou, and Y. Qian, 2014: Responses of East Asian summer monsoon to natural and anthropogenic forcings in the 17 latest CMIP5 models. *Geophys. Res. Lett.*, **41**. doi:10.1002/2013GL058705.

Sperber, K. R., J. M. Slingo, and H. Annamalai, 2000: Predictability and the relationship between subseasonal and interannual variability during the Asian summer monsoon. *Quart. J. Roy. Meteor. Soc.*, **126**, 2545–2574.

Sperber, K. R., C. Brankovic, M. Deque, C. S. Fredericksen, R. Graham, A. Kitoh, C. Kobayashi, T. Palmer, K. Puri, W. Tennent, and E. Volodin, 2001: Dynamical seasonal predictability of the Asian summer monsoon. *Mon. Wea. Rev.*, **129**, 2226–2248.

Sperber, K. R., 2003: Propagation and the vertical structure of the Madden-Julian Oscillation. *Mon. Wea. Rev.*, **131**, 3018–3037.

Sperber, K. R. and H. Annamalai, 2008: Coupled model simulations of boreal summer intraseasonal (30–50 day) variability, Part 1: Systematic errors and caution on use of metrics. *Clim. Dynam.*, **31**, 345–272. doi:10.1007/s00382-008-0367-9.

Sperber, K. R., H. Annamalai, I.-S. Kang, A. Kitoh, A. Moise, A. Turner, B. Wang, and T. Zhou, 2013: The Asian summer monsoon: An intercomparison of CMIP5 vs. CMIP3 simulations of the late 20th century. *Clim. Dynam.*, **41**, 2771–2744. doi:10.1007/s00382-012-1607-6.

Sperber, K. R. and H. Annamalai, 2014: The use of fractional accumulated precipitation for the evaluation of the annual cycle of monsoons. *Clim. Dynam.*, **43**, 3219–3244. doi:10.1007/s00382-014-2099-3.

Sultan, B. and S. Janicot, 2003: The West African monsoon dynamics. Part II: The "preonset" and "onset" of the summer monsoon. *J. Climate*, **16**, 3389–3406.

Taylor, K. E., R. J. Stouffer, and G. Meehl, 2012: An overview of CMIP5 and the experiment design. *Bul. Amer. Meteor. Soc.*, **93**, 485–498. doi:10.1175/BAMS-D-11-00094.1.

Tierney, J. E., J. E. Smerdon, K. J. Anchukaitis, and R. Seager, 2013: Multidecadal variability in East African hydroclimate controlled by the Indian Ocean. *Nature*, **493**, 389–392. doi:10.1038/nature11785.

Tory, K., S. Chand, R. Dare, and J. McBride, 2013: An assessment of a model-, grid-, and basin-independent tropical cyclone detection scheme in selected CMIP3 global climate models. *J. Climate*, **26**, 5508–5522.

Troup, A. J., 1961: Variations in upper tropospheric flow associated with the onset of the Australian summer monsoon. *Indian J. Meteor. Geophys.*, **12**, 217–230.

Turner, A. G., P. M. Inness, and J. M. Slingo, 2007: The effect of doubled CO_2 and model basic state biases on the monsoon-ENSO system. I: Mean response and interannual variability. *Quart. J. Roy. Meteor. Soc.*, **133**, 1143–1157.

Turner, A. G. and H. Annamalai, 2012: Climate change and the South Asian summer monsoon. *Nature Clim. Change*, **2**, 1–9. doi:10.1038/NCLIMATE1495.

Ueda, H., A. Iwai, K. Kuwako, and M. E. Hori, 2006: Impact of anthropogenic forcing on the Asian summer monsoon as simulated by eight GCMs. *Geophys. Res. Lett.*, **33**, L06703.

Vera, C., B. W. Gutowski, C. R. Mechoso, N. Goswami, C. Reason, C. D. Thorncroft, J. A. Marengo, B. Hewitson, H. Hendon, C. Jones, and P. Lionello, 2013: Understanding and predicting climate variability and change at regional scales. In *Climate Science for Serving Society: Research, Modeling and Prediction Priorities*, Eds. G. R. Asrar and J. W. Hurrell, Springer, pp. 273–306.

Waliser, D. E. and coauthors, 2003: AGCM simulations of intraseasonal variability associated with the Asian summer monsoon. *Clim. Dynam.*, **21**, 423–446.

Walker, G. T., 1924: Correlation in seasonal variations of weather, IV, A further study of world weather. *Memoirs of the Indian Meteorol. Dept.*, **24**, 275–332.

Wang, B., Q. Ding, X. Fu, I.-S. Kang, K. Jin, J. Shukla, and F. Doblas-Reyes 2005: Fundamental challenge in simulation and prediction of summer monsoon rainfall. *Geophys. Res. Lett.*, **32**, L15711. doi:10.1029/2005GL022734.

Wang, B. and Q. Ding, 2006: Changes in global monsoon precipitation over the past 56 years. *Geophys. Res. Lett.*, **33**, L06711. doi:10.1029/2005GL025347

Wang, B., S.-Y. Yim, J.-Y. Lee, J. Liu, and K.-J. Ha, 2014: Future change of Asian-Australian monsoon under RCP 4.5 anthropogenic warming scenario. *Clim. Dynam.*, **42**, 83–100. doi:10.1007/s00382-013-1769-x.

Webster, P. J., V. O. Magana, T. N. Palmer, J. Shukla, R. A. Tomas, M. Yanai, and T. Yasunari, 1998: Monsoons: Processes, predictability and the prospects for prediction. *J. Geophys. Res.*, **103**, 14 451–14 510.

Webster, P., A. M. Moore, J. P. Loschnig, and R. R. Leben, 1999: Coupled ocean-atmosphere dynamics in the Indian Ocean during 1997–98. *Nature*, **401**, 356–360.

Wheeler, M. C. and H. H. Hendon, 2004: An all-season real-time multivariate MJO index: Development of an index for monitoring and prediction. *Mon. Wea. Rev.*, **132**, 1917–1932.

Xue, Y. and coauthors, 2010: Intercomparison and analyses of the climatology of the West African Monsoon in the West African Monsoon Modeling and Evaluation project (WAMME) first model Intercomparison experiment. *Clim. Dynam.*, **35**, 3–27. doi:10.1007/s00382-010-0778-2.

Yasunari, T., 1979: Cloudiness fluctuations associated with the Northern Hemisphere summer monsoon. *J. Meteor. Soc. Japan*, **57**, 227–242.

Yasunari, T., 1981: Structure of an Indian monsoon system with around 40-day period. *J. Meteor. Soc. Japan*, **59**, 336–354.

Yin, L., R. Fu, E. Shevliakova, and R. E. Dickenson, 2013: How well can CMIP5 simulate precipitation and its controlling processes over tropical South America? *Clim. Dynam.*, **41**, 3127–3143. doi: 10.1007/s00382-012-1582-y.

Zhang, C. J. and H. Zhang, 2010: Potential impacts of East Asian winter monsoon on climate variability and predictability in the Australian summer monsoon region. *Theoretical and Applied Climatology*, **1**, 161–177. doi:10.1007/s00704-009-0246-2.

Zhang, H. and A. F. Moise, 2016: The Australian summer monsoon in current and future climate. In: *The Monsoons and Climate Change — Observations and Modeling*, Eds. L. M. V. de Carvalho and C. Jones, Springer, pp. 67–120.

Zhang, H., A. Moise, P. Liang, and L. Hanson, 2013: The response of summer monsoon onset/retreat in Sumatra-Java and tropical Australia region to global warming in CMIP3 models. *Clim. Dynam.*, **40**, 377–399.

Zhang, H., G. Dong, A. F. Moise, R. Colman, L. Hanson, and H. Ye, 2015: Uncertainty in CMIP5 model-projected changes in the onset/retreat of the Australian summer monsoon. *Clim. Dynam.*, **46**, 2371–2389. doi:10.1007/s00382-015-2707-x.

Zhang, L. and T. Zhou, 2011: An assessment of monsoon precipitation changes during 1901–2011. *Clim. Dynam.*, **37**, 279–296. doi:10.1007/s00382-011-0993-5.

Zou L., Y. Qian, T. Zhou, and B. Yang, 2014: Parameter tuning and calibration of RegCM3 with MIT-Emanuel cumulus parameterization scheme over CORDEX East Asia domain. *J. Climate*, **27**, 7687–7701. doi:10.1175/JCLI-D-14-00229.1.

Part II
Intraseasonal Oscillations

Chapter 8

The Planetary-Scale Structure of the Madden-Julian Oscillation

Ángel F. Adames* and John M. Wallace

Department of Atmospheric Sciences,
University of Washington, Seattle, WA, USA
**angelf88@uw.edu*

This chapter summarizes the most salient features of the Madden Julian Oscillation's (MJO) planetary wave structure. At upper tropospheric levels the MJO signature in wind (**V**) and geopotential height (Z) consists of equatorial Kelvin waves flanked by Rossby waves centered along 28°N/S, confined to the subtropics by the climatological-mean westerly jet streams and propagating all the way around the equatorial belt. At lower tropospheric levels the MJO resembles the response to a pulsating heat source over the Maritime Continent as in the solutions of Matsuno (1966) and Gill (1980). The contrasting upper and lower tropospheric structures can be interpreted as the superposition of a first internal baroclinic modal structure, which is dominant in the tropics and an equivalent barotropic field consisting mainly of extratropical wavetrains. The lower-tropospheric anomalies in wind and geopotential height induce a distinctive frictionally induced boundary layer convergence pattern, which is displaced to the east of the corresponding pattern in lower free tropospheric convergence. The contrasting divergence profiles in the boundary layer and free troposphere are responsible for the observed "bottom-up" evolution of the convection and the distinctive "swallowtail shaped" vertical velocity pattern.

1. Introduction

The Madden-Julian Oscillation (MJO, after Madden and Julian 1971, 1972) is the dominant mode of tropical intraseasonal variability. Its circulation in the equatorial plane was originally described in Fig. 16 of Madden and Julian (1972) as consisting of a succession of eastward-propagating centers of deep convection coupled with planetary-scale circulation cells that extend through the depth of the troposphere. As more data over the tropics became available, it was proposed that the planetary wave structure might be associated with equatorially-trapped Kelvin waves to the east of the enhanced convection and equatorial Rossby waves to the west, in accordance with Matsuno's (1966) solution for the response to an equatorial heat source. Webster (1972) and Gill (1980) extended Matsuno's shallow water equatorial wave analysis to three dimensions and offered analytic solutions to the forced equatorial waves that lend further insight into their dynamical interpretation.

When many individual events are averaged to form a composite or by performing regression analysis, the MJO-related outgoing longwave radiation (OLR) anomalies are found to propagate eastward with a phase speed of roughly $5\,\mathrm{m\,s^{-1}}$ across the equatorial Indian and western and central Pacific Oceans (see Fig. 1), with a period ranging from 30 to 90 days (Hendon and Salby 1994; Matthews 2000; Sperber 2003; Kiladis *et al.* 2005). The slowness of the propagation has led some authors to interpret the MJO as the response to a pulsating equatorial heat source (Yamagata and Hayashi 1984, Yamagata 1987). However, this interpretation does not explain what causes the oscillation to occur on an intraseasonal timescale or what

The Global Monsoon System: Research and Forecast (3rd Edition)
Edited by C. P. Chang et al.

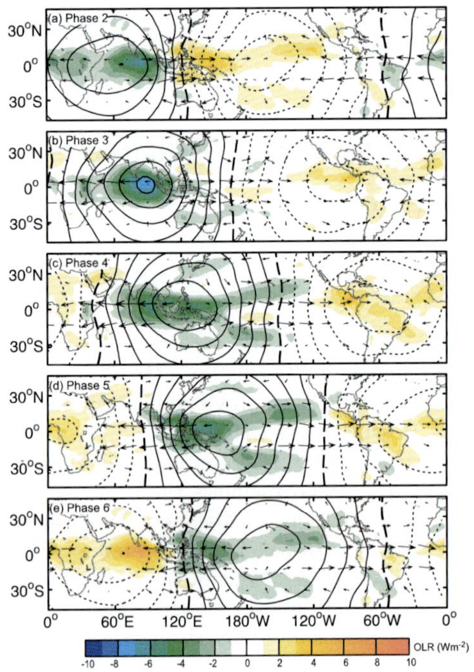

Fig. 1. Leading EOFs of 850 minus 150 hPa velocity potential $\Delta\chi$ with the corresponding regression patterns for OLR (colored shading) and the total 150 hPa wind vectors (arrows). The panels correspond to different linear combinations of PC1 and PC2 of $\Delta\chi$. (a) $-$PC2; (b) PC1/$\sqrt{2}$ $-$ PC2/$\sqrt{2}$(c) PC1, (d) PC1/$\sqrt{2}$ +PC2/$\sqrt{2}$; and (e) PC2. The MJO phases that correspond to each panel is depicted in the top left corner of each panel. Dashed contours denote negative $\Delta\chi$ anomalies, solid contours positive $\Delta\chi$ anomalies and the gray contour depicts the zero value of $\Delta\chi$. Contour interval $0.75 \times 10^6\ \mathrm{m^2\,s^{-1}}$. This and all subsequent figures are based on data for all calendar months.

causes the slow eastward propagation of the heat source.

Many questions still remain regarding the MJO's structure. For example, many studies have identified upper-tropospheric, off-equatorial cyclonic and anticyclonic gyres to the east and west of the MJO-related enhanced equatorial convection. However, descriptions differ from study to study with respect to the positioning and amplitude of the gyres. Some studies have portrayed these features as Rossby waves existing in the upper and lower troposphere with reversed phase, while others have portrayed them as upper tropospheric features only.

The horizontal structure of the MJO in vertical velocity, divergence and precipitation exhibits a "swallowtail" shape that is not captured in the shallow water wave solutions of Matsuno (1966) and Gill (1980) and the vertical velocity and divergence fields exhibit a westward phase tilt with height in the equatorial plane. Wang (1988), and Wang and Rui (1990) theorized that boundary layer friction acting upon the low level flow should induce boundary layer convergence to the east of the region of maximum ascent and divergence to the west of it. Some authors have interpreted both features in terms of the dynamics of equatorial Kelvin and Rossby waves (Wang and Rui 1990; Maloney and Hartmann 1998), others (Zhang and Ling 2012) have argued that these structures do not relate directly either type of equatorially-trapped wave.

This chapter describes the structure of the MJO and elucidates the dynamical processes that shape its most salient features. For this purpose we use daily, 1.5° longitude × 1.5° latitude horizontal resolution ERA-Interim anomaly fields (Dee *et al.*, 2011), and the daily interpolated. 2.5° longitude × 2.5° latitude horizontal resolution OLR field (Liebmann and Smith 1996) regressed onto an MJO index based on 850 minus 150 hPa velocity potential ($\Delta\chi$, see Fig. 1), as in Adames and Wallace (2014a,b), herein referred to as AW1 and AW2. A comparison of $\Delta\chi$ with other MJO indices is presented in the appendix of AW1. Results are based on year-round data and are not separated by season. Section 2 provides an overview of the variability associated with the MJO. Section 3 focuses on the MJO-related horizontal wind field wind (**V**) and geopotential height (Z) fields, which exhibit simple vertical structures, and serve as a basis for an initial description of the MJO. Section 4 discusses temperature (T), vertical velocity (ω) and divergence (Div), which exhibit more complex vertical structures with significant phase tilts. A short summary is given in Sec. 5.

2. Variability Associated with the MJO

First we consider the variability associated with the MJO from a global perspective. We calculate the root-mean-squared (rms) amplitude of the respective fields averaged over the MJO cycle, which is half the square root of the sum of the squared amplitude of the anomalies in regression maps (D) based on PC1 and PC2 of $\Delta\chi$:

$$D_{rms} = \frac{1}{2}\sqrt{D_1^2 + D_2^2}$$

The top panel of Fig. 2 shows the rms amplitude of the 400 hPa ω field layer superimposed upon the annual mean, climatological-mean zonal wind field averaged over the 100–300 hPa layer. The largest amplitudes are observed within the equatorial belt (<10°N/S) over the Indo-Pacific warm pool region (60°E–180°), but large rms values of ω can also be found over the eastern Pacific south of Central America. The tropical (<30°N/S) regions of large rms ω are also regions in which large OLR and precipitation anomalies are observed. Patches of enhanced ω are also observed ~30°–40°N/S, in regions of strong (>30 m s^{-1}) climatological-mean westerly winds.

The rms amplitude of the Z field averaged over the 100–300 hPa layer is shown in the middle panel. Among the most prominent features in this pattern are elongated equatorially-symmetric maxima centered along 28°N and 28°S, just equatorward of the climatological mean jet streams. The subtropical features are most clearly defined over the Indo-Pacific warm pool region but are apparent at other longitudes as well. The locations of the MJO-related variance maxima and their relation to the climatological-mean background flow in the middle panel of Fig. 2 suggest the existence of upper tropospheric Rossby waveguides centered along ~28°N/S, which we refer to as the *Rossby wave corridors*. Poleward of the Northern Hemisphere jet streams are a pair of

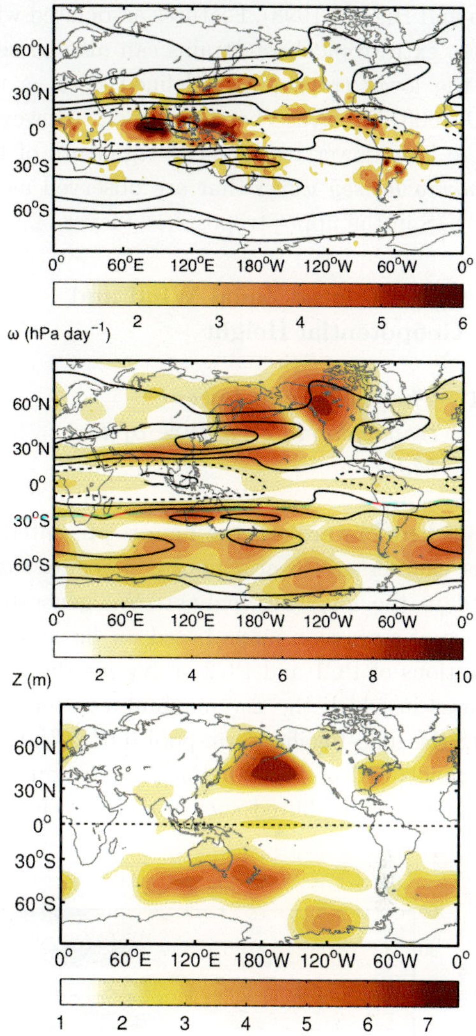

Fig. 2. Root mean squared (rms) amplitude of 400 hPa ω (top panel), the 100–300 hPa layer-mean Z (middle) averaged over the MJO cycle superimposed upon the annual mean, climatological mean 100–300 hPa u (contours). Shading interval 1 m beginning at 2 m. Solid contours denote annual mean westerlies dashed contours denote annual mean easterlies and the zero line is rendered as a dotted contour. Contour interval interval 10 m s^{-1}. Bottom panel: rms 500–1000 hPa layer-mean Z.

equatorially-symmetric wavetrains centered over the central Pacific reminiscent of the Pacific-North American pattern (PNA, Wallace and Gutzler 1981) and the Pacific-South American Pattern (PSA, Karoly 1989, Ghil and Mo 1991;

Mo and Higgins 1998). Features associated with these extratropical wavetrains can also be seen in the lower troposphere, defined here as the 500–1000 hPa layer (bottom panel). However, there is no lower tropospheric signature of the *flanking Rossby waves* that are observed along 28°N/S in the upper tropospheric layer.

3. MJO-Related Zonal Wind and Geopotential Height

3.1. *The upper and lower tropospheric structure of the MJO*

From here onward, all the figures presented in this chapter are "warm pool composites" created in the following manner: We generate a series of regression maps based on linear combinations of PC1 and PC2 of $\Delta\chi$ for the MJO phases in which the $\Delta\chi$ maximum is centered over the Indo-Pacific warm pool (60°E–180°, as in panels a–d of Fig. 1). Each regression map is then zonally shifted such that its "reference longitude" (0°) corresponds to the longitude at which $\Delta\chi$ is a maximum. These shifted maps are then composited, yielding a single map that describes the mean structure of MJO-related features relative to the reference longitude as they propagate eastward from 60°E to the Date Line. An analogous "non-warm pool composite" can be made based on all the longitude points that lie outside of the warm pool region (Date Line to 60°E).

Figure 3 shows a warm pool composite of Z and the horizontal wind field $\mathbf{V}$ composited over the warm pool region (left column) and over the remainder of the tropics (right panels), averaged over the upper troposphere (100–300 hPa, top panels) and the lower troposphere (500–1000 hPa, bottom panels). The reference longitude, depicted by a red circle, corresponds to the location at which $\Delta\chi$ is a maximum.

The equatorial (<10°N/S) wind field is predominantly zonal and of opposing sign in the upper and lower troposphere. Other aspects of the upper and lower-tropospheric patterns associated with the MJO are quite different.

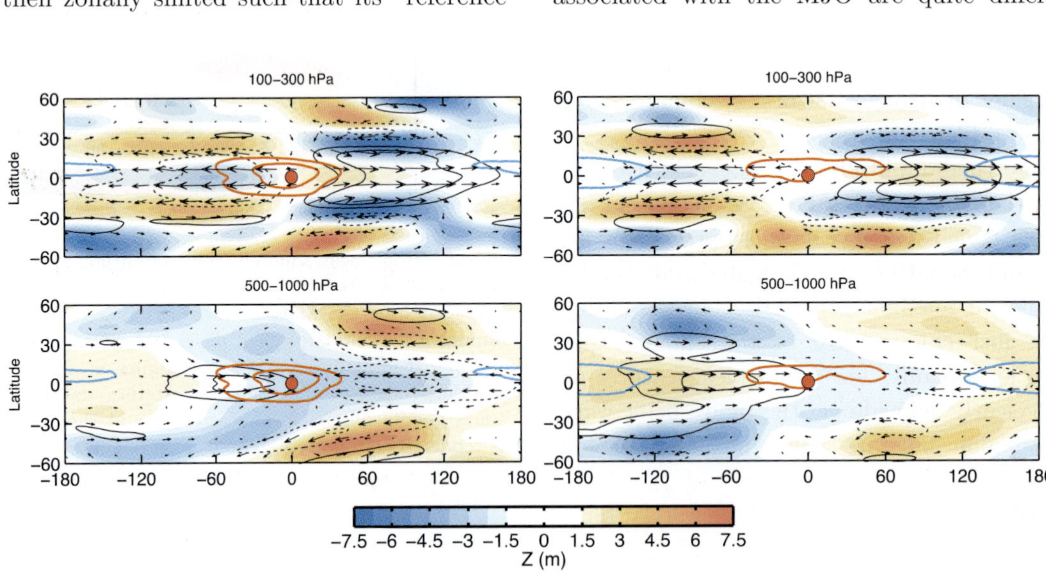

Fig. 3. Warm pool composite of $\mathbf{V}$ (arrows), Z (shading) averaged over the upper tropospheric layer (100–300 hPa, top panel) and lower tropospheric layer (500–1000 hPa, bottom panel) for the warm pool region (left, 60°–180°E) and the rest of the tropics (180°W–60°E). The orange contours depict negative OLR anomalies (contour interval 2 W m^{-2}), black contours depict zonal wind anomalies (interval 0.5 m s^{-1} in top panel, 0.25 m s^{-1} in bottom panel); negative contours (easterlies) dashed.

In the upper tropospheric layer, single, equatorially symmetric circulation couplets, centered ~28° from the equator are evident to the east and west of the reference longitude. These centers are responsible for the maxima in the rms Z in the middle panel of Fig. 2, and they correspond to the subtropical cyclonic and anticyclonic gyres pointed out by Murakami (1988), Knutson and Weickmann (1987), Rui and Wang (1990), Hendon and Salby (1994), and in subsequent studies. As suggested by Barlow (2012) and AW1, and confirmed in a modeling study by Monteiro *et al.* (2014), these waves owe their existence to the strong westerly winds observed in the upper troposphere associated with the climatological-mean jets located ~30°N/S.

The OLR minimum (orange contours) is zonally shifted slightly toward the west of the $\Delta\chi$ maxima, towards the region of upper level easterlies. Near the equator, u and Z are positively correlated, consistent with Kelvin wave dynamics, but upon close inspection it is evident that they are not exactly in phase.

The warm pool composite map for the lower troposphere (Fig. 3, lower left panel), exhibits an equatorial signature to the east of the reference longitude reminiscent of an equatorial Kelvin wave. This feature corresponds to the low-level pressure signal noted by Milliff and Madden (1996), Matthews (2000) and Sperber (2003). It is wider in meridional extent than the equatorial signature in the upper tropospheric layer. To the west of the reference longitude, a pair a cyclonic centers is located near 30°N/S, flanking a region of equatorial westerlies. These lower tropospheric Rossby waves exhibit a stronger meridional wind component along their eastern flank than the upper level flanking Rossby waves. The implications of this meridional wind component are discussed further in the following section. The pair of extratropical anticyclonic gyres located ~45°N/S, to the east of the reference longitude contributes to widening the region of easterlies well beyond the equatorial belt (see also Adames *et al.* 2014). The pattern in the lower tropospheric Z and wind field over the warm pool is reminiscent of the planetary wave response to an isolated, stationary heat source as described by Gill (1980).

Over the rest of the tropics (right column of Fig. 3), the low level Z signature is weaker and less clearly defined and less closely resembles the pattern described by Gill (1980). In contrast, only minor differences between the warm pool and non-warm pool composites are seen in the upper troposphere.

3.2. *Modal decomposition of u and Z*

Like many tropical atmospheric phenomena, the MJO exhibits a *"first internal" baroclinic* vertical structure in the u and Z fields. This baroclinic structure is approximately sinusoidal, with a phase reversal between the upper and lower troposphere with a node located near 400 hPa. In this subsection, we perform maximal covariance analysis (MCA) on the u field paired with Z field in a matrix composed of the merged regression patterns based on PC1 and PC2 of $\Delta\chi$. This operation is analogous to conventional MCA but with time replaced by pressure level. For comparison, EOF analysis is performed on these fields separately as well. The leading mode of the expansion is obtained using the formula:

$$S(x, y, p, t) = \sum_{m=1}^{2} \sum_{n=1}^{N} R_{mn}(x, y) P_m(t) Q_n(p)$$

where the Q_n are a set of modal structures that describe MJO-related variation in the vertical and the R_{mn} are the corresponding set of horizontal regression maps: one corresponding to PC1 of $\Delta\chi$ and the other to PC2. The leading vertical structure function $Q_1(p)$ derived from this analysis is shown in Fig. 4. Accounting for 80% of the squared covariance between the u and Z regression patterns, it corresponds to the "first internal" baroclinic mode discussed above. The largest amplitude in Q_1 is located near the 125 hPa level for both u and Z. The largest

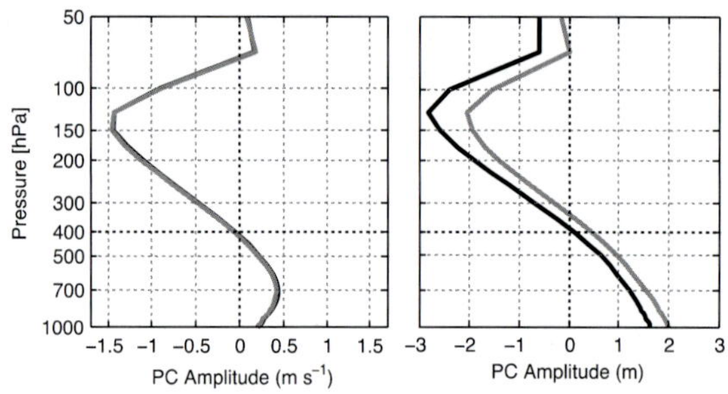

Fig. 4. Expansion coefficients (ECs, solid lines) of u (left) and Z (right) in the leading modal structure derived from the MCA of the vertical profiles of u and Z generated by regressing those fields on PC1 and PC2 of $\Delta\chi$. The analysis is based on grid points in the 10°S–10°N latitude belt. The gray lines are the same profiles but derived from EOF analysis instead. The ECs and PCs are scaled in proportion to their eigenvalues.

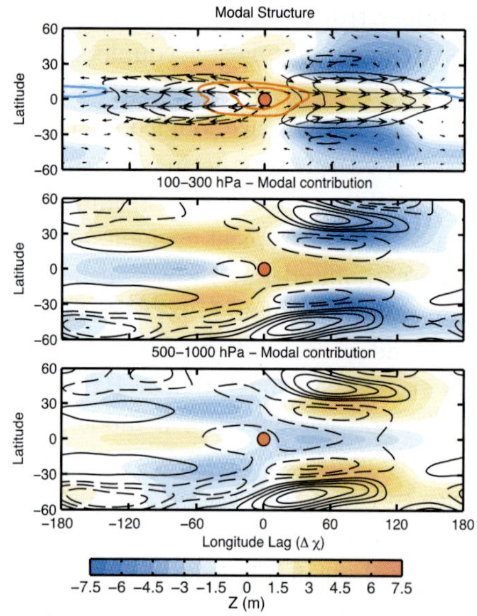

Fig. 5. Warm pool composites. *Top panel*: as in Fig. 3 but the fields presented correspond to horizontal maps obtained by regressing the modal structure in Fig. 4. All fields scaled to the 100–300 hPa layer mean value. *Middle and lower panels*: the colored shading shows the modal structure in Z, scaled to the 100–300 hPa layer in the middle panel, and to the lower tropospheric layer 500–1000 hPa in the bottom panel. Contour interval $0.5\,\mathrm{m\,s^{-1}}$. Contours in the middle and bottom panel are the residual Z field obtained by subtracting the modal structure from the regression maps in the respective layers. Contour interval 1 m.

amplitude in the lower troposphere is located near 750 hPa for u, and near the surface for Z. Both profiles exhibit a node near 400 hPa. The vertical structures obtained through empirical orthogonal function (EOF) analysis exhibit the same vertical structure as the expansion coefficients obtained from MCA, with the mode for u explaining 89% and the mode for Z explaining 64% of the variance in the stack of regression maps. Note that in the left panel, the structure obtained using EOF analysis and MCA are virtually identical.

A warm pool composite of the horizontal patterns in the modal structure is shown in Fig. 5 in a format analogous to those in the left column of Fig. 3. The patterns in the top panel are weighted in accordance with the $Q(p)$ profiles shown in Fig. 4. Comparing the top panel of Fig. 5 and the top left panel of Fig. 3 it is evident that at tropical latitudes the patterns associated with the modal structure derived from this analysis are very similar to the patterns in the upper tropospheric layer. Insight into the differences in structure between the upper and lower troposphere can be obtained by considering the residual patterns obtained by subtracting out the patterns corresponding to the modal structure.

The middle and bottom panels of Fig. 5 show the modal structure in Z as the shaded field, weighted in accordance with the mean value of $Q(p)$ in those layers. The contours in those panels show the residual Z field obtained by subtracting the modal structure from the regression fields shown in the left panels in Fig. 3. For both the upper and lower troposphere, it is evident that within 10° of the equator the modal field is dominant, while the residual field dominates the extratropical (>45°N/S) signature. Between these latitude belts, the residual field is instrumental in enhancing the wave-like signature evident in the upper-tropospheric layer while canceling the flanking Rossby waves in the lower tropospheric-layer, thereby widening the swath of low pressure to the east of the reference longitude. The similarity of the residual fields in the two layers indicates that its structure is predominantly barotropic. Hence, the vertical structure of the MJO can be thought of as being composed of a baroclinic modal structure and an equivalent barotropic residual.

Figure 6 shows a warm pool composite equatorial (10°S–10°N) longitude-height section of the u and Z fields. The total u and Z anomalies and the residuals formed by subtracting the leading mode in the MCA expansion from the regression fields are also shown. The regression profile and the modal structure are remarkably similar, while the residual is weak and equivalent barotropic. In the next section we will examine the structure of temperature, divergence, and vertical motion as revealed by warm pool and non-warm pool composites.

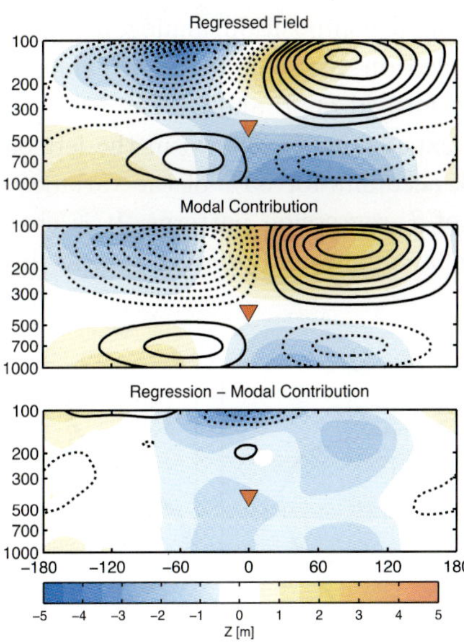

Fig. 6. Warm pool composite equatorial (10°S–10°N) cross section of u and Z: total field (top), modal contribution (center), and the residual obtained by subtracting the modal contribution from the total field (bottom). Shading denotes Z anomalies and contours denote u anomalies. Contour interval 0.25 m s^{-1}. Zero contours are omitted. The red triangle indicates the longitude of the $\Delta\chi$ maximum.

4. Temperature, Vertical Velocity and Divergence

4.1. *Overview*

In this section we discuss the structure of T, ω and *Div* in the MJO. Figure 7 shows longitude-height cross-sections of T (contoured), the mass circulation (ρu, ρw. arrows). and the mass divergence (ρDiv, shading), averaged over the equatorial belt 10°S to 10°N, composited over the

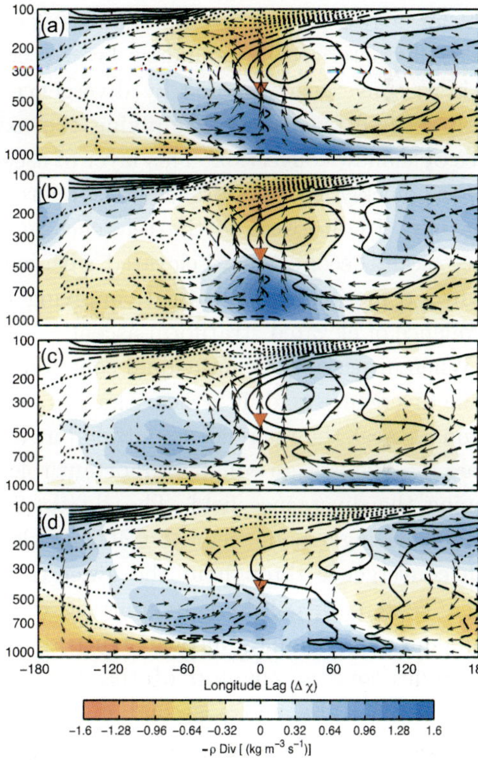

Fig. 7. Composite equatorial (10°S–10°N) longitude-height cross sections of temperature (contours), mass circulation ($\rho u, \rho w$, arrows), mass divergence (ρDiv, shading, panel a), $\rho\partial u/\partial x$ (panel b) and $\rho\partial v/\partial y$ (panel c) over the warm pool region, and ρDiv composited over the rest of the tropics (180°W–60°E, panel d). Contour interval 0.05°C. The red triangle indicates the longitude of the maximum in $\Delta\chi$. The ρw component has been multiplied by a factor of 250. The largest zonal flux vector is around 0.6 kg m^{-2} s^{-1}, and the largest vertical flux vector around 7×10^{-4} kg m^{-2} s^{-1}. The zero value of T is depicted as a dashed contour.

warm pool region (a) and over the rest of the tropics (d). Panels (b) and (c) are as in panel (a) but showing the zonal ($\rho\partial u/\partial x$, panel b) and meridional ($\rho\partial v/\partial y$, panel c) contributions to the mass divergence field. Over the warm pool region the MJO-related T are strongest ~30° to the east of the reference longitude at the 250 hPa level, and exhibit significant eastward tilt with height above that level, and westward tilt with height below it. This warm region extends up to 150° to the east of the reference longitude and exhibits a structure similar to that in an equatorial Kelvin wave (see Kiladis *et al.* 2009 for a review on this topic). This region is also characterized by boundary layer (1000–850 hPa) convergence capped by divergence in the lower-free troposphere, implying the existence of shallow convection. This layer of ascent deepens quickly as one approaches the reference longitude. To the west of the reference longitude, ascent becomes more elevated with weak downward motion underneath it, implying the prevalence of elevated stratiform ascent.

When inspecting the individual contribution to the divergence field, it is clear that $\partial u/\partial x$ is dominant near the reference longitude but elsewhere $\partial v/\partial y$ plays an equally important role in shaping the divergence field. The $\partial u/\partial x$ field exhibits a single node near the 400 hPa level with convergence throughout the lower troposphere centered at 800 hPa. In contrast, $\partial v/\partial y$ exhibits a more complex vertical structure with two nodes, one at the top of the boundary layer near 850 hPa and a second in the mid-troposphere. The vertical structure of $\partial v/\partial y$ is suggestive of an important contribution from friction in the boundary layer. Neither $\partial u/\partial x$ nor $\partial v/\partial y$ exhibits a vertical tilt but they are zonally shifted with respect to one another such that the total divergence field exhibits a substantial vertical tilt. We can therefore interpret the large-scale circulation of the MJO as the superposition of a deep overturning circulation ($\partial u/\partial x$) in the zonal plane and shallow, equatorially symmetric meridional circulation ($\partial v/\partial y$) that is at least partially driven by friction.

Over the rest of the tropics (180°W–60°E), the zonal winds are nearly as strong as over the warm pool, but the divergence, temperature and vertical velocity fields are substantially weaker, in agreement with the right panels of Fig. 3, and the overturning cells are more elongated, with centers separated by ~200° of longitude, versus ~120° over the warm pool region. Throughout the equatorial belt boundary layer convergence is observed to the east of the reference longitude, and divergence to the west. The same circulation features seen in panels (b) and (c) for the warm pool composite are also evident to this region as well (not shown).

4.2. *Modal decomposition of temperature and vertical velocity*

A modal decomposition analogous to the one performed on u and Z in the previous section was performed on T and ω. The leading PC of the T expansion, T_1, shown in the left panel of Fig. 8, accounts for 63% of the variance of the stack of T regression patterns. It is characterized by anomalies of opposing polarity in layers

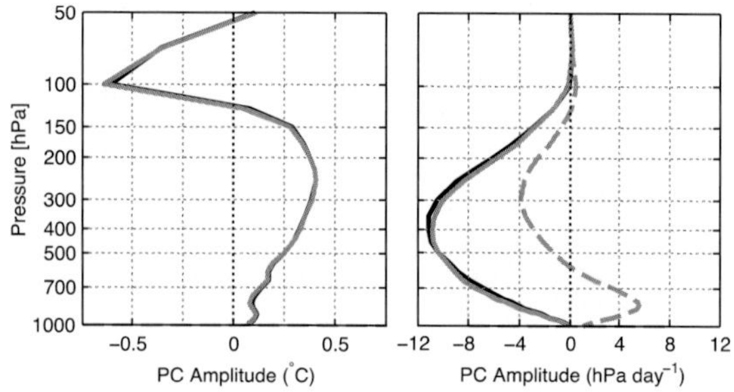

Fig. 8. Leading ECs (black) and PCs (gray) of T (left) and ω (right) derived from MC/EOF analysis of the vertical profiles of T and ω regressed on PC1 and PC2 of $\Delta\chi$ as in Fig. 4. The solid lines denote the first mode and the dashed line denotes the second mode. The PCs are scaled in proportion to their eigenvalues and expressed in units of °C for T and hPa day^{-1} for ω.

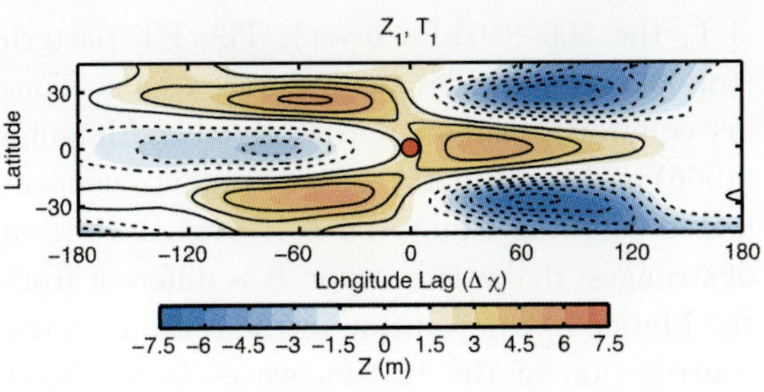

Fig. 9. Warm pool composite of T_1 (represented as contours) superimposed upon Z_1 (shading). Contour interval 0.05°C. T_1 scaled to the 250 hPa level, Z_1 scaled to the 100–300 hPa layer. The red circle indicates the longitude of the $\Delta\chi$ maximum.

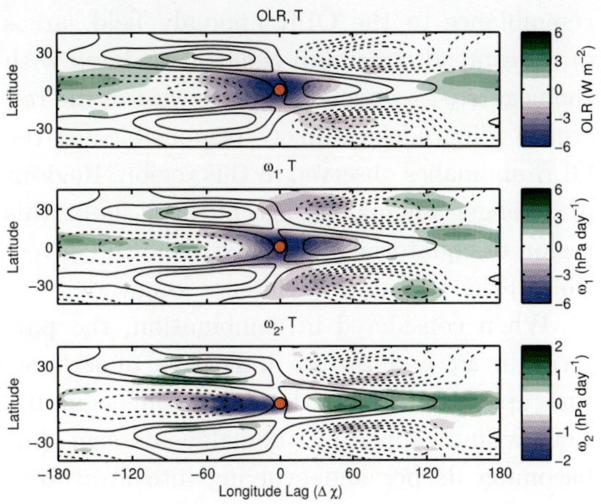

Fig. 10. Warm pool composite of T_1 (contours) superimposed upon different indicators of vertical velocity (shading): top panel OLR; middle panel ω_1; bottom panel ω_2. Contour interval 0.05°C. T_1 scaled to the 250 hPa level, ω_1 and ω_2 scaled to the 300 hPa level. The red circle indicates the $\Delta\chi$ maximum.

centered on the 250 and 100 hPa levels. The horizontal map of T_1 is nearly identical to the map of Z_1, as documented in Fig. 9. That the two maps are so similar confirms that T_1 is related to Z_1 through the hypsometric equation.

The leading mode of the vertical motion field, ω_1, which accounts for 72% of its variance is indicated by the solid line in the right panel of Fig. 8. It is of the same polarity at all levels and exhibits a maximum near 400 hPa. The second mode, ω_2, which accounts for another 18% of the variance, is characterized by perturbations of opposing polarity above and below 600 hPa, attaining its largest amplitudes near 850 and 300 hPa. The vertical profiles of ω_1 and ω_2 are analogous to linear combinations of the "deep convective" and "elevated stratiform" categories used by Houze (1982, 1989); Schumacher et al. (2004); Haertel *et al.* (2008); Zhang and Hagos (2008), among others to characterize latent heating profiles in tropical convective systems. T_1 and ω_1 can also be interpreted as convectively coupled modes since very similar profiles appear in leading mode of the MCA expansion of T and ω (the black lines in Fig. 8), which accouns for 72% of the squared covariance in the stack of regression maps. The MCA and EOF structures shown in Fig. 8 are virtually identical.

Warm pool composites of the horizontal maps associated with the modal structures in Fig. 8 are shown in Fig. 10. The negative ω_1 anomalies, shown in the middle panel, exhibit a structure to the OLR anomalies shown in the top panel but the east-west asymmetries in the shape of the area of enhanced ascent are more pronounced. To the east of the reference longitude anomalous ascent (negative ω_1) is bowed forward along the equator, and it widens and splits into two off-equatorial lines of anomalous ascent to the west of the reference longitude with edges near 18°N/S. This structure is reminiscent of the "swallowtail pattern" described by Zhang and Ling (2012).

Regions of anomalous ascent are also observed poleward and to the west of the flanking Rossby cyclones. No OLR anomalies are seen accompanying these regions of ascent since cooling associated with adiabatic lifting is approximately being balanced by anomalous horizontal temperature advetion (Barlow *et al.* 2005, Hoell *et al.* 2012, Fig. 5 of AW2).

The pattern for ω_2 is much more equatorially focused. It is positively correlated with the patterm in T_1 but shifted slightly eastward. While the horizontal pattern in ω_1 bears a stronger resemblance to the OLR anomaly field, areas of widespread stratiform clouds associated with the negative ω_2 anomalies to the west of the reference longitude are likely responsible for the OLR anomalies observed in this region. Regions of stronger anomalous ascent (ω_1) flank this region of equatorial elevated stratiform convection to the west of the reference longitude.

When considered in combination, the patterns in ω_1 and ω_2 are indicative of a "bottom up" evolution: with convection beginning shallow and capped by anomalous descent, but becoming deeper and evolving into stratiform convection with elevated bases.

4.3. *Relationship between vertical velocity and lower-tropospheric divergence*

We will now consider the structure of the lower-tropospheric divergence field and how it relates to the ω_1 and ω_2 modes described in the previous section. Figure 11 shows the structure of *Div*, **V** and Z within the boundary layer (BL, the 850–1000 hPa layer) and the lower free troposphere

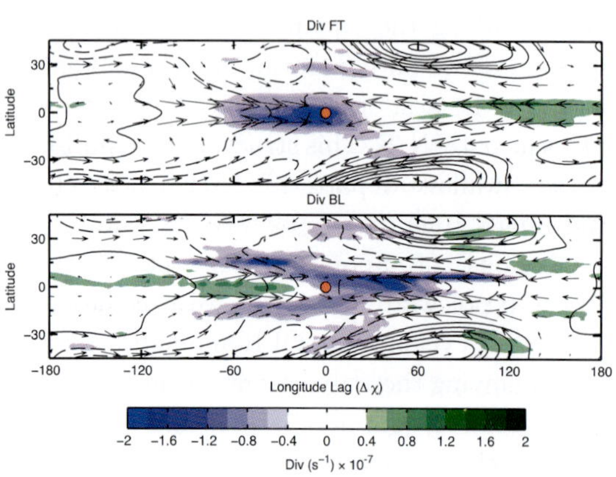

Fig. 11. *Top*: Warm pool composite of lower free tropospheric (FT) layer-averaged divergence (shading), wind (arrows) and Z anomalies (contours). *Bottom*: As in the top panel but averaged over the boundary layer. (BL) Contour interval 1 m. The largest wind vector is around 0.75 m s^{-1}. The red circle indicates the longitude of maximum $\Delta\chi$. Adapted from AW2.

(FT, the 500–850 hPa layer). The FT pattern (top panel) resembles the response to a stationary equatorial heat source described by Matsuno (1966) and Gill (1980), with divergence concentrated over the equatorial belt over the region of strongest diabatic heating. It is different from the Matsuno/Gill solution in the sense that the contribution of the Rossby waves to the west of the heating is not as well defined and the Z anomalies extend well beyond the equatorial belt. The FT near-equatorial winds exhibit a clearly discernible meridional component, with diffluence to the east of the reference longitude and confluence to the west. It is worth noting that Div_{FT} exhibits a structure similar to the OLR field seen in the top panel of Fig. 10.

The structure of the divergence field in the BL (Div_{BL}) is different from that of Div_{FT}. The maximum in convergence is shifted $\sim$35° of longitude to the east of the strongest FT convergence, as noted by Lau *et al.* (1988); Hendon and Salby (1994); and Maloney and Hartmann (1998). The wind field in this layer exhibits a meridional component that is of opposing sign to that in the FT. To the west of the region of maximum large-scale convergence is a region of equatorial BL divergence flanked by lines of off-equatorial convergence ($\sim$18°N/S).

The structure of the Div_{BL} can be represented in terms of the Laplacian of the BL geopotential field ($\phi = gZ$), augmented the divergence of the Coriolis parameter (i.e. $\nabla \cdot (f\mathbf{k} \times \mathbf{u})$, see Eq. (6.2$a$) of Wang and Li, 1994):

$$Div_{BL} \approx \frac{\varepsilon}{f^2 + \varepsilon^2}(\nabla\phi + \beta u + f\beta v\varepsilon^{-1})$$

where $\varepsilon = 2.6$ day^{-1} is the Raleigh friction coefficient, chosen to be consistent with results of studies by Deser (1993); Li and Wang (1994) and Stevens *et al.* (2002), and $\beta = 2.28 \times 10^{-11}$ m^{-1} s^{-1} is the beta parameter ($\beta = df/dy$). A warm pool composite of the estimated divergence field is shown in Fig. 12. That the estimated divergence field based on the previous equation closely matches the structure

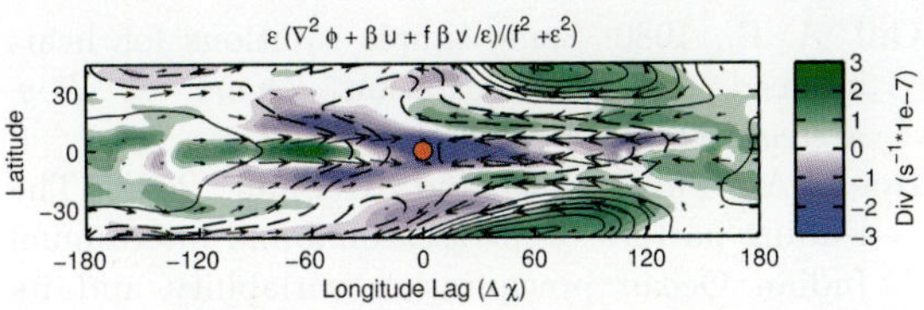

Fig. 12. Warm pool composite estimate of BL divergence based on applying Eq. (3) to the observed BL (850–1000 hPa layer averaged) Z, u and v fields. Contour interval 1 m. The largest wind vector is ~0.75 m s^{-1}. The red circle indicates the maximum in $\Delta\chi$.

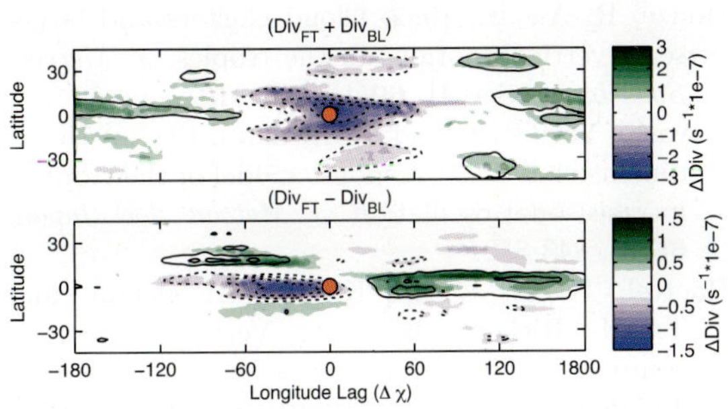

Fig. 13. *Top*: Warm pool composite of ω_1 (contours) and the sum of FT and BL divergence (shading). Contour interval 1.5 hPa day^{-1}. *Bottom*: warm pool composite of ω_2 (contours) and the difference between FT and BL divergence (shading). Contour interval 0.5 hPa day^{-1}. The red circle indicates the longitude of the maximum $\Delta\chi$. Adapted from AW2.

of the observed Div_{BL} field suggests that it is driven by planetary-scale wave motions in the presence of BL friction.

It is possible to represent the modal structures of ω in terms of linear combinations of BL and FT *Div* through the continuity equation. The deep convective profile, ω_1 can be represented as the sum of the divergence in the two layers. In the case of ω_2, descent should peak near 850 hPa and weaken above it. Expressions for ω_1 and ω_2 take the form:

$$\omega_1 \propto Div_{FT} + Div_{BL}$$

$$\omega_2 \propto Div_{FT} - Div_{BL}$$

The correspondence is nearly perfect, as seen in Fig. 13. This result implies that the different profiles in divergence in the BL and FT conspire to create the observed vertical velocity fields, its westward tilt with height and its characteristic swallowtail shape (as seen in Fig. 10).

5. Summary and Discussion

In this chapter we have described the three-dimensional structure of the u, Z, ω, T and *Div* fields in the MJO. The analysis reveals that In the lower troposphere the major features in these fields follow directly from the dynamics of equatorial planetary waves: i.e., an equatorial Kelvin wave prevails to the east of the region of maximum ascent and an equatorial Rossby wave to the west of it, consistent with the Matsuno-Gill solutions for the response to a steady-state equatorial heat source.

The influence of the mean flow causes the MJO to assume somewhat different structure in the upper troposphere. The Kelvin-wave like signature is restricted to a narrower latitude band than in the lower troposphere and the subtropical signature in Z is dominated by flanking Rossby waves located equatorward of the climatological-mean jet maxima near 28°N/S.

Despite these differences, much of the planetary wave structure of the MJO, as manifested in the zonal wind and geopotential height fields, can be interpreted in terms of a baroclinic modal structure that is dominant in the tropics <30°N/S and a barotropic mode composed of extratropical wavetrains that emanate from the upper-level flanking Rossby waves within the jet exit regions.

This baroclinic modal structure is also connected with the observed vertical velocity fields. In the presence of BL friction and the Coriolis force, the large-scale vertical velocity profile develops from the bottom up, with the deepest ascent, represented in this study by strong negative values of ω_1, assuming the form of a swallowtail pattern.

The bottom up evolution of vertical velocity can also be interpreted as the superposition of two circulation cells: a deep zonal overturning circulation in the zonal plane and a pair of shallow, equatorially symmetric circulation cells. Nether of the cells exhibits any tilt with height, but they are shifted zonally such that when they are added together they yield a pattern with a westward tilt with height, as observed in Fig. 7.

Acknowledgments

This research was supported by the National Science Foundation's Graduate Research Fellowship Program (NSF-GRFP) grant DGE-0718124. The work of the second author was funded by the National Science Foundation under grant ATM 1122989.

References

Adames, Á. F. and J. M. Wallace, 2014a: Three-dimensional structure and evolution of the MJO and its relation to the mean flow. *J. Atmos. Sci.*, **71**, 2007–2026.

Adames, Á. F. and J. M. Wallace, 2014b: Three-dimensional structure and evolution of the vertical velocity and divergence fields in the MJO. *J. Atmos. Sci.*, in press.

Adames, Á. F., J. Patoux, and R. C. Foster, 2014: The contribution of extratropical waves to the MJO wind field. *J. Atmos. Sci.*, **71**, 155–176.

Barlow, M., M. Wheeler, B. Lyon, and H. Cullen, 2005: Modulation of daily precipitation over Southwest Asia by the Madden-Julian Oscillation. *Mon. Weather Rev.*, **133**(12), 3579–3594.

Barlow, M., 2012: The Madden-Julian Oscillation influence on Africa and West Asia. *Intraseasonal Variability of the Atmosphere-Ocean Climate System*, 2nd ed., W. Lau and D. Waliser, Eds., Springer Praxis, 477–495.

Dee, D. P., *et al.*, 2011: The ERA-interim reanalysis: Configuration and performance of the data assimilation system. *Quart. J. Roy. Meteor. Soc.*, **137**(656), 553–597.

Deser, C., 1993: Diagnosis of the surface momentum balance over the tropical Pacific Ocean. *J. Climate*, **6**(1), 64–74.

Ghil, M. and K. Mo, 1991: Intraseasonal oscillations in the global atmosphere. Part II: Southern Hemisphere. *J. Atmos. Sci.*, **48**(5), 780–790.

Gill, A. E., 1980: Some simple solutions for heat-induced tropical circulation. *Quart. J. Roy. Meteor. Soc.*, **106**(449), 447–462.

Hoell, A., M. Barlow, and R. Saini, 2012: The leading pattern of intraseasonal and interannual Indian Ocean precipitation variability and its relationship with Asian circulation during the boreal cold season. *J. Climate*, **25**(21), 7509–7526.

Houze, R. A., 1989: Observed structure of mesoscale convective systems and implications for large-scale heating. *Quart. J. Roy. Meteor. Soc.*, **115**(487), 425–461.

Houze, R. A., Jr., 1982: Cloud clusters and large-scale vertical motions in the tropics. *J. Meteor. Soc. Japan.* Ser. II, **60**(1), 396–410.

Hayashi, Yoshikazu, and S. Miyahara, 1987: A three-dimensional linear response model of the tropical intraseasonal oscillation. *J. Meteor. Soc. Japan*, **65**(6), 843-852.

Haertel, P. T., G. N. Kiladis, A. Denno, and T. M. Rickenbach, 2008: Vertical-mode decompositions of 2-day waves and the Madden-Julian Oscillation. *J. Atmos. Sci.*, **65**(3), 813–833.

Hendon, H. H. and M. L. Salby, 1994: The life cycle of the Madden-Julian Oscillation. *J. Atmos. Sci.*, **51**, 2225–2237.

Karoly, D. J. 1989: Southern Hemisphere circulation features associated with El Niño–southern oscillation events. *J. Climate*, **2**: 1239–1251.

Kiladis, G. N., K. H. Straub, and P. T. Haertel, 2005: Zonal and vertical structure of the Madden-Julian Oscillation. *J. Atmos. Sci.*, **62**, 2790–2809.

Kiladis, G. N., M. C. Wheeler, P. T. Haertel, K. H. Straub, and P. E. Roundy, 2009: Convectively coupled equatorial waves. *Rev. Geophys.*, **RG2003**, 1–42.

Knutson, T. R. and K. M. Weickmann, 1987: 30–60 day atmospheric oscillations: Composite life cycles of convection and circulation anomalies. *Mon. Wea. Rev.*, **115**(7), 1407–1436.

Lau, N.-C., I. M. Held, and J. D. Neelin, 1988: The Madden-Julian Oscillation in an idealized general circulation model. *J. Atmos. Sci.*, **45**(24), 3810–3832.

Li, T. and B. Wang, 1994: A thermodynamic equilibrium climate model for monthly mean surface winds and precipitation over the tropical Pacific. *J. Atmos. Sci.*, **51**(11), 1372–1385.

Liebmann, B. and C. A. Smith, 1996: Description of a complete (interpolated) outgoing longwave radiation dataset. *Bull. Amer. Met. Soc.*, **77**, 1275–1277.

Madden, R. and P. Julian, 1971: Detection of a 40–50 day oscillation in the zonal wind in the tropical Pacific. *J. Atmos. Sci.*, **28**, 702–708,

Madden, R. and P. Julian, 1972: Description of global scale circulation cells in the tropics with a 40–50 day period. *J. Atmos. Sci.*, **29**, 1109–1123.

Maloney, E. D. and D. L. Hartmann, 1998: Frictional moisture convergence in a composite life cycle of the Madden-Julian Oscillation. *J. Climate*, **11**, 2387–2403.

Matsuno, T., 1966: Quasi-geostrophic motions in the equatorial area. *J. Meteor. Soc. Japan*, **44**, 25–43.

Matthews, A., 2000: Propagation mechanisms for the Madden-Julian oscillation. *Quart. J. Roy. Meteor. Soc.*, **126**, 2637–2651.

Milliff, R. F. and R. A. Madden, 1996: The existence and vertical structure of fast, eastward-moving disturbances in the equatorial troposphere. *J. Atmos. Sci.*, **53**, 586–597.

Mo, K. C. and R. W. Higgins, 1998: The Pacific-South American modes and tropical convection during the Southern Hemisphere winter. *Mon. Wea. Rev.*, **126**(6), 1581–1596.

Monteiro, J. M., Á. F. Adames, J. M. Wallace, and J. S. Sukhatme, 2014: Interpreting the upper-level structure of the Madden-Julian Oscillation, *Geophs. Res. Lett.*, revised.

Murakami, T., 1988: Intraseasonal Atmospheric Teleconnection Patterns during the Northern Hemisphere winter. *J. Climate*, **1**(2), 117–131.

Rui, H. and B. Wang, 1990: Development characteristics and dynamic structure of tropical intraseasonal convection anomalies. *J. Atmos. Sci.*, **47**(3), 357–379.

Schumacher, C., R. A. Houze, and I. Kraucunas, 2004: The tropical dynamical response to latent heating estimates derived from the TRMM precipitation radar. *J. Atmos. Sci.*, **61**(12), 1341–1358.

Sobel, A. H. and C. S. Bretherton, 2000: Modeling tropical precipitation in a single column. *J. Climate*, **13**(24), 4378–4392.

Sperber, K. R., 2003: Propagation and the vertical structure of the Madden-Julian Oscillation. *Mon. Wea. Rev.*, **131**, 3018–3037.

Stevens, B., J. Duan, J. C. McWilliams, M. Muñnich, and J. D. Neelin, 2002: Entrainment, Rayleigh friction, and boundary layer winds over the tropical Pacific. *J. Climate*, **15**(1), 30–44.

Wallace, J. M. and D. S. Gutzler, 1981: Teleconnections in the geopotential height field during the Northern Hemisphere winter. *Mon. Wea. Rev.*, **109**(4), 784–812.

Wang, B. and T. Li, 1994: Convective interaction with boundary-layer dynamics in the development of a tropical intraseasonal system. *J. Atmos. Sci.*, **51**(11), 1386–1400.

Webster, P. J., 1972: Response of the tropical atmosphere to local, steady forcing. *Mon. Wea. Rev.*, **100**(7), 518–541.

Yamagata, T., 1987: A simple moist model relevant to the origin of intraseasonal disturbances in the tropics. *J. Meteor. Soc. Japan*, **65**, 153–165.

Yamagata, T. and Y. Hayashi, 1984: A simple diagnostic model for the 30–50 day oscillation in the tropics. *J. Meteor. Soc. Japan*, **62**(5), 709–717.

Zhang, C. and S. M. Hagos, 2009: Bi-modal structure and variability of large-scale diabatic heating in the tropics. *J. Atmos. Sci.*, **66**(12), 3621–3640.

Zhang, C. and J. Ling, 2012: Potential vorticity of the Madden-Julian Oscillation. *J. Atmos. Sci.* **69**, 65–78.

Chapter 9

Simulation of the Madden-Julian Oscillation Using General Circulation Models

Daehyun Kim* and Eric D. Maloney†

*Department of Atmospheric Sciences, University of Washington, Seattle, WA 98195, USA
daehyun@uw.edu
http://www.atmos.washington.edu/~daehyun/

†Department of Atmospheric Science, Colorado State University, Fort Collins, CO 80521, USA
emaloney@atmos.colostate.edu
http://maloney.atmos.colostate.edu/

In this chapter, we summarize our experiences with respect to the representation of the Madden-Julian oscillation (MJO) in general circulation models (GCMs). Specifically, we focus on diagnostics that have been used in evaluating GCM simulations, the empirical relationships that have been reported between GCM configurations and MJO simulation skill, and the theoretical frameworks that have been used to interpret the empirical relationships. Popular MJO simulation diagnostics that capture salient features of the MJO were recently standardized by the U.S. CLIVAR MJO Working Group, which make tracking of improvements possible across different models and model generations. Simulation of the MJO is sensitive in particular to the details of the cumulus parameterization used in GCMs. Inhibiting too-frequent activation of vigorous deep convection, which can be achieved a number of ways, generally leads to enhanced MJO variability. Recent MJO theoretical developments have led to new frameworks in which to interpret MJO modeling results. Different, but not mutually exclusive, theoretical frameworks have been applied in different studies, reflecting the lack of a single, satisfactory theory of the MJO. Possible future directions for further improvement of the MJO in GCMs are also discussed.

1. Introduction

The Madden-Julian Oscillation (MJO, Madden and Julian 1972) is the dominant intraseasonal mode in the tropics. Anomalous (i.e. deviation from seasonal cycle) planetary-scale tropical convection that propagates eastward over the Indo-Pacific warm pool is the core ingredient of the MJO. The convective envelope is tightly coupled to the tropical circulation, and thereby affects other tropical phenomena and influences mid- to high-latitude circulation anomalies through upper-tropospheric Rossby wave propagation (see Lau and Waliser 2011 for comprehensive reviews on interactions between the MJO and other components of the climate system).

The non-negligible influence of the MJO on mid-latitude weather suggests that an accurate forecast of the MJO may improve mid-latitude weather forecasting. Currently, the prediction skill of the MJO using numerical weather forecast models, which is traditionally defined as the forecast lead-time when the anomaly correlation of an MJO index falls below 0.5, is about 20–25 days (Rashid *et al.* 2011, Vitart and Molteni 2010, Zhang and van den Dool 2012 and Wang *et al.* 2013; see also Lee and

The Global Monsoon System: Research and Forecast (3rd Edition)
Edited by C. P. Chang et al.

Wang and Wheeler *et al.* in this book) when assessed using the popular Real-time Multivariate MJO index of Wheeler and Hendon (2004). A recent predictability study at the European Centre for Medium-Range Weather Forecasts (ECMWF) (Vitart 2010), however, showed that MJO predictability is longer than 45 days, implying room for further improvement in MJO prediction skill with numerical models, and that model deficiencies in representing the MJO are limiting our capability to forecast the MJO.

Previous and ongoing efforts to improve the representation of the MJO in GCMs will be reviewed in this chapter. This task might seem daunting if we recall that the representation of the MJO in GCMs has been unsatisfactory in several multi-model intercomparison studies of the last two decades (Slingo *et al.* 1996; Waliser *et al.* 2003; Lin *et al.* 2006; Kim *et al.* 2009; Hung *et al.* 2013). Despite the continued MJO simulation deficiencies noted in these studies, we think that some progress has been made, and hence this article is a summary of what we have learned.

In this review, we will focus on three aspects of MJO modeling based on the typical evolution of an MJO modeling study: (i) diagnostics that have been used to evaluate the MJO (Sec. 2), (ii) sensitivity of the simulated MJO to model configurations (Sec. 3), and (iii) the theoretical frameworks that have been used to interpret/understand this sensitivity (Sec. 4). Perspectives on possible future research will be given in Sec. 5, which will be followed by a summary and conclusions in Sec. 6.

2. Diagnostics

2.1. *Statistical methods*

The fact that the MJO has a characteristic spatial scale of zonal wavenumbers 1-3, temporal scale of 30–60 days, and eastward propagation direction has encouraged people to use spectral analysis to assess MJO simulation fidelity. The variability or power of simulated eastward-propagating disturbances that have similar space- and time-scales to the observed MJO is often compared among different models, and to that of observed MJO (e.g. Slingo *et al.* 1996).

The eastward propagating nature of the MJO makes lag-correlation analysis a useful diagnostic tool. This is typically done using a band-pass (e.g. 20–100 days) filtered time series of a variable at a reference point (e.g. eastern equatorial Indian Ocean) and a meridionally averaged (e.g. 10°S–10°N) time series of the same variable or another physically related quantity. If the variable is dominated by eastward propagation, the lag-correlation map between two time-series would exhibit eastward propagation (e.g. Lin *et al.* 2006; Hung *et al.* 2013).

The dominance of the MJO in explaining intraseasonal time-scale variability of many variables that is coherent across the tropics stimulated people to use Empirical Orthogonal Function (EOF) analysis or variants of it to extract the MJO mode from data. Multivariate EOF analysis that includes both OLR and winds has also recently become a popular basis to diagnose the MJO (e.g. Wheeler and Hendon 2004). The 3-dimensional structure of the MJO can be constructed by compositing dynamical and thermodynamic variables based on the phase of the MJO, using the leading PCs as a basis for compositing. The large-scale structure can then be compared to that of observations (Maloney and Hartmann 2001; Sperber 2005; Maloney 2009; Maloney *et al.* 2010; Kim *et al.* 2012).

2.2. *Observable parameters*

The parameters used in validating MJO simulations have continued to evolve as new observations become available, moving from simple signatures or proxies of tropical deep convection such as OLR, to detailed vertical distribution of clouds, water vapor, and other geophysical quantities derived from recent satellite missions. Improved global reanalyses have also provided improved representation of the observed dynamical structure of the MJO for comparison to simulations.

2.3. *MJO Working Group diagnostics*

As reviewed above, a wide variety of diagnostics are now available to validate models, and no two studies have traditionally employed the same diagnostic analysis. Also, details of the diagnostics (e.g. period window for bandpass filtering) are often different even if a same statistical method is used. This inhomogeneity has made it difficult to systematically track progress on MJO simulation quality as models evolve. The U.S. CLIVAR MJO Working Group developed a package of code and scripts to evaluate the representation of the MJO in GCM simulations (CLIVAR MJO WG 2009; Kim *et al.* 2009) to overcome these issues of inhomogeneity. The package includes a hierarchy of diagnostic tools from time series power spectra to MJO life-cycle composites. The package is becoming a standard diagnostic tool in the modeling community (e.g. Liu *et al.* 2009; Fu and Wang 2009; Chikira and Sugiyama 2010; Kim *et al.* 2011; Weaver *et al.* 2011; Kim *et al.* 2012; Wang *et al.* 2012; Chikira and Sugiyama 2013).

3. Sensitivity to Model Configurations

Increasing horizontal resolution has been shown to be ineffective in improving the representation of the MJO in hydrostatic GCMs (e.g. Rajendran *et al.* 2008; Liu *et al.* 2009). Increasing the number of vertical levels, however, improves the fidelity of the MJO in some GCMs (e.g. Inness *et al.* 2001; Jia *et al.* 2009). With increased vertical resolution, GCMs are able to more reasonably simulate the tri-modal distribution of cloud-top height (Johnson *et al.* 1999), helping GCMs to better simulate the transition from shallow to deep convection, which is important in simulating the gradual deepening of cloud top height during the MJO development phase. The effects of air-sea coupling on MJO simulation have been mixed. Its effect has been shown to be positive (Flatau *et al.* 1997; Waliser *et al.* 1999; Inness *et al.* 2003), neutral (Maloney and Kiehl 2002; Sperber 2004), or even negative (Hendon 2000; Liess *et al.* 2004) depending on the study. Interestingly, Crueger *et al.* (2013) found that the effects of air-sea coupling on the representation of the MJO depend on horizontal resolution of the atmospheric GCM. They showed that improvement with air-sea coupling enabled is much greater when a relatively higher horizontal resolution is used (T127 vs. T63). The physical mechanism behind this empirical relationship is unclear.

Numerous studies have shown that the ability of a GCM to simulate the MJO is sensitive to details of the cumulus parameterization. Due to the complex interaction between the parameterizations in GCMs, however, it is sometimes hard to isolate the effect of cumulus parameterization on the MJO simulation. The MJO is an emergent property of a GCM that involves complex interactions among several different model parameterizations. Models often have different convection schemes, but their other parameterizations also differ, such as the boundary layer, large-scale condensation and radiation schemes. Therefore, in this review we confine our discussion to studies where only details of cumulus parameterization are changed in a single GCM.

In Table 1, specific features of cumulus parameterization that are suggested to be important for MJO simulation are listed along with desired changes to them for a better MJO simulation. The important aspects of the convection scheme include: the closure assumption, triggering function, strength of mixing between cloud parcel and environment, efficiency of convective rain re-evaporation, and convective momentum transport. While different aspects of the convection scheme have been

Table 1. Summary of the sensitivity of MJO simulation to details of convection scheme.

Aspect of convection scheme	Direction toward a better MJO simulation	References
Triggering function	A more strict triggering of deep convection	Tokioka *et al.* (1988), Wang and Schlesinger (1999), Lee *et al.* (2003), Lin *et al.* (2008), Zhang and Mu (2005), Hannah and Maloney (2011), Benedict *et al.* (2013)
Mixing assumption	Stronger mixing between cloud and environment	Tokioka *et al.* (1988), Bechtold *et al.* (2008), Fu and Wang (2009), Chikira and Sugiyama (2010), Kim *et al.* (2012a), Kim *et al.* (2012b), Hirons *et al.* (2012a), Hirons *et al.* (2012b)
Rain re-evaporation	More efficient rain re-evaporation	Maloney and Hartmann (2001), Hannah and Maloney (2011), Kim *et al.* (2012b)
Closure assumption	Making convection scheme remove less instability	Zhang and Mu (2005), Zhou *et al.* (2012), Benedict *et al.* (2013)
Convective momentum transport	Enabling convective momentum transport	Zhou *et al.* (2012), Ling *et al.* (2009)*

*Counter examples.

emphasized in different MJO modeling studies, one common relationship is that inhibiting too-frequent activation of vigorous deep convection improves MJO variability in GCMs. For example, applying a more stringent condition for triggering deep convection has resulted in enhanced MJO variability in many studies. We note that using superparameterization in a GCM (Grabowski 2001) or using cloud-system resolving resolution globally (Satoh *et al.* 2008) that presumably includes more realistic representation of the processes in Table 1 engenders an improved MJO (e.g. Benedict and Randall 2009 see also Tomoki *et al.* in this book).

Table 1 demonstrates that we have learned how to work with convection schemes to improve MJO simulations. When the modeling results are used to understand the dynamics of the MJO, theories of the MJO have provided frameworks for it. In the next section, we attempt to summarize those theoretical frameworks that have been used in MJO modeling studies.

4. Theoretical Frameworks

In Table 2, the theories of the MJO that have been suggested, the salient mechanisms or processes of these theories crucial to the MJO, and the modeling studies that assess these theories are summarized.

Among other factors important in the wave-CISK theory (Lau and Peng 1987; Chang and Lim 1988), the vertical profile of diabatic heating, which determines the phase speed of the propagating wave, has been often emphasized in MJO modeling studies. Following wave-CISK, a bottom-heavy heating profile is preferred for a simulation of correct phase speed of the MJO. There is some controversy, however, in the importance of the heating structure. For example, in Zhang and Song (2009) and Li *et al.* (2009), a bottom-heavy heating profile was shown to simulate a better MJO, while Seo and Wang (2010) suggested a top-heavy heating structure is better.

The frictional wave-CISK theory (Wang 1988; Wang and Rui 1990) emphasizes the role of frictional convergence in the planetary boundary layer to the east of MJO-associated convection. This feature is often found in modeling studies when a GCM simulate a good MJO. However, results from the so-called mechanism-denial experiments suggest that this might not be a crucial process in MJO simulation (Chao and Chen 2001; Kim *et al.* 2011b). For example, in Kim *et al.* (2011b), the simulated MJO was

Table 2. Summary of theories that have been used in MJO modeling studies.

Theory/View	Important physical process	MJO Modeling studies
Wave-CISK (Lau and Peng 1987; Chang and Lim 1988)	Bottom-heavy vertical profile of diabatic heating	Park *et al.* (1990), Wang and Schlesinger (1999), Li *et al.* (2009), Seo and Wang (2010)*
Frictional wave-CISK (Wang 1988; Wang and Rui 1990)	Existence of frictional convergence in the boundary layer to the east of enhanced MJO convection	Lau and Lau (1986), Lau and Chan (1988), Sperber *et al.* (1997), Waliser *et al.* (1999), Wang and Schlesinger (1999), Maloney and Hartmann (2001), Lee *et al.* (2003), Seo and Wang (2010), Chao and Chen (2001)*, Kim *et al.* (2011b)*
Wind-induced surface heat exchange (Emanuel 1987; Neelin *et al.* 1987)	Enhanced surface latent heat flux associated with wind anomalies	Maloney and Sobel (2004), Sobel *et al.* (2008; 2010), Kim *et al.* (2011b), Landu and Maloney (2011)
Air-sea coupling (Wang and Xie 1998)	SST changes and their impacts on surface latent heat fluxes	Waliser *et al.* (1999), Hendon (2000)*, Watterson (2002), Bernie *et al.* (2008), Klingaman *et al.* (2011)
Cloud-radiation interaction (Raymond 2001)	Anomalous longwave radiative heating associated with the MJO	Kim *et al.* (2011b), Landu and Maloney (2011), Anderson and Kuang (2012), Chikira (2014), Lee *et al.* (2001)*
Discharge-recharge mechanism (Blade and Hartmann 1993)	Gradual build-up of moist static energy anomaly from surface	Wang and Schlesinger (1999), Maloney (2002), Sperber *et al.* (2005), Benedict and Randall (2009), Thayer-Calder and Randall (2009), Cai *et al.* (2013)
Moisture mode (Raymond 2001; Sobel and Maloney 2012; 2013)	Moisture-convection coupling, details of the column integrated moisture/moist static energy budget	Maloney (2009), Benedict and Randall (2009), Maloney *et al.* (2010), Anderson and Kuang (2012), Wu and Deng (2013), Chikira (2014)
Multiscale theory (Majda and Stechmann 2009b; Wang and Liu 2011; Liu and Wang 2012)	Upscale momentum and energy transfer	Deng and Wu (2010; 2011), Zhou *et al.* (2012)

*Counter examples or opposite results.

insensitive to the exclusion of the anomalous surface friction and associated boundary layer convergence, suggesting that the boundary layer frictional convergence may not be a necessary condition for proper simulation of the MJO.

The wind-induced surface heat exchange (WISHE) mechanism stresses the role of surface latent heat flux anomalies produced by the anomalous surface wind. Although observations have cast doubt on the details of the original WISHE hypothesis in which WISHE operated in an low-level easterly basic state flow (Emanuel 1987; Neelin *et al.* 1987), recent modeling studies suggest that WISHE may still an important process in modeling the MJO (e.g. Maloney and Sobel 2004). They have argued that the anomalous surface latent heat flux associated with the MJO, which slightly lags convection but has overlap with enhanced convection, provides energy to the convecting column and acts as a destabilization mechanism.

Air-sea coupling was emphasized in Wang and Xie (1998) who showed that Sea Surface Temperature (SST) changes associated with the MJO and their impacts on surface latent heat fluxes play an important role on the phase speed of the MJO. Several GCM studies found that enabling air-sea coupling improves MJO simulation (e.g. Waliser *et al.* 1999), although counter examples exist (e.g. Hendon 2000).

Raymond (2001) suggested the importance of cloud-radiation interaction to MJO dynamics. Anomalous longwave radiative heating, which provides a positive feedback to enhanced deep convection, is crucial in this theory. Consistent with this argument, recent studies have found that the anomalous radiative heating associated with the MJO convection is a primary destabilization mechanism of the MJO (Andersen and Kuang 2012; Chikira 2014), and the simulated MJO weakens significantly when cloud-radiation interactions are suppressed (Kim *et al.* 2011b; Landu and Maloney 2011), although counter-example also exists in the literature (e.g. Lee *et al.* 2001). In discharge-recharge theory (e.g. Bladé and Hartmann 1993; Benedict and Randall 2007), the time scale of MJO is determined by the time scale required for build-up and decay of lower tropospheric moist static energy (MSE) anomalies. Consistent with this theory, modeling studies have found a gradual deepening of the MSE anomaly in the GCM simulations with a reasonable MJO. It is not clear, however, whether this feature is a crucial physical mechanism for the existence of the MJO or just a consequence of the existence of the MJO. Also, the discharge-recharge theory is incapable of explaining the eastward propagation of the MJO as well as its scale selection.

A recent view of the MJO, often referred to as a "moisture mode" theory, cites column integrated moisture or MSE anomalies as fundamental to the development and propagation of MJO (Neelin and Yu 1994; Raymond 2001; Sobel *et al.* 2001; Majda and Stechmann 2009a; Sobel and Maloney 2012; 2013; Adames and Kim 2016). In these theories, anomalous convection associated with the MJO is assumed to be closely tied to anomalous column integrated water vapor (or MSE for weak tropical temperature gradients). Therefore, explaining the growth and propagation of the convective anomalies associated with the MJO is equivalent to explaining those of the column integrated water vapor. How strongly the simulated convection is coupled to the column integrated moisture (e.g. Kim *et al.* 2012), and how accurately the GCM simulates the column integrated MSE budget are the crucial aspects that GCMs need to mimic (e.g. Maloney 2009; Maloney *et al.* 2010; Andersen and Kuang 2012).

In the multiscale theory of the MJO (Majda and Stechmann 2009), the upscale transfer of heat and momentum from synoptic-scale waves to the MJO is an important process for the maintenance of the MJO. In some recent modeling studies (Deng and Wu 2010; 2011; Zhou *et al.* 2012) the role of convective momentum transport in the simulation of the MJO is emphasized, supporting the multiscale theory.

5. Perspectives on Future Research

5.1. *MJO-mean state trade-off*

An apparent conflict exists in MJO modeling. On one hand, we know how to improve the MJO in GCM simulations (Sec. 3), while on the other hand GCM simulations of the MJO have been always poor (Sec. 1). Using 10 AGCM simulations, Kim *et al.* (2011a) showed that the seasonal-mean precipitation is degraded when the MJO is enhanced through the modifications to the convection scheme. If any changes in convection scheme that enhance MJO sacrifice the mean state, the changes could be easily and justifiably rejected by model development teams, because the mean state has a higher priority in model development. Therefore, the MJO-mean state trade-off might explain why in intercomparison studies, where official versions of GCMs are used, GCMs' ability to simulate the MJO has been poor.

The MJO-mean state trade-off might imply that the method of improving MJO variability is not correct, but also that the right model mean state could be derived for the wrong reasons. It is also possible that there are some common missing processes. Further research is needed to explain and overcome this trade-off, including further process-oriented diagnosis of GCMs, as discussed in the next section.

5.2. *Process-oriented MJO simulation diagnostics*

The diagnostics discussed in Sec. 2 are useful in evaluating the MJO in one or multiple GCM simulations. They do not tell us, however, why some models simulate a better MJO than others and what is physically important to improve the GCM. It would be useful to develop diagnostics that could inform about the strength or weakness of a parameterization in simulating the MJO. Such "process-oriented MJO simulation diagnostics" would be able to provide insights on desired changes to physical processes in the GCM for a better MJO simulation, but would also provide relevant information on MJO dynamics.

Kim *et al.* (2014) suggest that a diagnostic that represents the strength of coupling between lower free tropospheric relative humidity and convection could be a suitable process-oriented MJO simulation diagnostic. Using 28 CMIP3 and CMIP5 simulations, they show that a measure of moisture sensitivity of simulated convection explains about 50% of inter-model variance in MJO simulation capability. Also, Benedict *et al.* (2014) showed that models in which convection and associated divergent circulations too strongly discharge MSE from the column tend to have a difficult time destabilizing the MJO.

The diagnostics used in Kim *et al.* (2014) and Benedict *et al.* (2014) are also closely related to the characteristics of moist physics (e.g. fractional entrainment rate and efficiency of rain re-evaporation), and therefore could help their improvement.

We expect that there are processes other than the moisture sensitivity of convection that are important in GCMs' capability to simulate the MJO. Identifying those processes and developing diagnostics that represents the processes need to be done in the future.

5.3. *MJO in a warmer climate?*

An increasing area of research involves how the MJO will change in a future warmer climate. Several modeling studies and observational studies have suggested that MJO activity will increase in a future climate (e.g. Slingo *et al.* 1999; Jones and Carvalho 2006; Pohl and Matthews 2007; Jones and Carvalho 2010; Liu 2012; Schubert *et al.* 2013; Arnold *et al.* 2013). However, many of these studies use observational analogues to future warming (e.g. ENSO) that are imperfect in the context of MJO dynamics, apply globally uniform warming patterns to assess changes in MJO activity although future warming is expected to be inhomogeneous (e.g. Xie *et al.* 2010), or derive results from a single model. Maloney and Xie (2013) argued using an aqua-planet GCM that changes in future MJO activity are very sensitive to the pattern of SST warming. They also argued that changes in MJO wind variability will likely not scale with precipitation variability, but will instead be mediated by changes in tropical dry static stability. A more extensive analysis of how MJO precipitation and wind variability is projected to change in future climate is needed.

6. Summary and Conclusions

Our experiences with respect to the representation of the MJO in the GCMs were summarized in this review article. Although the simulation of the MJO is still a strict test for any contemporary GCM, we note some progress in all parts of MJO modeling: diagnostics for characterizing the simulated MJO, empirical knowledge about the sensitivity of MJO simulation to model configuration, and the framework for interpretation and understanding of the results.

Popular MJO diagnostics have been standardized, enabling a tracking of MJO simulation skill as a community. Observations are continually becoming more sophisticated in time: evolving from upper tropospheric winds to OLR to surface rain rate and recently to satellite datasets providing vertical profiles of clouds. New observations and global reanalysis datasets have enhanced our understanding of the MJO as well as assessment of model performance.

MJO modeling is challenging because the MJO in models is very sensitive to most uncertain part of the GCM: the representation of cumulus convection. The use of limited-domain high-resolution models (e.g. cloud-system resolving models and large-eddy simulations) and more observations will deepen our understanding of the physics of cumulus convection. This enhanced understanding, especially that related to the representation of entrainment and detrainment, will eventually be beneficial to MJO modeling.

No single satisfactory theory of the MJO exists. Theories have provided frameworks to interpret modeling results, but at the same time GCMs have provided an excellent test bed for the theories. Development of theories and improvement of GCMs will complement each other for a better simulation and understanding of the MJO.

Currently, we have a number of GCMs that simulate at least the basic characteristics of the MJO, although many of them suffer from the MJO-mean state trade-off syndrome. A closer examination of these simulations is needed, including sophisticated process-oriented diagnosis that will allow us to assess reasons for model MJO improvements. The changes of the MJO in a warmer climate can be assessed using the GCMs that best simulate the MJO.

References

Adames, Á. F. and D. Kim, 2016: The MJO as a dispersive, convectively coupled moisture wave: Theory and observations. *J. Atmos. Sci.*, **73**, 913–941.

Andersen, J. A. and Z. Kuang, 2012: Moist static energy budget of MJO-like disturbances in the atmosphere of a zonally symmetric aquaplanet. *J. Climate*, **25**, 2782–2804.

Arnold, N. P., Z. Kuang, and E. Tziperman, 2013: Enhanced MJO-like variability at high SST. *J. Climate*, **26**, 988–1001.

Benedict, J. J. and D. A. Randall, 2007: Observed characteristics of the MJO relative to maximum rainfall. *J. Atmos. Sci.*, **64**, 2332–2354.

Benedict, J. J. and D. A. Randall, 2009: Structure of the Madden-Julian Oscillation in the superparameterized CAM. *J. Atmos. Sci.*, **66**, 3277–3296.

Benedict, J. J., E. D. Maloney, A. H. Sobel, and D. M. W. Frierson, 2014: Gross moist stability and MJO simulation skill in three full-physics GCMs. *J. Atmos. Sci.*, **71**, 3327–3349.

Benedict, J. J., E. D. Maloney, A. H. Sobel, D. M. Frierson, and L. J. Donner, 2013: Tropical intraseasonal variability in Version 3 of the GFDL atmosphere model. *J. Climate*, **26**, 426–449.

Bernie, D. J., E. Guilyardi, G. Madec, J. M. Slingo, S. J. Woolnough, and J. Cole, 2008: Impact of resolving the diurnal cycle in an ocean-atmosphere GCM. Part 2: A diurnally coupled CGCM, *Clim. Dyn.*, **31**, 909–925.

Bladé, I. and D. L. Hartmann, 1993: Tropical intraseasonal oscillations in a simple nonlinear model. *J. Atmos. Sci.*, **50**, 2922–2939.

Cai, Q., G. J. Zhang, and T. Zhou, 2013: Impacts of shallow convection on MJO simulation: A moist static energy and moisture budget analysis. *J. Climate*, **26**, 2417–2431.

Chang, C.-P. and H. Lim, 1988: Kelvin-wave CISK: A possible mechanism for the 30–50-day oscillation. *J. Atmos. Sci.*, **45**, 1709–1720.

Chao, W. C. and B. Chen, 2001: The role of surface friction in tropical intraseasonal oscillation. *Mon. Wea. Rev.*, **129**, 896–904.

Chikira, M. and M. Sugiyama, 2010: A cumulus parameterization with state-dependent entrainment rate. Part I: Description and sensitivity to temperature and humidity profiles. *J. Atmos. Sci.*, **67**, 2171–2193.

Chikira, M. and M. Sugiyama, 2013: Eastward-propagating intraseasonal oscillation represented by Chikira-Sugiyama cumulus parameterization. Part I: Comparison with observation and reanalysis. *J. Atmos. Sci.*, **70**, 3920–3939.

Chikira, M., 2014: Eastward-propagating intraseasonal oscillation represented by Chikira–Sugiyama cumulus parameterization. Part II: Understanding moisture variation under weak temperature gradient balance. *J. Atmos. Sci.*, **71**, 615–639.

CLIVAR Madden-Julian Oscillation Working Group; D. Waliser, K. Sperber, H. Hendon, D. Kim, E. Maloney, M. Wheeler, K. Weickmann, C. Zhang, L. Donner, J. Gottschalck, W. Higgins, I.-S. Kang, D. Legler, M. Moncrieff, S. Schubert, W. Stern, F. Vitart, B. Wang, W. Wang, and S. Woolnough, 2009: MJO simulation diagnostics. *J. Climate*, **22**, 3006–3030.

Crueger, T., B. Stevens, and R. Brokopf, 2013: The Madden-Julian Oscillation in ECHAM6 and the introduction of an objective MJO metric. *J. Climate*, **26**, 3241–3257.

Deng, L. and X. Wu, 2010: Effects of convective processes on GCM simulations of the Madden-Julian Oscillation. *J. Climate*, **23**, 352–377.

Deng, L. and X. Wu, 2011: Physical mechanisms for the maintenance of GCM-simulated Madden-Julian Oscillation over the Indian Ocean and Pacific. *J. Climate*, **24**, 2469–2482.

Emanuel, K. A., 1987: An air-sea interaction-model of intraseasonal oscillations in the tropics, *J. Atmos. Sci.*, **44**, 2324–2340.

Flatau, M., P. J. Flatau, P. Phoebus, and P. P. Niiler, 1997: The feedback between equatorial convection and local radiative and evaporative processes: The implications for intraseasonal oscillations. *J. Atmos. Sci.*, **54**, 2373–2386.

Fu, X. and B. Wang, 2009: Critical roles of the stratiform rainfall in sustaining the Madden-Julian Oscillation: GCM experiments. *J. Climate*, **22**, 3939–3959.

Grabowski, W. W., 2001: Coupling cloud processes with the large-scale dynamics using the Cloud-Resolving Convection Parameterization (CRCP). *J. Atmos. Sci.*, **58**, 978–997.

Hannah, W. M. and E. D. Maloney, 2011: The role of moisture-convection feedbacks in simulating the Madden-Julian Oscillation. *J. Climate*, **24**, 2754–2770.

Hendon, H. H., 2000: Impact of air–sea coupling on the Madden-Julian Oscillation in a general circulation model. *J. Atmos. Sci.*, **57**, 3939–3952.

Hirons L. C., P. Inness, F. Vitart, and P. Bechtold, 2012a: Understanding advances in the simulation of intraseasonal variability in the ECMWF model. Part I: The representation of the MJO. *Q. J. R. Meteorol. Soc.* DOI:10.1002/qj.2060.

Hirons L. C., P. Inness, F. Vitart, and P. Bechtold, 2012b: Understanding advances in the simulation of intraseasonal variability in the ECMWF model. Part II: The application of process-based diagnostics. *Q. J. R. Meteorol. Soc.* DOI:10.1002/qj.2059.

Hung, M.-P., J.-L. Lin, W. Wang, D. Kim, T. Shinoda, and S. J. Weaver, 2013: MJO and convectively coupled equatorial waves simulated by CMIP5 climate models. *J. Clim.*, **26**, 6185–6214.

Inness, P. M., J. M. Slingo, S. J. Woolnough, R. B. Neale, and V. D. Pope, 2001: Organization of tropical convection in a GCM with varying vertical resolution; implications for the simulation of the Madden-Julian Oscillation. *Clim. Dyn.*, **17**, 777–793.

Inness, P. M., J. M. Slingo, E. Guilyardi, and J. Cole, 2003: Simulation of the Madden-Julian Oscillation in a coupled general circulation model. Part II: The role of the basic state. *J. Climate*, **16**, 365–382.

Jia, X., C. Li, J. Ling, and C. Zhang, 2008: Impacts of a GCM's resolution on MJO simulation. *Adv. Atmos. Sci.*, **25**, 139–156.

Johnson, R. H., T. M. Rickenbach, S. A. Rutledge, P. E. Ciesielski, and W. H. Schubert, 1999: Trimodal characteristics of tropical convection. *J. Climate*, **12**, 2397–2418.

Jones, C. and L. M. V. Carvalho, 2006: Changes in the activity of the Madden-Julian Oscillation during 1958–2004. *J. Climate*, **19**, 6353–6370.

Jones, C. and L. M. V. Carvalho, 2010: Will global warming modify the activity of the Madden-Julian Oscillation? *Quart. J. Roy. Meteor. Soc.*, **137**, 544–552.

Kiladis, G. N., K. H. Straub, and P. T. Haertel, 2005: Zonal and vertical structure of the Madden-Julian Oscillation. *J. Atmos. Sci.*, **62**, 2790–2809.

Kim, D. and coauthors, 2009: Application of MJO simulation diagnostics to climate models, *J. Climate*, **22**, 6413–6436.

Kim, D., A. H. Sobel, D. M. W. Frierson, E. D. Maloney, and I.-S. Kang, 2011a: A systematic relationship between intraseasonal variability and mean state bias in AGCM simulations. *J. Climate*, **24**, 5506–5520.

Kim, D., A. H. Sobel, and I.-S. Kang, 2011b: A mechanism denial study on the Madden-Julian Oscillation. *J. Adv. Model. Earth Syst.*, **3**, M12007.

Kim, D. and I.-S. Kang, 2012: A bulk mass flux convection scheme for climate model — description and moisture sensitivity. *Clim. Dyn.*, **38**, 411–429.

Kim, D., A. H. Sobel, A. D. Del Genio, Y. Chen, S. J. Camargo, M.-S. Yao, M. Kelley, and L. Nazarenko, 2012: Tropical intraseasonal variability simulated in the NASA GISS general circulation model. *J. Climate*, **25**, 4641–4659.

Kim, D. and coauthors, 2014: Process-oriented MJO simulation diagnostic: moisture sensitivity of simulated convection. *J. Climate*, **27**, 5379–5395.

Klingaman, N. P., S. J. Woolnough, H. Weller, and J. M. Slingo, 2011: The impact of finer-resolution air-sea coupling on the intraseasonal oscillation of the Indian summer monsoon, *J. Climate*, **24**, 2451–2468.

Landu, K. and E. D. Maloney, 2011: The intraseasonal moist static energy budget in reanalysis data. *J. Geophys. Res.*, **116**, D21117.

Lau, K. M. and L. Peng, 1987: Origin of low-frequency, intraseasonal oscillations in the tropical atmosphere. Part I: Basic theory. *J. Atmos. Sci.*, **44**, 950–972.

Lau, W. K. and D. E. Waliser, 2012: Intraseasonal variability in the atmosphere-ocean climate system. Springer.

Lau, N. C. and K. M. Lau, 1986: The structure and propagation of intraseasonal oscillations appearing in a GFDL general circulation model. *J. Atmos. Sci.*, **43**, 2023–2047.

Lee, J.-Y., X. Fu, and B. Wang, 2016: Real-time forecasting of modes of tropical intraseasonal variability: The Madden Julian and Boreal Summer Intraseasonal Oscillations. *Global Monsoon System*, 3rd ed., edited by C. P. Chang *et al.*

Lee, M. I., I. S. Kang, J. K. Kim, and B. E. Mapes, 2001: Influence of cloud-radiation interaction on simulating tropical intraseasonal oscillation with an atmospheric general circulation model, *J. Geophys. Res.*, **106**, 14,219–14,233.

Lee, M.-I., I.-S. Kang, and B. E. Mapes, 2003: Impacts of cumulus convection parameterization on aqua-planet AGCM simulations of tropical intraseasonal variability. *J. Meteor. Soc. Japan*, **81**, 963–992.

Liess, S., L. Bengtsson, and K. Arpe, 2004: The intraseasonal oscillation in ECHAM4 Part I: Coupled to a comprehensive ocean model. *Clim. Dyn.*, **22**, 653–669.

Li, C., X. Jia, J. Ling, W. Zhou, and C. Zhang, 2009: Sensitivity of MJO simulations to diabatic heating profiles. *Clim. Dyn.*, **32**, 167–187.

Lin, J.-L. and coauthors, 2006: Tropical intraseasonal variability in 14 IPCC AR4 climate models. Part I: Convective signals. *J. Climate*, **19**, 2665–2690.

Lin, J.-L., M.-I. Lee, D. Kim, I.-S. Kang, and D. M. W. Frierson, 2008: The impacts of convective parameterization and moisture triggering on AGCM-simulated convectively coupled equatorial waves. *J. Climate*, **21**, 883–909.

Ling, J., C. Li, and X. Jia, 2009: Impacts of cumulus momentum transport on MJO simulation. *Adv. Atmos. Sci.*, **26**, 864–876.

Liu, P. and coauthors, 2009: Tropical intraseasonal variability in the MRI-20km60L AGCM. *J. Climate*, **22**, 2006–2022.

Liu, P., 2013: Changes in a modeled MJO with idealized global warming. *Clim. Dyn.*, **40**, 761–773.

Liu, F. and B. Wang, 2012: A frictional skeleton model for the Madden-Julian Oscillation. *J. Atmos. Sci.*, **69**, 2749–2758.

Madden, Roland A. and Paul R. Julian, 1972: Description of global-scale circulation cells in the tropics with a 40–50 day period. *J. Atmos. Sci.*, **29**, 1109–1123.

Majda, A. J. and S. N. Stechmann, 2009a: The skeleton of tropical intraseasonal oscillations. *Proc. Natl. Acad. Sci.*, **106**, 8417–8422.

Majda, A. J. and S. N. Stechmann, 2009b: A simple dynamical model with features of convective momentum transport. *J. Atmos. Sci.*, **66**, 373–392.

Maloney, E. D. and D. L. Hartmann, 2001: The sensitivity of intraseasonal variability in the NCAR CCM3 to changes in convective parameterization. *J. Climate*, **14**, 2015–2034.

Maloney, E. D., 2002: An intraseasonal oscillation composite lifecycle in the NCAR CCM3.6 with modified convection. *J. Climate*, **15**, 964–982.

Maloney, E. D. and J. T. Kiehl, 2002: Intraseasonal eastern Pacific precipitation and SST variations in a GCM coupled to a slab ocean model. *J. Climate*, **15**, 2989–3007.

Maloney, E. D. and A. H. Sobel, 2004: Surface fluxes and ocean coupling in the tropical intraseasonal oscillation. *J. Climate*, **17**, 4368–4386.

Maloney, E. D., 2009: The moist static energy budget of a composite tropical intraseasonal oscillation in a climate model. *J. Climate*, **22**, 711–729.

Maloney, E. D., A. H. Sobel, and W. M. Hannah, 2010: Intraseasonal variability in an aquaplanet general circulation model. *J. Adv. Model. Earth Syst.*, **2**, Art. #5, 24 pp.

Maloney, E. D. and S.-P. Xie, 2013: Sensitivity of MJO activity to the pattern of climate warming. *J. Adv. Modeling Earth Sys.*, **5**, 32–47, doi:10.1029/2012MS000171.

Neelin, J. D., I. M. Held, and K. H. Cook, 1987: Evaporation-wind feedback and low-frequency variability in the tropical atmosphere. *J. Atmos. Sci.*, **44**, 2341–2348.

Neelin, J. D. and J. Yu, 1994: Modes of tropical variability under convective adjustment and the Madden-Julian Oscillation. Part I: Analytical theory. *J. Atmos. Sci.*, **51**, 1876–1894.

Park, C. K., D. M. Straus, and K.-M. Lau, 1990: An evaluation of the structure of tropical intraseasonal oscillations in three general circulation models. *J. Meteor. Soc. Japan*, **68**, 403–417.

Pohl, B. and A. J. Matthews, 2007: Observed changes in the lifetime and amplitude of the Madden-Julian Oscillation associated with interannual ENSO sea surface temperature anomalies. *J. Climate*, **20**, 2659–2674.

Rashid, H. A., H. H. Hendon, M. C. Wheeler, and O. Alves, 2011: Prediction of the Madden-Julian Oscillation with the POAMA dynamical prediction system. *Clim. Dyn.*, **36**, 649–661.

Rajendran, K., A. Kitoh, R. Mizuta, S. Jajani, and T. Nakazawa, 2008: High resolution simulation of mean convection and its intraseasonal variability over the tropics in MRI/JMA 20-km mesh AGCM. *J. Climate*, **21**, 3722–3739.

Raymond, D. J., 2001: A new model of the Madden-Julian Oscillation, *J. Atmos. Sci.*, **58**, 2807–2819.

Satoh, M., T. Matsuno, H. Tomita, H. Miura, T. Nasuno, and S. Iga, 2008: Nonhydrostatic icosahedral atmospheric model (NICAM) for global cloud resolving simulations. *J. Comput. Phys.*, **227**, 3486–3514.

Schubert, J. J., B. Stevens, and T. Crueger, 2013: Madden-Julian Oscillation as simulated by the Max Planck Institute for Meteorology Earth System Model: Over the last and into the next millennium, *J. Adv. Model. Earth Syst.*, **5**, 71–84, doi:10.1029/2012MS000180.

Seo, K. H. and Wang, W., 2010: The Madden-Julian Oscillation simulated in the NCEP climate forecast system model: The importance of stratiform heating. *J. Climate*, **23**, 4770–4793.

Slingo, J. M. and coauthors, 1996: Intraseasonal oscillations in 15 atmospheric general circulation models: Results from an AMIP diagnostic subproject. *Clim. Dyn.*, **12**, 325–357.

Slingo, J. M., D. P. Rowell, K. R. Sperber, and F. Nortley, 1999: On the predictability of the interannual behaviour of the Madden-Julian Oscillation and its relationship with El Nino. *Quart. J. Roy. Meteor. Soc.*, **125**, 583–609.

Sobel, A. H., J. Nilsson, and L. M. Polvani 2001: The weak temperature gradient approximation and balanced tropical moisture waves. *J. Atmos. Sci.*, **58**, 3650–3665.

Sobel, A. H., E. D. Maloney, G. Bellon, and D. M. Frierson, 2008: The role of surface heat fluxes in tropical intraseasonal oscillations. *Nature Geoscience*, **1**, 653–657.

Sobel, A. H., E. D. Maloney, G. Bellon, and D. M. Frierson, 2010: Surface fluxes and tropical intraseasonal variability: A reassessment. *J. Adv. Model. Earth Syst.*, **2**, doi:10.3894/JAMES.2010.2.2.

Sobel, A. H. and E. D. Maloney, 2012: An idealized semi-empirical framework for modeling the Madden-Julian Oscillation. *J. Atmos. Sci.*, **69**, 1691–1705.

Sobel, A. H. and E. D. Maloney, 2013: Moisture modes and the eastward propagation of the MJO. *J. Atmos. Sci.*, **70**, 187–192.

Sperber, K. R., J. M. Slingo, P. M. Inness, and W. M. Lau, 1997: On the maintenance and initiation of the intraseasonal oscillation in the NCEP/NCAR reanalysis and in the GLA and UKMO AMIP simulations. *Clim. Dyn.*, **13**, 769–795.

Sperber, K. R., 2004: Madden-Julian variability in NCAR CAM2.0 and CCSM2.0. *Clim. Dyn.*, **23**, 259–278.

Sperber, K. R., S. Gualdi, S. Legutke, and V. Gayler, 2005: The Madden-Julian Oscillation in ECHAM4 coupled and uncoupled general circulation models. *Clim. Dyn.*, **25**, 117–140.

Thayer-Calder, K. and D. A. Randall, 2009: The role of convective moistening in the Madden-Julian Oscillation. *J. Atmos. Sci.*, **66**, 3297–3312.

Tokioka, T., K. Yamazaki, A. Kitoh, and T. Ose, 1988: The equatorial 30–60-day oscillation and the Arakawa–Schubert penetrative cumulus parameterization. *J. Meteor. Soc. Japan*, **66**, 883–901.

Miyakawa, T., *et al.*, 2016: Review: Global cloud/cloud-system resolving models and the Madden-Julian Oscillation. *Global Monsoon System*, 3rd ed., edited by C. P. Chang *et al.*

Vitart, F. and F. Molteni, 2010: Simulation of the Madden-Julian Oscillation and its teleconnections in the ECMWF forecast system. *Quart. J. Roy. Meteor. Soc.*, **136**, 842–855.

Xie, S.-P., C. Deser, G. A. Vecchi, J. Ma, H. Teng, and A. T. Wittenberg, 2010: Global warming pattern formation: Sea surface temperature and rainfall. *J. Climate*, **23**, 966–986.

Waliser, D. E., K. M. Lau, and J.-H. Kim, 1999: The influence of coupled sea surface temperatures on the Madden-Julian Oscillation: A model perturbation experiment. *J. Atmos. Sci.*, **56**, 333–358.

Waliser, D. E. and coauthors, 2003: AGCM simulations of intraseasonal variability associated with the Asian summer monsoon. *Clim. Dyn.*, **21**, 423–446.

Wang, B., 1988: Dynamics of tropical low-frequency waves: An analysis of the moist Kelvin wave. *J. Atmos. Sci.*, **45**, 2051–2065.

Wang, B. and H. Rui, 1990: Dynamics of the coupled moist Kelvin-Rossby wave on an equatorial-plane. *J. Atmos. Sci.*, **47**, 397–413.

Wang, B. and X. Xie, 1998: Coupled modes of the warm pool climate system. Part I: The role of air-sea interaction in Madden-Julian Oscillation. *J. Climate*, **11**, 2116–2135.

Wang, B. and F. Liu, 2011: A model for scale interaction in the Madden-Julian Oscillation. *J. Atmos. Sci.*, **68**, 2524–2536.

Wang, J., W. Wang, X. Fu, and K.-H. Seo, 2012: Tropical intraseasonal rainfall variability in the CFSR. *Clim. Dyn.*, **38**, 2191–2207.

Wang W., M. P. Hung, S. J. Weaver, A. Kumar, and X. Fu, 2013: MJO prediction in the NCEP Climate Forecast System version 2. *Clim. Dyn.*, 10.1007/s00382-013-1806-9.

Wang, W. Q. and M. E. Schlesinger, 1999: The dependence on convection parameterization of the tropical intraseasonal oscillation simulated by the UIUC 11-layer atmospheric GCM. *J. Climate*, **12**, 1423–1457.

Watterson, I. G., 2002: The sensitivity of subannual and intraseasonal tropical variability to model ocean mixed layer depth. *J. Geophys. Res.*, **107**, 4020. doi:10.1029/2001JD000671.

Weaver, S. J., W. Wang, M. Chen, and A. Kumar, 2011: Representation of MJO variability in the NCEP Climate Forecast System. *J. Climate*, **24**, 4676–4694.

Wheeler, M. C. and H. H. Hendon, 2004: An all-season real-time multivariate MJO index: Development of an index for monitoring and prediction. *Mon. Wea. Rev.*, **132**, 1917–1932.

Wheeler, M. C., H.-J. Kim, J.-Y. Lee, and J. C. Gottschalck, 2016: Real-time forecasting of modes of tropical intraseasonal variability: The Madden Julian and Boreal Summer Intraseasonal Oscillations. *Global Monsoon System*, 3rd ed., edited by C. P. Chang *et al.*

Wu, X. and L. Deng, 2013: Comparison of moist static energy and budget between the GCM-Simulated Madden-Julian Oscillation and observations over the Indian Ocean and Western Pacific. *J. Climate*, **26**, 4981–4993.

Zhang, G. J. and M. Mu, 2005: Simulation of the Madden-Julian Oscillation in the NCAR CCM3 using a revised Zhang–McFarlane convection parameterization scheme. *J. Climate*, **18**, 4046–4064.

Zhang, G. J. and X. L. Song, 2009: Interaction of deep and shallow convection is key to Madden-Julian Oscillation simulation. *Geophys. Res. Lett.*, **36**, L09708. doi:10.1029/2009GL037340.

Zhang Q. and H. van den Dool, 2012: Relative merit of model improvement versus availability of retrospective forecasts: The case of Climate Forecast System MJO prediction. *Wea. Forecast.* **27**, 1045–1051.

Zhou, L., B. R. Neale, M. Jochum, and R. Murtugudde, 2012: Improved Madden-Julian Oscillations with improved physics: The impact of modified convection parameterizations. *J. Climate*, **25**, 1116–1136.

Chapter 10

Real-Time Forecasting of Modes of Tropical Intraseasonal Variability: The Madden-Julian and Boreal Summer Intraseasonal Oscillations

Matthew C. Wheeler

Centre for Australian Weather and Climate Research, Bureau of Meteorology, GPO Box 1289, Melbourne, VIC 3001, Australia
m.wheeler@bom.gov.au
www.cawcr.gov.au/staff/mwheeler/

Hae-Jeong Kim

APEC Climate Center, Busan, 612-020, Korea
shout@apcc21.org

June-Yi Lee

Global Monsoon Climate Laboratory, Pusan National University, Busan, 609-735, Korea
juneyi@pusan.ac.kr

Jon C. Gottschalck

Climate Prediction Center, NOAA/National Weather Service, College Park, MD 20740, USA
jon.gottschalck@noaa.gov

We review the progress of two related activities on real-time forecasting of modes of large-scale tropical intraseasonal variability. The modes are the Madden Julian Oscillation (MJO) and the Boreal Summer Intraseasonal Oscillation (BSISO). The MJO forecasting activity has reached a mature status and is judged a success based on: (a) the number of user hits its web page receives; (b) the application of the MJO forecasts for user-oriented forecast products; and (c) the number of research papers for which the forecasting and verification methods have been used. The BSISO forecasting activity is still developing and in 2013 had five different model contributions, with another to be added in 2014. Verification of the BSISO forecasts during 2013 show a similar level of skill for the main BSISO mode (BSISO1) compared to the MJO, but less for the 2nd mode (BSISO2). There are plans to analyze the relation of the BSISO modes to impacts in the East Asian monsoon, and also to provide a multi-model ensemble forecast.

1. Introduction

The Madden-Julian Oscillation (MJO) is the primary mode of intraseasonal variability existing year-round in most tropical regions whereas the Boreal Summer Intraseasonal Oscillation (BSISO) is the primary mode of intraseasonal variability existing during the boreal summer in the Asian monsoon region. At the last workshop of the international monsoon community (IWM-IV in 2008), a review of then recent research on intraseasonal forecasting (Goswami *et al.* 2011) introduced to the community a coordinated activity for monitoring forecasts of the

The Global Monsoon System: Research and Forecast (3rd Edition)
Edited by C. P. Chang et al.

MJO in operational dynamical models. At that time this MJO forecasting activity was still in its infancy. The activity was seen as an important step for the further realization of intraseasonal predictions, particularly in the tropics but also indirectly of the low-frequency extra-tropical weather variability influenced by the MJO. It has since been deemed a significant success (e.g. Wheeler *et al.* 2013; see further below). This review provides an update on the progress and maturation of this MJO activity as well as introducing a new activity to apply a similar idea and approach to forecasts of the BSISO.

2. MJO Forecasting Activity

The motivation for this activity was that by closely monitoring the state of the MJO in observations and forecast model output we should increase understanding of: (a) the MJO itself; (b) the MJO's relation to weather impacts; and (c) its prediction skill in models. Using real-time forecasts increases the relevancy of the information and has allowed its application for user-oriented products. As described in the earlier review chapter of Goswami *et al.* (2011), the two-component Real-time Multivariate MJO (RMM; Wheeler and Hendon 2004) index is used as a measure of the MJO in both observations and forecasts. The RMM index is computed for the model analyses and each day of model forecast output and displayed in the two-dimensional phase space defined by the index (e.g. Fig. 1). Such a display allows for an easy visual comparison between different models and ensemble members as well as for comparison to the verifying observations. As can be seen from the number of different models to choose from in the web display in Fig. 1 (as taken from http://www.cpc.ncep.noaa.gov/products/precip/CWlink/MJO/CLIVAR/clivar_wh.shtml) there are now 16 different model streams contributing to this activity.

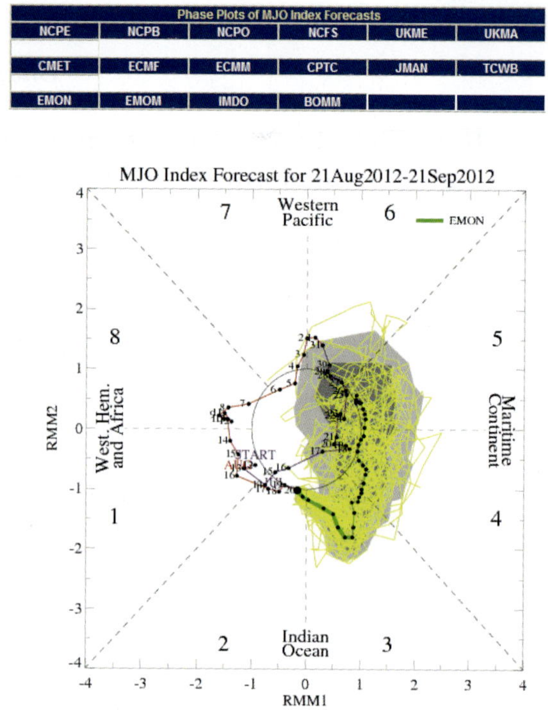

Fig. 1. Snapshot of real-time MJO forecast display as hosted on the CPC-NCEP web page. Shown are the observations (purple and red lines) and forecasts (thick/thin green line for ensemble mean during/after first 7 days, yellow line for individual members) of the RMM index as displayed in the phase space defined by RMM1 and RMM2. Further explanation of the diagram and acronyms can be found at http://www.cpc.ncep.noaa.gov/products/precip/CWlink/MJO/CLIVAR/clivar_ wh.shtml.

To reach this high level of model participation required substantial international collaboration and coordination. The initial concept of a coordinated activity using forecasts from multiple models was discussed in late 2007 at a meeting of the Climate Variability and Predictability (CLIVAR) MJO Working Group (Sperber and Waliser 2008). The help of the Working Group on Numerical Experimentation (WGNE) was called upon to solicit participation by operational modelling centres worldwide. Further, the Climate Prediction Center (CPC) of the National Centers for Environmental Prediction (NCEP) volunteered to host the data collection and display, which they have continued to maintain at the web-site listed above.

Details of the model MJO index calculation and forecast verification method can be found in Gottschalck *et al.* (2010). Briefly, the method involves projecting the daily output of the numerical models onto a pair of Empirical Orthogonal Functions (EOFs) computed using 15°S–15°N averages of three fields: outgoing longwave radiation (OLR); 850 hPa zonal wind; and 200 hPa zonal wind. The three fields are a function of longitude only, and the observed EOFs are used for the model output as well. An advantage of the RMM index is that the majority of the MJO signal can be extracted from daily data with essentially no time filtering. It has become the most widely used index of the MJO, although some newer alternatives have been developed for different needs (Ventrice *et al.* 2013; Kiladis *et al.* 2014). For verification and skill comparisons a standard method has been defined that uses the bivariate correlation (Lin *et al.* 2008) and root-mean-square error between the observed and forecast RMM indices.

An example of the verification of all the real time MJO forecasts received at NCEP between November 2008 and October 2012 is provided in Fig. 2. It shows the bivariate correlation between the forecast and observed bivariate RMM index for each forecast model as a function of lead time. Some of these are ensemble systems (NCEP, NCBC, UKME, CANM, ECMF, ECMM, JMAN, EMON, EMOM, CFSM), and some aren't (UKMA, GFSO, TCWB). For the ensemble systems the ensemble mean is used in this calculation of the bivariate correlation. For example, UKME uses the ensemble mean from the UK Met Office system whereas UKMA is from a single member with a higher resolution "deterministic" system. This plot shows that for the forecasts provided, the skill of the lower resolution UKME ensemble is greater than the skill of the higher resolution UKMA deterministic run. This is generally the case for other systems

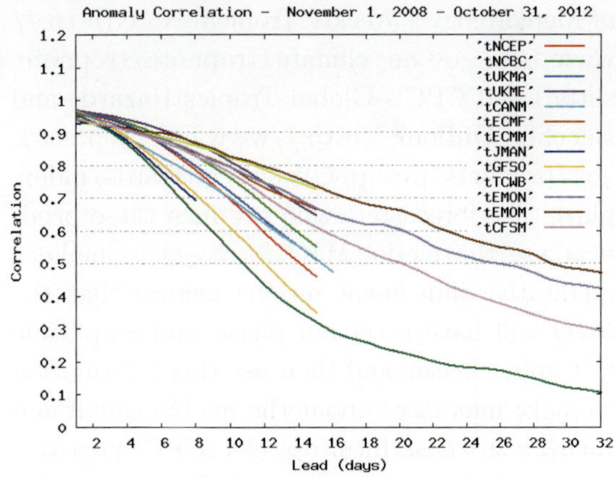

Fig. 2. Bivariate correlation skill between forecast and observed values of the bivariate RMM index as a function of forecast lead time.

as well, showing the advantage of an ensemble for MJO forecasting.

All models provide total fields from which the NCEP reanalysis climatology is removed. However, some models also have a long hindcast dataset from which they can compute anomalies with respect to their hindcast climatology. For example, EMOM is the ECMWF monthly forecast system with their hindcast climatology removed, whereas EMON is the same system but the NCEP climatology has been used. The skill in this sample shows slightly higher skill for EMOM, showing the advantage of having a hindcast climatology for removal of model bias.

Success of this MJO forecast activity can be measured in several ways. The first is by the number of hits being received by the CPC web page that displays the forecasts. For example, during the month of March 2012 there were over 283,000 hits giving the page a rank of 22nd most popular at CPC. March 2012 corresponded to a time of strong MJO activity, but even during times of weak MJO during mid-2012 the site still received over 20,000 hits per month.

A second measure of success is the incorporation of information from the MJO forecasts into user-oriented forecast products. Two important examples of this are the Australian Bureau of Meteorology's Weekly Tropical Note (http://www.bom.gov.au/climate/tropnote/tropnote.shtml) and CPC's Global Tropics Hazards and Benefits Outlook (http://www.cpc.ncep.noaa.gov/products/precip/CWlink/ghazards/index.php). The forecasters who produce these products use the model MJO forecasts to make a qualitative judgement on the chance that the MJO will have a certain phase and amplitude in coming weeks, and then use this information to make inferences about the likely rainfall and weather in those future weeks. CPC have verified that this approach has provided positive skill in most locations.

A third measure of success is the number of research papers for which aspects of this MJO forecast activity, and the methods used, have since been used. A sample of the more relevant papers are Vitart *et al.* (2010), Rashid *et al.* (2011), Fu *et al.* (2011), Lin and Brunet (2011), Matsueda and Endo (2011), Marshall *et al.* (2011), Zhang and Van den Dool (2012), and Fu *et al.* (2013). It thus appears that the initial goal as discussed by the MJOWG in 2007, of establishing a uniform method for quantifying MJO prediction skill, has been realized.

3. BSISO Forecasting Activity

One shortcoming of the MJO forecasting activity is that the MJO, as measured by the RMM index, accounts for a smaller fraction of intraseasonal variability existing during boreal summer compared to austral summer (Lee *et al.* 2013). In particular, the RMM index only captures a fraction of the northward propagating intraseasonal variability that is observed in the Asian summer monsoon (ASM). Further, the extent to which forecast models simulate northward propagation cannot be accurately diagnosed with the RMM. Therefore, it became a goal of the successor to the MJO Working Group, the international MJO Task Force (MJO-TF), to produce a new index and research activity based around the intraseasonal variability that is particular to the ASM region.

Work on this new activity began in 2010 through discussions at a workshop in Korea (Hendon *et al.* 2011). The initial work concentrated on defining a similar index or set of indices to the RMM index that may capture a greater fraction of the intraseasonal variance within the ASM. Instead of using 15°S–15°N averaged data as input to the EOF analysis, maps of data were used over the domain 10°S–40°N, 40–160°E, resulting in the definition of 2 modes of variability each consisting of a pair of spatial EOFs (Figs. 3 and 4). Only two variables were used in the analysis (OLR and 850 hPa zonal wind) and for the months of May–October only. We call the modes BSISO1 (consisting of EOFs 1 and 2; Fig. 3) and BSISO2 (consisting of EOFs 3 and 4; Fig. 4), with BSISO1 representing the canonical northward propagating variability that often occurs in conjunction with the eastward MJO with quasi-oscillating periods of 30–60 days, and BSISO2 representing the northward and northwestward propagating variability with periods of 10–30 days existing primarily during the pre-monsoon and monsoon onset season. Full details of the BSISO indices and their properties are provided in Lee *et al.* (2013).

In late 2012 a request was sent by the MJO-TF to the Director of the APEC Climate Center to host a data collection and display of these new BSISO indices using model forecasts from the operational centres, which was graciously accepted. Then, with the help of WGNE, a new letter was sent to the operational centres in early 2013 asking for their contribution of forecasts for a new web page based on the BSISO forecasts at APCC. During the 2013 boreal summer there were 5 model forecast streams contributing to the display at http://www.apcc21.org/eng/service/bsiso/fore/japcc030601.jsp (Table 1 and Fig. 5): from the Australian Bureau of Meteorology (labelled as BOM in Fig. 5); the NCEP Coupled Forecast System (CFS) and Global Forecast System (GFS); the European Centre

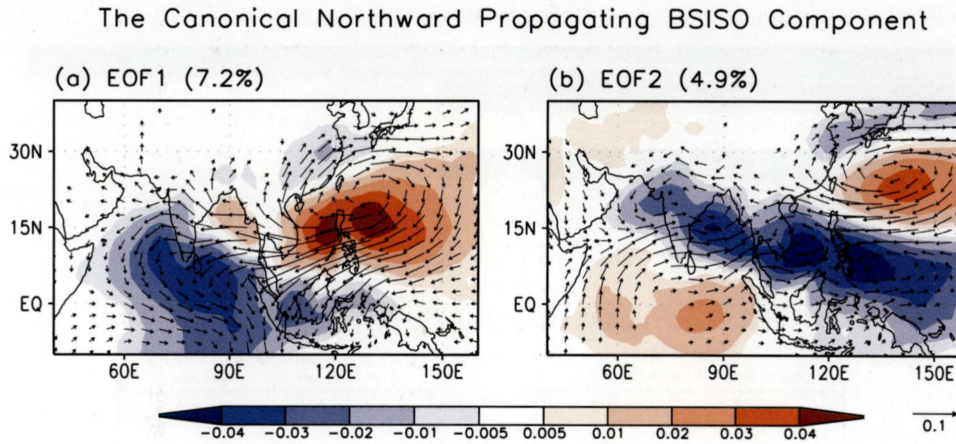

Fig. 3. Spatial structure of BSISO1 mode as represented by EOFs 1 and 2 of the analysis of Lee *et al.* (2013) using OLR (shading; normalized by a value of 33.04 W m^{-2}) and 850-hPa zonal wind (x-component of vector; normalized by a value of 4.01 m s^{-1}). Note that the meridional component of the wind vector is obtained through regression. The percentages provide their respective percent of variance explained.

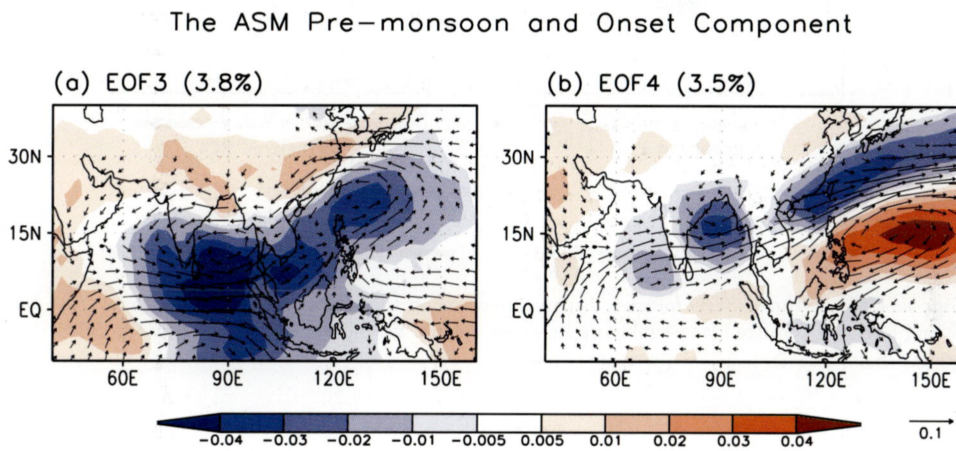

Fig. 4. Spatial structure of BSISO2 mode as represented by EOFs 3 and 4 of the analysis of Lee *et al.* (2013) using OLR (shading; normalized by a value of 33.04 W m^{-2}) and 850-hPa zonal wind (x-component of vector; normalized by a value of 4.01 m s^{-1}). Note that the meridional component of the wind vector is obtained through regression. The percentages provide their respective percent of variance explained.

Table 1. Contributing models to the BSISO forecasting activity during 2013. In 2014 a contribution from the Central Weather Bureau of Taiwan is to be added.

Institute	Model	Ensemble size	Forecast period	Update frequency	Resolution
NCEP	Climate Forecast System	4	40 days	Once a day	T126 L64
	Global Forecast System	1	16 days	Once a day	T574, T190 L64
Australian Bureau of Meteorology	POAMA2.4 multi-week model	33	40 days	Twice per week	T47 L17
ECMWF	ECMWF Ensemble Prediction System	51	32 days	Twice per week	T639, T319 L62
UK Met Office	MOGREPS-15	24	15 days	Once a day	60 km L70

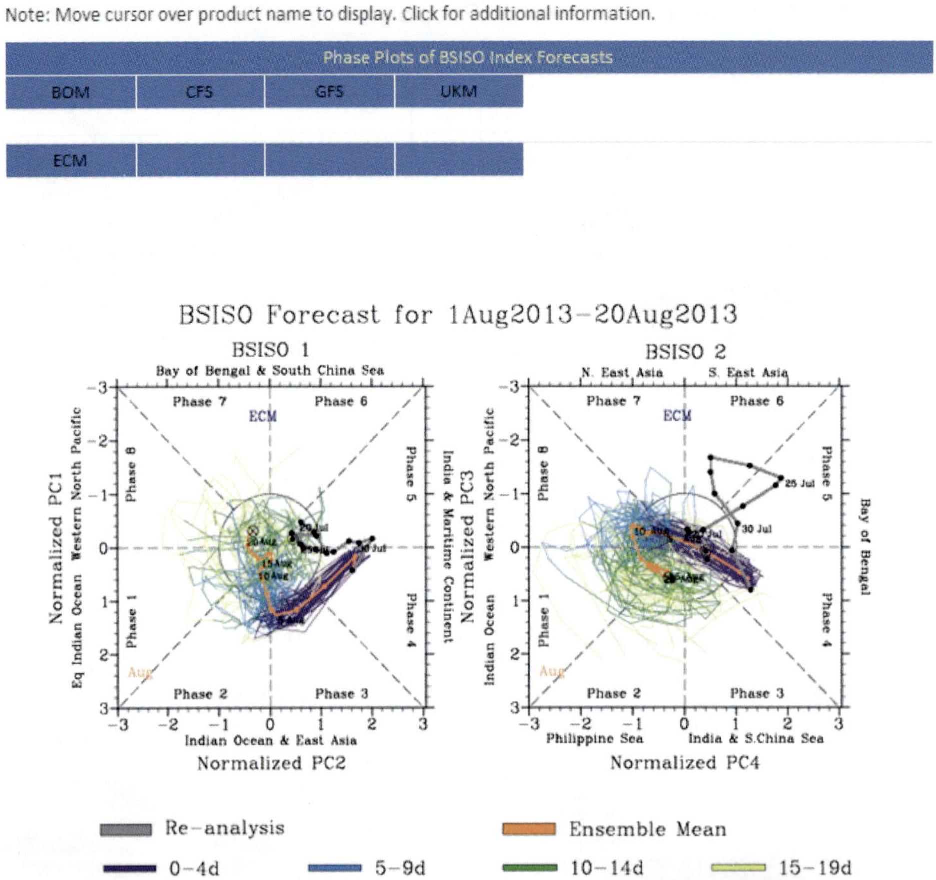

Fig. 5. Snapshot of real-time BSISO forecast display as hosted on the APCC web page. Shown are the observations (grey line) and forecasts (orange for ensemble mean, other colours for individual members) of the BSISO1 (left) and BSISO2 (right) indices.

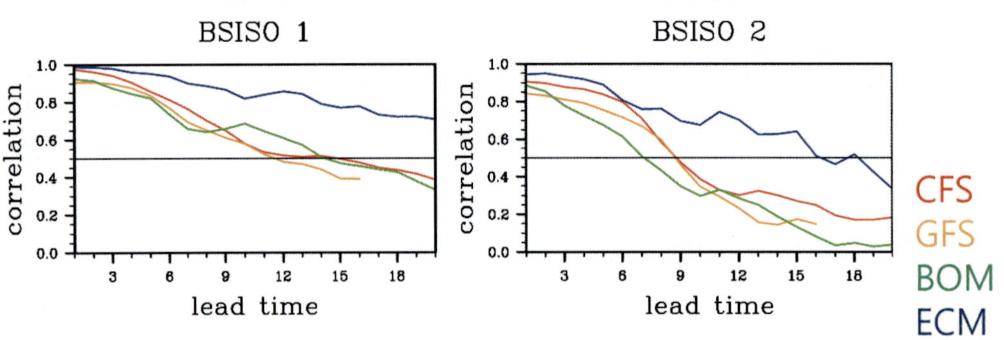

Fig. 6. As in Fig. 2 except for the skill at predicting BSISO1 (left panel) and BSISO2 (right panel) using all forecasts received during May–October 2013 (except UKMO). Refer to text for model acronyms.

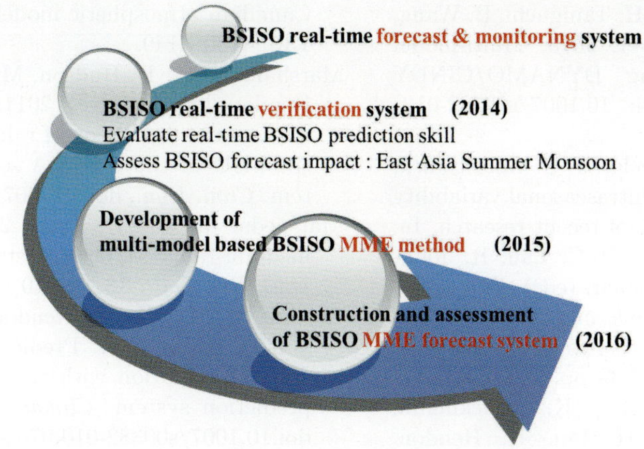

Fig. 7. Summary of the development plan for the BSISO forecasting activity at APCC.

for Medium Range Weather Forecasts (ECM); and from the UK Met Office (UKM). Since then, a commitment has also been received from the Central Weather Bureau of Taiwan to provide forecast data in 2014.

Application of these forecasts (e.g. for more user-oriented products like those listed above for the MJO) is still under development, but will likely require some time for the wider forecast community to become aware. Verification statistics of the forecasts received during May–October 2013 are provided in Fig. 6. Like in Fig. 2, the verification measure used is the bivariate correlation skill, noting that BSISO1 and BSISO2 are each composed of 2 components (i.e. they are both bivariate). The correlations skill for the ECM model is higher for both BSISO1 and BSISO2 compared to the other models, and the skill at predicting BSISO1 is higher than that for BSISO2 across all models. This higher skill for BSISO1 is to be expected given its longer characteristic time scale. Interestingly, the skill of the ECM model for predicting BSISO1 (correlations skill of 0.7 at a lag of 20 days in Fig. 6) is higher than that for any model at predicting the MJO at the same lag (Fig. 2). However, we must caution that this result is based only on forecasts received for the BSISO indices in 2013, and it is currently unknown how generally they will apply to other years.

Future plans include construction of a real-time verification system, assessing BSISO impacts in the East Asian Summer Monsoon, and the production of a multi-model ensemble (MME) forecast system (Fig. 7). It is hoped that this activity may eventually attain a similar level of success as listed above for the MJO activity.

Acknowledgments

We thank the members of the MJO Task Force for their inputs to these activities, especially Duane Waliser, Harry Hendon, Kenneth Sperber, and Eric Maloney. We also thank the forecast contributions that have been facilitated by a number of individuals including Andrew Marshall, Wanqiu Wang, Ann Shelly, and Frederic Vitart. Aurel Moise and Andrew Marshall provided internal reviews.

References

Fu, X., B. Wang, J.-Y. Lee, W. Wang, and L. Gao, 2011: Sensitivity of dynamical intraseasonal prediction skills to different initial conditions. *Mon. Wea. Rev.*, **139**, 2572–2592.

Fu, X., J.-Y. Lee, P.-C. Hsu, H. Taniguchi, B. Wang, W. Wang, and S. Weaver, 2013: Multi-model MJO forecasting during DYNAMO/CINDY period. *Clim. Dyn.*, doi: 10.1007/s00382-013-1859-9.

Goswami, B. N., M. C. Wheeler, J. C. Gottschalck, and D. E. Waliser, 2011: Intraseasonal variability and forecasting: A review of recent research. In: C.-P. Chang, Y. H. Ding, N.-C. Lau, R. Johnson, B. Wang, and T. Yasunari (eds), *The Global Monsoon System: Research and Forecast (2nd edition)*. World Scientific Series on Asia-Pacific Weather and Climate, Vol. **5**, pp. 389–408.

Gottschalck, J., M. Wheeler, K. Weickmann, F. Vitart, N. Savage, H. Lin, H. Hendon, D. Waliser, K. Sperber, M. Nakagawa, C. Prestrelo, M. Flatau, and W. Higgins, 2010: A framework for assessing operational Madden-Julian Oscillation forecasts: A CLIVAR MJO Working Group project. *Bull. Amer. Meteor. Soc.*, **91**, 1247–1258.

Hendon, H. H., K. R. Sperber, D. E. Waliser, and M. C. Wheeler, 2011: Modelling monsoon intraseasonal variability: From theory to operational forecasting. *Bull. Amer. Meteor. Soc.*, **92**, ES32–ES35. doi:10.1175/2011BAMS3164.1.

Kiladis, G., J. Dias, K. Straub, M. Wheeler, S. Tulich, K. Kikuchi, K. Weickmann, and M. Ventrice, 2014: A comparison of OLR and circulation based indices for tracking the MJO. *Mon. Wea. Rev.*, doi:10.1175/MWR-D-13-00301.1, in press.

Lee, J.-Y., B. Wang, M. C. Wheeler, X. Fu, D. E. Waliser, and I.-S. Kang, 2013: Real-time multivariate indices for the boreal summer intraseasonal oscillation over the Asian summer monsoon region. *Clim. Dyn.*, **40**, 493–509.

Lin, H. and G. Brunet, 2011: Impact of the North Atlantic Oscillation on the forecast skill of the Madden-Julian Oscillation. *Geophys. Res. Lett.*, **38**, doi: 10.1029/2010GL046131.

Lin, H., G. Brunet, and J. Derome, 2008: Forecast skill of the Madden-Julian Oscillation in two Canadian atmospheric models. *Mon. Wea. Rev.*, **136**, 4130–4149.

Marshall, A. G., D. Hudson, M. C. Wheeler, H. H. Hendon, and O. Alves, 2011: Assessing the simulation and prediction of rainfall associated with the MJO in the POAMA seasonal forecast system. *Clim. Dyn.*, doi:10.1007/s00382-010-0948-2.

Matsueda, M. and H. Endo, 2011: Verification of medium-range MJO forecasts with TIGGE. *Geophys. Res. Lett.*, **38**, doi: 10.1029/2011GL047480.

Rashid, H. A., H. H. Hendon, M. C. Wheeler, and O. Alves, 2011: Prediction of the Madden-Julian Oscillation with the POAMA dynamical prediction system. *Climate Dyn.*, **36**, 649–661. doi:10.1007/s00382-010-0754-x.

Sperber, K. R. and D. E. Waliser, 2008: New approaches to understanding, simulating, and forecasting the Madden–Julian Oscillation. *Bull. Amer. Meteor. Soc.*, **89**, 1917–1920.

Ventrice, M. J., M. C. Wheeler, H. H. Hendon, C. J. Schreck, and C. D. Thorncroft, 2013: A modified multivariate Madden Julian Oscillation index using velocity potential. *Mon. Wea. Rev.*, **141**, 4197–4210.

Vitart, F., A. Leroy, and M. C. Wheeler, 2010: A comparison of dynamical and statistical predictions of weekly tropical cyclone activity in the Southern Hemisphere. *Mon. Wea. Rev.*, **138**, 3671–3682.

Wheeler, M. C. and H. H. Hendon, 2004: An all-season real-time multivariate MJO Index: Development of an index for monitoring and prediction. *Mon. Wea. Rev.*, **132**, 1917–1932.

Wheeler, M., E. Maloney, and MJO Task Force, 2013: Madden-Julian Oscillation (MJO) Task Force: A joint effort of the climate and weather communities. *CLIVAR Exchanges*, **61**, 9–12.

Zhang, Q. and H. Van den Dool, 2012: Relative merit of model improvement versus availability of retrospective forecasts: The case of climate forecast system MJO prediction. *Wea. Forecast.*, **27**, 1045–1051.

Chapter 11

Propagation Characteristics of Tropical-Subtropical Intraseasonal Variations during Boreal Summer and Their Mechanisms

In-Sik Kang* and Min-Seop Ahn

School of Earth Environment Sciences,
Seoul National University, Seoul, 151–742, Korea
**kang@climate.snu.ac.kr*
http://climate.snu.ac.kr

This paper is aimed to summarize the previous findings on various propagation characteristics of intraseasonal variations (ISV) over the Asian Monsoon region during boreal summer and reviews and suggests major mechanisms associated with the propagation characteristics. The ISV consists of various components whose characteristics are different in terms of propagation direction and its time scale. Three major components include the eastward propagating component with time scales of 30–60 days, which propagates mainly along the tropical belt, the westward propagating component with shorter time scales of a week to a month in the subtropical Western Pacific and Indochina, and the northward propagating component with time scales of about 40 days in the Indian Ocean and Western Pacific region. It is demonstrated that the essential mechanism for the eastward and westward propagations is rooted in the convectively coupled Kelvin-Rossby waves with frictional moisture convergence. The propagation direction and time scales are shown to be determined by the horizontal structures of surface boundary conditions, particularly the sea surface temperature. The northward propagation, on the other hand, appears to be related to the strong vertical wind shear in the South Asian Monsoon region, as appears only during boreal summer. The wind shear mechanisms include the vorticity titling mechanism and the cumulus momentum mixing (CMT) and the CMT is discussed as a major mechanism in this paper.

1. Introduction

The most distinctive feature of the intraseasonal oscillation (ISO) in the tropics is the Madden and Julian Oscillation (MJO), which is characterized by the eastward propagation along the equatorial region with a time scale of 30–60 days (Madden and Julian 1972, and others). Not only the eastward propagation, but the westward moving disturbances with intraseasonal time scales are also seen in the subtropical western Pacific and Indian region (Wang and Rui 1990b) and the northward propagation is also a distinctive feature of ISO over the Indian Ocean and western Pacific during boreal summer (Yasunari 1979; Sikka and Gadgil 1980; Goswami 1998; Annamalai *et al.* 1999; Lawrence and Webster 2002; Hsu *et al.* 2004). The present paper reviews the propagation characteristics of the ISO and the associated mechanisms.

The intraseasonal oscillation is more-or-less confined in the tropical Indian Ocean and western Pacific during boreal winter, which is characterized by the eastward propagation with a time scale of 30–60 days. This ISO is often referred to as the MJO and the mechanisms governing the MJO have been proposed by a number of studies. In contrast, the summer ISO has a variety of propagation characteristics. It has the eastward and westward propagation signals, which affect the summer climate in the Asian Monsoon region. The northward propagation of the ISO also strongly affects the onset and withdrawal of Asian summer monsoon in both Indian and East

The Global Monsoon System: Research and Forecast (3rd Edition)
Edited by C. P. Chang et al.

Asian regions. The mechanism of the northward propagation was proposed by a number of studies (Wang and Xie 1997; Kemball-Cook and Wang 2001; Lawrence and Webster 2002; Hsu *et al.* 2004; Annamalai and Sperber 2005). The dynamics of the northward propagation, solely appearing in the boreal summer, may require a mechanism which interacts with the Monsoon climatological mean flow whose summer mean is distinctively different from the winter. In this sense, the theories related to the strong vertical shear of summer-mean zonal flow are plausible to explain the northward propagating ISO. Those theories include the vorticity titling mechanism (Jiang *et al.* 2004; Drabohlav and Wang 2005) and the cumulus momentum mixing (Kang *et al.* 2010).

The present paper is aimed to review previous papers, which described the propagation characteristics of observed ISO during boreal summer and the related mechanisms proposed. Also this paper suggests the most plausible mechanisms separately for the eastward-westward and northward propagations of ISO.

2. Observed Characteristics

The propagation speed of the MJO depends on the region. It is relatively slow in the warm western Pacific and relatively fast in the eastern equatorial Pacific where the sea surface temperature is relatively cold. The different MJO phase speeds are due to the regional differences in the moist static stability, which affects the Kelvin wave speed. The eastward propagation of ISO is mostly clearly seen along the equator. Off the equator in 10–30N, on the other hand, the westward propagating ISO signal appears particularly during boreal summer (Lau *et al.* 1988). They showed that this westward propagating ISO has relatively fast time scales (10–30 days) compared to that of the tropical ISO. Tanaka *et al.* (1992) found that the 20–25 day mode exists north of 13°N while the 30–60 day mode is active south of 13°N. Chen *et al.* (2000) identified the presence of two well-separated spectral peaks with periods around 30–60 days and 12–24 days in the South China Sea and the Western Pacific. Also, Murakami (1980) found 20–30-day perturbations propagating westward along 10°N–20°N in the western Pacific and northward over the South China Sea.

A distinctive characteristic of intraseasonal variations during boreal summer is the northward propagation in the Indian Ocean and western Pacific region in addition to the eastward propagation near the equator (Krishnamurti and Subrahmanyam 1982; Chen and Murakami 1988; Goswami *et al.* 1998; Annamalai *et al.* 1999; Kemball-Cook and Wang 2001; Lawrence and Webster 2002; Hsu *et al.* 2004). To distinguish the boreal summer intraseasonal oscillation from the wintertime MJO, Kemball-Cook and Wang (2001) began to call it as boreal summer intraseasonal oscillation (BSISO). The BSISO is closely related to the "active" and "break" period of the summer monsoon (Yasunari 1979; Annamalai and Slingo 2001). The propagation characteristics of the summer ISO in each grid points of the Asian Monsoon domain can be clearly seen in Fig. 1. As mentioned above, the summer ISO is dominated by eastward propagation in the equatorial region, with some westward signal appearing in the subtropics north of 20°N, and northward propagation clearly seen in the Bay of Bengal and the western Pacific off the equator.

3. Mechanisms for the Eastward and Westward Propagations

Although a number of theories have been proposed for the MJO, most of those mechanisms include the interaction between large-scale moisture dynamics and convection, which seems an essential dynamical process of the MJO (Wang 2005). In particular, the convectively coupled Kelvin wave proposed by Lau and Peng (1987) is able to explain several aspects of the observed

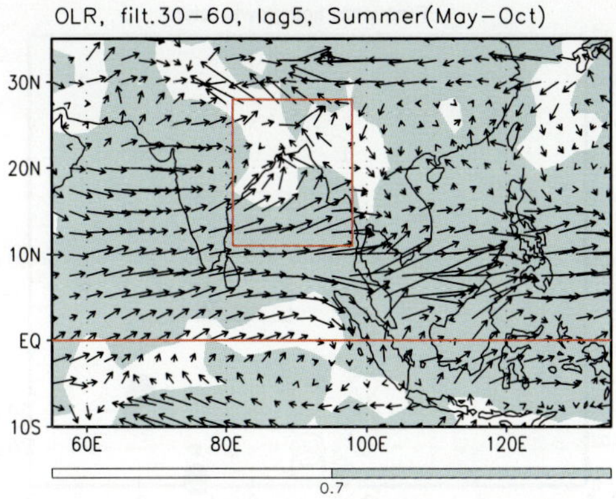

Fig. 1. Propagation tendency vectors of 30–60 day filtered OLR, obtained by the east/west and north/south ratio of 5 day lagged correlation coefficient. The 5 day lagged correlation is computed between each grid point (i.e. base point) and surrounding grid points with the 30–60 day OLR during boreal summer (May-Oct). The vector denotes the most probable direction of propagation at every base points within the 5 days. Shading represents the 5 day lagged autocorrelation coefficient greater than 0.7.

MJO properties (Wheeler and Kiladis 1999), such as the eastward propagating planetary-scale waves, although it is faster than that of the MJO. In the frictional wave-CISK theory, the so-called "moist coupled Kelvin-Rossby wave" theory, the Ekman pumping of moisture at the top of planetary boundary layer (PBL) provides additional moisture supply to the convection by a surface frictional convergence, which slows down the eastward propagation speed. This frictional convergence plays an essential role in coupling the moist Kelvin and Rossby wave modes (Wang and Rui 1990a). This theory has been supported by several observational studies (Hendon and Salby 1994; Maloney and Hartmann 1998; Matthews 2000) and modeling studies (Sperber *et al.* 1997; Waliser *et al.* 1999; Lee *et al.* 2003). Kang *et al.* (2013) revisited this moist coupled Kelvin-Rossby wave theory to understand the mechanism for the propagation characteristics of eastward and westward propagating ISO, in particular the slow eastward propagation of the ISO near the equator (MJO) and the relatively fast ISO signal propagating westward off the equator in the subtropical Monsoon region. They highlighted that the SST structure determines the propagation characteristics of ISO using various aqua-planet experiments similar to Maloney *et al.* (2010) and using a simple model with moisture dynamics.

Some of the content of Kang *et al.* (2013) is introduced in this review paper, since it provides a basic mechanism for both eastward and westward propagations of ISO. The governing equation of the simple model, that they used, is based on the linearized shallow water equations with moisture physics and a frictional surface boundary-layer, similar to those of Wang and Rui (1990a). It is a one and a half level model, which includes the first baroclinic mode of the free troposphere and a barotropic PBL mode. Interactive moisture physics is added in the model with convective heating resulting from the vertical moisture flux in the lower layer. In the model, the precipitation and the convective heating are induced by both the large-scale horizontal moisture convergence in the lower layer and the frictional convergence in the planetary boundary layer. A unique feature of this model is the control of SST for determining the atmospheric basic-state moisture field and the associated static stability, which is critical for determining the propagation characteristics of the ISO. The formula of the SST that they used is as follows.

$$SST = SST_0 \exp[-((y - y_0)/y_L)^2/2] \qquad (1)$$

where $SST_0 = 29.5°\text{C}$, y_0 is the SST center latitude, and y_L is the meridional e-folding scale of SST.

Firstly, they specified the SST center at the equator, $y_0 = 0$, and then the meridional SST scale, y_L, was changed to examine how the change of SST modifies the coupling of the Kelvin and Rossby waves. Figure 2a clearly shows that this model selects the long waves to become unstable and the most unstable wave has a wavenumber 1. The long-wave selection

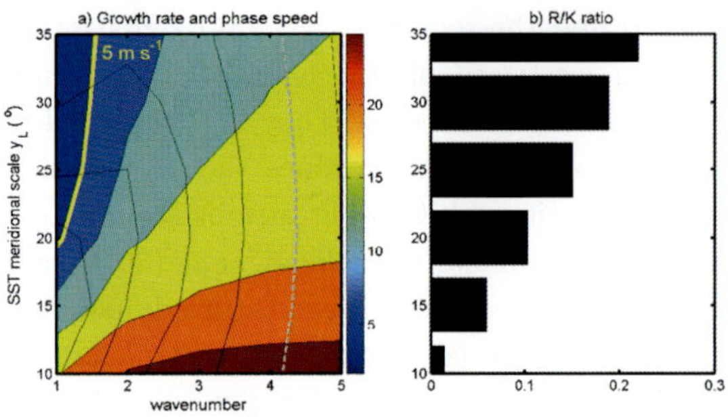

Fig. 2. (a) Growth rate (contour) and phase speed (Shading) as functions of the wavenumber and meridional SST scale y_L. Positive (negative) contours are denoted by solid (dashed) lines. (b) Ratios of the Rossby to Kelvin (R/K) component as functions of the meridional SST scale. Here, $y_0 = 0$. See more details in Kang *et al.* (2013).

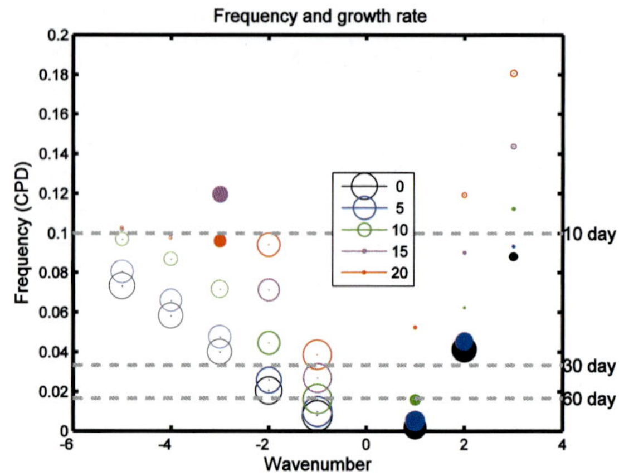

Fig. 3. Frequency and growth rate as a function of the wavenumber. Unstable and damped modes are represented by the dots and circles, respectively. The growth rates are proportional to the diameters, with respect to the maximum growth and damping rate of 0.13 day^{-1} and -0.6 day^{-1}, respectively. Results with different SST centers are denoted by different colors, as indicated in the figure. Here $y_L = 25°$. From Kang *et al.* (2013).

of instability is partly due to large surface friction and strong momentum diffusion of the short waves in the free atmosphere, indicating the importance of the Ekman pumping mechanism for the tropical intraseasonal variations. When the warm SST is confined near the equator, $y_L = 10°$, this model presents a fast eastward propagating mode, which has a phase speed of 20 m s^{-1} for wavenumber 1, and shorter waves propagate eastward even faster.

Figure 2b shows that for the narrow SST ($y_L = 10°$), the R/K ratio is very small and the model presents a strong eastward propagating Kelvin-type mode. When the SST becomes broader, this model presents stronger Rossby component (Fig. 2b), meanwhile, the phase speeds also decrease to the range of the observed MJO, about 5 m s^{-1}. It is noted that the present analytic model is based on a linearized shallow water equation in an equatorial beta plane, and therefore the equatorially trapped waves are emphasized and off-equatorial waves may not be well represented in the model. In spite of its simplicity, the model produces the most unstable modes, whose spatial patterns are similar to the observed counterparts, particularly in the tropics (see Kang *et al.* 2013). The results clearly show that the narrow SST only enhances the fast eastward Kelvin waves, while the broad SST provides enough off-equatorial moisture for the growth of the Rossby component, which couples with the Kelvin component and slows down the eastward propagating mode similar to the observed MJO.

The role of the SST center location on the coupled Kelvin-Rossby waves can be examined with Fig. 3. In the boreal summer time, the ITCZ shifts northward, and this ITCZ shift is associated with a shift of SST center off the equator to the NH subtropics, particularly in the western Pacific. The off-equatorial SST center can be represented by $y_0 > 0$ in Eq. (1). When the SST is centered at the equator, this model presents strong unstable eastward but damped westward modes. With a broad SST, the slow eastward mode has a considerable Rossby component, although it is still smaller than the eastward Kelvin component. When the SST center moves further to the north, the eastward modes become less unstable and eventually are damped as seen in Fig. 3, since the equatorial

SST is relatively cold and the moisture content is relatively small to support the growth of the Kelvin component. On the contrary, the westward modes obtain enough off-equatorial moisture for the Rossby component, and even become unstable when the SST center locates further to the north of 15°N. It is pointed out that these solutions are based on the meridional SST profile given by Eq. (1). In reality, during boreal summer, the SST at the equator is still very warm, slightly cooler than off-equatorial SST. In such a situation, the SST broadening effect in addition to the center shift affects the phase speed of the coupled Kelvin-Rossby waves, and therefore, the eastward phase speed is slowed down and the westward propagating waves can also be excited. When the SST is centered at 10°N, an off-equatorial rain band associated with the eastward modes (not shown) appears to be similar to the summer rain band of the intraseasonal oscillation during the boreal summer in India and the western Pacific region. When the SST center moves further to the north (such SST displacement occurs in boreal summer), this simple model presents unstable westward Rossby-type disturbances (see the details in Kang *et al.* 2013).

4. Mechanisms for the Northward Propagation

The mechanism of the northward propagation may not be explained in terms of the tropical wave dynamics. The dynamics of the northward propagation, solely appearing in the season of boreal summer, may require a mechanism which interacts with the Monsoon climatological mean flow whose summer mean is distinctively different from the winter. In this sense, the recent studies (Jiang *et al.* 2004; Annamalai and Sperber 2005; and Kang *et al.* 2010) are promising to explain the northward propagation during the summer. Those studies emphasize the role of mean easterly wind shear which is an important feature of summer Monsoon in South Asia and Indian Ocean region and the tropical Western Pacific. Jiang *et al.* (2004) demonstrated that the vorticity tilting term, in the presence of strong vertical wind shear, creates barotropic vorticity in the north of the convection and the vorticity induces frictional boundary-layer moisture convergence and therefore eventually the convection moves to the north. On the other hand, Kang *et al.* (2010) proposed the cumulus momentum transport (CMT) as an essential mechanism for the northward propagation of the ISO. The importance of the CMT on the northward movement of the BSISO can be inferred from the multi-model analysis of ISO, documented by Sperber and Annamalai (2008). They showed that the simulated ISOs are mostly weaker than the observed and that half the models fail to simulate the northward propagation. Kang *et al.* (2010) counted the models with and without the CMT among those used by Sperber and Annamalai (2008). As shown by Kang *et al.* (2010), the models with CMT (7 out of 8) simulate the northward propagation, but all of the models without CMT do not show the northward propagation signal. This result implies that the CMT is an important mechanism that should be included in GCMs to better simulate the northward propagation characteristics. Kang *et al.* (2010) also showed that a version of Seoul National University's ocean-atmosphere coupled GCM simulated northward propagation when the CMT was implemented, and the northward propagation disappeared when the CMT was disabled. And then, they used an axially symmetric shallow water model with a parameterized CMT to understand the role of CMT in the northward propagation of ISO.

It is noted that a large mean vertical shear exists in South Asian region during boreal summer with the west-southwesterly Monsoon flow in the lower troposphere and the strong easterlies in the upper-troposphere associated with the Tibetan anticyclonic circulation. In the mean state with a large easterly vertical shear, the convective mixing provides easterly momentum

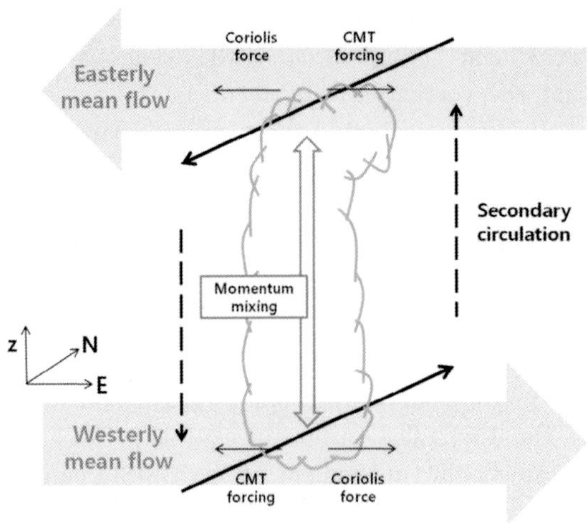

Fig. 4. Schematic diagram of the three-dimensional circulation and convection in the presence of the CMT mechanism under the mean condition of a large vertical shear in the zonal wind, represented by the large grey arrows.

in the lower layer and westerly momentum in the upper layer. Such zonal wind tendencies by the CMT should be balanced by a secondary meridional circulation with the Coriolis force (there is no zonal pressure gradient in the axially symmetric model). As a result, as seen in Fig. 4, the anomalous southerly and northerly winds are induced in the lower and upper layers, respectively. The anomalous meridional winds induce the lower-layer convergence and upper-level divergence to the north of the convection, resulting in the northward propagation of the convective system.

5. Summary

In the Asian Monsoon region, the surface boundary condition has a complicated spatial structure with land, ocean, and mountains. All of these components contribute to the detailed structure of the ISO. For simplicity, the broader SST can be considered as the tropical warm pool over the Indian Ocean and western Pacific, where the MJO has a strong eastward propagation component with a slow time scale, but the MJO speed becomes faster in the eastern Pacific, whose SST structure is similar to that of the narrower SST. It is expected that the monsoon resulting from a land-sea contrast also strongly influences the MJO particularly during a boreal summer, when the mean precipitation and large moisture shifted to the subtropical western Pacific and continental side. Those effects of monsoons, a warm pool, and the cold tongue of the tropical SST on the MJO could be a topic of future study.

It should also be mentioned that in Kang *et al.* (2013), the convection (precipitation) is assumed to be determined by a steady state balance between precipitation and the moisture convergence. However, by adding the tendency term of the moisture equation, Majda and Stechmann (2009) and Sobel and Maloney (2012) showed the existence of a moisture mode which mimics the observed MJO. In their studies, however, the governing equations of the circulation system, determining the moisture convergence, were crudely treated. It may be interesting to examine how the results of Kang *et al.* (2013) can be modified by considering a comprehensive moisture equation with parameterized precipitation processes. This subject will be a topic of future study.

Acknowledgments

This work was supported by the National Research Foundation of Korea (NRF) grant funded by the Korean government (MEST) (NRF-2012M1A2A2671775) and by the Brain Korea 21 Plus.

References

Annamalai, H., J. M. Slingo, K. R. Sperber, and K. Hodges, 1999: The mean evolution and variability of the Asian summer monsoon: Comparison of ECMWF and NCEP/NCAR reanalyses. *Mon. Wea. Rev.*, **127**, 1157–1186.

Annamalai, H. and J. Slingo, 2001: Active/break cycles: Diagnosis of the intraseasonal variability of the Asian summer monsoon, *Clim. Dyn.*, **18**, 85–102.

Annamalai, H. and K. R. Sperber, 2005: Regional heat sources and the active and break phases of boreal summer intraseasonal (30–50 day) variability. *J. Atmos. Sci.*, **62**, 2726–2748.

Chen, T.-C. and M. Murakami, 1988: The 30–50 day variation of convective activity over the western Pacific Ocean with emphasis on the northwestern region. *Mon. Wea. Rev.*, **116**, 892–906.

Chen, T.-C., M.-C. Yen, and S.-P. Weng, 2000: Interaction between the summer monsoons in East Asia and the South China Sea: Intraseasonal monsoon modes. *J. Atmos. Sci.*, **57**, 1373–1392.

Drbohlav, H. and B. Wang, 2005: Mechanism of the northward-propagating intraseasonal oscillation: Insights from a zonally symmetric model. *J. Climate*, **18**, 952–972.

Goswami, B., 1998: Interannual variations of Indian summer monsoon in a GCM: External conditions versus internal feedbacks. *J. Climate*, **11**, 501–522.

Jiang, X., T. Li, and B. Wang, 2004: Structures and mechanisms of the northward propagating boreal summer intraseasonal oscillation. *J. Climate*, **17**, 1022–1039.

Hendon, H. H. and M. L. Salby, 1994: The life cycle of the Madden-Julian Oscillation. *J. Atmos. Sci.*, **51**, 2225–2237.

Hsu, H., C. Weng, and C. Wu, 2004: Contrasting characteristics between the northward and eastward propagation of the intraseasonal oscillation during the boreal summer. *J. Climate*, **17**, 727–743.

Kang, I.-S., D. Kim, and J.-S. Kug, 2010: Mechanism for northward propagation of boreal summer intraseasonal oscillation: Convective momentum transport. *Geophys. Res. Lett.*, **37**, L24804.

Kang I.-S., L. Fei, M.-S. Ahn, Y.-M. Yang, and B. Wang, 2013: The role of SST structure in convectively coupled Kelvin–Rossby waves and its implication for MJO formation. *J. Climate*, **26**, 5915–5930.

Kemball-Cook, S. and B. Wang, 2001: Equatorial waves and air-sea interaction in the boreal summer intraseasonal oscillation. *J. Climate*, **14**, 2923–2942.

Krishnamurti, T. and D. Subrahmanyam, 1982: The 30–50 day mode at 850 mb during MONEX. *J. Atmos. Sci.*, **39**, 2088–2095.

Lau, K. M. and L. Peng, 1987: Origin of low-frequency (intraseasonal) oscillations in the tropical atmosphere. Part I: Basic theory. *J. Atmos. Sci.*, **44**, 950–972.

Lau, K. M., G. J. Yang, and S. H. Shen, 1988: Seasonal and intraseasonal climatology of summer monsoon rainfall over East-Asia. *Mon. Wea. Rev.*, **116**, 18–37.

Lawrence, D. M. and P. J. Webster, 2002: The boreal summer intraseasonal oscillation: Relationship between northward and eastward movement of convection. *J. Atmos. Sci.*, **59**, 1593–1606.

Lee, M.-I., I.-S. Kang, and B. E. Mapes, 2003: Impacts of cumulus convection parameterization on aqua-planet AGCM simulations of tropical intraseasonal variability. *J. Meteor. Soc. Japan*, **81**, 963–992.

Madden, R. A. and P. R. Julian, 1972: Description of global-scale circulation cells in the tropics with a 40–50 day period. *J. Atmos. Sci.*, **29**, 1109–1123.

Majda, A. J. and S. N. Stechmann, 2009: A simple dynamical model with features of convective momentum transport. *J. Atmos. Sci.*, **66**, 373–392.

Maloney, E. D. and D. L. Hartmann, 1998: Frictional moisture convergence in a composite life cycle of the Madden-Julian Oscillation. *J. Climate*, **11**, 2387–2403.

Maloney, E. D., A. H. Sobel, and W. M. Hannah, 2010: Intraseasonal variability in an aquaplanet general circulation model. *J. Adv. Model. Earth Syst.*, **2**, 5–24.

Matthews, A. J., 2000: Propagation mechanisms for the Madden-Julian Oscillation. *Quart. J. Roy. Meteor. Soc.*, **126**, 2637–2651.

Murakami, T., 1980: Temporal variation of satellite-observed outgoing longwave radiation. Part I: Long-period (15–30-day) observations. *Mon. Wea. Rev.*, **108**, 408–426.

Sikka, D. R. and S. Gadgil, 1980: On the maximum cloud zone and the ITCZ over Indian, longitudes during the Southwest Monsoon. *Mon. Wea. Rev.*, **108**, 1840–1853.

Sobel, A. and E. Maloney, 2012: An idealized semi-empirical framework for modeling the Madden-Julian Oscillation. *J. Atmos. Sci.*, **69**, 1691–1703.

Sperber, K. R., J. M. Slingo, P. M. Inness, and W. K. M. Lau, 1997: On the maintenance and initiation of the intraseasonal oscillation in the NCEP/NCAR reanalysis and in the GLA and UKMO AMIP simulations. *Clim. Dyn.*, **13**, 769–795.

Sperber, K. R. and H. Annamalai, 2008: Coupled model simulations of boreal summer intraseasonal (30–50 day) variability, Part 1: Systematic errors and caution on use of metrics. *Clim. Dyn.*, **31**, 345–372.

Tanaka, M., 1992: Intraseasonal oscillation and the onset and retreat dares of the summer monsoon over east, southeast Asia and the western Pacific region using GMS high cloud amount data. *J. Meteor. Soc. Japan*, **70**, 613–629.

Waliser, D. E., K. M. Lau, and J.-H. Kim, 1999: The influence of coupled sea surface temperatures on the Madden-Julian Oscillation: A model perturbation experiment. *J. Atmos. Sci.*, **56**, 333–358.

Wang, B. and H. Rui, 1990a: Dynamics of the coupled moist Kelvin–Rossby wave on an equatorial β-plane. *J. Atmos. Sci.*, **47**, 397–413.

Wang, B. and H. Rui, 1990b: Synoptic climatology of transient tropical intraseasonal convective anomalies: 1975–1985. *Meteor. Atmos. Phys.*, **44**, 43–61.

Wang, B. and X. Xie, 1997: A model for the boreal summer intraseasonal oscillation. *J. Atmos. Sci.*, **54**, 72–86.

Wang, B., 2005: Theory, Chapter 10 in *Tropical Intraseasonal Oscillation in the Atmosphere and Ocean*. Eds. William K.-M. Lau and D. E. Waliser. Praxis Publishing.

Wheeler, M. and G. N. Kiladis, 1999: Convectively coupled equatorial waves: Analysis of clouds and temperature in the wavenumber–frequency domain. *J. Atmos. Sci.*, **56**, 374–399.

Yasunari, T., 1979: Cloudiness fluctuations associated with the Northern Hemisphere summer monsoon. *J. Meteor. Soc. Japan*, **57**, 227–242.

[illegible] radiation dataset. Bull. Amer. Meteor. Soc., 77, [illegible]

[illegible] heating estimates derived from the TRMM [illegible]

[illegible]

Chapter 12

Predictability and Prediction of the Madden-Julian Oscillation: A Review on Progress and Current Status

June-Yi Lee

Research Center for Climate Sciences,
Pusan National University, Busan, 46241, Korea
juneyi@pusan.ac.kr
gmcl.pusan.ac.kr

Xiouhua Fu* and Bin Wang†

International Pacific Research Center and Department of Meteorology,
University of Hawaii, Honolulu, 96822, USA
**xfu@hawaii.edu*
†wangbin@hawaii.edu

The progress and current status of prediction activities on the Madden-Julian Oscillation (MJO) are reviewed. We have witnessed the significant progress in dynamical MJO prediction from 7-day limit using the (NCEP)-1 reanalysis vintage model during 1990s to about 30-day limit using the latest version of European Centre for Medium-Range Weather Forecasts (ECMWF) model during 2010s. Motivated by significant societal demands for reliable subseasonal prediction, several international efforts have been coordinated including real-time MJO and boreal summer intraseasonal oscillation (BSISO) prediction intercomparison by MJO Task Force and Working Group on Numerical Experimentation (WGNE), and Intraseasonal Variability Hindcast Experiment (ISVHE). Analysis of 12 climate models' hindcast participated in ISVHE showed that the MJO can be potentially predictable up to 35∼45 days and the multi-model ensemble has a useful skill for the Realtime Multivariate MJO (RMM) index up to 26∼28 days for 1989–2008 during boreal winter, demonstrating capability of current models for seamless weather-intraseasonal-seasonal prediction. It is also noted that the current MJO prediction depends on initial and target MJO phase and El Niño and Southern Oscillation (ENSO) phase. The dynamical models' common biases in simulating the MJO are slower eastward propagation, underestimation of MJO amplitude, and the Maritime Continent barrier. Key factors to further improve the MJO prediction include improvement of models' intrinsic MJO mode, better initialization, and proper representation of air-sea coupling.

1. Introduction

The Madden-Julian Oscillation (MJO) in honor of the discoverers (Madden and Jullian 1971, 1994) is the dominant intraseasonal variability in the global climate system with quasi-oscillating period of 30–60 days. The atmospheric circulation anomalies associated with the MJO have significant impacts on the local and remote aspects of weather and climate through upscale/downscale modulations and tropical-extratropical teleconnection (Donald *et al.* 2006). The MJO even affects El Niño and Southern Oscillation (ENSO) and its teleconnection (Kessler and Kleeman 2000; McPhaden *et al.* 2006; Hoell *et al.* 2014). It has been noted that the MJO represents an essential source of predictability at the subseasonal time scale (Lau and Waliser 2005; Waliser 2011; Zhang 2013; Zhang *et al.* 2013).

The Global Monsoon System: Research and Forecast (3rd Edition)
Edited by C. P. Chang et al.

The MJO convective anomalies initiate over the equatorial Africa and the western equatorial Indian Ocean (Wang and Rui 1990a) and propagate eastward together with the associated circulation system as a Kelvin-Rossby wave couplet (Wang and Rui 1990b, Hendon and Salby 1994) in involvement with multi-scale interactions (Majda and Biello 2004; Wang and Liu 2011). During boreal winter when the MJO convection mainly propagates eastward along the equator, there is significant modulation to Australian summer monsoon (Hendon and Liebmann 1990; MacBride *et al.* 1995) as well as the North Pacific-North America circulation (Higgins and Mo 1997; Moon *et al.* 2012), resulting in extreme precipitation in California (Jones 2000), Oregon and Washington (Bond and Vecchi 2003; Moon *et al.* 2012). During boreal summer, when the main centers of convection variability associated with the MJO are shifted away from the equator to off 10–20°N, there is significant modulation to Asian monsoon (Annamalai and Sperber 2005; Kajikawa and Yasunari 2005; Kikuchi *et al.* 2012; Lee *et al.* 2013; Lee *et al.* 2015), North American monsoon (Higgins and Shi 2001; Lorenz and Hartmann 2006; Neena *et al.* 2014b), and tropical cyclones (Maloney and Hartmannn 2000; Molinari and Vollaro 2000; Bessafi and Wheeler 2006; Klotzbach 2010).

Societal demands have been increasing for the extended forecasts with time scales longer than weeks. The recurrent nature of the MJO with a period of 30–60 days offers an opportunity to bridge the forecasting gap between medium-range weather forecast (~one week) and seasonal prediction (longer than one month) (e.g. Waliser 2005; Fu *et al.* 2008; Brunet *et al.* 2010; Hoskins 2012). Most global operational and research weather/climate models, however, still face a variety of challenges to realistically simulate and accurately predict the MJO (Lin *et al.* 2006; Vitart *et al.* 2007; Wang and Seo 2009; Gottschalck *et al.* 2010; Rashid *et al.* 2010; Fu *et al.* 2011; Weaver *et al.* 2011; and Matsueda and Endo 2011). Achieving a better understanding of the MJO dynamics, predictability, and its associated weather and climate impacts is of critical importance for extending weather forecast beyond its current limit and developing seamless suite of weather and climate prediction.

This chapter reviews not only the progress and the current status on the MJO prediction using both empirical and dynamical models but also the estimation of potential predictability of the MJO using various approaches suggested. Particular attention is also paid to pressing issues in improving the MJO prediction. There are five related review articles in this book. Kim and Maloney (Chapter 9) review the progress and current status of MJO simulation by general circulation models. Wheeler *et al.* (Chapter 10) review the progress of real-time forecasting of the MJO and boreal summer intraseasonal oscillation (BSISO) under international collaboration among operational centers. Tomoki and Sato (Chapter 14) address the status of the MJO prediction using a cloud-resolving coupled general circulation model. Zhang (Chapter 27) describes the CINDY/DYNAMO field campaign for studying the initiation of MJO and its prediction. Lastly, Klingaman *et al.* (Chapter 13) review details of diabatic processes in the MJO. Table 1 lists the acronyms of the institutions, models, programs, and projects mentioned in the text.

2. The Progress of MJO Prediction

During recent two decades, climate scientists have made substantial efforts on developing and improving the MJO prediction models empirically and dynamically. Although deficiencies still remain in global climate models, we have witnessed significant improvement in prediction of the MJO. The progress can be divided into three phases: Phase I as empirical era, Phase II as dynamical era, and Phase III as multi-model era.

Table 1. Acronym names of institutions, models, programs, and projects used in the text.

Acronym	Full names
ABOM	Australia Bureau of Meteorology
CCCma	Canadian Centre for Climate Modeling and Analysis
CFS	Coupled forecast system model
CFSR	Climate Forecast System Reanalysis
CINDY	Cooperative Indian Ocean Experiment on Intraseasonal Variability
CLIVAR	Variability and Predictability of the Ocean-Atmosphere System
CMCC	Coupled Model of Euro-Mediterranean Center on Climate Center
DYNAMO	Dynamics of the Madden-Julian Oscillation
EC	Environmental Canada
ECMWF	European Centre for Medium-Range Weather Forecasts
ERA	ECMWF re-analysis
GEM	Global Environmental Multiscale
GFDL	Geophysical Fluid Dynamics Laboratory
ISVHE	Intraseasonal Variability Hindcast Experiment
JMA	Japan Meteorological Agency
NCAR	National Center for Atmospheric Research
NCEP	National Centers for Environmental Prediction
NOAA	National Oceanic and Atmospheric Administration
POAMA	Predictive Ocean Atmosphere Model for Australia
RPN	Recherche en Prevision Numerique
SNU	Seoul National University
UHHCM	University of Hawaii Hybrid Coupled Model
WCRP	World Climate Research Program
WWRP	World Weather Research Program

2.1. *Phase I: Empirical era (until early 2000s)*

Until early 2000s, empirical forecasts had better performance than dynamical models on MJO prediction. Two review papers by Waliser (2005, 2006) well summarized the status of empirical and dynamical forecasts during the Phase I. At that time, several statistical forecast showed a useful skill up to 15 days for rainfall, OLR, and 200-hPa velocity potential field on intraseasonal time scales by using empirical approaches including principal oscillating patterns (von Storch and Xu 1990), singular value decomposition (Waliser *et al.* 1999), EOF and regression (Lo and Hendon 2000; Jones *et al.* 2004), Fourier filtering (Wheeler and Weickmann 2001), singular spectrum analysis (Mo 2001), multi-variable physically based Bayesian model (Webster and Hoyos 2004), linear inverse model (Newman *et al.* 2003), and vector autoregressive model (Maharaj and Wheeler 2005). On the other hand, dynamical forecasts of the MJO using the atmospheric-only model of the NCEP-1 reanalysis vintage showed a useful skill only up to about 7 days for boreal winter season (Hendon *et al.* 1999; Jones *et al.* 2000).

2.2. *Phase II: Dynamical era (until early 2010s)*

In the past decade, dynamical models have improved greatly in predicting the MJO and have been outperforming statistical schemes (e.g. Sperber and Waliser 2008; Goswami *et al.* 2011; Kang and Kim 2010; Vitart 2014; Kim *et al.* 2014a). Besides the advance in dynamical models at each modeling groups, international collaborations on developing uniform metrics in monitoring, evaluating, and predicting the MJO also contributed to the extensive MJO studies and then further facilitated the improvement on the MJO prediction. The MJO diagnostics and metrics were developed by MJO Working Group (MJOWG) to provide a means to quantitatively compare MJO forecast skill across the centers and to facilitate the development of a multi-model forecast of the MJO (CLIVAR MJO WG 2009; Kim *et al.* 2009; Gottschalck *et al.* 2010).

Since Wheeler and Hendon (2004) developed the 2-component Real-time Multivariate MJO (RMM) index (RMM1 and RMM2), it has been served as an objective measure of the MJO to compare MJO forecast skill among prediction models and widely used for real-time monitoring and evaluating forecast applications (e.g. Leroy and Wheeler 2008; Wheeler *et al.* 2009; Gottschalck *et al.* 2010; Rashid *et al.* 2011). It is defined by the first two principal component time series of the multivariate empirical orthogonal function (EOF) modes of the equatorial mean (between 15°S and 15°N) outgoing longwave radiation and zonal winds at 850 and 200 hPa. The equatorial symmetric nature of the RMM index makes it an excellent measure of the equatorial eastward propagating MJO. Since no temporal filtering is required, the index is extremely useful for real-time applications.

For the purpose of defining model forecast skill, Lin *et al.* (2008) suggested to use the bivariate correlation coefficient and root-mean square error (RMSE) between the observed and forecast two-component RMM index. A correlation skill of 0.5, which is approximately corresponding to a RMSE of $\sqrt{2}$, has been regarded as a threshold for a useful skill according to Rashid *et al.* (2011). Several atmosphere-ocean coupled models had the bivariate correlation skill exceeding 0.5 beyond 20 days. In particular, Vitart (2014) showed that the latest version of ECMWF model has the useful skill beyond 27 days. GFDL coupled model (CM) also exhibits the correlation skill of greater than 0.5 at a lead time of 27 days (Xiang *et al.* 2015). Table 2 summarizes performance of dynamical models' long-term *hindcast* in predicting the RMM index in terms of the day when the model's bivariate correlation skill reaches 0.5. Most of recent coupled models such as ECMWF Cy38r1, GFDL CM, GloSea5, NCEP CFSv2, POAMA 1.5b, SNU CGCM, show the useful bivariate correlation skill of 0.5 for the RMM index up to 20~27 days. Kang and Kim (2010) suggested that dynamical predictions now exceed statistical predictions for the RMM index, showing that the useful skill limit of statistical models is at 14~17 days and 8~12 days for the multi regression model and wavelet/SSA based model, respectively. Jiang *et al.* (2008) showed that a multivariate

Table 2. Summary of dynamical models' forecast ability for the MJO in terms of forecast lead day when a model's bivariate correlation skill for the 2-component RMM index drops below 0.5.

Model	Atmospheric initial condition	$\text{Day}_{0.5\ \text{skill}}$	Reference
CCCma GCM3	NCEP/NCAR R1	6 days	Lin *et al.* 2008
RPN GEM	NCEP/NCAR R2	10 days	Lin *et al.* 2008
ECMWF Cy32r3	ERA-40, operational analysis	23 days	Vitart and Molteni 2010
ECMWF Cy38r1	ERA-Interim	above 27 days	Vitart 2014; Kim *et al.* 2014b
GFDL CM	NCEP GFS	27 days	Xiang *et al.* 2015
GloSea5	ERA-Interim	15~20 days	MacLachlan *et al.* 2015
NCEP CFSv1	NCEP/NCAR R2	10~15 days	Seo *et al.* 2009
NCEP CFSv2	CFSR	20~21 days	Wang *et al.* 2014
POAMA 1.5b	ERA-40 relaxation	21 days	Rashid *et al.* 2011
SNU CGCM	NCEP/NCAR R2	20 days	Kang and Kim 2010

lag-regression statistical model has the correlation skill of greater than 0.5 for the RMM index at a lead time of 15 days.

Despite the significant societal and environmental demands for accurate prediction of MJO and intraseasonal variability, *operational* prediction of MJO is still limited and its achievement is seen as a great challenge faced by operational weather forecast centers. Zhang *et al.* (2013) showed that the majority of operational forecast models have the useful real-time forecast skill of 0.5 throughout the first 15 days, with several maintaining above the skill of 0.7 for all daily forecasts from 1 Sep 2011–31 Mar 2012.

2.3. *Phase III: International coordination and multi-model era (2010s)*

Several internationally coordinated activities were recently formed to meet the great challenges to better understanding and predicting the MJO. Zhang *et al.* (2013) summarizes the activities from the perspectives of numerical simulation, prediction, diagnostics, and field campaigns. The international collaborations include MJO Task Force (MJOTF), IntraSeasonal Variability Hindcast Experiment (ISVHE), Year of Tropical Convection (YOTC), MJO Vertical Structure and Physical Processes Project (Jiang *et al.* 2014), WWRP-WCRP Subseasonal to Seasonal prediction (S2S) project, CINDY/DYNAMO Field Campaign (Yoneyama *et al.* 2013), and the second phase of North American Multimodel Ensemble (Kirtman *et al.* 2014).

The WWRP-WCRP MJOTF was initially formed in 2010 following the earlier success of the US CLIVAR MJOWG (Wheeler *et al.* 2013, 2014). It aims to improve MJO representation in weather and climate models, and enabling more skillful predictions of the MJO and related phenomena. MJOTF and the Working Group on Numerical Experimentation (WGNE) have coordinated real-time forecast activities for the MJO hosted in NCEP Climate Prediction Center since 2010 and boreal summer intraseasonal oscillation (BSISO) hosted in APEC Climate Center since 2013 with participation of major operational modeling centers. Wheeler *et al.* (Chapter 10 in this book) describe details in the activities and the current status of real-time operational forecast skill for the MJO and BSISO.

While the Multi-Model Ensemble (MME) has proven to be an effective technique to deliver better seasonal prediction (e.g. Yoo and Kang 2005; Wang *et al.* 2009 and many others), it has not been demonstrated whether a MME can improve MJO forecasts. Therefore, the possible/expected benefit of an MME to the subseasonal problem (and in turn a more seamless prediction suite) has not been realized. Kang and Kim (2010) and Fu *et al.* (2013) demonstrated potential for improving the MJO prediction through the MME approach. The former (the latter) developed a hybrid statistical-dynamical model (MME with CFSv2 and UH coupled model) that extends a forecast limit up to 3 days (1 week) compared to individual models. A major stumbling block to such an effort has been the lack of a quality long-term multi-model hindcast dataset to develop effective MME schemes suitable for MME prediction.

Motivated by significant societal demands for reliable subseasonal prediction, the coordinated ISVHE was launched in 2009 (Mani *et al.* 2014a, b; Lee *et al.* 2014) as a second phase of Climate Prediction and its Application to Society (CliPAS) project (Wang *et al.* 2009; Lee *et al.* 2010). It has been supported by APEC Climate Center, and NOAA Climate Test Bed, and endorsed by the CLIVAR Asian-Australian Monsoon Panel, YOTC/MJOTF, and the Scientific Steering Committee of Asian Monsoon Years (2007–2012). It aims to determine intraseasonal predictability and realized prediction skill in the current dynamical models. The ISVHE project is the first attempt to produce a long-term hindcast dataset that specifically targets the needs and themes associated with intraseasonal prediction research. The objectives of the ISVHE are to:

(1) Better understand the physical basis for intraseasonal prediction and estimate the potential and practical predictability of intraseasonal variability, including the MJO, in a multimodel framework;
(2) Develop optimal strategies for an MME ISV prediction system, including optimal initialization schemes and quantification of prediction skill with forecast metrics under operational conditions;
(3) Discover new physical mechanisms associated with intraseasonal variability that cannot be obtained from analyses of a single model;
(4) Identify model deficiencies in prediction of intraseasonal variability in a multimodel framework and suggest ways to improve models.

Currently, hindcast datasets produced by 12 models have been collected from six operational centers and six research groups covering the period from January 1989 to December 2008 initiated either every 10 days in each calendar month or every first day of each month for at least a 45-day integration. Neena *et al.* (2014a) and Lee *et al.* (2015) introduced details for the hindcast experiment, participating models, and hindcast specification.

3. The Current Status of MJO Prediction in ISVHE

The MJO prediction using individual models and their MME are evaluated using the ISVHE data. Table 3 shows the bivariate correlation skill for the 2-component RMM index at the first forecast day and the threshold day when the correlation skill drops at 0.5 in individual models. The integration period is 1999–2010, 2005–2008, and 1989–2008 for CFSv2 initiated every 5 days, UHHCM initiated every 5 days, and other models initiated first day of every month, respectively, from November to April. Individual model performance shows that the skill at the first forecast day ranges from 0.76 to 0.93 and the forecast limit varies from 8 to 26 days. It is noted that model improvement positively affects the MJO forecast skill. The forecast limit is extended from 18 days in POAMA 1.5 to 25 days POAMA 2.4 and from 12 days in CFSv1 to 23 days in CFSv2. Zhang and v. d. Dool (2012) also showed improvement for the MJO prediction of CFS v2 from CFS v1.

Table 3. The bivariate correlation skill for the 2-component RMM index at day 1 and the last day when the skill reaches at 0.5 in dynamical model predictions participated in the ISVHE.

Institute	Model	1-day skill	$\text{Day}_{0.5\,\text{skill}}$
ABOM	POAMA 1.5	0.9	18 days
	POAMA 2.4	0.93	25 days
CMCC	CMCC	0.84	21 days
EC	GEM	0.92	18 days
ECMWF	ECMWF	0.92	26 days
GFDL	CM2	0.76	8 days
JMA	JMA CGCM	0.92	21 days
NCEP	CFSv1	0.91	12 days
	CFSv2	0.91	23 days
SNU	SNU CGCM	0.91	22 days
UH	UHHCM	0.9	22 says

The MME is constructed using eight models initiated from the first day of month from November to April for the 20 years of 1989–2008. The total forecast sample size is 120. The MME has the useful forecast skill of 0.5 up to 27 days for the 2-component RMM index and 27 and 26 days for the RMM1 and RMM2, respectively (Fig. 1). The MME is better than individual models for most of forecast lead time. However, ECMWF model beats the MME in some lead time because the inter-model spread is considerable, thus suggesting necessity for developing optimal MME method to further extend the forecast limit for the MJO. Nonetheless, the

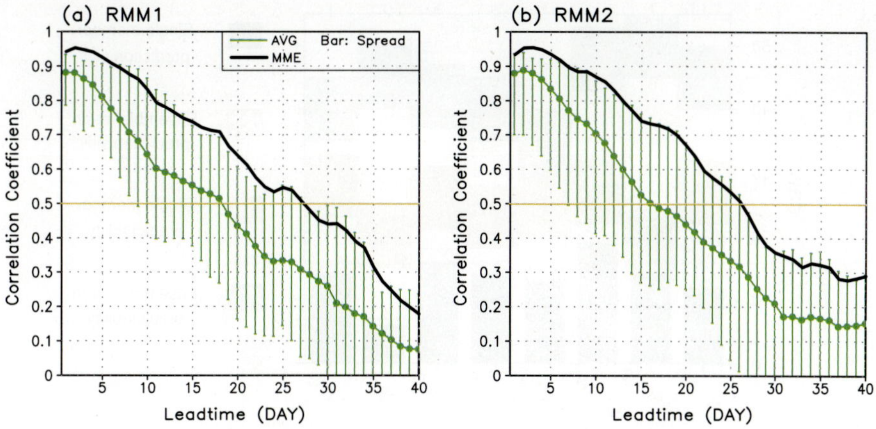

Fig. 1. The Anomaly Correlation Coefficient skill for the (a) RMM1 and (b) RMM2 as a function of forecast lead time predicted by 12 climate models and their MME participated in ISVHE initiated from the first day of month from November to April for the 21 years of 1989–2008 modified from Zhang *et al.* (2013).

MME efficiency defined as the distance from an averaged skill of individual model to a MME skill (Wang *et al.* 2009) is significantly large in the case of the MJO. For example, the useful forecast limit for the RMM1 (RMM2) is much extended from 18 (17) days in the average of individual models to 27 (26) days in the MME. It is further noted that the MJO prediction skill depends on the initial phase of MJO and ENSO phase. Many coupled models still suffer the prediction barrier over the Maritime Continent. Thus, coupled models tend to have better skill with initial MJO phase of 2 and 3 (Vitart 2013; Wang *et al.* 2013; Kim *et al.* 2014). The MJO forecast skill tends to be higher during La Nina than El Nino (Seo 2009) attributable to drier condition over Maritime Continent and faster MJO propagation during El Nino than La Nina (Pohl and Matthews 2007).

4. Predictability of the MJO

Considerable effort has been put forth to measure the MJO predictability. Potential predictability of the MJO has been estimated using the perfect model approach using dynamical models' ensemble simulation (Waliser *et al.* 2003; Reichler and Roads 2005; Pegion and Kirtman 2008; Neena *et al.* 2014) and the nonlinear local Lyapunov exponent (NLLE) method based on observation.

In the perfect model approach, there are two approaches for measuring the predictability of the MJO in hindcast. In the first method, so-called the single-member estimate, one of any given hindcast ensemble member are considered as the control forecast and those from every other ensemble member are considered as perturbed forecast (Waliser *et al.* 2003). In the second method, so-called ensemble estimate, the error growth is estimated for a single ensemble member (control) and the ensemble mean over all the other ensemble members (perturbed) (Pegion and Kirtman 2008).

Neena *et al.* (2014a) measured the MJO predictability based on the two perfect model approaches using the ISVHE hindcast. Figure 2 shows two MJO predictability estimates against respective MJO prediction skill estimates (single member and ensemble mean) in eight models. On average, the MJO predictability limit is 20$\sim$30 days and 35$\sim$45 days in single-member and ensemble mean estimate, respectively. It is noted that the ensemble-mean based predictability is higher than the single-member based predictability in all models. Neena *et al.*

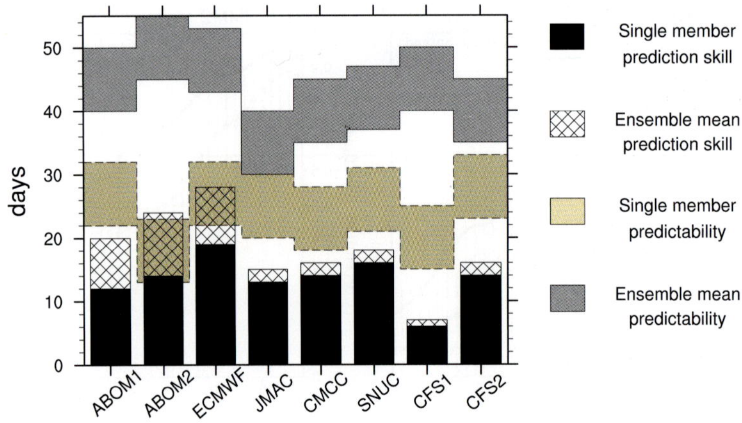

Fig. 2. The single member prediction skill (black bar) and ensemble mean prediction skill (hatched bar) estimates for MJO for the eight models are shown along with their respective, single member (wheat shaded area) and ensemble mean (grey shaded area) estimates of MJO predictability (+/5 range) from Neena *et al.* (2014).

(2014a) further showed that the predictability with the initial MJO phase of 2 and 3 is higher than that with other phases in most models and predictability for secondary MJO events is higher than that for primary events.

Figure 2 showed that the predictability estimate based on the perfect model approach strongly depends on model being used. The estimation is 44~54 days by ABOM2 but 30~40 days by JMAC. To avoid the model dependency, Ding *et al.* (2010) estimated the MJO predictability based on the NLLE using observational data only. The general idea of the algorithm is to find local analogs of evolution pattern, so-called local dynamical analog, from observational time series. The NLLE is estimated by calculating the mean exponential divergence rate of initially analogous points. The predictability can be quantitatively determined by investigating the average growth of initial errors derived from the NLLE. The theoretical potential predictability by NLLE exceeds 35 days.

5. Discussion

It is encouragingly to see that the MJO can be predictable up to 45 days and the state-of-the-art climate models have the useful skill of 0.5 for the two-component RMM index up to 26~28 days during boreal winter. However, most global operational and research weather/climate models still face a variety of challenges to accurately predict the MJO. Key factors to improve the MJO prediction include improvement of models' intrinsic MJO mode, initialization, and proper representation of air-sea coupling. The dynamical models' common biases are slower eastward propagation, underestimation of amplitude, and the Maritime Continent barrier. It is also noted that the current MJO prediction depends on initial and target MJO phase and on ENSO phase.

Physical basis for the intraseasonal prediction lies in both atmospheric initial conditions and external boundary forcings in the state of the ocean, land, and sea ice. Fu *et al.* (2011) demonstrated that the coupled model prediction for the MJO is sensitive to atmospheric initial condition and the ISO signal-recovered initial condition tends to improve the MJO prediction.

Air-sea coupling plays an important role on initiation, strength, and propagation of the MJO as well as quadrature precipitation-SST phase relationship (Fu *et al.* 2007, 2008, 2013; Woolnough *et al.* 2007; Pegion and Kirtman 2008). Although substantial observational and modeling evidence points out the importance of interactive air-sea coupling on MJO dynamics, our knowledge on the degree of the impacts of the coupling (Watterson 2002; Zhang *et al.* 2006; Hsu and Li 2012), on the feedback processes (Wang and Xie 1998; Marshall *et al.* 2008; Fu *et al.* 2008), and our capability of faithfully representing these processes in the state-of-the-art coupled models are still very limited (Lin *et al.* 2006; Wu *et al.* 2008; Hung *et al.* 2013). Fu *et al.* (2013) demonstrated that air-sea coupling extends the MJO skill by about 1 week and atmosphere-only forecast driven by forecasted daily SST have a similar skill as the coupled forecast during the DYNAMO/CINDY period from October 2011 to March 2012.

Continuous improvement in prediction models is one of key to improve the MJO prediction. Vitart (2014) showed evolution of ECMWF operational sub-seasonal forecast skill for the RMM index from April 2002 to March 2012 using a common set of hindcast for 1995–2001. The MJO prediction has improved significantly from 2002 (22 days) to 2012 (30 days) with an average gain of about 1 day of prediction skill per year. It is demonstrated that the amplitude of the MJO in ECMWF system has become more realistic, although the model still tends to have weaker MJOs than observation.

Most coupled models tend to have slower propagation and weaker amplitude of the MJO than observation that leads to degrade the MJO forecast skill especially at long lead (Janowiak *et al.* 2010; Rashid *et al.* 2010; Matsueda and Endo 2011; Fu *et al.* 2013; Vitart 2014; Wang *et al.* 2014). Rashid *et al.* (2010) showed that phase speed error in POAMA 1.5 tends to be increased as forecast lead time increases because the model has slower MJO propagation intrinsically. Viart (2014) showed that the latest version of ECMWF operational system still has MJO amplitude and phase speed error. Thus, it is important for model to have realistic MJO propagation and amplitude in order to improve its prediction.

We have observed that understanding and prediction of the MJO have been significantly advanced during the last decade. Although the skill level achieved using the state-of-the-art coupled models is not sufficient yet for practical application, it clearly demonstrates potential for extending forecast beyond its current weather limit and developing seamless suite of weather and climate prediction.

Acknowledgments

This work was supported by the National Research Foundation of Korea (NRF) through a Global Research Laboratory (GRL) Grant (MEST 2011-0021927) and APEC Climate Center. We thank all participating modeling groups in Intraseasonal Hindcast Experiments.

References

Annamalai, H. and K. R. Sperber, 2006: Regional heat sources and the active and break phases of boreal summer intraseasonal (30–50 day) variability. *J. Atmos. Sci.*, **62**, 2725–2748.

Bessafi, M. and M. C. Wheeler, 2006: Modulation of south Indian Ocean tropical cyclones by the Madden-Julian Oscillation and convectively coupled equatorial waves. *Mon. Wea. Rev.*, **134**, 638–656.

Bond, N. A. and G. A. Vecchi, 2003: The influence of the Madden-Julian Oscillation on precipitation in Oregon and Washington. *Wea. Forecasting*, **18**, 600–613.

Donald, A., H. Meinke, B. Power, A. de H. N. Maia, M. C. Wheeler, N. White, R. C. Stone, and J. Ribbe, 2006: Near-global impact of the Madden-Julian Oscillation on rainfall. *Geophys. Res. Lett.*, **33**, L09704.

Fu, X., J.-Y. Lee, P.-C. Hsu, H. Taniguchi, B. Wang, W. Wang, and S. Weaver, 2013: Multi-model MJO forecasting during DYNAMO/CINDY period. *Clim. Dyn.*, **41**, 1067–1081.

Fu, X., B. Wang, J.-Y. Lee, W. Q. Wang, and L. Gao, 2011: Sensitivity of dynamical intraseasonal prediction skills to different initial conditions. *Mon. Wea. Rev.*, **139**, 2572–2592.

Fu, X., B. Wang, D. E. Waliser, and L. Tao, 2007: Impact of atmosphere-ocean interaction on the predictability of monsoon intraseasonal oscillation. *J. Atmos. Sci.*, **64**, 157–174.

Fu, X., B. Yang, Q. Bao, and B. Wang, 2008: Sea surface temperature feedback extends the predictability of tropical intraseasonal oscillation. *Mon. Wea. Rev.*, **136**, 577–597.

Gottschalck, J., M. C. Wheeler *et al.*, 2010: A framework for assessing operational Madden-Julian Oscillation forecasts: A CLIVAR MJO working group project. *Bull. Am. Meteorol. Soc.*, **91**, 1247–1258.

Hendon, H. H. and B. Liebmann, 1990: A composite study of the onset of the Australian summer monsoon. *J. Atmos. Sci.*, **48**, 2909–2923.

Hendon, H. H., B. Liebmann, M. Newman, J. D. Glick, and J. Schemm, 1999: Medium range forecast errors associated with active episodes of the MJO. *Mon. Wea. Rev.*, **128**, 69–86.

Hendon, H. H. and M. L. Salby, 1994: The life cycle of the Madden-Julian Oscillation. *J. Atmos. Sci.*, **51**, 2225–2237.

Higgins, R. W. and K. C. Mo, 1997: Persistent North Pacific Circulation Anomalies and the tropical intraseasonal oscillation. *J. Climate*, **10**, 223–244.

Higgins, W. and W. Shi, 2001: Intercomparison of the principal modes of interannual and intraseasonal variability of the North American monsoon system. *J. Climate*, **14**, 403–417.

Hoell, A., M. Barlow, M. C. Wheeler, and C. Funk, 2014: Disruptions of El Nino-Southern Oscillation teleconnections by the Madden-Julian Oscillation. *Geophys. Res. Lett.*, **41**(3), 998–1004.

Hsu, P.-C. and T. Li, 2012: Role of the boundary layer moisture asymmetry in causing the eastward propagation of the Madden-Julian Oscillation. *J. Climate*, **25**, 4914–4931.

Hung, M., J. Lin, W. Q. Wang, D. Kim, T. Shinoda, and S. Weaver, 2013: MJO and convectively coupled equatorial waves simulated by CMIP5 climate models. *J. Climate*, **26**, 6185–6214.

Jiang, X., D. E. Waliser, M. C. Wheeler *et al.*, 2008: Assessing the skill of an all-season statistical forecast model for the Madden-Julian Oscillation. *Mon. Wea. Rev.*, **136**, 1940–1596.

Jiang, X., D. E. Waliser, P. K. Xavier *et al.*, 2014: Vertical structure and physical processes of the Madden-Julian Oscillation: Exploring key model physics in climate simulations. *J. Geophys. Res.*, submitted.

Jones, C., 2000: Occurrence of extreme precipitation events in California and relationships with the Madden-Julian Oscillation. *J. Climate*, **13**, 3576–3587.

Jones, C., L. M. Carvalho, R. W. Higgins *et al.*, 2004: A statistical forecast model of tropical intraseasonal convective anomalies. *J. Climate*, 17, 2078–2095.

Jones, C., D. E. Waliser, J.-K. E. Schemm, and W. K.-M. Lau, 2000: Prediction skill of the Madden-Julian Oscillation in dynamical extended range forecasts. *Clim. Dyn.*, **16**, 273–289.

Kajikawa, Y. and T. Yasunari, 2005: Interannual variability of the 10–25- and 30–60-day variation over the South China Sea during boreal summer. *Geophys. Res. Lett.*, **32**, L04710.

Kang, I.-S. and H.-M. Kim, 2010: Assessment of MJO predictability for boreal winter with various statistical and dynamical models. *J. Climate*, **23**, 2368–2378.

Kessler, W. S. and R. Kleeman, 2000: Rectification of the MJO into the ENSO cycle. *J. Climate*, **13**, 3560–3575.

Kikuchi, K., B. Wang, and Y. Kajikawa, 2012: Bimodel representation of the tropical intraseasonal oscillation. *Clim. Dyn.*, **38**, 1989–2000.

Kim, D. *et al.* 2009: Application of MJO simulation diagnostics to climate models. *J. Climate*, **22**, 6413–6436.

Kim, H.-M., J. Belanger, E. Blake *et al.*, 2014a: Chapter 5.3: Beyond synoptic timescale/ Intraseasonal timescales, Documentation, *8th International Workshop on Tropical Cyclone*, WMO.

Kim, H.-M., P. J. Webster, V. E. Toma, and D. Kim, 2014b: Predictability and prediction skill of the MJO in two operational forecasting systems. *J. Climate*, **27**(14), 5364–5378.

Kirtman, B. *et al.*, 2014: The North American multimodel ensemble: Phase-1 seasonal-to-interannual prediction; Phase-2 toward developing intraseasonal prediction. *Bull. Am. Meteorol. Soc.*, **95**, 585–601.

Klotzbach, P. J., 2010: On the Madden-Julian Oscillation-Atlantic hurricane relationship. *J. Climate*, **23**, 282–293.

Lau, W. K. M. and D. E. Waliser (Eds.), 2005: *Intraseasonal Variability of the Atmosphere-Ocean Climate System*, 474 pp., Springer, Heidelberg, Germany.

Lee, J.-Y., B. Wang, M. C. Wheeler, X. Fu, D. E. Waliser, and I.-S. Kang, 2013: Real-time multivariate indices for the boreal summer intraseasonal oscillation over the Asian summer monsoon region. *Clim. Dyn.*, **40**, 493–509.

Lee, J.-Y., B. Wang, I.-S. Kang, J. Shukla *et al.*, 2010: How are seasonal prediction skills related to models' performance on mean state and annual cycle? *Clim. Dyn.*, **35**, 267–283.

Lee, S.-S., B. Wang, D. E. Waliser, J. M. Neena, and J.-Y. Lee, 2015: Predictability and prediction skill of the boreal summer intraseasonal oscillation in the Intraseasonal Variability Hindcast Experiment. *Clim. Dyn.*, **45**, 2123–2135.

Leroy, A. and M. C. Wheeler, 2008: Statistical prediction of weekly tropical cyclone activity in the Southern Hemisphere. *Mon. Wea. Rev.*, **136**, 3637–3654.

Lin, H., G. Brunet, and J. Derome, 2008: Forecast skill of the Madden-Julian Oscillation in two Canadian atmospheric models. *Mon. Wea. Rev.*, **136**, 4130–4149.

Lin, J. L. *et al.* 2006: Tropical intraseasonal variability in 14 IPCC AR4 climate models. Part I: Convective signals. *J. Climate*, **19**, 2665–2690.

Lo, F. and H. H. Hendon, 2000: Empirical extended-range prediction of the Madden-Julian Oscillation. *Mon. Wea. Rev.*, **128**, 2528–2543.

Lorenz, D. J. and D. L. Hartmann, 2006: The effect of the MJO on the North American monsoon. *J. Climate*, **19**, 333–343.

MacLachlan, C., A. Arribas, K. A. Peterson *et al.*, 2015: Global seasonal forecast system version 5 (GloSea5): A high-resolution seasonal forecast system. *Quart. J. R. Meteorol. Soc.*, **141**, 1072–1084.

Madden, R. A. and P. R. Julian, 1971: Detection of a 40–50 day oscillation in the tropics. *J. Atmos. Sci.*, **43**, 3138–3158.

Madden, R. A. and P. R. Julian, 1994: Observations of the 40–50-Day Tropical Oscillation — a Review. *Mon. Wea. Rev.*, **122**, 814–837.

Maharaj, E. A. and M. C. Wheeler, 2005: Forecasting an index of the Madden-Oscillation. *Int. J. Climatol.*, **25**, 1611–1618.

Majda, A. J. and J. A. Biello, 2004: A multiscale model for the intraseasonal oscillation. *Proc. Natl. Acad. Sci.*, **101**, 4736–4741.

Maloney, E. D. and D. L. Hartmann, 2000: Modulation of eastern North Pacific hurricanes by the Madden-Julian Oscillation. *J. Climate*, **13**, 1451–1460.

Mani, N. J., J.-Y. Lee, D. Waliser, B. Wang, and X. Jiang, 2014a: Predictability of the Madden Julian Oscillation in the Intraseasonal Variability Hindcast Experiment (ISVHE). *J. Climate*, **27**, 4531–4543.

Mani, N. J., X. Jiang, D. E. Waliser, J.-Y. Lee, and B. Wang, 2014b: Eastern Pacific intraseasonal variability: A predictability perspective. *J. Climate*, **27**, 8869–8883.

Marshall, A. G., O. Alves, and H. H. Hendon, 2008: An enhanced moisture convergence-evaporation feedback mechanism for MJO air-sea interaction. *J. Atmos. Sci.*, **65**, 970–986.

Matsueda, M. and H. Endo, 2011: Verification of medium-range MJO forecasts with TIGGE. *Geophys. Res. Lett.*, **38**, L11801.

McBride, J. L., N. E. Davidson, K. Puri, and G. C. Tyrell, 1995: The flow during TOGA COARE as diagnosed by the BMRC tropical analysis and prediction system. *Mon. Wea. Rev.*, **123**, 717–736.

McPhaden, M. J., X. Zhang, H. H. Hendon, and M. Wheeler, 2006: Large scale dynamics and MJO forcing of ENSO variability. *Geophys. Res. Lett.*, **33**, L16702.

Mo, K. C., 2001: Adaptive filtering and prediction of intraseasonal oscillations. *Mon. Wea. Rev.*, **129**, 802–817.

Molinary, J. and D. Vollaro, 2000: Planetary- and synoptic-scale influences on eastern Pacific tropical cyclogenesis. *Mon. Wea. Rev.*, **128**, 3296–3307.

Moon, J.-Y., B. Wang, and K.-J. Ha, 2012: Modulation on 2009/10 winter snowstorms in the United States. *J. Climate*, **25**, 978–991.

Newman, M., P. D. Sardeshmukh, C. R. Winkler, and J. S. Whitaker, 2003: A study of subseasonal predictability. *Mon. Wea. Rev.*, **131**, 1715–1732.

Palmer, T. N. *et al.*, 2004: Development of a European multi-model ensemble system for seasonal to interannual prediction (DEMETER). *Bull. Am. Meteorol. Soc.*, **85**, 853–872.

Pegion, K. and B. P. Kirtman, 2008: The impact of air-sea interactions on the predictability of the Tropical Intraseasonal Oscillation. *J. Climate*, **21**, 5870–5886.

Rashid, H. A., H. H. Hendon, M. C. Wheeler, and O. Alves, 2011: Prediction of the Madden-Julian Oscillation with the POAMA dynamical prediction system. *Clim. Dyn.*, **36**, 649–661.

Reichler, T. and J. O. Roads, 2005: Long-range predictability in the tropics. Part II: 30–60-day variability. *J. Climate*, **18**, 634–650.

Slingo, J. M., P. M. Inness, and K. R. Sperber, 2005: Modeling, in Intraseasonal Variability of the Atmosphere-Ocean Climate System, edited by W. K. M. Lau and D. E. Waliser, Springer, pp. 361–388.

Sperber, K., J. Slingo, and P. Inness, 2011: Modeling intraseasonal variability, in *Intraseasonal Variability of the Atmosphere-Ocean Climate System*, 2nd ed., edited by W. K. M. Lau and D. E. Waliser, Springer, p. 399.

Sperber, K. R. and D. E. Waliser, 2008: New approaches to understanding, simulating, and forecasting the Madden-Julian Oscillation, *Bull. Am. Meteor. Soc.*, **89**, 1917–1920.

Vitart, F., 2014: Evolution of ECMWF sub-seasonal forecast skill scores. *Quart. J. Roy. Meteor. Soc.*, **140**, 1889–1899.

Vitart, F. and F. Molteni, 2010: Simulation of the Madden-Julian Oscillation and its teleconnections in the ECMWF forecast system. *Quart. J. Roy. Meteor. Soc.*, **136**, 842–855.

Von Storch, H. and J. Xu, 1990: Principal oscillation pattern analysis of the 30- to 60-day oscillation in the tropical troposphere. *Clim. Dyn.*, **4**, 175–190.

Waliser, D. E., C. Jones, J. K. E. Schemm, and N. E. Graham, 1999: A statistical extended-range tropical forecast model based on the slow evolution of the Madden-Julian Oscillation. *J. Climate*, **12**, 1918–1939.

Waliser, D. E., 2005: Predictability and forecasting. *Intraseasonal Variability in the Atmosphere-Ocean Climate System*, Chap. 12, edited by W. K. M. Lau and D. E. Waliser, Springer, pp. 389–423.

Waliser, D. E., 2006: Predictability of tropical intraseasonal variability. *Predictability of Weather and Climate*, Chap. 11. T. N. Palmer and R. Hagedorn, Eds., Cambridge Univ. Press, pp. 275–305.

Waliser, D. E., 2011: Predictability and Forecasting. Chapter 12, *Intraseasonal Variability of the Atmosphere-Ocean Climate System*. 2nd ed., edited by W. K. M. Lau and D. E. Waliser, Springer, p. 613.

Waliser, D. E., K. M. Lau, W. Stern, and C. Jones, 2003: Potential predictability of the Madden-Julian Oscillation. *Bull. Amer. Meteor. Soc.*, **84**, 33–50.

Wang, B. and F. Liu, 2011: A model for scale interaction in the Madden-Julian Oscillation. *J. Atmos. Sci.*, **68**, 2524–2536.

Wang, B. and H. Rui, 1990a: Synoptic climatology of transient tropical intraseasonal convection anomalies: 1975–1985. *Meteor. Atmos. Phys.*, **44**, 43–61.

Wang, B. and H. Rui, 1990b: Dynamics of coupled moist Kelvin-Rossby waves on an equatorial beta-plane. *J. Atmos. Sci.*, **47**, 397–413.

Wang, B., J.-Y. Lee, J. Shukla, I.-S. Kang, C.-K. Park *et al.*, 2009: Advance and prospectus of seasonal prediction: Assessment of APCC/CliPAS 14-model ensemble retrospective seasonal prediction (1980–2004). *Clim. Dyn.*, **33**, 93–117.

Wang, B. and X. Xie, 1998: Coupled modes of the warm pool climate system. Part I: The role of air-sea interaction in maintaining Madden-Julian Oscillation. *J. Climate*, **11**, 2116–2135.

Wang, W. Q., M.-P. Hung, S. J. Weaver, A. Kumar, and X. H. Fu, 2013: MJO prediction in the NCEP climate forecast system version 2. *Clim. Dyn.*, **42**, 2509–2520.

Watterson, I. G., 2002: The sensitivity of subannual and intraseasonal tropical variability to model ocean mixed layer depth. *J. Geophys. Res.*, **107**, D24020.

Weaver, S. J., W. Q. Wang, M. Y. Chen, and A. Kumar, 2011: Representation of MJO variability in the NCEP climate forecast system. *J. Climate*, **24**, 4676–4694.

Webster, P. J. and C. Hoyos, 2004: Forecasting monsoon rainfall and river discharge variability on 20–25 day time scales. *Bull. Amer. Meteorol. Soc.*, **85**, 1745–1765.

Wheeler, M. C., E. Maloney, and MJO Task Force, 2013: Madden-Julian Oscillation (MJO) Task Force: A joint effort of the climate and weather communities, CLIVAR Exchanges (61), 9–12.

Wheeler, M. C. and H. H. Hendon, 2004: An all-season real-time multivariate MJO index: Development of an index for monitoring and prediction. *Mon. Wea. Rev.*, **132**, 1917–1932.

Wheeler, M. C., H. H. Hendon *et al.* 2009: Impacts of the Madden-Julian Oscillation on Australian rainfall and circulation. *J. Climate*, **22**, 1482–1498.

Wheeler, M. C. and K. M. Weickmann, 2001: Real-time monitoring and prediction of modes of coherent synoptic to intraseasonal tropical variability. *Mon. Wea. Rev.*, **129**, 2677–2694.

Woolnough, S. J., F. Vitart, and M. A. Balmaseda, 2007: The role of the ocean in the Madden-Julian Oscillation: Implications for MJO prediction. *Quart. J. Roy. Meteor. Soc.*, **133**, 117–128.

Wu, R., B. P. Kirtman, and K. Pegion, 2008: Local rainfall-SST relationship on subseasonal time scales in satellite observations and CFS. *Geophys. Res. Lett.*, **35**, L22706.

Xiang, B., M. Zhao, X. Jiang *et al.* 2015: The 3–4 week MJO prediction skill in a GFDL coupled model. *J. Climate*, **28**, 5351–5364.

Yoneyama, K., C. Zhang, and C. N. Long, 2013: Tracking pulses of the Madden-Julian Oscillation. *Bull. Amer. Meteorol. Soc.*, **94**, 1871–1891.

Yoo, J.-H. and I.-S. Kang, 2005: Theoretical examination of a multi-model composite for seasonal prediction. *Geophys. Res. Lett.*, **32**, L15711.

Zhang, C., 2005: Madden-Julian Oscillation. *Rev. Geophys.*, **43**, RG2003. doi:10.1029/2004RG000158.

Zhang, C., 2013: Madden-Julian Oscillation: Bridging weather and climate. *Bull. Amer. Meteorol. Soc.*, **94**, 1849–1870.

Zhang, C. and J. Gottschalck, 2002: SST anomalies of ENSO and the Madden-Julian Oscillation in the equatorial Pacific. *J. Climate*, **15**, 2429–2445.

Zhang, C., J. Gottschalck, E. D. Maloney *et al.*, 2013: Cracking the MJO nut. *Geophys. Res. Lett.*, **40**, 1223–1230.

Zhang, C., M. Dong, S. Gualdi *et al.*, 2006: Simulations of the Madden-Julian oscillation in four pairs of coupled and uncoupled global models. *Clim. Dyn.*, **27**, 573–592.

Zhang, C. and H. v. d. Dool, 2012: Relative Merit of Model improvement versus availability of retrospective forecasts: The case of Climate Forecast System MJO prediction. *Wea. Forecast.*, **27**, 1045–1051.

Zhou, S. and A. J. Miller, 2005: The interaction of the Madden-Julian Oscillation and the Artic Oscillation. *J. Climate*, **18**, 143–159.

Chapter 13

Vertical Structure and Diabatic Processes of the Madden-Julian Oscillation

Nicholas P. Klingaman*,§, Xianan Jiang,† Prince K. Xavier,‡
Jon Petch,‡ Duane Waliser,† Steven J. Woolnough*

*National Centre for Atmospheric Science-Climate,
University of Reading, Reading, Berkshire, RG6 6BB, UK

§n.p.klingaman@reading.ac.uk

http://yotc.ucar.edu/mjo/vertical-structure-and-diabatic-processes-mjo

†Joint Institute for Regional Earth System Science and Engineering,
University of California, Los Angeles, California, USA

‡Met Office, Exeter, UK

The "Vertical Structure of Diabatic Processes of the Madden-Julian Oscillation" global-model evaluation project developed a novel experimental framework, which produces a complete characterization of models' abilities to simulate the Madden-Julian oscillation (MJO). The three components of the project comprise 2-day and 20-day hindcasts and 20-year simulations; each obtained heating, moistening and momentum tendencies from the models' sub-grid parameterizations. Thirty-five centers provided output for at least one component; nine centers provided data for all three. The models vary greatly in MJO fidelity in climate and hindcast experiments, yet fidelity in one was not correlated with fidelity in the other. In 20-year simulations, strong MJO models demonstrated heating, vertical-velocity and zonal-wind profiles that tilted westward with height, as in reanalysis data. The 20-day hindcasts showed no correspondence between the shape of the heating profile and hindcast skill. Low-to-mid-level moistening at moderate rain rates was a consistent feature of high-skill models and absent from low-skill models, suggesting a role for boundary-layer and congestus clouds in the MJO transition, which was confirmed by timestep data from the 2-day hindcasts. These hindcasts revealed a poor simulation of the MJO transition phase, even at short leads, with large mid-tropospheric dry biases and discrepancies in radiative-heating profiles.

1. Introduction

The Madden-Julian Oscillation (MJO; Madden and Julian, 1971) is the leading mode of sub-seasonal (30–70 day) tropical variability. The active (suppressed) phase of the MJO consists of a large-scale envelope of deep (suppressed) convection closely followed by strong (weak) low-level westerly winds (Lawrence and Webster, 2002). MJO events often form in the equatorial Indian Ocean, before propagating east through the Maritime Continent into the West Pacific at approximately a $5\,\mathrm{ms}^{-1}$ (Madden and Julian, 1994). As the MJO circumnavigates the tropics, it modulates regional precipitation and the large-scale circulation, including the Australian (Wheeler *et al.*, 2009), African (Lavender and Matthews, 2009), South Asian (Pai *et al.*, 2011) and North American (Lorenz and Hartmann, 2006) monsoons. Such modulations also affect tropical cyclogenesis in the Indian, Pacific and Atlantic basins (Camargo *et al.*, 2009). The circulation response to the MJO-associated diabatic heating extends to the extra-tropics, producing a teleconnection to the North Atlantic Oscillation (Cassou, 2008). Therefore, the MJO provides a significant source of sub-seasonal predictability globally.

Although many studies have documented the impacts of the MJO, the fundamental physics

The Global Monsoon System: Research and Forecast (3rd Edition)
Edited by C. P. Chang et al.

of its generation, maintenance and propagation are intensely debated. Proposed theories focus on instabilities arising from interactions between components of the diabatic heating and the large-scale circulation. These theories emphasize different sources of heating that drive moisture convergence at different heights, including free-tropospheric moisture convergence from low-level heating (Lau and Peng, 1987), frictional moisture convergence in the boundary layer (Wang and Rui, 1990), as well as radiative cooling that drives instability and tropospheric moisture convergence (Stephens *et al.*, 2004). In modelling studies, Li *et al.* (2009) highlighted low-level heating from shallow convection prior to the onset of deep convection, while Fu and Wang (2009) found that stratiform, mid-level heating was essential. This uncertainty is further complicated by disagreement in profile shape and amplitude among observations and reanalysis, including among products processed from the same Tropical Rainfall Measuring Mission (TRMM) instruments (Ling and Zhang, 2011; Jiang *et al.*, 2011). Moistening from detraining convection likely also plays a role in sustaining the suppressed phase of the MJO. Woolnough *et al.* (2010) found that models which produced more precipitation and net drying below the freezing level in the suppressed phase were unable to adequately transition to the active phase. As a result of this lack of physical understanding, most general circulation models (GCMs) exhibit biases in diabatic heating and moistening processes, as well as incorrect or weak feedbacks between that heating and the large-scale circulation. These biases may lead to the substantial errors found in GCM representations of the MJO (e.g. Kim *et al.*, 2009), which in turn limit sub-seasonal skill.

2. The "Vertical Structure and Diabatic Processes" of the MJO Global-Model Evaluation Project

We are conducting a novel global-model evaluation project to assess the vertical structure and diabatic processes of the MJO in GCMs. The project is endorsed by the Global Atmospheric Systems Studies (GASS) panel, the Years of Tropical Convection (YoTC) and the World Climate Research Programme–World Weather Research Programme (WCRP–WWRP) MJO Task Force. The project has developed a framework through which model developers can improve their sub-gridscale physical parameterisations, by linking errors in the simulation of the MJO to errors in GCM vertical profiles of heat, moisture and momentum. The MJO provides a rigorous test for GCM physics schemes, as well as the interactions between the physics and the large-scale dynamics. Due to the many teleconnections from the MJO, reducing model MJO biases should also improve the simulation of other tropical and extra-tropical phenomena. A key advantage of this evaluation project is the acquisition of temperature, moisture and momentum tendencies from each of the model sub-grid physics schemes; these data are not available from previous, more general intercomparisons, such as the Coupled Model Intercomparison Project (CMIP), particularly not at the temporal frequency obtained here.

The three project components are designed to take advantage of known links between biases in short-range (e.g. NWP) forecasts and climate simulations (Boyle *et al.*, 2008):

Component 1: Twenty-year climate simulations with either atmosphere-only or coupled GCMs. These simulations assess the overall fidelity of each model's representation of the MJO, relative to the climatology and mean-state biases of that model, as well as teleconnections from the MJO to monsoon systems and extra-tropical variability. Six-hourly output of all sub-grid tendencies has been obtained for the full twenty-year period.

Component 2: Two-day hindcasts, initialized daily during the active phases of two strong MJO events (20 October–10 November 2009 and 20 December–10 January 2010) in the Indian

Ocean within the YoTC period (Cases E and F), initialized from the European Centre for Medium-range Weather Forecasts (ECMWF) YoTC analyses. Tendencies are obtained every timestep on GCM horizontal and vertical grids, over the Warm Pool region. This enables detailed, comprehensive comparisons of parameterizations when models are close to the initial state and contain an active MJO.

Component 3: Twenty-day hindcasts, initialized daily during the two events above, but for a wider range of start dates to capture the MJO genesis and lysis at lead times beyond ten days. Three-hourly output is used to diagnose changes in GCM behavior and skill with lead time, as GCMs move from the near-observed state — identified in component 2 — towards their preferred climate — identified in component 1. The aim is to link the degradation in a model's representation of the MJO with forecast lead time to the growth of biases in diabatic heating and moistening profiles.

Thirty-five centers provided data for at least one component; nine provided data for all three components. A complete list of participating centers, which experiments each centre performed and details of the model configurations are available from our project website: http://yotc.ucar.edu/mjo/vertical-structure-and-diabatic-processes-mjo. All data collected are available to the community through a link on the above page.

3. Component 1: Twenty-Year Climate Simulations

Model fidelity in representing the eastward propagation of the MJO in 26 GCM simulations is assessed by lag-regression and wavenumber–frequency analysis. In Fig. 1, Hovmöller diagrams of rainfall evolution based on observed and GCM-simulated rainfall are derived by lag-regression of 20–100-day band-pass filtered anomalous rainfall against itself averaged over an Indian Ocean box. The observed systematic eastward propagation of the MJO is captured in only about one-quarter of models. Based on pattern correlations of Hovmöller diagrams between each model simulation and TRMM, two groups of seven GCMs are identified as the top (red labels) and bottom (blue labels) 25% GCMs for representing the MJO. Similar skill scores can be derived by applying the "east/west variance ratio" approach based on wavenumber- frequency analysis (Kim *et al.*, 2009).

Various process-oriented metrics [e.g. those suggested by the MJO Task Force (CLIVAR Madden-Julian Oscillation Working Group, 2009)] have been examined to identify key processes essential for faithful MJO representation. These include rainfall fraction from large-scale condensation, the rainfall probability density function (PDF), vertical relative humidity (RH) profiles as a function of rain rate and the 850 hPa mean zonal wind over the equatorial eastern Indian Ocean and western Pacific. Analyses suggest that the processes responsible for reliable MJO simulations in the top 25% GCMs are complex; no single process yet examined is strongly linked to the improved MJO in these models. For example, of five selected top 25% GCMs, the total rainfall in two is dominated by the convective component, while the convective and large-scale components play comparable roles in the other three. We note that air–sea coupling improves the MJO in several models, particularly the CNRM GCM, as previously reported (e.g. Inness and Slingo, 2003).

To further elucidate key processes for realistic MJO simulation, the vertical structures of anomalous zonal wind, temperature, vertical velocity, diabatic heating, and specific humidity corresponding to intra-seasonal rainfall variability over the Indian Ocean are analyzed for the top and bottom 25% of GCMs; the results are compared to ERA-Interim (Fig. 2). TAMU-modCAM4 was excluded from the top 25% composite, since that GCM prescribes an observed diabatic-heating profile based on MJO phase;

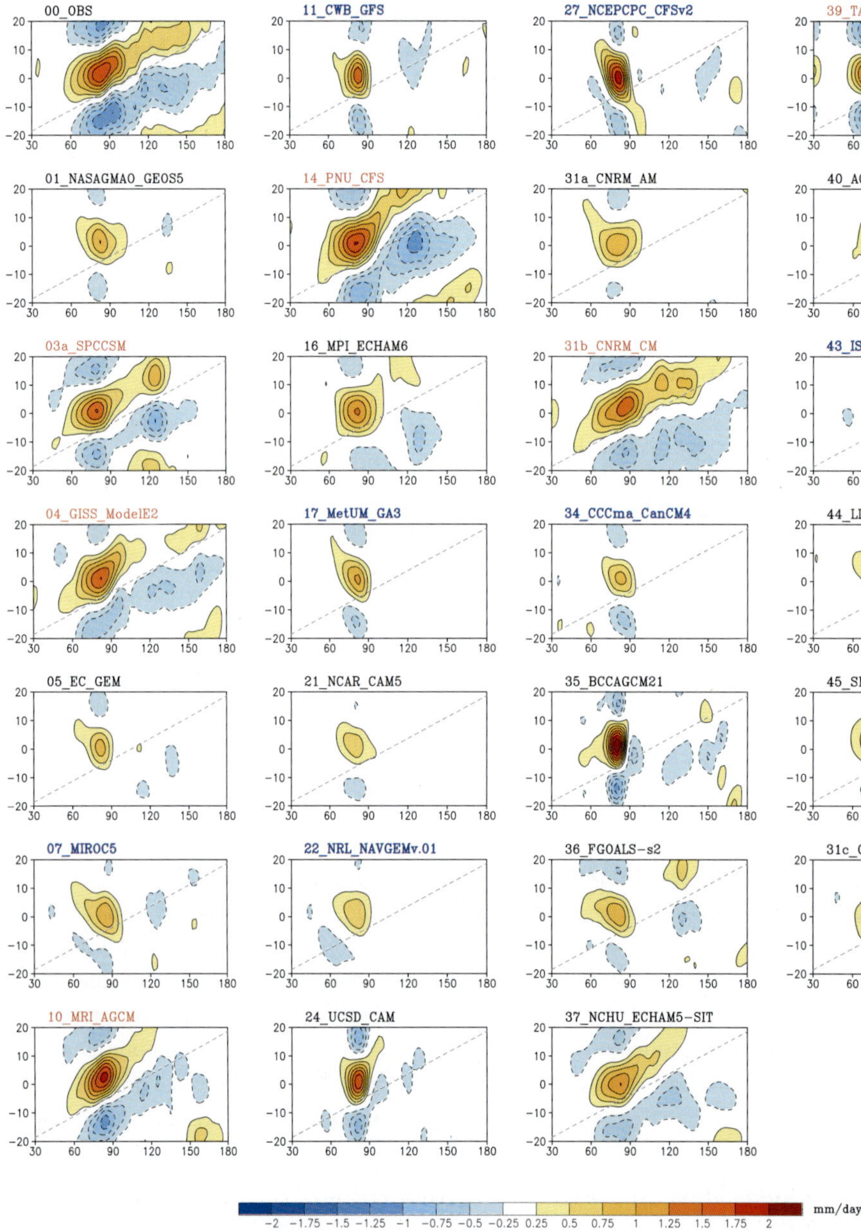

Fig. 1. Longitude (°E)-time (day) diagrams of latitude-averaged (10°S–10°N) rainfall regressed on area-averaged (75–85°E, 5°S–5°N) rainfall. Rainfall was first 20–100 day bandpass-filtered. Regressions are scaled by the standard deviation of area-averaged rainfall; units are mm day^{-1}. The dashed line is 5 m s^{-1} eastward propagation. The top 25% (red) and bottom 25% (blue) GCMs are identified by pattern correlation with TRMM (OBS; top left).

this made very small differences to the composites. The most prominent observed features in the vertical profiles of these fields are well-captured in the top 25% composite. In contrast, deficiencies are seen in the profiles from the bottom 25% GCMs: the first baroclinic responses in zonal wind associated with enhanced convection are not well-defined; and the westward tilt with

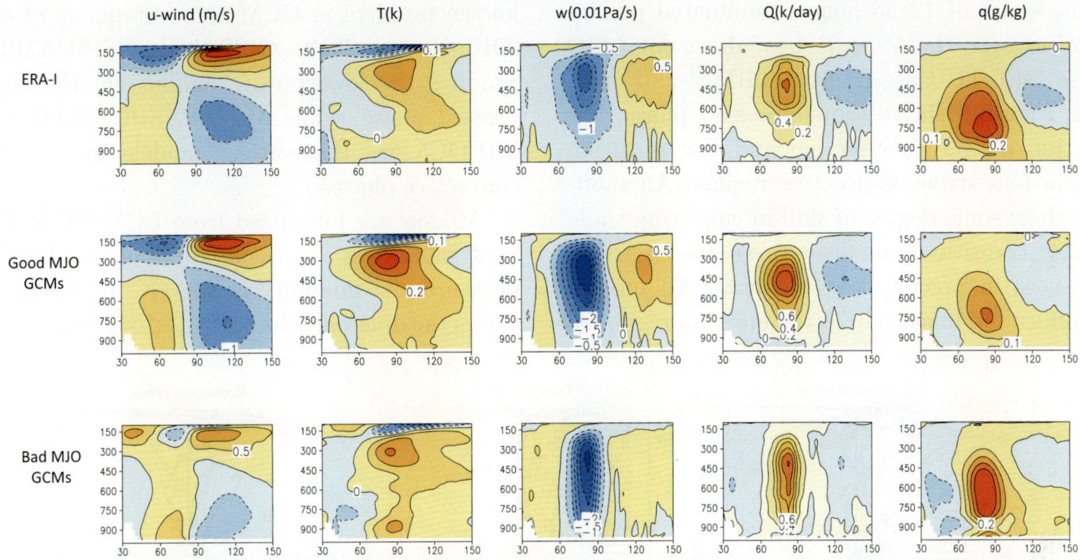

Fig. 2. Pressure (hPa)–longitude (°E) profiles of lag-regressed zonal wind, temperature, vertical velocity, diabatic heating, and specific humidity onto 20–100-day bandpass-filtered, area-averaged Indian Ocean rainfall (75–85°E; 5°S-5°N) based on ERA-Interim (top) and the composites of the top 25% GCMs (red labels in Fig. 1, excluding TAMU-modCAM4) and the bottom 25% GCMs (blue labels in Fig. 1). All variables are averaged 10°S–10°N.

height in vertical velocity, diabatic heating, and moisture fields seen in reanalysis and the top 25% GCMs is not clearly evident. This may indicate a lack of key pre-conditioning processes for the eastward-propagating MJO.

More detailed analyses have been carried out to understand the essential differences in the vertical structures associated with intraseasonal rainfall variability between these two groups of GCMs as shown in Fig. 2, including energetic conversion terms (Holloway *et al.*, 2013), the moist static energy budget (Maloney, 2009; Andersen and Kuang 2012), and normalized gross moist stability (Raymond *et al.*, 2009; Benedict *et al.*, 2014). These will be reported separately.

4. Component 2: Short-Range Hindcasts

In the 48-hour hindcasts, 12–36 hours is used as the time window for analysis that minimizes model spin-up effects and differences in

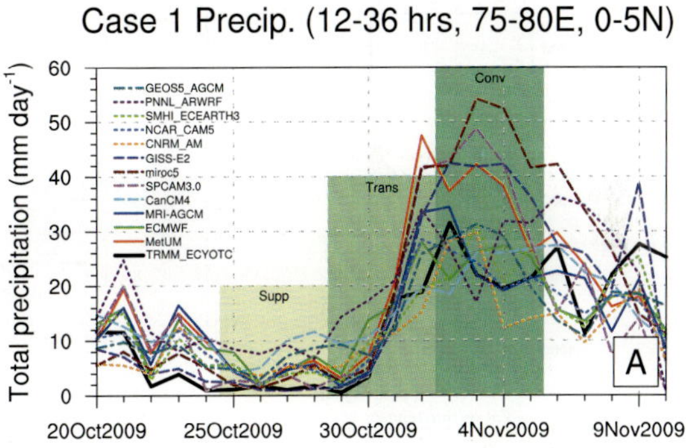

Fig. 3. Time series of 12–36 hours total precipitation from TRMM (solid black) and models over 75–80°E, 0°–5°N for YoTC Case E ("Case 1"). Three phases of the convective transition are marked as suppressed, transition and convective depending on the rainfall amounts.

the evolution of the large-scale dynamics. We focus on understanding the transition of MJO convection over a sufficiently large sub-domain that can represent the large-scale convective transition. A 5° × 5° domain is chosen (75–80°E, 0°–5°N) that is large enough to include information from the coarser resolution models (e.g. SPCAM3.0, CanCM4). Figure 3 shows time series of 12–36 hour accumulated precipitation over the 5° × 5° box above for MJO case 1 (YoTC Case E) from TRMM and models. Figure 3 shows the suppressed phase, initiation of MJO convection and its transition towards a strong convective regime. All models show some degree of skill in capturing these phase transitions, however there are large differences in precipitation amounts. All models produce too much rain in the low-rain regime, a known problem in GCMs (e.g. Stephens *et al.*, 2010; Xavier, 2012). GISS-E2 and SP-CAM3.0 produce the lowest rain amounts in the suppressed phase, while MetUM, SPCAM3.0 and MIROC5 are the wettest models during the convective phase.

Models are initialized from ECMWF-YoTC analysis; biases represent adjustment of temperature and humidity in the GCMs. Figure 4 shows the difference between 12–36 hour average

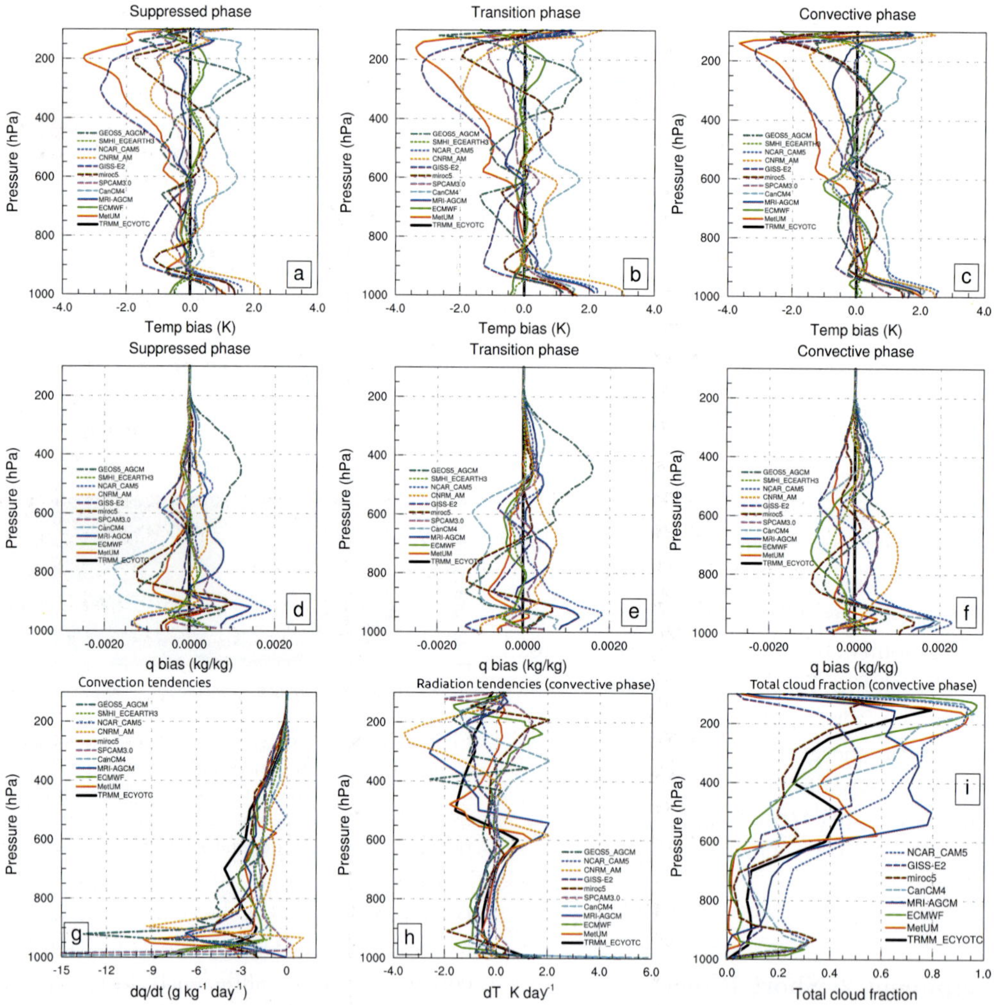

Fig. 4. Difference between (a–c) 12–36 hour average temperature and (d–f) specific humidity from the corresponding ECMWF YoTC 3–24-hour forecasts. Biases for suppressed (left panels), transition (middle) and convective phases (right panels) for YoTC Case E are shown (phases are defined in Fig. 3). (g) shows moisture tendencies due to convection during the transition phase; (h) shows the temperature tendency due to the radiation scheme in the convective phase; (i) shows the the total cloud fraction during convective phase.

temperature (a–c) and specific humidity (q, d–f) from the corresponding ECMWF-YoTC 0–23-hour forecasts for the three phases. Most models have warm biases near the surface in all three phases. Temperatures remain fairly close to the YoTC forecasts up to about the freezing level; a few models (e.g. MetUM, GISS-E2, CNRM AM) develop large cold biases above this which peak in the upper troposphere (Fig. 4a–c). Humidity (q) biases are much more varied for models, although many models show mid-level dry biases in the suppressed and transition phases.

An important feature of the MJO suppressed–convective transition is gradual mid-level moistening due to shallow cumulus and cumulus congestus clouds. This moistening and associated latent heating play a role in preconditioning the atmosphere for intense convection (Benedict and Randall, 2007). Figure 4g shows the moisture tendencies from the convection scheme during the transition phase. The thick black line is the tendency from ECMWF YoTC 0–23-hour forecasts. The YoTC tendencies are strongly a function of ECMWF model parameterizations, despite better constrained dynamics than the models. We treat the YoTC tendencies as a broad guideline for evaluating diabatic processes. The models' moisture tendencies are shown in colors. The removal of moisture due to convection at mid-levels is often a signature of cumulus congestus. Models produce varying amounts of shallow convective drying, but with relatively low mid-level values, which may reduce moistening prior to deep convection.

Temperature biases in the models have larger spread above 600 hPa compared to the lower-mid levels. Temperature tendencies from physics and dynamics from the models (not shown) suggest that even though there are large differences among models in the tendency terms, there is general agreement on the sign of the tendencies, except for the radiation tendencies during the convective phase (Fig. 4h). Temperature tendencies due to radiation schemes show large (relative to their mean value) spread above 600 hPa. Some of these large uncertainties may be related to the representation of cloud properties in the models. Figure 4i shows the total cloud fraction during the convective phase of MJO. Above the freezing level, models tend to have large differences in their cloud fraction. For example, the ECMWF model produces 20% cloud fraction at around 500 hPa while MRI-AGCM produces almost 80%. There are large spreads in cloud-liquid and ice profiles (not shown), which may affect the upper-level heating differences and large-scale temperature biases. However, it remains a challenge to extract a precise relationship between radiative tendencies and MJO performance. There is no truth available to estimate the biases in radiative-heating and cloud profiles. A lack of clear and consistent clustering of models in temperature, radiative or cloud profiles at short range makes it difficult to relate to the MJO performance in the other components, given that the short-range results are based on a single MJO event and on a relatively smaller region.

Nevertheless this framework emphasizes the general behavior of model physics in terms of their uncertainties and interactions with large-scale dynamics. The inter-model spread in diabatic heating and moistening processes highlighted here calls for more accurate representation of shallow cumulus/congestus and cloud-radiative properties. More focused analysis is required to understand the relationship between heating, moistening and cloud properties at short-range and the key biases relevant for MJO simulation in NWP and climate models.

5. Component 3: Long-Range Hindcasts

In the 20-day hindcasts, model "skill" is evaluated by computing the Wheeler and Hendon (2004) Real-time Multi-variate MJO (RMM) indices from each model, using the method in Gottschalck *et al.* (2010), and comparing against observed RMM indices from satellite-derived outgoing longwave radiation (OLR) and ECMWF YoTC analysis zonal winds. "Skill" is a loose term here, since only two events were simulated using a total of 94 start dates. This experiment aims not to provide a thorough assessment of skill, but rather to distinguish between the diabatic processes of those models that simulate these MJO cases well from those that do so poorly.

The 13 GCMs and two statistical models — linear inverse models, "LM" and "LS" in Fig. 5 — display a wide range of skill for these events (Fig. 5a). Several models, such as the two versions of the Community Atmospheric Model, version 5 (CAM5; "C5" and "CZ"), show greater than 20 days' skill at a bivariate correlation of 0.7; others are similar to or worse than a persistence forecast ("PE"), which implies that the MJO displays little or no propagation. All GCMs show greater skill for RMM2, which represents the anti-correlation in convection between the Indian Ocean and the West Pacific, than in RMM1, which represents variability over the Maritime Continent (Fig. 5b). For the 11 GCMs that also supplied results to Component 1, the daily climatology of the 20-year simulations was used to determine how quickly the model drifted from the initial conditions towards its attractor, using fields such as winds and OLR in the deep tropics (not shown). Those models that drifted more (less) quickly displayed lower (higher) skill; lower-skill GCMs tend to damp sub-seasonal anomalies and return to their mean climates.

Several process-oriented diagnostics were applied, such as vertical profiles of specific humidity and vertical-velocity anomalies as functions of rainfall (e.g. Thayer-Calder and Randall, 2009). There were no significant correlations between fidelity in these profiles, compared to observations, and model skill (not shown). These profiles did not vary with lead time, indicating that they are robust features of the GCM, regardless of MJO strength.

Diabatic-heating profiles from each model were computed by compositing on quartiles of rainfall rate, after first re-gridding the heating and rainfall to a 10° × 10° horizontal grid to ensure only broad-scale features were cap-

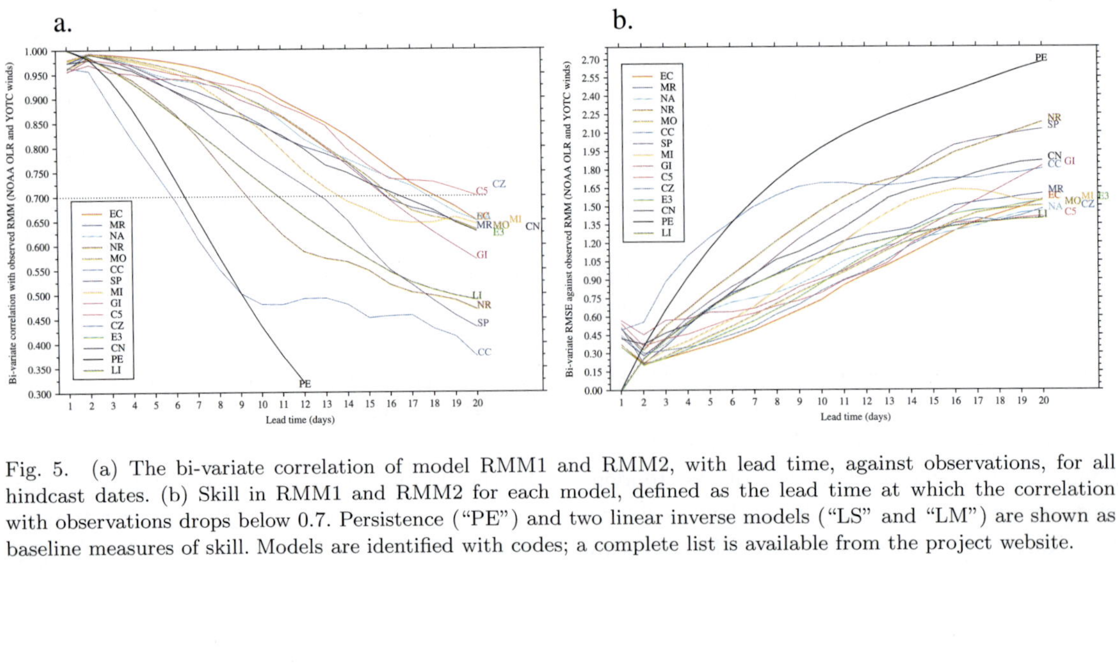

Fig. 5. (a) The bi-variate correlation of model RMM1 and RMM2, with lead time, against observations, for all hindcast dates. (b) Skill in RMM1 and RMM2 for each model, defined as the lead time at which the correlation with observations drops below 0.7. Persistence ("PE") and two linear inverse models ("LS" and "LM") are shown as baseline measures of skill. Models are identified with codes; a complete list is available from the project website.

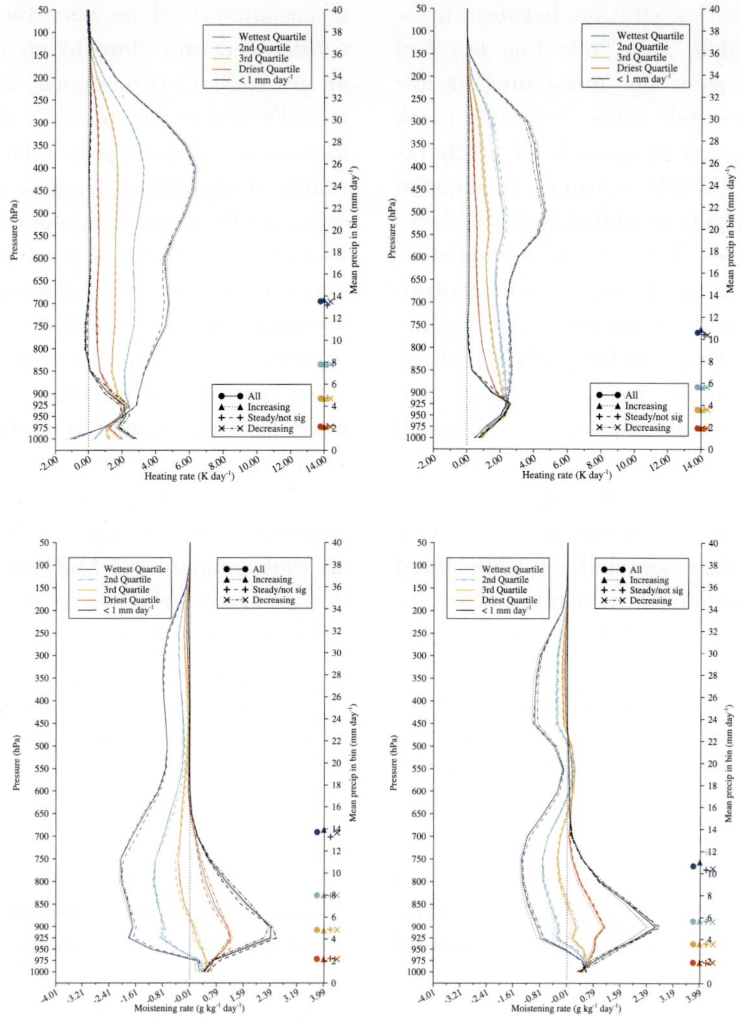

Fig. 6. For 3-hr (a, b) diabatic heating from physics ("Q1") and (c, d) total moistening (dq/dt), composite profiles by rain-rate quartile for (a, c) CanCM4 and (b, d) CAM5-ZM). Quartiles are constructed at each grid point, from all rain rates $> 1\,\mathrm{mm\ day^{-1}}$; profiles are averaged over all ocean points in 60°–90°E, 10°S–10°N, using all start dates and lead times. Composite profiles for rain rates $\leq 1\,\mathrm{mm\ day^{-1}}$ are in black. Symbols on the right-hand axis give the mean rain rate ($\mathrm{mm\ day^{-1}}$) for each quartile.

tured (Fig. 6). These profiles also largely failed to segregate those models with a high predictive skill from those with lower skill. One poor-performing model, NavGEM ("NR" in Fig. 5; profiles not shown), displayed unrealistically bottom-heavy heating profiles. Another low-skill GCM, CanCM4 ("CC"; Fig. 6a), had a top-heavy profile similar to the high-skill CAM5 (Fig. 6b), as well as a similar evolution of heating profiles with increasing rain rate. A top-heavy heating profile may be a necessary, but not a sufficient condition, for accurate MJO predictions. The composite heating profiles were almost constant with lead time, suggesting that the timestep profiles in Component 2 are not unduly influenced by the short lead time.

A diagnostic based on total moistening (i.e. $\delta q/\delta t$), as a function of rain-rate quartile, produced a cleaner separation between high- and low-skill models. CanCM4 (Fig. 6c) and CAM5 (Fig. 6d) show net moistening at low levels for light rainfall rates (red and black lines), but in the second quartile of precipitation (orange line) CAM5 continues to moisten in the lower troposphere while CanCM4 shows near-zero tendencies. The transition from shallow moistening during the suppressed phase to mid-level and then deep moistening during the active phase was found to be much smoother, and much closer to the YoTC 3–24 hour forecasts, in the high-skill models. The high-skill models all simulated net moistening above the boundary layer in the second quartile, while all of the low-skill models displayed negative or negligible $\delta q/\delta t$. The low-skill models showed a too-quick transition, with increasing precipitation, from low-level moistening and upper-level drying to deep moistening and low-level drying. This suggests a key role for boundary-layer clouds and congestus in the simulation of MJO initiation and propagation. In GCMs, the $\delta q/\delta t$ profile is often driven by the counterbalance between dynamics and physics: at high (low) rain rates, the model dynamics moistens (dries) and physics dries (moistens). However, at moderate rain rates (3–6 mm day^{-1}), several high-skill models had mid-tropospheric moistening from both advection and physics, suggesting an accurate transition from suppressed to active MJO conditions may require moistening by both processes.

6. Summary and Conclusions

The "Vertical structure and diabatic process of the MJO" global-model evaluation project has established a framework for analyzing biases in model physical parameterizations associated with tropical convection, which affect the simulation and prediction of weather and climate phenomena globally. The three components of the framework provide a robust characterization of MJO activity in a GCM, from the behavior of parameterizations when the model is highly constrained and close to an initial state with an active MJO (Component 2), through to the model's ability to generate and maintain an intra-seasonal oscillation relative to its climatology (Component 1). The third component examines the degradation of the model from the former to the latter, identifying how biases in diabatic heating and moistening profiles may accelerate that degradation.

Analysis of the three components has identified several over-arching conclusions. First, many process-oriented MJO diagnostics derived from previous studies based on one or several models, when applied to the 35 models in this project, fail to distinguish between those models that simulate the MJO well and those that do so poorly, whether in initialized hindcasts or decadal-length simulations. It appears that no single process holds the key to the MJO in these models; rather, there are many processes that are necessary, but not sufficient conditions for a reliable simulation of tropical convection. Second, there is little correlation between a model's skill in predicting the MJO — based only on the two YoTC cases examined — and its ability to generate sub-seasonal variability in a free-running simulation. Several of the best-performing models in Component 3 (e.g. CAM5 and the U. K. Met Office Unified Model) performed poorly in Component 1, while some of the best models in that component (e.g. the Super-Parameterized CAM and the Goddard Institute for Space Studies ModelE) produced poor hindcasts. Finally, all components have demonstrated that reliable representations of moistening, particularly at low and mid-levels during the transition between suppressed and active phases, may be critical to simulating the MJO (Figs. 2, 4 and 6), perhaps more so than the representation of convective heating.

This project has produced a rich dataset, which is open to the community via: http://yotc.ucar.edu/mjo/vertical-structure-and-diabatic-processes-mjo. We have only "scratched the surface" of this dataset. Although we have focused on the MJO, the data could be used to examine many other phenomena (e.g. monsoons, equatorial waves, tropical cyclones and tropical–extra-tropical teleconnections); many variables from Components 1 and 3 are available globally, with tendencies available 50°N–50°S. We hope that further studies will develop and extend these results, to continue to test and refine the physical parameterizations critical to the simulation of tropical sub-seasonal variability.

While this book chapter was in press, further and updated conclusions from this project were published in the peer-reviewed literature. The interested reader is referred to Klingaman *et al.* (2015a,b), Jiang *et al.* (2015) and Xavier *et al.* (2015).

References

Andersen, J. A. and Z. Kuang, 2012: Moist static energy budget of MJO-like disturbances in the atmosphere of a zonally symmetric aquaplanet. *J. Climate*, **25**, 2782–2804

Benedict, J. J. and D. A. Randall, 2007: Observed characteristics of the MJO relative to maximum rainfall. *J. Atmos. Sci.*, **64**, 2332–2354.

Boyle, J. S., S. Klein, G. Zhang, S. Xie, and X. Wei, 2008: Climate model forecast experiments for TOGA COARE. *Mon. Wea. Rev.*, **136**, 808–832.

Benedict, J. J., A. H. Sobel, E. D. Maloney, and D. M. Frierson, 2014: Gross moist stability and MJO simulation skill in three full-physics GCMs. *J. Atmos. Sci.*, **71**, 3327–3349.

Camargo, S. J., M. C. Wheeler, and A. H. Sobel, 2009: Diagnosis of the MJO modulation of tropical cyclogenesis using an empirical index. *J. Atmos. Sci.*, **66**, 3061–3074.

Cassou, C., 2008: Intraseasonal interaction between the Madden-Julian oscillation and the North Atlantic oscillation. *Nature*, **455**, 523–527.

CLIVAR Madden-Julian Oscillation Working Group, 2009: MJO simulation diagnostics. *J. Climate*, **22**, 3006–3030.

Fu, X. and B. Wang, 2009: Critical roles of the stratiform rainfall in sustaining the Madden-Julian Oscillation: GCM experiments. *J. Climate*, **22**, 3939–3959.

Gottschalck, J. M. Wheeler, K. Weickmann, F. Vitart, N. Savage, H. Lin, H. Hendon, D. Waliser, K. Sperber, M. Nakagawa, C. Prestrelo, M. Flatau, and W. Higgins, 2010: A framework for assessing operational Madden-Julian Oscillation forecasts: A CLIVAR MJO Working Group project. *Bull. Amer. Meteor. Soc.*, **91**, 1247–1258.

Holloway, C. E., S. J. Woolnough, and G. M. S. Lister, 2013: The effects of explicit versus parameterized convection on the MJO in a large-domain high-resolution tropical case study. Part I: Characterization of large-scale organization and propagation. *J. Atmos. Sci.*, **70**, 1342–1369.

Inness, P. M. and J. M. Slingo, 2003: Simulation of the Madden-Julian Oscillation in a coupled general circulation model. Part I: Comparison with observations and an atmosphere-only GCM. *J. Climate*, **16**, 345–364.

Jiang, X., D. E. Waliser, W. S. Olson, W. K. Tao, T. S. L'Ecuyer, K. F. Li, Y. L. Yung, S. Shige, S. Lang, and Y. N. Takayabu, 2011: Vertical diabatic heating structure of the MJO: Intercomparison between recent reanalyses and TRMM estimates. *Mon. Wea. Rev.*, **139**, 3208–3223.

Jiang, X., D. E. Waliser, P. K. Xavier, J. Petch, N. P. Klingaman, S. J. Woolnough, B. Guan, G. Bellon, T. Crueger, C. DeMott, C. Hannay, H. Lin, W. Hu, D. Kim, C.-L. Lappen, M.-M. Lu, H.-Y. Ma, T. Miyakawa, J. A. Ridout, S. D. Schubert, J. Scinocca, K.-H. Seo, E. Shindo, X. Song, C. Stan, W.-L. Tseng, W. Wang, T. Wu, X. Wu, K. Wyser, G. J. Zhang, and H. Zhu, 2015: Vertical structure and physical processes of the Madden-Julian Oscillation: exploring key model physics in climate simulations. *J. Geophys. Res.*, **120**, 4718–4748.

Kim, D., K. Sperber, W. Stern, D. Waliser, I. S. Kang, E. Maloney, W. Wang, K. Weickmann, J. Benedict, M. Khairoutdinov, M. I. Lee, R. Neale, M. Suarez, K. Thayer-Calder, and G. Zhang, 2009: Application of MJO simulation diagnostics to climate models. *J. Climate*, **22**: 6413–6436.

Klingaman, N. P., X. Jiang, P. K. Xavier, J. Petch, D. E. Waliser, and S. J. Woolnough, 2015a: Vertical structure and physical processes of the Madden-Julian Oscillation: synthesis and summary. *J. Geophys. Res.*, **120**, 4671–4689.

Klingaman, N. P., S. J. Woolnough, X. Jiang, D. E. Waliser, P. K. Xavier, J. Petch, M. Caian,

C. Hannay, D. Kim, H.-Y. Ma, W. J. Merryfield, T. Miyakawa, M. Pritchard, J. A. Ridout, R. Roehrig, E. Shindo, F. Vitart, H. Wang, N. R. Cavanaugh, B. E. Mapes, A. Shelly and G. J. Zhang, 2015b: Vertical structure and physical processes of the Madden-Julian oscillation: Linking hindcast fidelity to simulated diabatic heating and moistening. *J. Geophys. Res.*, **120**, 4690–4717.

Lau, K. M. and L. Peng, 1987: Origin of low-frequency (intraseasonal) oscillations in the tropical atmosphere. Part I: basic theory. *J. Atmos. Sci.*, **44**, 950–972.

Lavender, S. L. and A. J. Matthews, 2009: Response of the West African monsoon to the Madden-Julian Oscillation. *J. Climate*, **22**, 4097–4116.

Lawrence, D. M. and P. J. Webster, 2002: The boreal intraseasonal oscillation: Relationship between northward and eastward movement of convection. *J. Atmos. Sci.*, **59**, 1593–1606.

Li, C. Y., X. L. Jia, J. Ling, W. Zhou, and C. Zhang, 2009: Sensitivity of MJO simulations to diabatic heating profiles. *Clim. Dyn.*, **32**, 167–187.

Ling, J. and C. Zhang, 2011: Structural evolution in heating profiles of the MJO in global reanalyses and TRMM retrievals. *J. Climate*, **24**, 825–842.

Lorenz, D. J. and D. L. Hartmann, 2006: The effect of the MJO on the North American monsoon. *J. Climate*, **19**, 333–343.

Madden, R. A. and P. R. Julian, 1971: Detection of a 40–50 day oscillation in the zonal wind in the tropical Pacific. *J. Atmos. Sci.*, **28**, 702–708.

Madden, R. A. and P. R. Julian, 1994: Observations of the 40–50 day tropical oscillation — a review. *Mon. Wea. Rev.*, **122**, 814–837.

Maloney, E. D., 2009: The moist static energy budget of a composite tropical intraseasonal oscillation in a climate model. *J. Climate*, **22**, 711–729.

Pai, D. S., J. Bhate, O. P. Sreejith, and H. R. Hatwar, 2011: Impact of MJO on the intraseasonal variation of summer monsoon rainfall over India. *Clim. Dyn.* **36**, 41–55.

Raymond, D. J., S. S. Sessions, A. H. Sobel, and Z. Fuchs, 2009: The mechanics of gross moist stability. *J. Adv. Model. Earth Sys.*, **1**, 1–19.

Stephens, G. L., P. J. Webster, R. H. Johnson, R. Engelen, and T. S. L'Ecuyer, 2004: Observational evidence for the mutual regulation of the tropical hydrological cycle and tropical sea surface temperatures. *J. Climate*, **17**, 2213–2224.

Stephens, G. L., T. L'Ecuyer, R. Forbes, A. Gettlemen, J.-C. Golaz, A. Bodas-Salcedo, K. Suzuki, P. Gabriel, and J. Haynes, 2010: Dreary state of precipitation in global models. *J. Geophys. Res.*, **115** (D24), D24,211.

Thayer-Calder, K. and D. A. Randall, 2009: The role of convective moistening in the Madden-Julian Oscillation. *J. Atmos. Sci.*, **66**, 3297–3312.

Wang, B. and H. Rui, 1990: Dynamics of the coupled moist Kelvin–Rossby wave on an equatorial β-plane. *J. Atmos. Sci.*, **47**, 397–413.

Wheeler, M. C. and H. H. Hendon, 2004: An all-season real-time multivariate MJO index: Development of an index for monitoring and prediction. *Mon. Wea. Rev.*, **132**, 1917–1932.

Wheeler, M. C., H. H. Hendon, S. Cleland, H. Meinke, and A. Donald, 2009: Impacts of the Madden-Julian Oscillation on Australian rainfall and circulation. *J. Climate*, **22**, 1482–1498.

Woolnough, S. J., P. N. Blossey, K.-M. Xu, P. Bechtold, J.-P. Chaboureau, T. Hosomi, S. F. Iacobellis, Y. Luo, J. C. Petch, R. Y. Wong, and S. Xie, 2010: Modelling convective processes during the suppressed phase of a Madden-Julian Oscillation. *Quart. J. Roy. Meteor. Soc.*, **136**, 333–353.

Xavier, P. K., 2012: Intraseasonal convective moistening in CMIP3 Models, *J. Climate*, **25**, 2569–2577.

Xavier, P. K., J. C. Petch, N. P. Klingaman, S. J. Woolnough, X. Jiang, D. E. Waliser, M. Caian, J. Cole, S. M. Hagos, C. Hannay, D. Kim, T. Miyakawa, M. S. Pritchard, R. Roehrig, E. Shindo, F. Vitart, and H. Wang, 2015: Vertical structure and physical processes of the Madden-Julian oscillation: Biases and uncertainties at short range. *J. Geophys. Res.*, **120**, 4749–4763.

Chapter 14

Global Cloud/Cloud-System Resolving Models and the Madden-Julian Oscillation

Tomoki Miyakawa* and Tomonori Sato†

**Atmosphere and Ocean Research Institute, The University of Tokyo, Kashiwa, 277-8568, Japan*
miyakawa@aori.u-tokyo.ac.jp

†Faculty of Earth Environmental Sciences, Hokkaido University, Sapporo, 060-0810, Japan
t_sato@ees.hokudai.ac.jp

In this chapter, we provide an overview of Madden-Julian oscillation (MJO) simulations with global high-resolution models that require large amounts of computational resources. After introducing the recent activities regarding such models, we focus on the Nonhydrostatic Icosahedral Atmospheric Model (NICAM), in which we are involved, and summarize our experiences with respect to the representation of the MJO in NICAM. State-of-the-art supercomputers have carried us to a point where global sub-kilometer grid systems are applicable for short simulations of several days, and sub-10-km grid systems can be applied for simulations that are several years long. The high resolution and the more explicit physics compared to those of conventional general circulation models allow a higher degree of freedom in the modeled state of the atmosphere. The increased degree of freedom not only provides the model a higher potential to closely represent the real atmosphere but it also provides more reasons for the model to be different from reality. Use of a highly complex model requires a great amount of effort to determine numerous parameters in physics schemes. Although simulation of MJO events by NICAM today is successful overall, the question remains whether it can maintain realistic climatology and intra-seasonal variability in climate simulations. The development of an efficient procedure to appropriately confine the range of these parameters is one of the main issues for the future. Moreover, it is important to keep in mind that higher complexity could increase the difficulty of interpreting the results and improving scientific understanding.

1. Introduction

Considering that the Madden-Julian Oscillation (MJO, Madden and Julian 1972) is a planetary-scale phenomenon and that the convective envelope that determines its activity is composed of numerous clouds of various scales, the use of global cloud/cloud-system resolving models (GCRMs) is a very straightforward approach to study the MJO. However, the use of such models relies largely on the available computational resources. As of 2014, only a very limited number of modeling centers can afford to run ample MJO simulations with such models.

Two model groups led the global cloud-resolving activity in the early 2000s. One of the two groups developed the Nonhydrostatic Icosahedral Atmospheric Model (NICAM, Fig. 1 and Sec. 2), which can directly express the microphysical properties of clouds on a global, high-resolution three-dimensional mesh called the "icosahedral grid system" (Fig. 2). The idea of the grid system enables computationally efficient high-resolution simulations by avoiding

The Global Monsoon System: Research and Forecast (3rd Edition)
Edited by C. P. Chang et al.

Fig. 1. Snapshot of an MJO simulation by NICAM (7 km mesh) run on the K computer (SPIRE field 3, project ID hp120313). Precipitation (colors) and outgoing longwave radiation (white shades). Visualized by Daisuke Matsuoka, CEIST, JAMSTEC.

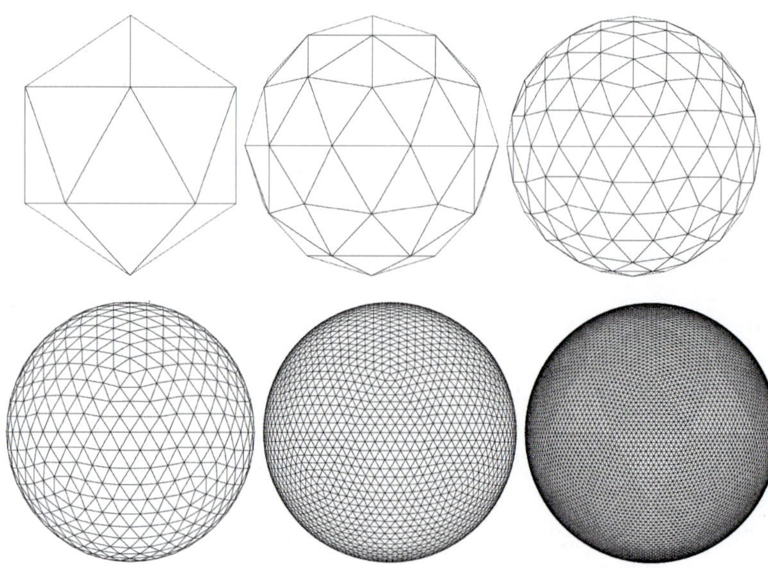

Fig. 2. The "icosahedral" grid system used in NICAM (Satoh *et al.* 2014).

distortion of grid size and grid shape. Apart from exceptional high-resolution experiments at 1.7 km and 870 m meshes, NICAM is mostly run at 3.5, 7, or 14 km meshes. The other group developed a unique hybrid of a conventional general circulation model (GCM) and a regional two-dimensional cloud-resolving model (CRM), as shown in Fig. 3. The concept of implementing CRMs to substitute for a convective scheme (Grabowski 2001, Khairoutdinov and Randall

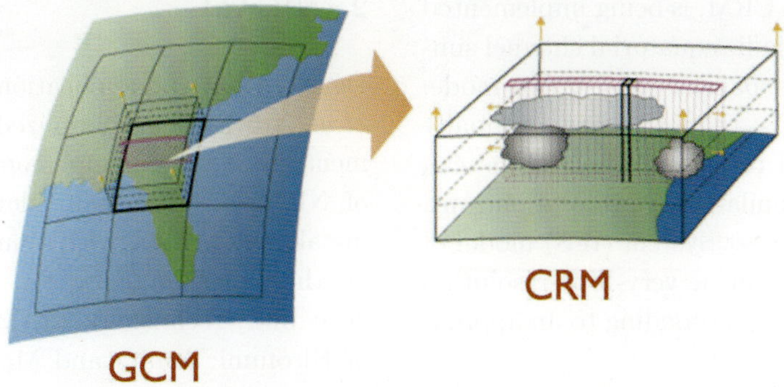

Fig. 3. A schematic describing the concept of "super parameterization." Two-dimensional cloud-resolving models are implemented in each grid-cell of the GCM. Figure reproduced from the National Monsoon Mission webpage (http://www.tropmet.res.in/monsoon/files/model_development.php).

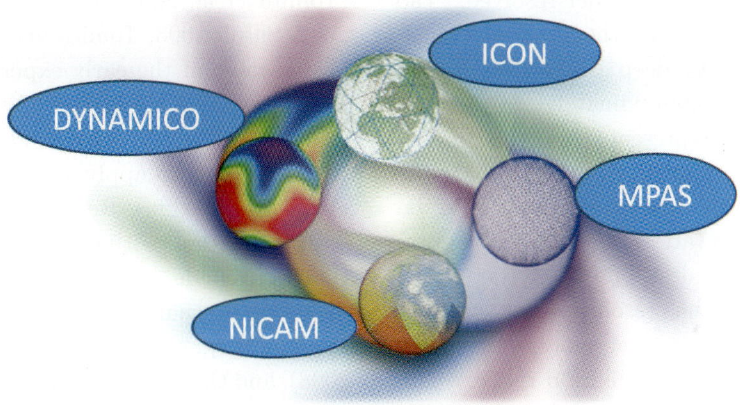

Fig. 4. Collaborative work by modeling groups that apply "Icosahedral" grid systems for global high-resolution simulations. Edited figure originally by Ryuji Yoshida and the ICOMEX consortium.

2001, Arakawa 2004) is now widely accepted as the "super parameterization" (SP) or the "multi-scale modeling framework". The SP is implemented in the Community Atmospheric Model (CAM) and its ocean-coupled version, and has been used in various recent studies. Both groups targeted the MJO as one of their primary interests, and had significant success stories. Miura *et al.* (2007b) successfully reproduced the eastward migration of the convective envelope of an MJO event. Benedict and Randall (2009) showed that SPCAM can simulate atmospheric variability on intraseasonal space and time (30–60 day) scales that are very similar to those of the observed MJO structures.

The success of these models, as well as the increasing availability of computational resources, stimulated the motivation of GCRM development. Skamarock *et al.* (2014) recently reported that the Model for Prediction Across Scales (MPAS) now runs at a 3 km mesh for global experiments. Additional global high-resolution models based on icosahedral grid systems are being developed, such as ICON (Zängl *et al.* 2014) and DYNAMICO (Dubos *et al.* 2015). These modeling groups built a consortium named ICOMEX (Zängl *et al.* 2011) for a collaborative study targeting an improvement in global high-resolution models with icosahedral grid systems (Fig. 4). The WRF model, an established regional CRM, is being implemented as an option for periodic equatorial channel simulations and even expanded to a global mode. Furthermore, GCMs that are not originally designed for "cloud resolving" simulations can now be capable of similarly high resolution runs. The Integrated Forecast System (IFS) model of the ECMWF is testing a very-high-resolution simulation (T7999, corresponding to an approximately 2 km mesh).

GCRMs and similarly high-resolution GCMs are now exploring global simulations at approximately 1–20 km meshes. However, certain convective features with horizontal scales less than 1 km are still probably under-resolved. Tao *et al.* (2009) argued that the gradual transition from shallow to deep convection in the tropics requires an approximately 250 m mesh to be sufficiently represented in the model. The use of conventional convective parameterizations at these resolutions may also be questioned, as they are designed to be representative of larger scales (O(100)–O(1000) km), corresponding to the resolution of conventional GCMs. Many global models currently occupy an intermediate resolution range where it is unclear whether they should explicitly represent clouds/cloud-systems or apply cumulus parameterizations. As for NICAM, it performs surprisingly well without convective parameterization in terms of reproducing the eastward migration of the MJO convective envelope, even when run with the 14 km mesh.

Hereafter, we mainly focus on the activity of NICAM, in which the authors are directly involved. In Sec. 2, we provide an overview of the development and applications of NICAM. The MJO performance of NICAM is summarized in Sec. 3. Section 4 describes various issues remaining and the difficulties encountered with global high-resolution experiments. Section 5 discusses strategies and future perspectives for GCRM simulation. Section 6 summarizes and concludes the chapter.

2. NICAM

Development and application of NICAM (Satoh *et al.* 2014) was synchronized with the improvement of Japanese Supercomputers. The basis of NICAM was initially developed eyeing the installment of the Earth Simulator (ES) in 2002 at the Japan Agency for Marine-Earth Science and Technology (JAMSTEC). The duo of Hirofumi Tomita and Masaki Satoh led the project, which was promoted by Taro Matsuno. They developed a nonhydrostatic dynamical core based on the "Icosahedral" grid system further optimized by a "spring-grid" method (Tomita *et al.* 2001, Satoh 2002, Tomita *et al.* 2002, Satoh 2003, Tomita and Satoh 2004). ES hosted most of the early experiments including the aqua-planet simulation (Tomita *et al.* 2005, Miura *et al.* 2005, Nasuno *et al.* 2007, Mapes *et al.* 2008) that apply high-resolution meshes (14, 7, and 3.5 km). Research targets expanded after simulations with realistic land-sea distribution were made available (Miura *et al.* 2007a), including diurnal cycles (Sato *et al.* 2007, Noda *et al.* 2012), cloud populations (Inoue *et al.* 2008), and the MJO (Miura *et al.* 2007b, Nasuno *et al.* 2009, Liu *et al.* 2009, Miyakawa *et al.* 2012).

The recent installment of the K computer at the RIKEN Advanced Institute for Computational Science (AICS) pushed the application of the GCRM to a new stage. The maximum resolution improved to 870 m mesh (Miyamoto *et al.* 2013), marking the opening of the much-anticipated global sub-1-km mesh simulation era (Heffernan 2008). The increased computational power can also allow longer simulations with a low-resolution version or increasing the sample sizes of short-range simulations. In fact, an AMIP-like climate simulation of over 30 years and future climate simulations have been completed with 14 km mesh, and are currently being intensively investigated (Kodama *et al.* 2015, Yamada *et al.* 2013). A series of 40 day MJO simulations that consist of 54 members has been conducted at 14 km resolution (Miyakawa *et al.* 2014), as described in Sec. 3. Short- to middle-range (<1 year) simulations at 7 and 14 km resolutions are now routinely executed.

The model is being continuously upgraded and various features of the outputs are being increasingly closely investigated. Another useful application of NICAM arises from grid-transformation systems that enables the model resolution to be locally enhanced (Tomita 2008a, Iga and Tomita 2014), allowing the model to be run like a regional atmospheric model but without a lateral boundary, which lowers the computational requirements so that NICAM runs become available with ordinary cluster machines (e.g. Nasuno 2013). The microphysics scheme has been upgraded to handle six categories of water (Tomita 2008b, Roh and Satoh 2014), and also given an option of double-moment microphysics (Seiki *et al.* 2014). Although a mixed-layer ocean model was originally available in NICAM to reflect the effect of atmosphere–ocean interactions for oceanic cloud systems, a dynamically coupled version is currently under operation checking, and is scheduled to be completed during 2015. Various sub-grid turbulence schemes have been implemented and tested (Noda *et al.* 2013). A satellite simulator has been implemented and the output compared with various satellite data (e.g. Hashino *et al.* 2013, Ham *et al.* 2013, Dodson *et al.* 2013). Because of the advantage of strict mass conservation in NICAM, the model is also utilized as a tracer transport model (e.g. Niwa *et al.* 2011, Goto 2014), and chemical schemes are planned to be implemented similarly to the CHASER model (Yashiro *et al.* 2011) in the near future.

3. MJO Performance

The high-resolution output data of GCRMs are often "good looking" and highly appealing to the community. However, GCRMs easily become needlessly costly, complex, and a problematic tool compared to conventional GCMs if thoughtlessly used. It is important to design GCRM simulations using strategies that exploit the advantages of the model.

Extended-range (>2 weeks) prediction of the MJO is an ideal target for GCRM simulations. Figure 5 shows a snapshot of OLR and precipitation simulated by NICAM, together with a schematic summarizing the appearance of cloud characteristics. It is apparent from the figure that the tropics (approximately 25S–25N) and higher latitudes are dominated by very different cloud regimes. The higher latitudes exhibit eastward-moving vortices, corresponding

Fig. 5. A snapshot of OLR and precipitation simulated by NICAM (left panel) and a schematic summarizing the convective characteristics observed in the snapshot (right panel).

to synoptic high/low pressure systems fueled by the baroclinic instability along the jet stream. Conventional theoretical studies tell us that it is practically impossible to predict the high/low pressure patterns much further than two weeks in advance because of non-linear growth of small errors. Therefore, it is less likely that GCRMs, no matter how accurate they may become, will have much impact in improving prediction of day-to-day patterns of the synoptic pressure systems. In contrast, a "perfect model test" performed by the ECMWF IFS model suggests that the potential predictability of the MJO reaches approximately 45 days (Vitart and Molteni 2011). The ECMWF reports that the predictability at the intraseasonal time scale (>2 weeks) in the mid-latitudes largely depends on the predictability of the MJO (Vitart and Molteni 2010). These findings imply that realistic representation of the atmosphere by numerical models may enable MJO predictions valid for longer than a month and that the improvement may possibly extend to extra-tropical regions because of the influence of the MJO on mid-latitude regions.

The simulation of a 2006 winter event (Miura *et al.* 2007b) by NICAM successfully reproduced the eastward migration of a convective envelope realistically and displayed the promisingly high potential of GCRMs in terms of MJO studies. After the K computer greatly reinforced the available computational resources, Miyakawa *et al.* (2014) conducted a case-sweep experiment targeting all the winter-MJO events identified from 2003–2012, and statistically proved that the model is currently capable of providing valid prediction of MJOs at a lead time of approximately 4 weeks. However, the physical reason why the GCRM successfully simulates MJO remains an open question. Given that the model performs well despite the lack of fine structures of cloud systems in the initial data, which is a simple interpolation from a lower-resolution dataset, subtle details of individual cloud systems are unlikely to be the sources of the high predictability. Statistical behaviors of feedbacks from cloud systems in realistic relation with large-scale conditions are the key issues, improved here by explicit calculation of the cloud systems.

Finer representation of generation/diffusion and redistribution of heat, mass, and momentum by the cloud systems are possible contributors to the improvement of the statistical behavior. Miyakawa *et al.* (2012) discussed this on the basis of an analysis of the MJO case simulated by Miura *et al.* (2007b), and found that the convective momentum transport, a crudely regarded component in GCMs, could play a role in altering the eastward speed of the MJO, and/or the tilting of its vertical structure, and in enhancement of surface evaporation. Detailed cloud microphysics (e.g. condensation/evaporation, freezing/melting, distinction of ice/snow/graupel, and aggregation) is also a likely contributor that probably leads to a better representation of the balance between convective heating and cloud radiative cooling of the MJO. Prichard *et al.* (2014) similarly suggested that in the CRM of SPCAM, the impact of resolving the detailed structure of cloud systems may be secondary compared to the application of detailed microphysics.

4. Moving Forward

The major tasks in today's MJO researches are: (a) to understand the MJO; (b) to improve MJO forecasts (initial-value problem); (c) to make reliable long-term future projections (boundary-value problem); and (d) to evaluate the impacts of the MJO on local weather/climate. These tasks will benefit from improvement of MJO representation in GCRMs. However, it is unclear whether further increases in resolution and/or complication of physics schemes will lead to such improvement. Furthermore, it is not easy to evaluate how effective the upgrades of resolution or physics schemes are in terms of model performance. Even if the upgrades improve the model's potential to reproduce the atmosphere realistically, this may not become apparent until the model is suitably tuned for the new model setup. There is no guarantee that a successful set of parameters at a certain resolution will remain a suitable choice at higher resolutions. Another problematic issue is that good model performances in short simulations (initial-value problems) do not guarantee model fidelity in climate simulations (boundary problems), as schematically shown in Fig. 6. This issue is magnified in GCRMs compared to regional CRMs and conventional GCMs, because GCRMs need to cope with both problems simultaneously at a wider range of temporal and spatial scales using a common framework. The best set of cloud microphysics parameters that produce the most realistic cold pools are often not the set that result in a better global radiative balance, perhaps because the best choice of parameters only converge at an infinite resolution. GCRM studies are currently facing two important challenges: (1) to find an efficient way to tune the model at higher resolutions or with more complex physics schemes; and (2) to prevent drifts in long-term simulations. Sophisticated theories that address the resolution dependencies of various phenomena and effectively limit the ranges of parameters will become crucial for coping with these problems within an affordable amount of time as we proceed to higher resolutions.

With regard to improving MJO predictions using GCRMs, another major issue is how to obtain appropriate initial atmospheric states. Although the MJO prediction skills are already impressively high in spite of the lack of detailed structures of cloud/cloud-system scales in the initial states derived as a simple interpolation of coarser-resolution reanalysis datasets, there remains a concern that the initial shock shifts the large-scale atmospheric state significantly away from reality and degrades the model performance. The initial shock is likely caused by

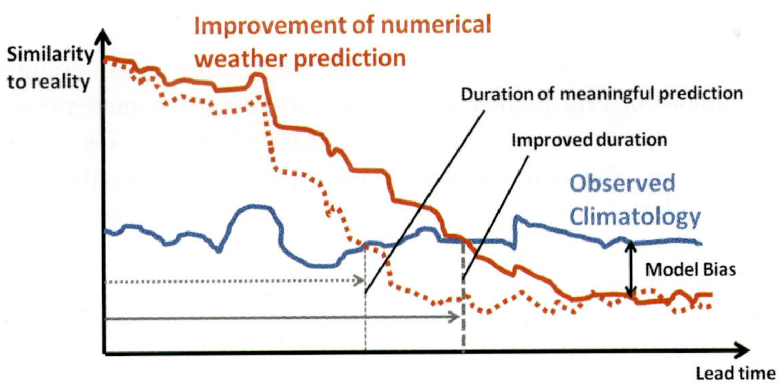

Fig. 6. A conceptual schematic describing the difference between improving short-term weather predictions and improving long-term climate simulations. The similarity between the reality and the observed climatology (blue plot) perturbs, but will stay constant on average unless there is a trend in the climate. The similarity between the reality and model simulations (red plots) initiated from observed states become degraded with time, and become worthless after the observed climatology becomes more similar to reality. The degrading will eventually stop when the model settles to perturb about its own climatology. Model biases are determined as the statistical differences between the observed climatology and model climatologies after the memory of the initial conditions have disappeared. Weather predictions are an initial-value problem, and the skills for which depend strongly on the accuracy of the initial conditions and the precision of individual processes. Climate simulations are a boundary problem, and the skills for which depend on how well the large-scale balances between multiple processes are represented (e.g. convective heating and radiative cooling). Elongated duration of a meaningful weather prediction does not guarantee smaller biases between the observed and modeled climatologies. The requirement to improve both in a common framework is a major challenge in current GCRM activities.

the imbalance between water vapor and cloud condensation and by the lack of kinetic energy at lower scales. A data assimilation system using the GCRM itself is needed to appropriately resolve this issue; however, it will require a large amount of effort and computational resources. If we disregard the computational costs in the hope that the computational resources will continue to increase, applying an ensemble prediction method using perturbed initial conditions is a promising way to further enhance prediction skills, as proven by many operational forecast models. Constructing efficient sets of ensemble members is another challenge to be faced in the near future.

5. Technical Issues

Extremely large computations using state-of-the-art supercomputers raise unique technical issues. In this section we note our experiences regarding such issues, in the hope of providing foresight for those who will participate in high-performance computing activities. Most of the issues can be attributed to the large data size, the large number of computational nodes involved, the large number of grid points, and the inflexible schedules.

The 870 m mesh run by NICAM requires 6 TB of input data, and the output data is approximately 500 TB for a 2 day simulation saved once every 30 min. When the simulation is executed with 20,480 nodes, nearly 1/4 of the entire system, the number of I/O files is on the order of 100,000 files. The scale of I/O required for atmospheric simulations are orders larger than most other HPC fields, e.g. orbital path simulations. The number of required I/O files is 100 times larger for older versions of NICAM, which by far exceeds the limit set on the K computer system. As waiting for the limit to be enlarged is not an option given the rigid schedule, the file format system of NICAM had to be quickly replaced so that many of the files can be merged. Although this enabled the K computer system to handle large simulations, the individual file-sizes increased to the order of 100 GB, which resulted in elongation of the already very long post-processing times. The NICAM output is on the icosahedral grid system, and is usually converted into lat–lon grids in the post-processing procedure. The increase in time required for seeking the target address in the huge files greatly slowed down the data conversion, and the post-processing time became much longer than the time required for executing the simulations. The disk space allowed on the K computer system quickly overflows unless the data is promptly converted and transferred to external archive disks. Thus, the data processing is often the bottle-neck of experiments on the K computer, raising the risk of losing enormous amounts of computational resources as a result of expiration. The validity check of executed simulations also became time-consuming, impeding long-term simulations that require many restarts. Tools for data processing and quick-looks are under redevelopment to reduce the required time. The large amount of output data also increases the difficulty of archiving the dataset. The required time and the risk of corruption increase upon data transfer to external archive disks, and disk space for backups is often unavailable. Therefore, we usually re-run the simulations if we lose data due to server errors, but the regenerated dataset may not be numerically identical to the former dataset if there is a major update to the K computer system.

The use of a large number of computational nodes increases the load of communications, which often ruins scalability, i.e. the computational efficiency per node decreases. Kodama *et al.* (2014) challenged this issue by developing a communication system based on a unique topology optimized for the K computer system. The use of many nodes also increases the risk of encountering node failures.

The increase in the number of grid points covering the globe was such that it resulted in a requirement of higher precision for variables used in grid formation, data processing, and visualization tools. The number of grids also exceeded the limits for certain external libraries and software. For example, the 870 m mesh outputs need to be divided into 20 panels if they are to be displayed using the widely used graphical tool GrADS without decreasing the resolution. The number of grid points also exceeds the number of pixels on a normal display.

The inflexible schedule resulting from the expiration limit of computational resources magnifies these issues. The waiting time for jobs submitted to the supercomputer can change drastically depending on the conditions of other research groups, making the situation even worse. The optimal number of nodes to speed up the simulations also alters depending on the jobs requested by other groups. We cannot avoid encountering unexpected troubles when we accept the challenge to run large simulations that exceed our past experiences. It often happens that we must quickly take measures that suit the various situations, under the pressure that a large amount of computational resources may be lost in the case of failure. To reduce some risk and stress, it may be advisable for research groups to set multiple levels for their "goals" and avoid "all-or-nothing" types of experimental designs if possible, in case they cannot complete their originally intended simulations.

6. Concluding Remarks

GCRMs will deservingly remain a frontier topic in the next decade. The rapidly increasing computational resources make various kinds of upgrades to the models and experiments available, e.g. higher resolution, higher-order closures, more complicated physics, longer durations, and larger sample sizes, providing a promising future for GCRM studies. The "post-K computer" project is already in progress, and is anticipated to accomplish 1 exa-flops, i.e. 100 times faster than the K computer. However, it is essential that we design the experiments carefully and avoid needless over-complication of scientific problems. Although the increased computational power enables us to apply model configurations with higher degrees of freedom, it is important to keep in mind that an essential part of science is the practice of simplification. A facile complication of the model would easily lead to increased difficulty in understanding the phenomena in question. Furthermore, increased complexity can also occasionally deteriorate model performances. Here we draw an analogy between simulation of the atmospheric state and simulation of the roll of an unevenly weighed, fiercely thrown dice. Increasing the precision of bounces of the dice in a numerical simulation will improve the early trajectory and angles of the dice for each attempt, but after several bounces the prediction will become essentially random. The probability distribution of the final upward surface will converge to a statistically derived value determined by the weight balance of the dice. If we improve the bouncing process but distort the weight balance of the dice in exchange, the early trajectory may be improved but the probability distribution of the final upward surface will become erroneous. Here the early trajectory, the probability distribution of the final upward surface, the bouncing process, and the weight balance of the dice are analogies for short-term weather forecasts, climatological statistics, cloud processes, and cloud statistics, respectively. Although the bouncing process and the weight balance are perhaps separate issues in the simulation of a dice roll, the cloud processes and cloud statistics in atmospheric simulations are closely related and therefore difficult issues. Furthermore, improvement of the precision of cloud processes requires increased degrees of freedom, which makes keeping control of cloud statistics more complicated and difficult. It is important to keep in mind the difference between increasing the precision of individual processes and improving the overall statistics, and avoid getting lost in uncontrollably increased degrees of freedom.

Acknowledgments

We thank the reviewer for comments that significantly improved the clarity of this chapter. Contents of the chapter are based on the NICAM-team members' experiences from simulations run on the Earth Simulator (JAMSTEC) and the K computer (RIKEN AICS). The drafting process was financially supported by Strategic Programs for Innovative Research of the MEXT, Japan, and by JSPS KAKENHI Grant Number 15K17757.

References

Arakawa, A., 2004: REVIEW ARTICLE The cumulus parameterization problem: Past, present, and future. *J. Climate*, **17**, 2493–2525.

Benedict, J. J. and D. A. Randall, 2009: Structure of the Madden-Julian Oscillation in the superparameterized CAM. *J. Atmos. Sci.*, **66**, 3277–3296.

Dodson, J. B., D. A. Randall and K. Suzuki, 2013: Comparison of observed and simulated tropical cumuliform clouds by CloudSat and NICAM. *J. Geophys. Res. Atmos.*, **18**, 1852–1867.

Dubos, T., S. Dubey, M. Tort, R. Mittal, Y. Meurdesoif, and F. Hourdin, 2015: DYNAMICO-1.0, an icosahedral hydrostatic dynamical core designed for consistency and versatility. *Geosci. Model Dev.*, **8**, 3131–3150.

Madden, R. A. and P. R. Julian, 1972: Description of global-scale circulation cells in the tropics with a 40–50 day period. *J. Atmos. Sci.*, **29**, 1109–1123.

Grabowski, W. W., 2001: Coupling cloud processes with the large-scale dynamics using the cloud-resolving convection parameterization (CRCP). *J. Atmos. Sci.*, **58**, 978–998.

Goto, D., 2014: Modeling of black carbon in Asia using a global-to-regional seamless aerosol-transport model. *Environ. Pollution*, **195**, 330–335.

Ham, S.-H., B.-J. Sohn, S. Kato, and M. Satoh, 2013: Vertical structure of ice cloud layers from CloudSat and CALIPSO measurements and comparison to NICAM simulations. *J. Geophys. Res.*, **118**, 9930–9947.

Hashino, T., M. Satoh, Y. Hagihara, T. Kubota, T. Matsui, T. Nasuno, and H. Okamoto, 2013: Evaluating global cloud distribution and microphysics from the NICAM against CloudSat and CALIPSO. *J. Geophys. Res.*, **118**, 7273–7292.

Heffernan, O., 2008: They say they want a revolution. *Nature*, **453**, 268–269.

Iga, S. and H. Tomita, 2014: Improved smoothness and homogeneity of icosahedral grids using the spring dynamics method. *J. Comp. Phys.*, **258**, 208–226.

Inoue, T., M. Satoh, H. Miura, and B. Mapes (2008): Characteristics of cloud size of deep convection simulated by a global cloud resolving model. *J. Meteor. Soc. Japan*, **86A**, 1–15.

Khairoutdinov, M. F. and D. A. Randall, 2001: A cloud resolving model as a cloud parameterization in the NCAR community climate system model: Preliminary results. *Geophys. Res. Lett.*, **28**, 3617–3620.

Kodama, C., M. Terai, A. T. Noda, Y. Yamada, M. Satoh, T. Seiki, S. Iga, H. Yashiro, H. Tomita, and K. Minami, 2014: Scalable rank-mapping algorithm for an icosahedral Grid system on the massive parallel computer with a 3-D torus network. *Parallel Computing*, **40**, 362–373.

Kodama, C., Y. Yamada, A. T. Noda, K. Kikuchi, Y. Kajikawa, T. Nasuno, T. Tomita, T. Yamaura, H. G. Takahashi, M. Hara, Y. Kawatani, M. Satoh, and M. Sugi, 2015: A 20-year climatology of a NICAM AMIP-type simulation. *J. Meteor. Soc. Japan*, **93**, 393–424.

Liu, P., M. Satoh, B. Wang, H. Fudeyasu, T. Nasuno, T. Li, H. Miura, H. Taniguchi, H. Masunaga, X. Fu, and H. Annamalai, 2009: A MJO simulated by the NICAM at 14-km and 7-km resolutions. *Mon. Wea. Rev.*, **137**, 3254–3268.

Mapes, B., S. Tulich, T. Nasuno, and M. Satoh, 2008: Predictability aspects of global aqua-planet simulations with explicit convection. *J. Meteor. Soc. Japan*, **86A**, 175–185.

Miura, H., M. Satoh, and M. Katsumata, 2009: Spontaneous onset of a Madden-Julian Oscillation event in a cloud-system-resolving simulation. *Geophys. Res. Lett.*, **36**, L13802.

Miura, H., H. Tomita, T. Nasuno, S. Iga, M. Satoh, and T. Matsuno, 2005: A climate sensitivity test using a global cloud resolving model under an aqua planet condition. *Geophys. Res. Lett.*, **32**, L19717.

Miura, H., M. Satoh, H. Tomita, T. Nasuno, S. Iga, and A. T. Noda 2007a: A short-duration global cloud-resolving simulation with a realistic land and sea distribution. *Geophys. Res. Lett.*, **34**, L02804.

Miura, H., M. Satoh, T. Nasuno, A. T. Noda, and K. Oouchi, 2007b: A Madden-Julian Oscillation event realistically simulated by a global cloud-resolving model. *Science*, **318**, 1763–1765.

Miyamoto, Y., Y. Kajikawa, R. Yoshida, T. Yamaura, H. Yashiro, and H. Tomita, 2013: Deep moist atmospheric convection in a subkilometer global simulation, *Geophys. Res. Lett.*, **40**, 4922–4926.

Nasuno, T., H. Tomita, S. Iga, H. Miura, and M. Satoh, 2007: Multi-scale organization of convection simulated with explicit cloud processes on an aquaplanet. *J. Atmos. Sci.*, **64**, 1902–1921.

Nasuno, T., 2013: Forecast skill of Madden-Julian Oscillation events in a global nonhydrostatic model during the CINDY2011/DYNAMO observation period, SOLA, **9**, 69–73.

Nasuno, T., H. Miura, M. Satoh, A. T. Noda, and K. Oouchi, 2009: Multi-scale organization of convection in a global numerical simulation of the December 2006 MJO event using explicit moist processes. *J. Meteor. Soc. Japan*, **87**, 335–345.

Noda, A. T., K. Nakamura, T. Iwasaki, and M. Satoh, 2013: A numerical study of a stratocumulus-topped boundary-layer: Relations of decaying clouds with stability parameters across inversion. *J. Meteor. Soc. Japan*, **91**, 727–746.

Noda, A. T., K. Oouchi, M. Satoh, and H. Tomita, 2012: Quantitative assessment of diurnal variation of tropical convection simulated by a global nonhydrostatic model without cumulus parameterization. *J. Climate*, **25**, 5119–5134.

Miyakawa, T., M. Satoh, H. Miura, H. Tomita, H. Yashiro, A. T. Noda, Y. Yamada, C. Kodama, M. Kimoto, and K. Yoneyama, 2014: Madden-Julian Oscillation prediction skill of a new-generation global model. *Nature Commun.*, **5**, 3769.

Miyakawa, T., Y. N. Takayabu, T. Nasuno, H. Miura, M. Satoh, and M. W. Moncrieff, 2012: Convective momentum transport by rainbands within a Madden-Julian Oscillation in a global nonhydrostatic model with explicit deep convective processes. Part I: Methodology and general results. *J. Atmos. Sci.*, **69**, 1317–1338.

Miyamoto, Y., Y. Kajikawa, R. Yoshida, T. Yamaura, H. Yashiro, and H. Tomita, 2013: Deep moist atmospheric convection in a subkilometer global simulation. *Geophys. Res. Lett.*, **40**, 4922–4926.

Niwa, Y., H. Tomita, M. Satoh, and R. Imasu, 2011: A three-dimensional icosahedral grid advection scheme preserving monotonicity and consistency with continuity for atmospheric tracer transport. *J. Meteor. Soc. Japan*, **89**, 255–268.

Prichard, M. S., C. S. Bretherton, and C. A. DeMott, 2014: Restricting 32–128 km horizontal scales hardly affects the MJO in the Superparameterized Community Atmosphere Model v.3.0 but the number of cloud-resolving grid columns constrains vertical mixing. *J. Adv. Model. Earth Syst.*, **6**, 723–739. doi:10.1002/2014MS000340.

Roh, W. and Satoh, M., 2014: Evaluation of precipitating hydrometeor parameterizations in a single-moment bulk microphysics scheme for deep convective systems over the tropical open ocean. *J. Atmos. Sci.*, **71**, 2654–2673.

Sato, T., T. Yoshikane, M. Satoh, M. Miura, and H. Fujinami, 2008: Resolution dependency of the diurnal cycle of convective clouds over the Tibetan Plateau in a mesoscale model. *J. Meteor. Soc. Japan*, **86A**, 17–31.

Sato, T., H. Miura, and M. Satoh, 2007: Spring diurnal cycle of clouds over Tibetan Plateau: Global cloud-resolving simulations and satellite observations. *Geophys. Res. Lett.*, **34**, L18816.

Satoh, M., 2002: Conservative scheme for the compressible non-hydrostatic models with the horizontally explicit and vertically implicit time integration scheme. *Mon. Wea. Rev.*, **130**, 1227–1245

Satoh, M., 2003: Conservative scheme for a compressible nonhydrostatic model with moist processes. *Mon. Wea. Rev.*, **131**, 1033–1050.

Satoh, M., H. Tomita, H. Yashiro, H. Miura, C. Kodama, T. Seiki, A. T. Noda, Y. Yamada, D. Goto, M. Sawada, T. Miyoshi, Y. Niwa, M. Hara, T. Ohno, S. Iga, T. Arakawa, T. Inoue, and H. Kubokawa, 2014: The non-hydrostatic icosahedral atmospheric model: Description and development. *Progress in Earth and Planetary Science*, **1**, 18.

Seiki, T., M. Satoh, H. Tomita, and T. Nakajima, 2014: Simultaneous evaluation of ice cloud microphysics and non-sphericity of the cloud optical properties using hydrometeor video sonde and radiometer sonde in-situ observations. *J. Geophys. Res. Atmos.*, **119**, 6681–6701.

Skamarock, W. C., S.-H. Park, J. B. Klemp, and C. Snyder, 2014: Atmospheric kinetic energy spectra from global high-resolution nonhydrostatic simulations. *J. Atmos. Sci.*, in press.

Tao, W.-K., D. Anderson, J. Chern, J. Estin, A. Hou, P. Houser, R. Kakar, S. Lang, W. Lau, C. Peters-Lidard, X. Li, T. Matsui, M. Rienecker, M. R. Schoeberl, B.-W. Shen, J.-J. Shi, and X. Zeng, 2009: Goddard multi-scale modeling systems with unified physics. *Annales Geophysics*, **27**, 3055–3064.

Tomita, H., M. Tsugawa, M. Satoh, and K. Goto, 2001: Shallow water model on a modified icosagedral geodesic grid by using spring dynamics. *J. Comp. Phys.*, **174**, 579–613.

Tomita, H., M. Satoh, and K. Goto, 2002: An optimization of icosahedral grid by using spring dynamics. *J. Comp. Phys.*, **183**, 307–331.

Tomita, H. and M. Satoh, 2004: A new dynamical framework of nonhydrostatic global model using the icosahedral grid. *Fluid Dyn. Res.*, **34**, 357–400.

Tomita, H., H. Miura, S. Iga, T. Nasuno, and M. Satoh, 2005: A global cloud-resolving simulation: Preliminary results from an aqua planet experiment. *Geophys. Res. Lett.*, **32**, L08805. doi:10.1029/2005GL022459.

Tomita, H., 2008a: A stretched grid on a sphere by new grid transformation. *J. Meteor. Soc. Japan*, **86A**, 107–119.

Tomita, H., 2008b: New microphysical schemes with five and six categories by diagnostic generation of cloud ice. *J. Meteor. Soc. Japan*, **86A**, 121–142.

Vitart, F. and F. Molteni, 2010: Simulation of the Madden-Julian Oscillation and its teleconnections in the ECMWF forecast system. *Quart. J. Roy. Meteor. Soc.*, **136**, 842–855.

Vitart, F. and F. Molteni, 2011: Simulation of the Madden-Julian Oscillation and its impact over Europe in ECMWF's monthly forecasts. *ECMWF Newsletter*, **126**, 12–17.

Yamada, Y. and M. Satoh, 2013: Response of ice and liquid water paths of tropical cyclones to global warming simulated by a global nonhydrostatic model with explicit cloud microphysics. *J. Climate*, **26**, 9931–9945.

Yashiro, H., K. Sudo, S. Yonemura, and M. Takigawa, 2011: The impact of soil uptake on the global distribution of molecular hydrogen: Chemical transport model simulation. *Atmos. Chem. Phys.*, **11**, 6701–6719.

Zängl, G., H. Tomita, M. Satoh, T. Ludwig, L. Linardakis, J. Thuburn, and T. Dubos, 2011: ICOsahedral-grid models for EXascale Earth system simulations. *IS-ENES workshop*, Lecce, 14–16 Dec. 2011.

Zängl, G., D. Reinert, P. Rípodas, and M. Baldauf, 2014: The ICON (ICOsahedral Non-hydrostatic) modelling framework of DWD and MPI-M: Description of the non-hydrostatic dynamical core. *Quart. J. Roy. Meteor. Soc.* doi: 10.1002/qj.2378.

Part III
Extreme Weather

Chapter 15

Potential Vorticity Structure and Propagation Mechanism of Indian Monsoon Depressions

William R. Boos

Department of Geology and Geophysics, Yale University, New Haven, CT 06511, USA
billboos@alum.mit.edu

Brian E. Mapes

Rosenstiel School of Marine and Atmospheric Sciences, University of Miami, Miami, FL 33149, USA

Varun S. Murthy

Department of Geology and Geophysics, Yale University, New Haven, CT 06511, USA

Indian monsoon depressions are synoptic-scale storms that form primarily over the Bay of Bengal and propagate westward over the subcontinent, producing a large fraction of India's total summer precipitation. We recently showed that, contrary to long-standing ideas, the westward propagation of Indian monsoon depressions is accomplished primarily by horizontal adiabatic advection of potential vorticity (PV), not by vortex stretching or diabatic PV generation that occurs in the region of quasi-geostrophic ascent southwest of the vortex center. This chapter extends that work by using several reanalysis products to examine case studies of Indian monsoon depressions. In all reanalyses examined, monsoon depressions have maximum PV in the middle troposphere, at higher altitudes than the level of maximum relative vorticity. The horizontal structure of mid-tropospheric PV suggests that the axial asymmetry of the vortex that produces the nonlinear westward advection may result at least partly from diabatic heating. Thus, although storm motion is produced primarily by horizontal adiabatic advection, diabatic heating can play an indirect role by shaping the PV field that produces this advection.

1. Introduction

Tropical monsoon circulations contain synoptic-scale cyclonic vortices that produce a large fraction of the annual mean rainfall in many monsoon regions (Yoon and Chen 2005; Berry *et al.* 2012; Hurley and Boos 2015). In the Indian region these vortices are commonly classified as monsoon lows, monsoon depressions, deep depressions, and cyclonic storms based on their surface wind speeds and surface pressure anomalies (Ramage 1971; India Meteorological Department 2011). Monsoon depressions (hereafter simply referred to as depressions) are the subset of these vortices that have surface wind speeds of 8–13 $\mathrm{m\,s^{-1}}$ and typical central surface pressure anomalies of 4–10 hPa. Each year, an average of 5–7 depressions are observed during June–September in the Indian monsoon region. These typically form over the Bay of Bengal and travel northwest over the Indian subcontinent, lasting for 3–6 days (Mooley and

The Global Monsoon System: Research and Forecast (3rd Edition)
Edited by C. P. Chang et al.

Shukla 1987; Sikka 1977). Peak precipitation rates in the average depression are about 50 mm day^{-1} (Yoon and Chen 2005, Boos *et al.* 2015).

Although monsoon depressions are cyclonic vortices located in the tropics, they are distinct from tropical cyclones as traditionally defined. Monsoon depressions have a cold core in the lower troposphere and loosely organized convection with no central core or eyewall (e.g. Sikka 1977; Godbole 1977; Yoon and Chen 2005). In contrast, the U.S. National Hurricane Center defines a tropical cyclone as a warm-core vortex with a closed surface wind circulation about a well-defined center (NHC 2015). Furthermore, monsoon depressions frequently pass over land surfaces without loss of intensity, indicating that, unlike tropical cyclones, they do not maintain their intensity by extracting heat from the ocean via surface evaporative fluxes.

The mechanisms of growth and propagation of Indian monsoon depressions have long been associated with the strong vertical wind shear of the monsoon mean state, or equivalently with that state's strong poleward temperature gradient. The amplification of monsoon depressions to mature intensity has been most commonly attributed to some sort of baroclinic instability, or stable non-modal baroclinic growth, heavily modified by the diabatic effects of moist convection (e.g. Krishnamurti *et al.* 1983, Moorthi and Arakawa 1985, Farrell 1985, Krishnamurti *et al.* 2013). The vertical shear of the monsoon mean state has also been invoked to explain the apparent upstream propagation of Indian monsoon depressions against the low-level monsoon westerlies. Quasi-geostrophic lifting downshear (i.e. to the west) of the low-level vorticity maximum produces convectively coupled ascent and vortex stretching that was argued to shift the vortex westward (Rao and Rajamani 1970; Sanders 1984; Chen *et al.* 2005). For tropical cyclones viewed as potential vorticity (PV) maxima, this propagation mechanism has been called "direct generation of a positive PV tendency by asymmetric heating" (Wu and Wang 2001), because the downshear precipitation provides a diabatic PV source. Observations have shown that peak precipitation and ascent do occur west of the vortex center in monsoon depressions (Godbole 1977), although the peak ascent typically occurs southwest of the vortex center while storm propagation is to the northwest.

Boos *et al.* (2015) examined the PV tendencies involved in the propagation of Indian monsoon depressions and showed that direct generation of PV by diabatic sources is less important than the horizontal adiabatic advection of the depression's central core of PV by azimuthally asymmetric flow. However, Boos *et al.* (2015) largely assumed that the asymmetric flow accomplishing the horizontal advection originated from interaction of the depression with the planetary vorticity gradient — the so-called "beta drift" that is also thought to play a role in tropical cyclone motion (see Chan 2005 for a review). Boos *et al.* (2015) did not consider the possibility that diabatic heating might generate the azimuthally asymmetric flow that then adiabatically advects the depression to the northwest. That indirect role of diabatic heating was discussed by Wu and Wang (2001) in the context of tropical cyclone motion.

This chapter considers these diabatic and adiabatic mechanisms for the axial asymmetry and thus for the propagation mechanism of Indian monsoon depressions, in the context of two case studies. PV structures are compared across multiple reanalysis products, some of which provide estimates of diabatic heating rates. Our analyses extend the results of Boos *et al.* (2015) by showing that diabatic generation of PV may help produce the asymmetries of the PV field that are responsible for the horizontal adiabatic advection of PV in monsoon depressions.

2. Data and Methods

General characteristics of the propagation of Indian monsoon depressions are first established by examining a large ensemble of storms as represented in the European Centre for Medium-Range Weather Forecasts (ECMWF) Interim Reanalysis (ERA-Interim; Dee *et al.* 2011) for 1979–2012. ERA-Interim uses a spectral T255 atmospheric model constrained by observations, with 60 vertical levels and gridded output available every 6 hours at a horizontal resolution of roughly $0.7 \times 0.7°$. Depressions are identified and tracked in the ERA-Interim dataset using an automated algorithm (Hodges 1995) to identify propagating 850 hPa relative vorticity maxima having surface wind speeds and sea level pressure anomalies consistent with the India Meteorological Department (IMD) definition of a monsoon depression. Full details of this tracking and selection algorithm are provided in Hurley and Boos (2015). Weaker storms (monsoon lows) and stronger storms (deep depressions and cyclonic storms) are filtered out by these selection criteria.

Individual case studies make use of the ECMWF Year of Tropical Convection (EC-YOTC) analysis (Waliser *et al.* 2012), which is available for May 2008–April 2010. This dataset uses a spectral T799 atmospheric model with 97 vertical levels and gridded output available every 6 hours at a horizontal resolution of $0.25 \times 0.25°$. The EC-YOTC dataset provides tendencies associated with model parameterizations and dynamics during 24-hour forecasts initialized at 1200 UTC daily.

For case studies, comparisons have also been made with other reanalysis products: the NCEP-NCAR reanalysis (Kalnay *et al.* 1996), the Climate Forecast System Reanalysis (CFSR, Saha *et al.* 2010), and the MERRA reanalysis (Reinecker *et al.* 2011). Case study data from these sources were obtained from online aggregations and displayed together using the free Integrated Data Viewer (IDV) software. IDV "bundle" files for reproducing these analyses, freely usable from any networked computer, are available on request from the second author (mapes@miami.edu). Operational analyses from the NCEP Global Forecast System and the coarse 20th Century reanalysis were also examined, but are not presented here for brevity.

3. Results

3.1. *General propagation characteristics*

The tracking algorithm identified 159 vorticity maxima with track origins in the region 50–100°E, 0–30°N during June–Sept. 1979–2012 that can be classified as depressions. Consistent with previous studies (e.g. Mooley and Shukla 1987), genesis occurs primarily over the northern Bay of Bengal, and depressions propagate to the west-northwest with tracks terminating over northern India and surrounding regions (Fig. 1). A fair number of genesis points do fall over land, primarily in and around the eastern Indo-Gangetic plain, and some storms also form over the Arabian Sea. The joint distribution of the zonal and meridional components of

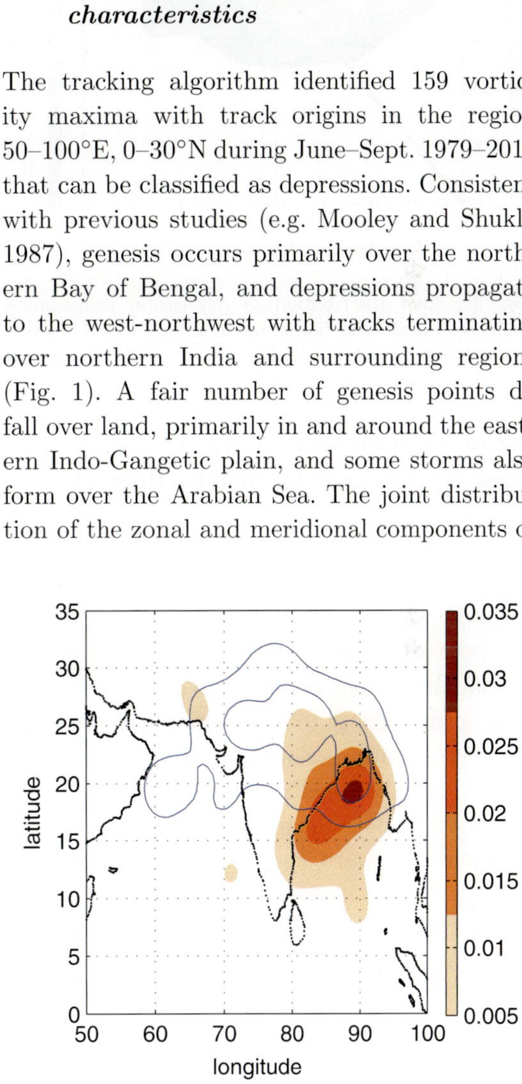

Fig. 1. Shading shows number of depression genesis points per square degree in ERA-Interim per summer season (June–Sept.) after smoothing with a Gaussian kernel having an isotropic standard deviation of 2°. Blue contours show the same quantity but for the endpoints of depression tracks. Contour intervals are 0.01 starting at 0.005 storms per summer per square degree.

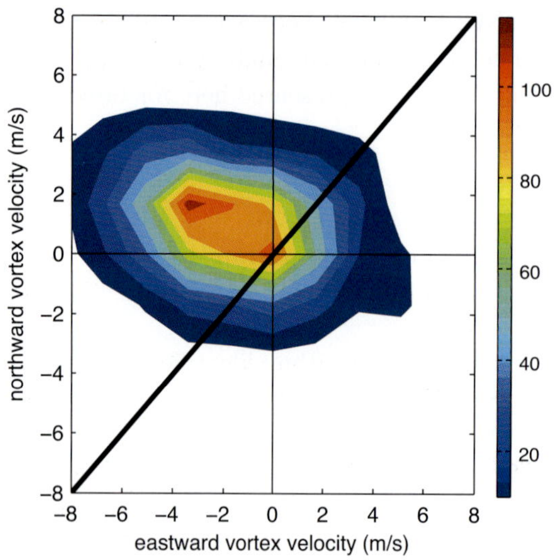

Fig. 2. Joint distribution of depression propagation velocities for all time points in the middle third of every track. Contour interval is a count of 10 starting at 10. Thick black line is the 1:1 line.

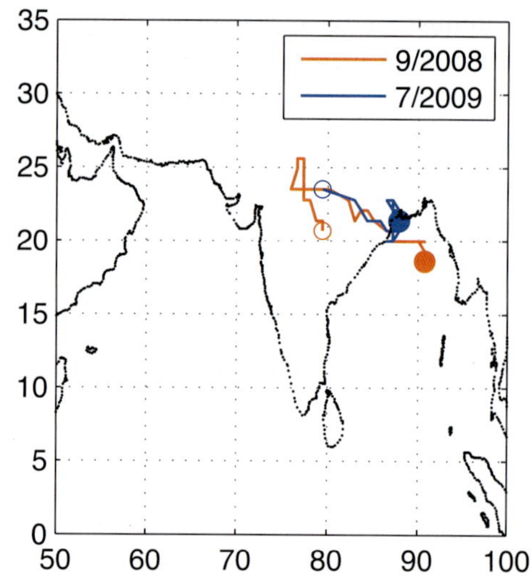

Fig. 3. Tracks of two depressions with genesis points (filled circles) on Sept. 13, 2008 (red) and July 8, 2009 (blue). Track termination points are shown by open circles and occur on Sept. 21, 2008 (red) and July 16, 2009 (blue).

the propagation velocity vector, sampled once every 6 hours over the middle third of every depression track, has a mode at a zonal velocity of $-3.4\,\mathrm{m\,s^{-1}}$ and a meridional velocity of $1.7\,\mathrm{m\,s^{-1}}$, which corresponds to propagation to the west-northwest at $3.8\,\mathrm{m\,s^{-1}}$ (Fig. 2). However, the propagation vector varies substantially in magnitude and direction, with some storms even propagating to the southeast, and the mean propagation velocity being $-1.5\,\mathrm{m\,s^{-1}}$ and $1.0\,\mathrm{m\,s^{-1}}$ in the zonal and meridional directions, respectively. Both the mode and mean are slower than the westward speed of 5° longitude/day ($5.8\,\mathrm{m\,s^{-1}}$) that is sometimes cited for the propagation speed of Indian monsoon depressions (e.g. Yoon and Huang 2012).

3.2. *Vorticity and PV structures*

We now examine the vorticity and PV distributions of two individual depressions, one occurring in Sept. 2008 and the other in July 2009. These storms were chosen because they occur during the period of the EC-YOTC dataset, have intensities typical of Indian monsoon depressions, and propagate in the classic west-northwest direction from their origin points over the northern Bay of Bengal (Fig. 3). When the maximum surface wind speed achieved within 500 km of the storm center is sampled every 6 hours along each track, the 2008 depression is found to have a distribution centered at slightly lower speeds than that associated with all depressions, while the 2009 storm is somewhat stronger (not shown).

The September 2008 case is shown in Fig. 4 near the time it crossed India's east coast. A second cyclone on the west coast was also present, so the transect was positioned to display them both. Relative vorticity (lower panels) has maximum values in the lower troposphere (the color scale saturates in Fig. 4c), although the vorticity column does extend into the upper troposphere. An enormous increase in detail is seen in the 0.25-degree EC-YOTC product when compared with the 2.5 degree NCEP-NCAR product. Figure 5 shows enlarged transects through the east coast depression of Fig. 4, for relative vorticity (top) and PV (bottom) from

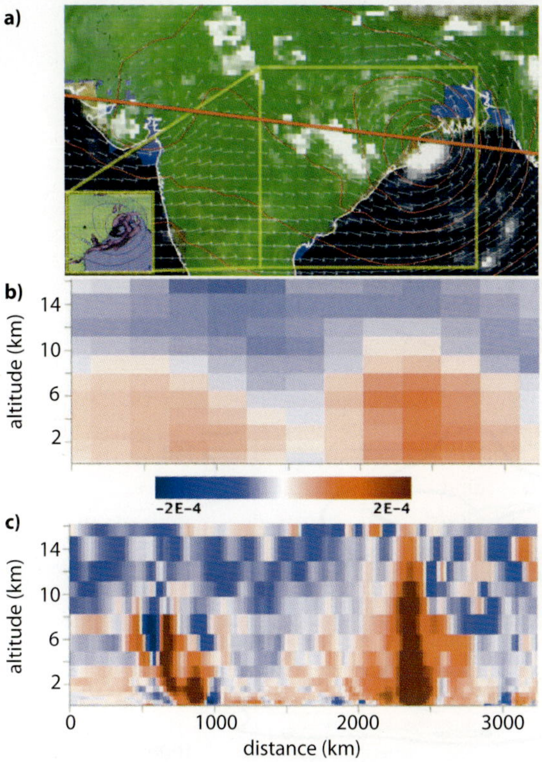

Fig. 4. Depictions of the monsoon cyclone (Bay of Bengal coast) and a secondary cyclone (Arabian Sea coast) on Sept. 16, 2008 at 1200 UTC. (a) Map view with EC-YOTC winds and geopotential contours at 850 hPa, overlain by TRMM 3B42 rainrate product (white shaded pixels). Inset (yellow box) shows a top view of the PV = 1 PVU isosurface in the EC-YOTC analysis (magenta). Red line shows the location of the cross sections below. (b) Relative vorticity section in the NCEP-NCAR reanalysis (on a 2.5 degree grid). (c) As in (b) but for the EC-YOTC analysis (on a 0.25 degree grid). Color bar is for both (b) and (c) and has units of s^{-1}.

three different reanalyses. In all three products (EC-YOTC, CFSR, and MERRA), the vertical structure of PV is more top-heavy than that of relative vorticity. This is indicative of the larger static stability in the mid-troposphere, near the melting level, since for monsoon depressions the PV is approximately equal to the product of the vertical component of the absolute vorticity vector and the climatological mean stratification (Boos *et al.* 2015).

The July 2009 case (Fig. 6) is somewhat weaker, but largely confirms the lessons of Figs. 4 and 5. In both cases, precipitation west of the cyclone is seen at the analysis time (1200 UTC; top panels of Figs. 4 and 6). Note that although the July 2009 case had a weaker PV anomaly than the Sept. 2008 case, its peak surface wind speeds were stronger, as represented in ERA-Interim (not shown).

These vorticity and PV structures are typical of those seen in monsoon depressions. To demonstrate this, we show in Fig. 7 the composite mean PV and zonal wind, averaged in a storm-centered coordinate system along the middle third of each of the 159 tracks in ERA-Interim. The maximum PV is clearly centered at 500 hPa, and a weaker secondary maximum is centered around 725 hPa. The PV maximum is located just slightly above the climatological mean zero zonal wind line, although the circulation of the storm is so strong that there is large horizontal shear across the PV maximum. The climatological baroclinicity is evident in the easterly vertical wind shear at the storm center, especially at the highest levels of the troposphere south of the center of the composite depression. Several conclusions can be drawn from these structures: (1) from a PV perspective, one would view depressions as being centered near 500 hPa, well above the lower troposphere where the relative vorticity peaks and the climatological mean flow is eastward; (2) the depression circulation is strong compared to the mean flow (i.e. it is far from the limit of a weak linear anomaly), so that any horizontal advection may be highly nonlinear and it may not be useful to think of depressions being advected by a mean flow. The rest of this chapter is devoted to examination of the mechanism of propagation.

3.3. *The central question: is beta drift sufficient?*

Boos *et al.* (2015) showed that the mid-tropospheric PV maximum of Indian monsoon

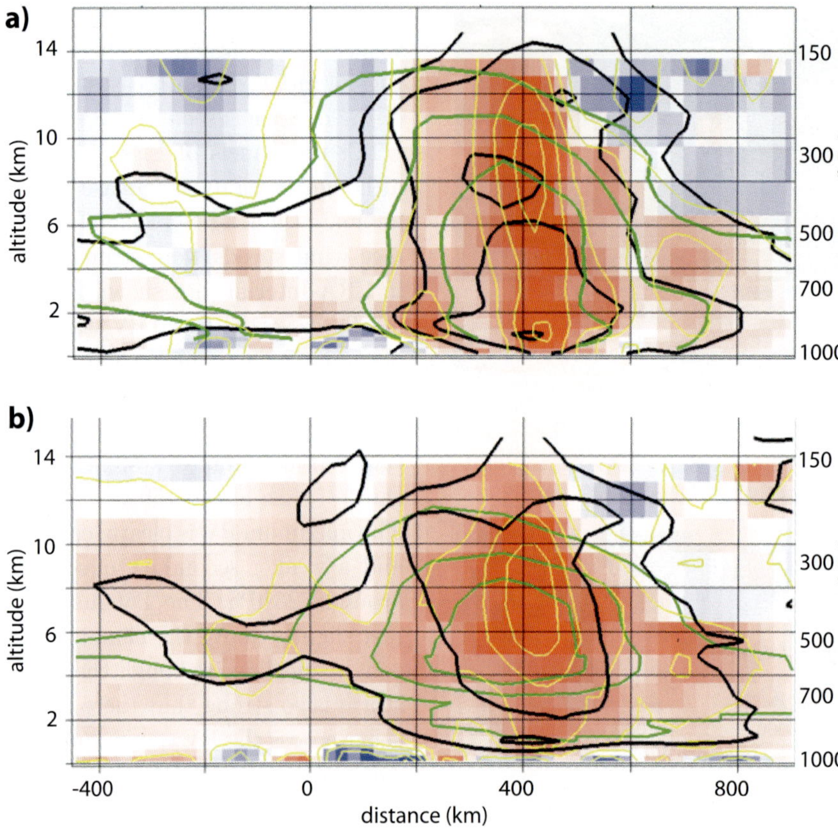

Fig. 5. Cross sections through the Bay of Bengal cyclone shown in Fig. 4, along the same transect. (a) Relative vorticity (contour interval $5 \times 10^{-5}\,s^{-1}$). (b) Potential vorticity (contour interval 0.5 PVU). Color shading (red is positive and blue negative) and yellow contours both show the EC-YOTC analysis, black contours show CFSR, and green contours show MERRA.

depressions moves westward by horizontal adiabatic advection. They demonstrated that, for more than 100 monsoon depressions, the storm motion had a direction and speed consistent with that of the PV-weighted horizontal wind at 500 hPa. Only the total horizontal wind was strong enough to explain the storm motion; the time-mean wind was much too weak at 500 hPa and was in the wrong direction in the lower troposphere. They also used a PV budget for the same Sept. 2008 storm examined here to show that the net Eulerian time tendency of PV was produced primarily by horizontal adiabatic advection, with diabatic PV generation playing only a minor role. Yet how, exactly, does this horizontal, nonlinear advection produce the net storm motion?

Nonlinear self-advection of vortices on rotating spheres has been well studied in the context of tropical cyclones, where this mechanism of propagation is known as "beta drift" (see reviews by Wang *et al.* 1998, Chan 2005). Beta drift occurs when a cyclonic vortex advects the background vorticity field, which has a poleward gradient, to produce an anomalous cyclone to the west of the original vortex and an anomalous anticyclone to the east. These induced anomalies, called beta gyres, have a circulation that interferes constructively to produce a net wind that advects the center of the

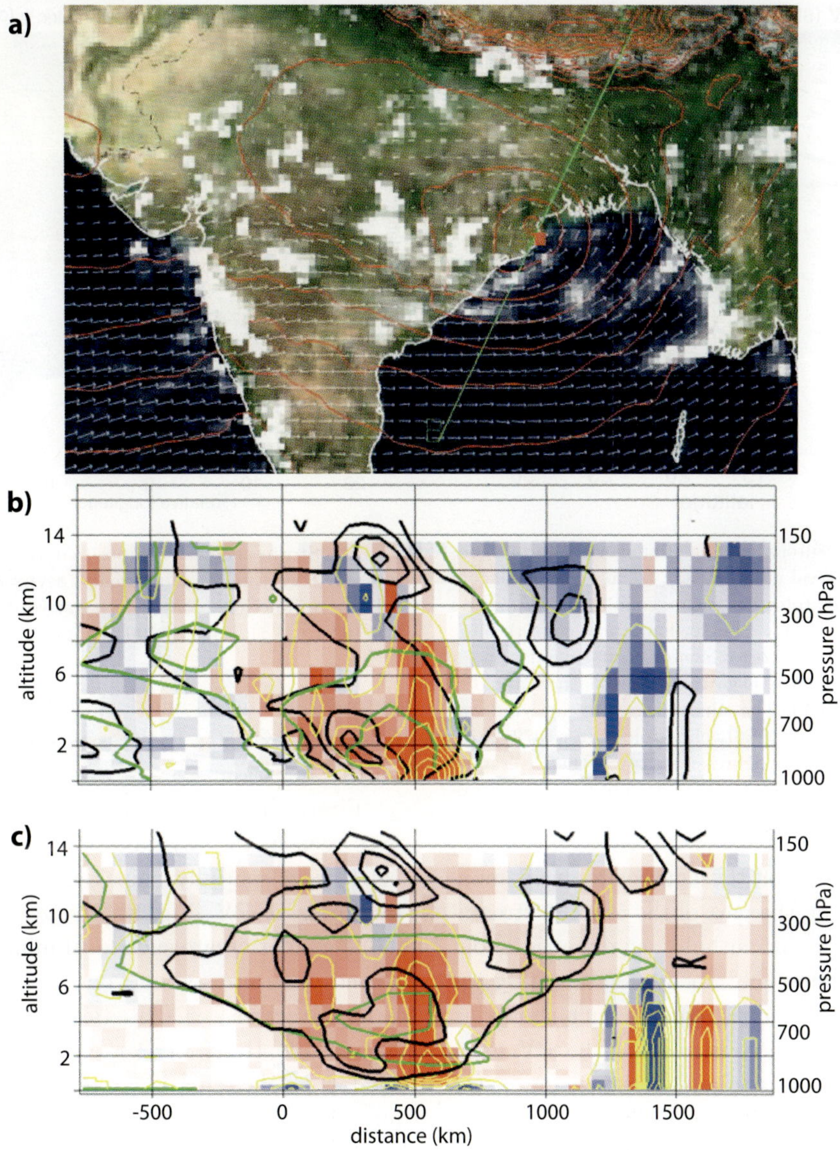

Fig. 6. For the monsoon depression of July 14, 2009 at 1200 UTC, (a) as in Fig. 4a, with the green line marking the transect for which vertical profiles are shown in the next two panels. (b) Relative vorticity, as in Fig. 5b. (c) Potential vorticity, as in Fig. 5c. The lower right region of (b) and (c) lies below the surface of the Himalayas and should be disregarded as it represents non-physical data in the EC-YOTC product.

original vortex. The anomalous gyres are typically rotated by the azimuthally symmetric flow to produce northwestward flow across the original vortex, although there is some sensitivity to the baroclinicity of the vortex, background flow, and other factors.

The existence of asymmetric gyres that advect monsoon depressions to the northwest can be demonstrated through decomposition of the horizontal streamfunction into components that are azimuthally symmetric and asymmetric about the vortex center (e.g. Fiorino and

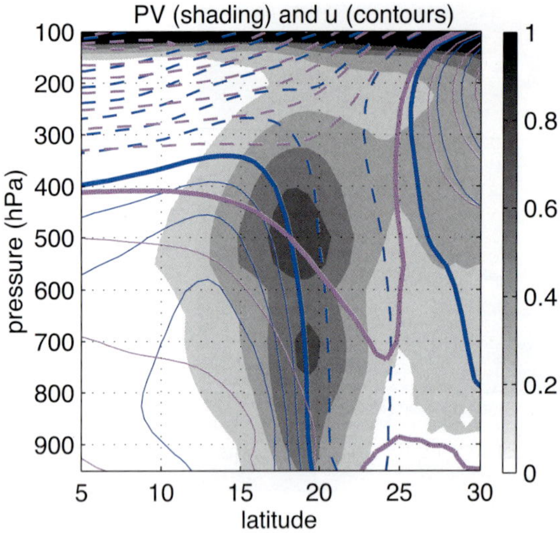

Fig. 7. Storm-centered composite mean PV (gray shading, contour interval of 0.2 PVU), zonal wind (blue contours), and climatological mean zonal wind (magenta contours), all along a single longitude line running through the 850 hPa vortex center. Zonal wind contour interval is 4 m/s with negative values dashed and the zero contour in bold. The climatological mean wind was obtained by sampling a daily climatology (for 1979–2012) along each point of the depression's track, then averaging in a storm-centered coordinate system to form a composite mean.

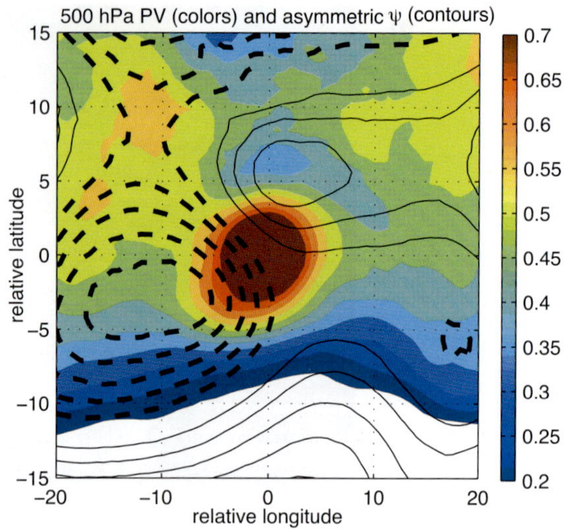

Fig. 8. Storm-centered composite mean PV (color shading, in PVU), and azimuthally asymmetric horizontal streamfunction (contours, negative dashed, interval $5 \times 10^5\,\mathrm{m^2\,s^{-1}}$), both at 500 hPa.

Elsberry 1989). For the composite-mean monsoon depression, the part of the 500 hPa streamfunction that is azimuthally asymmetric about the PV maximum (obtained by subtracting the azimuthally symmetric component) shows an anomalous cyclone southwest of the PV maximum and an anomalous anticyclone to the northeast (Fig. 8). These gyres together produce flow toward the northwest over the PV maximum, and have a structure qualitatively consistent with that seen in idealized models of barotropic beta drift (e.g. Chan 2005). The depression does lie in a region of poleward PV gradient, but this gradient seems fairly weak at the latitude of the depression. *This motivates the central question of this chapter: can the beta effect alone produce the azimuthally asymmetric vorticity needed for the northwestward advection of the 500 hPa PV structure?* Since composite averaging may obscure detailed relationships between gradients and local maxima in the PV field, we next turn to a case study to shed light on this question.

First, though, we note that the anticyclonic gyre seen in the asymmetric streamfunction at 500 hPa is spatially coincident with a relative minimum of PV (Fig. 8). Horizontal advection alone cannot create such local minima in a conserved quantity, so one wonders if friction, vertical advection, or diabatic heating might be responsible. Hunt and Parker (2016) suggest that the low-level circulation of monsoon depressions may interact barotropically with Himalayan topography to produce northwestward motion of the depressions; they frame this mechanism in terms of image vortices, but one might relatedly argue that friction with the Himalayas produces this PV minimum. It is unclear whether this interaction could operate at 500 hPa (average plateau surface pressures are around 550 hPa); destruction of PV in the dry boundary layer over the Tibetan Plateau might also play a role. It is also unclear whether interaction with topography could explain the cyclonic gyre to the southwest of the PV maximum.

3.4. *Case study of asymmetric gyre creation*

At 500 hPa, the streamfunction minimum of the Sept. 2008 depression is centered slightly southwest of the PV maximum, and the PV field around the depression is relatively disorganized without a clear and strong meridional gradient at the latitude of the depression's center (Fig. 9a). The offset between the streamfunction minimum and PV maximum is more clearly seen in the component of the streamfunction that is azimuthally asymmetric about the PV maximum; that field shows an anomalous cyclone southwest of the PV maximum and an anomalous anticyclone to the northeast (Fig. 9b), highly similar to the structure seen in the composite mean (Fig. 8). The total wind at 500 hPa is indeed directed toward the northwest (blue vector in Fig. 9b), and has a magnitude and direction only slightly different from that of the storm propagation (red vector). The fact that the time mean wind (the green vector) is about an order of magnitude weaker and directed toward the southwest indicates that advection by the time mean wind has little relevance for translation of the 500 hPa PV maximum. All of these features are consistent with the composite-mean behavior discussed above, and with the idea that horizontal, nonlinear advection produces the northwestward motion of the depression.

Chan *et al.* (2002) employed similar reasoning in the context of tropical cyclones, and called this the "advection of symmetric PV by the asymmetric flow [AASPV, which includes, but is not limited to, the environmental 'steering flow' and the beta-induced circulation (the so-called ventilation flow)]". Diabatic PV generation (and its subsequent vertical advection) shift the storm propagation vector only slightly to the southwest of the position it would have due to horizontal advection alone (Boos *et al.* 2015), indicating that the direct generation of PV by asymmetric heating plays little role in shifting the PV maximum directly.

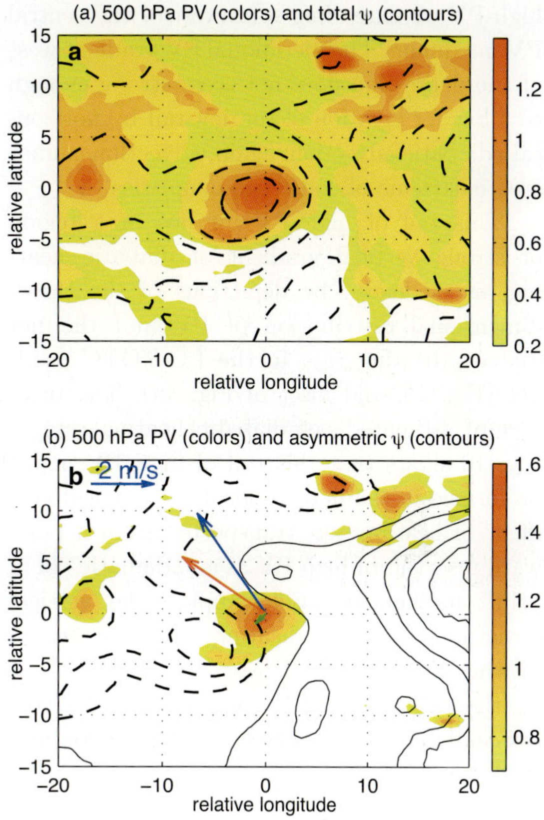

Fig. 9. For the 2008 depression at 1800 UTC on Sept. 16, in a storm-centered coordinate system: (a) total horizontal streamfunction at 500 hPa (contours, dashed negative, interval of 10^6 m^2 s^{-1}) and 500 hPa PV (color shading); (b) as in (a) but the azimuthally asymmetric part of the streamfunction, with a contour interval of 5×10^5 m^2 s^{-1}. Blue arrow is the total horizontal wind vector, green arrow is the climatological mean horizontal wind vector, and red arrow is the storm propagation velocity; all wind vectors are PV-weighted averages taken over a $2 \times 2°$ box centered on the PV maximum. Color shading in both panels shows the same PV field, but with a larger minimum contour in (b) to highlight the maxima.

However, we now suggest that this diabatic PV generation creates asymmetries in the flow field that allow for the adiabatic advection of the depression to the northwest. The relatively weak anticyclonic gyre northeast of the vortex center is not obviously associated with any negative PV anomaly, while the stronger cyclonic gyre to the southwest is roughly coincident with a "tail" of high-PV air extending southwest of the central PV maximum. The meridional PV gradient west of the depression does not seem strong enough to allow for creation of the PV tail by horizontal adiabatic advection: again, advection cannot create extrema of conserved quantities.

This tail of high PV air lies in a region of strong vertical gradient of diabatic heating, as evidenced by depictions of the three-dimensional distribution of PV and the net diabatic heating rate in the EC-YOTC product (Fig. 10, and inset of Fig. 4a). The transect of estimated net diabatic heating rate in Fig. 10 shows that this tail of high PV might be formed in the mid-tropospheric vertical gradient of heating, not transported from a poleward reservoir of high PV as an adiabatic "beta drift" mechanism would imply. The vertical structure of diabatic heating is better seen in heating rates averaged over a region centered on the tail of high PV southwest of the vortex center (Fig. 11a). The large vertical gradient in diabatic heating at 500 hPa results primarily from a stratiform heating structure (Houze 1997) represented by grid-scale condensation and evaporation handled by the microphysics scheme in the EC-YOTC model. In contrast, the model's parameterized convective heating peaks at 500 hPa, and has a vertical gradient that generates positive PV between the surface and 500 hPa, and negative PV between 500 hPa and the tropopause. The net PV tendency due to diabatic processes strongly resembles that due to the stratiform heating, with maximum PV generation at 500 hPa (Fig. 11b). Diabatic heat sources other than stratiform and convective heating (e.g. radiation) are comparatively small. There is net negative diabatic heating below 550 hPa, much of which results from the strength of the stratiform-like evaporative cooling.

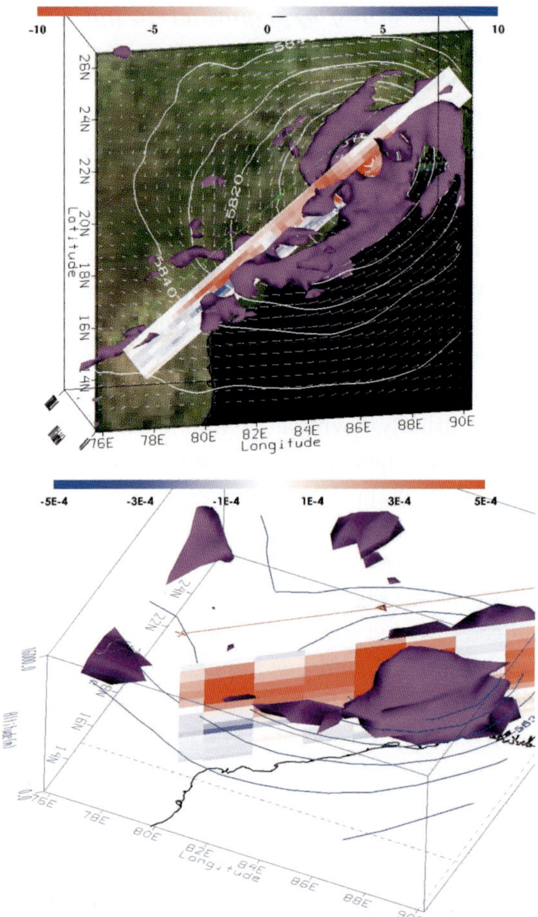

Fig. 10. Diabatic heating rate estimates from all physical processes in cross section (shading, red positive and blue negative, units K/s) through the rainy southwest quadrant of the Sept. 16, 2008 depression. The 1 PVU isosurface (magenta) is depicted from EC-YOTC (top) and MERRA (bottom), along with geopotential height contours at 500 hPa in both figures. In both depictions, the total heating is estimated to be negative in the lower troposphere (blue), indicative of a strong midlevel heating gradient (and associated PV source). In animations of both products (not shown), air with PV = 1 PVU appears to be formed in the midlevel vertical heating gradient and then advected downwind, not simply transported into the section from regions of high planetary vorticity to the north as an adiabatic "beta effect".

To have the net diabatic heating be negative in the lower troposphere during tropical precipitation is not very typical of tropical convection, even in mesoscale systems where the heating profile is known to be top-heavy due to stratiform precipitation contributions (e.g. Houze 1989, 1997). An especially strong stratiform cooling at low levels perhaps results from dry lower tropospheric conditions over India

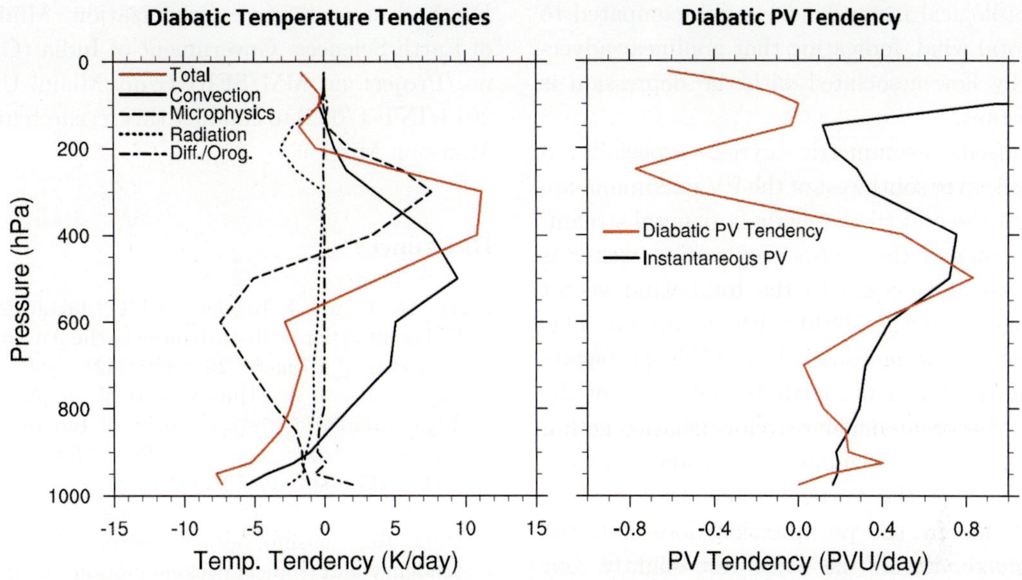

Fig. 11. Vertical profiles of EC-YOTC diabatic temperature tendencies (left) and diabatic PV tendency (right) in the region of high PV southwest of the vortex center during the monsoon depression of September 2008. The diabatic temperature tendency profiles are horizontally averaged in the region 15–19°N, 80–86°E at 1800 UTC on September 16, 2008. The diabatic PV tendency is the product of the absolute vorticity (not shown) and the vertical gradient of the diabatic temperature tendency; other components of the vorticity and potential temperature gradient vectors make a negligible contribution.

that allow for especially large evaporation of falling condensate. It is also possible that the ascent forced by the dynamics of the depression interacting with its sheared baroclinic environment, thought to be the reason for rain in the southwest quadrant, is more top-heavy than in typical deep-convectively driven stratiform precipitation. In a time-mean averaged over the center of the depression, rather than just its southwestern quadrant, the net diabatic heating is positive at nearly all levels even though the evaporative cooling is about 5 K day^{-1} throughout much of the lower troposphere (not shown).

Additional cases should be examined to determine how representative these analyses are of all monsoon depressions, and field data beyond gridded analyses would be helpful to add confidence to these results. Nevertheless, the distributions of PV, horizontal streamfunction, and diabatic heating in this case study suggest that diabatic heating rather than adiabatic advection of planetary vorticity is needed to create the asymmetries in the flow field that adiabatically advect depressions to the northwest.

4. Conclusions

Indian monsoon depressions have traditionally been thought of as lower tropospheric relative vorticity maxima embedded in mean low-level westerlies, requiring some special mechanism to explain their apparent upstream, westward propagation. We have shown that these depressions can alternatively be viewed as PV maxima with peak amplitude in the middle troposphere. Since PV, unlike relative vorticity, is a conserved quantity that can be altered only by non-conservative diabatic heating and mechanical forcing, these PV maxima might be expected to be advected to the west by the mid- to upper-level easterly mean flow. However, the climatological mean wind is weak compared to the total wind, indicating that nonlinear advection by flow associated with the depression is important.

Indeed, asymmetric gyres, especially a cyclonic gyre southwest of the PV maximum, are seen in the mid-tropospheric horizontal streamfunction, and the storm propagation vector is approximately equal to the total wind vector averaged over the 500 hPa PV maximum. This indicates that monsoon depressions propagate to the northwest primarily by advection of the vortex by an anomalous cyclone that, together with the peak precipitation, spreads off to the southwest of the vortex center. Diabatic PV generation by the precipitation does shift the PV maximum to the southwest slightly, but this has a comparatively minor effect on vortex motion, as shown by Boos *et al.* (2015). In the terminology of Wu and Wang's (2001) analysis of tropical cyclone motion, the "direct generation" of PV plays only a minor role in vortex motion, and instead adiabatic advection "by heating-induced asymmetric flow" may be the key effect of precipitation-related heating. Three-dimensional structures of PV and diabatic heating in a case study were shown to suggest a scenario in which strong stratiform heating generates azimuthally asymmetric PV and associated balanced flow that produces the northwestward advection of the main center of the depression. Further examination of monsoon depression dynamics in observations and numerical models would help to determine whether this propagation mechanism operates more generally in monsoon depressions in India and other regions.

Acknowledgments

WRB and VSM acknowledge support from Office of Naval Research Young Investigator Program award N00014-11-1-0617 and National Science Foundation award AGS-1253222. BEM acknowledges financial support given by the Earth System Science Organization, Ministry of Earth Sciences, Government of India (Grant no./Project no MM/SERP/Univ_Miami_USA/2013/INT-1/002) to conduct this research under Monsoon Mission.

References

Berry, G. J., M. J. Reeder, and C. Jakob, 2012: Coherent synoptic disturbances in the Australian monsoon. *J. Climate*, **25**, 8409–8421.

Boos, W. R., J. V. Hurley, and V. S. Murthy, 2015: Adiabatic westward drift of Indian monsoon depressions. *Quart. J. Roy. Meteor. Soc.*, **141**, 1035–1048. doi:10.1002/qj.2454.

Chan, J. C. L., F. M. F. Ko, and Y. M. Lei, 2002: Relationship between potential vorticity tendency and tropical cyclone motion. *J. Atmos. Sci.*, **59**, 1317–1336.

Chan, J. 2005: The physics of tropical cyclone motion. *Ann. Rev. Fluid Mechanics*, **37**, 99–128.

Chen, T.-C., J.-H. Yoon, and S.-Y. Wang, 2005: Westward propagation of the Indian monsoon depression. *Tellus A*, **57**, 758–769.

Dee, D. and coauthors, 2011: The ERA-Interim reanalysis: Configuration and performance of the data assimilation system. *Quart. J. Roy. Meteor. Soc.*, **137**, 553–597.

Farrell, B., 1985: Transient growth of damped baroclinic waves. *J. Atmos. Sci.*, **42**, 2718–2727.

Fiorino, M. and R. Elsberry, 1989: Some aspects of vortex structure related to tropical cyclone motion. *J. Atmos. Sci.*, **46**, 975–990.

Godbole, R. V., 1977: The composite structure of the monsoon depression. *Tellus*, **29**, 25–40.

Hodges, K., 1995: Feature tracking on the unit sphere. *Mon. Wea. Rev.*, **123**, 3458–3465.

Houze, R. A., 1989: Observed structure of mesoscale convective systems and implications for large-scale heating. *Quart. J. Roy. Meteor. Soc.*, **115**, 425–461.

Houze, R. A., Jr., 1997: Stratiform precipitation in regions of convection: A meteorological paradox? *Bull. Amer. Meteor. Soc.*, **78**, 2179–2196.

Hunt, K. M. R. and D. J. Parker, 2016: The movement of Indian monsoon depressions by interaction with image vortices near the Himalayan wall. *Quart. J. Roy. Meteor. Soc.*, **142**, 2224–2229. doi:10.1002/qj.2812.

Hurley, J. V. and W. R. Boos, 2015: A global climatology of monsoon low-pressure systems.

Quart. J. Roy. Meteor. Soc., **141**, 1049–1064. doi:10.1002/qj.2447.

India Meteorological Department, 2011: Tracks of cyclones and depressions over North Indian Ocean (from 1891 onwards). Technical Note, Regional Meteorological Centre, Chennai.

Kalnay, E. and coauthors, 1996: The NCEP/NCAR 40-year reanalysis project. *Bull. Amer. Meteor. Soc.*, **77**, 437–470.

Krishnamurti, T. N., R. J. Pasch, H.-L. Pan, S.-H. Chu, and K. Ingles, 1983: Details of low latitude medium range numerical weather prediction using a global spectral model I: Formation of a monsoon depression. *J. Meteorol. Soc. Japan*, **61**, 188–206.

Krishnamurti, T. N., A. Martin, R. Krishnamurti, A. Simon, A. Thomas, and V. Kumar, 2013: Impacts of enhanced CCN on the organization of convection and recent reduced counts of monsoon depressions. *Clim. Dyn.* **41**: 117–134.

Mooley, D. and J. Shukla, 1987: Characteristics of the westward-moving summer monsoon low pressure systems over the Indian region and their relationship with the monsoon rainfall. Tech Rep., University of Maryland, Department of Meteorology, Center for Ocean-Land-Atmosphere Interactions.

Moorthi, S. and A. Arakawa, 1985: Baroclinic instability with cumulus heating. *J. Atmos. Sci.*, **42**, 2007–2031.

National Hurricane Center, 2015: Glossary, accessed 5 April 2015, http://www.nhc.noaa.gov/about gloss.shtml.

Ramage, C. S., 1971: *Monsoon Meteorology*. Academic Press, New York.

Rao K. and S. Rajamani, 1970: Diagnostic study of a monsoon depression by geostrophic baroclinic model. *Indian J. Meteor. Geophys.*, **21**, 187–194.

Reinecker, M. M. and coauthors, 2011: MERRA: NASA's Modern-Era Retrospective Analysis for Research and Applications. *J. Climate*, **24**, 3624–3648.

Saha, S. and coauthors, 2010: The NCEP Climate Forecast System Reanalysis. *Bull. Amer. Meteor. Soc.*, **91**, 1015–1057.

Sanders, F., 1984: Quasi-geostrophic diagnosis of the monsoon depression of 5–8 July 1979. *J. Atmos. Sci.*, **41**, 538–552.

Sikka, D., 1977: Some aspects of the life history, structure and movement of monsoon depressions. *Pure Appl. Geophys.*, **115**, 1501–1529.

Waliser, D. E. and coauthors, 2012: The "Year" of Tropical Convection (May 2008–April 2010), *Bull. Amer. Meteor. Soc.*, **93**, 1189–1218.

Wang, B., R. L. Elsberry, W. Yuqing, and W. Liguang, 1998: Dynamics in tropical cyclone motion: A review. *Chinese J. Atmos. Sci.*, **22**, 416–434.

Wu, L. and B. Wang, 2001: Effects of convective heating on movement and vertical coupling of tropical cyclones: A numerical study. *J. Atmos. Sci.*, **58**, 3639–3649.

Yoon, J. H. and W.-R. Huang, 2012: Indian monsoon depression: Climatology and variability. In *Modern Climatology*, S.-Y. Wang, Ed., InTech, chap. 2, pp. 45–72.

Yoon, J.-H. and T.-C. Chen, 2005: Water vapor budget of the Indian monsoon depression. *Tellus A*, **57**, 770–782.

Chapter 16

Recent Progress on Squall Lines in East China

Zhiyong Meng
Department of Atmospheric and Oceanic Department,
Peking University, Beijing, China
zymeng@pku.edu.cn
http://www.atmos.pku.edu.cn/zymeng/

This chapter reviewed the main results of four recent published papers on squall lines in east China covering the radar characteristics, structure and predictability of squall lines under different environmental flow systems including tropical cyclone, westerly trough, and cold vortex. The general features of squall lines in east China was first examined based on two-year's mosaic of radar reflectivity in 2008 and 2009 in comparison to the U.S. squall lines. In particular, the general features of squall lines preceding landfall tropical cyclones were documented in details. Following that, the structure and predictability of squall lines associated with a westerly trough and a cold vortex were further examined. Results showed that the rear inflow that formed through the consolidation of three pairs of bookend vortices played an essential role in the formation of a large bowing structure within a squall line that formed in front of a westerly trough. Through a detailed damage survey, radar and radiosonde analyses, the radar features and formation environment of a tornadic supercell developing in a cold-vortex-associated squall line was examined. This chapter also examined the predictability of a squall line associated with a westerly trough. It was found that squall-line simulation could be very sensitive to model error from horizontal resolution and uncertainties in physical parameterization schemes. A linear improvement in the performance of squall line simulation was observed when the initial error was decreased gradually, with the largest contribution from initial moisture field.

1. Introduction

Squall lines are linear or quasi-linear mesoscale convective systems (MCSs) that can produce disastrous heavy rains, strong winds, hail, and sometimes even tornadoes. Features of squall lines are best represented by radar observation. Due to the advanced radar observation network, squall lines in the United States have been well documented (e.g. Bluestein and Jain 1985; Wyss and Emanuel 1988; Houze *et al.* 1990; Geerts 1998; Parker and Johnson 2000; Jirak *et al.* 2003) with a good knowledge of their general characteristics being obtained such as the spatial and temporal distributions, morphology, and formation environment. Since squall lines result from complicated multi-scale interaction with apparent impact from boundary layer features and underlying surface condition, they may take on different morphologies or produce different hazards depending on the background environmental conditions present (Chen and Chou 1993; Meng and Zhang 2012, hereafter MZ12).

Squall lines are very common mesoscale convective systems in wide area of China from early spring to late autumn. Due to the previously limited radar observation network, studies on the general features of squall lines in large areas of China, their climatological features and differences from squall lines in other areas such as the United States, as well as their

The Global Monsoon System: Research and Forecast (3rd Edition)
Edited by C. P. Chang et al.

detailed structure and predictability have been very few.

After more than ten years' effort, Doppler-radar network in China is becoming mature. The radar network in east China has become dense enough to examine radar climatology of squall lines (Fig. 1a). The development of radar observation capability also facilitates examinations of detailed features of squall lines with the aid of high-resolution numerical simulation. This chapter is targeted at summarizing four recent papers (MZ12; Meng *et al.* 2012; 2013; Meng and Yao 2014) that illustrated radar characteristics, structure, and predictability of squall lines in east China.

2. Radar-based Survey on the General Features of Squall Lines in East China

2.1. *Squall lines in all environmental background*

Based on the mosaic of composite radar reflectivity in 2008 and 2009, Meng *et al.* (2013) performed a comprehensive survey on the basic features of squall lines in east China including their temporal and spatial distributions, length, intensity, duration, orientation, movement, formation, dissipation and organizational modes, and environmental features in comparison to the U.S. squall lines. The criterion used to identify a squall line was that a linear or quasi-linear contiguous 40-dBZ band of no shorter than 100 km last at least 3 h.

Results showed that squall lines had a maximum frequency of formation in north China near the boundaries between Shandong, Henan, Anhui, and Jiangsu Provinces ("B" in Fig. 1a). Three secondary maxima in the frequency of occurrence were located in Hebei, Guangdong,

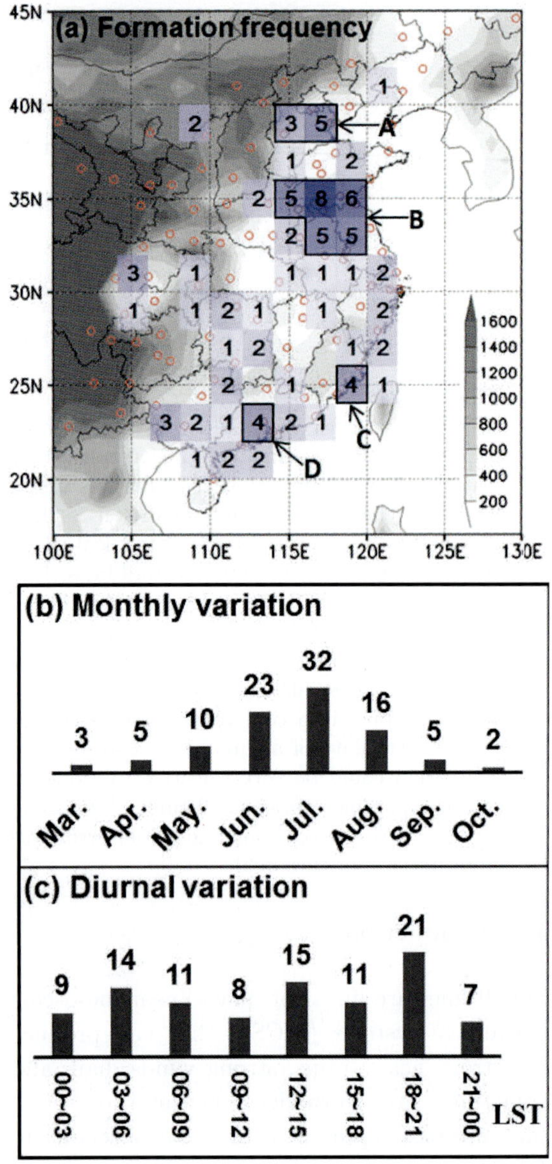

Fig. 1. The geographical distribution [a, blue shading, darker color denotes higher frequency. The numbers denote the number of times that the centroids of squall lines lie within the grid square that spans 2° × 2° area at their formation times. Capital letters A–D denote high frequency areas. The gray shading denotes the terrain height (every 200 m until 1600 m)], monthly (b) and diurnal (c) variations of the frequency of squall line formation. (Adapted from Meng *et al.* 2013).

and Fujian Provinces. The squall lines formed from March to October with a peak in July (Fig. 1b). Their diurnal variation had a major peak in the early evening and two minor peaks in the early morning and early afternoon (Fig. 1c). The squall lines were mostly southwest–northeast orientated. On average, they had an eastward motion at a speed of $14.4\,\mathrm{m\,s^{-1}}$, a maximum length of 243 km, a maximum intensity of 58–63 dBZ, and a life cycle of 4.7 h. The squall lines showed a dominant formation mode of broken-line (Bluestein and Jain 1985) and a dominant organization mode of trailing-stratiform (Parker and Johnson 2000). Most squall lines dissipated in the reversed broken-line mode (continuous line broke into separated cells and they individually dissipated). Likely due to the rich moisture in east China, a higher percentage of squall lines formed in an embedded-areal pattern (Bluestein and Jain 1985) relative to the U.S. cases.

Composite rawinsonde analyses in the area with high frequency of formation (A-D in Fig. 1a) showed that the pre-formation environment of the squall lines in mid-latitude east China was moister with weaker vertical wind shear relative to their U.S. counterparts (Table 1). Examination on the environment of squall lines with different organizational modes showed that the largest instability and the most humid environment were observed in the squall lines with a leading stratiform in east China rather than with trailing stratiform in the U.S. situation.

The squall lines in east China mainly formed in five synoptic weather patterns by identifying their dominant ambient weather systems including pre-trough, cold vortex, subtropical high, tropical cyclone (TC), and post-trough (Fig. 2). A cold vortex is a quasi-stationary upper troposphere low pressure system that often causes series of severe convective weather in warm season in north China. Favorable conditions for deep, moist convection of various patterns were examined. The pre-trough pattern was the dominant one which was characterized by adequate moisture supply, a low-level jet, and moderate 1000–700-hPa vertical wind shear. The cold-vortex and post-trough patterns generally had larger environmental instability. The main contributors in subtropical high pattern were rich moisture transported by southerly flows, short-wave disturbances, and moderate 1000–700-hPa vertical wind shear. The TC pattern was mainly featured by rich moisture, high conditional instability, and low-level convergence. This type of squall lines has not been documented worldwide in literature before MZ12.

2.2. *Squall lines preceding landfalling tropical cyclone*

The basic features associated with squall lines ahead of landfall tropical cyclones (Fig. 3) in China between 2007 and 2009 were examined in comparison to previous works on "standard" mid-latitude squall lines (MZ12). It was found

Table 1. A comparison between the averages of the derived properties of the environmental rawinsondes of the squall lines in midlatitude east China and those in previous studies. Also given for squall lines in east China are the standard deviations (in parentheses) and ranges of various properties. (Adapted from Meng *et al.* 2013.)

Sounding properties		CAPE ($\mathrm{J\,kg^{-1}}$)	CIN ($\mathrm{J\,kg^{-1}}$)	LI (K)	LCL (hPa)	PW (cm)
Bluestein and Jain (1985)		2260	33			2.8
Wyss and Emanuel (1988)		1208	76			
Parker and Johnson (2000)		1605		−5.4	831	3.4
Chen and Chou (1993)		1330				
Meng and Zhang (2012)		1548	67	−3.6	899	6.1
Squall lines in east China	Avg (std dev)	1480 (644)	77 (59)	−4.3 (1.4)	909 (55)	5.6 (1.2)
at midlatitudes	Range	730 to 2583	10 to 204	−2.2 to −6.3	780 to 965	3.6 to 7

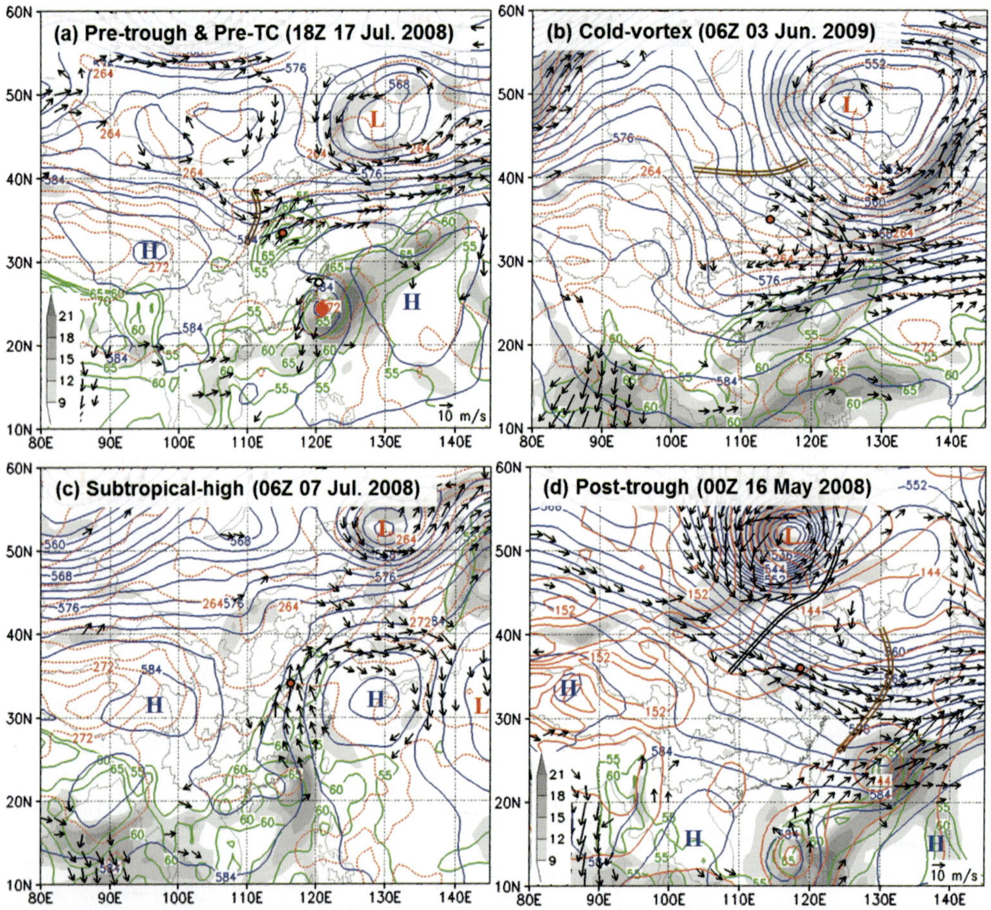

Fig. 2. Examples of various environmental flow patterns for (a) pre-trough and pre-TC, (b) cold vortex, (c) subtropical-high, and (d) post-trough. Plotted are geopotential height [blue contours, 10 geopotential meters (gpm)], temperature (dashed red contours, K) on 500 hPa, column-integrated precipitable water larger than 55 kg m^{-2} (green contours, kg m^{-2}), vertical shear vector between 700 and 1000 hPa with the magnitude larger than 8 m s^{-1} (the representative wind vector is given in the lower-right corners of panel a), and wind speed larger than 9 m s^{-1} at 850 hPa (shaded, every 3 m s^{-1}). The double solid brown lines denote the trough on 500 hPa. The red solid contour and the double solid black lines in (d) respectively denotes the geopotential height [10 geopotential meters (gpm)] and trough on 850 hPa. The places where the convection of the squall line was initiated are marked by red dots. The white dots indicate the areas that also have a possibility of squall line formation in the ambience of a TC. "H" and "L" denotes the high and low pressure centers at 500 hPa, respectively. The TC centers are given by the red typhoon marks in (a). (Adapted from Meng *et al.* 2013.)

that about 43% of landfall TCs in China in these three years were associated with pre-TC squall lines, half of which had multiple squall lines. They mostly formed in August and from late afternoon to midnight.

The highest formation frequency was observed in the flat area of Guangdong province.

The basic radar characteristics of pre-TC squall lines were summarized in Fig. 4. They tended to form about 600 km ahead of their parent TC to the right of the TC track with a dominant broken-line formation mode and a trailing-straitiform organization pattern. Relative to their mid-latitude counterparts and squall lines in Taiwan strait area, pre-TC squall lines had a shorter life span. They were generally stronger than squall lines in Taiwan strait area, but had a weaker cold pool than mid-latitude

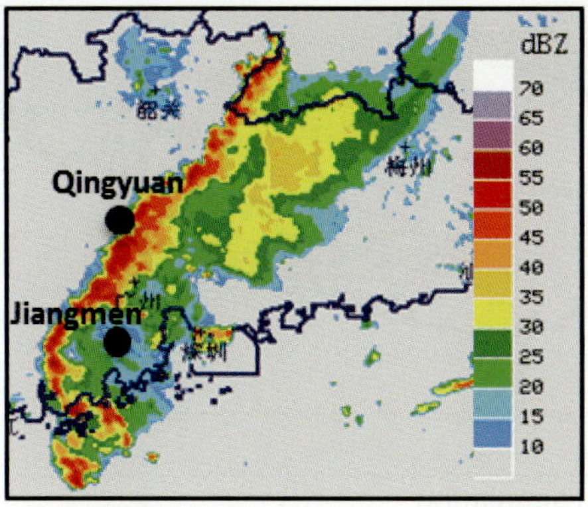

Fig. 3. An example of pre-TC squall lines. It was a squall line preceding severe tropical storm Kammuri (2008). The color shading is the composite radar reflectivity at 1320 UTC 4 Aug. 2008. (Adapted from MZ12.)

squall lines with slower movement and smaller changes of temperature, pressure and wind speed across the squall lines.

The formation environment of pre-TC squall lines was also examined. They tended to form in the transition area between the parent TC and subtropical high (Fig. 5). In comparison to their mid-latitude counterparts and squall lines in Taiwan Strait area (Chen and Chou 1993), pre-TC squall lines tended to form in a moister environment with a weaker low-level vertical wind shear. Their pre-formation environment had a smaller conditional instability than that of the spring-season squall lines in Oklahoma (Bluestein and Jain 1985) but a large conditional instability than those of prefrontal squall lines across the whole continental US (Wyss and Emanuel 1988) and Taiwan Strait area (Chen and Chou 1985) (Table 1).

In-depth case study through observational analysis and model simulation illustrated the role that the TC played in creating a local environment favorable for squall line development. The parent TC might have contributed to the formation of their preceding squall lines by increasing environmental instability and low-level moisture supply, as well as facilitating the linear organization of discrete convection through low-level convergent or confluent flows. It was found that changes to TC size and

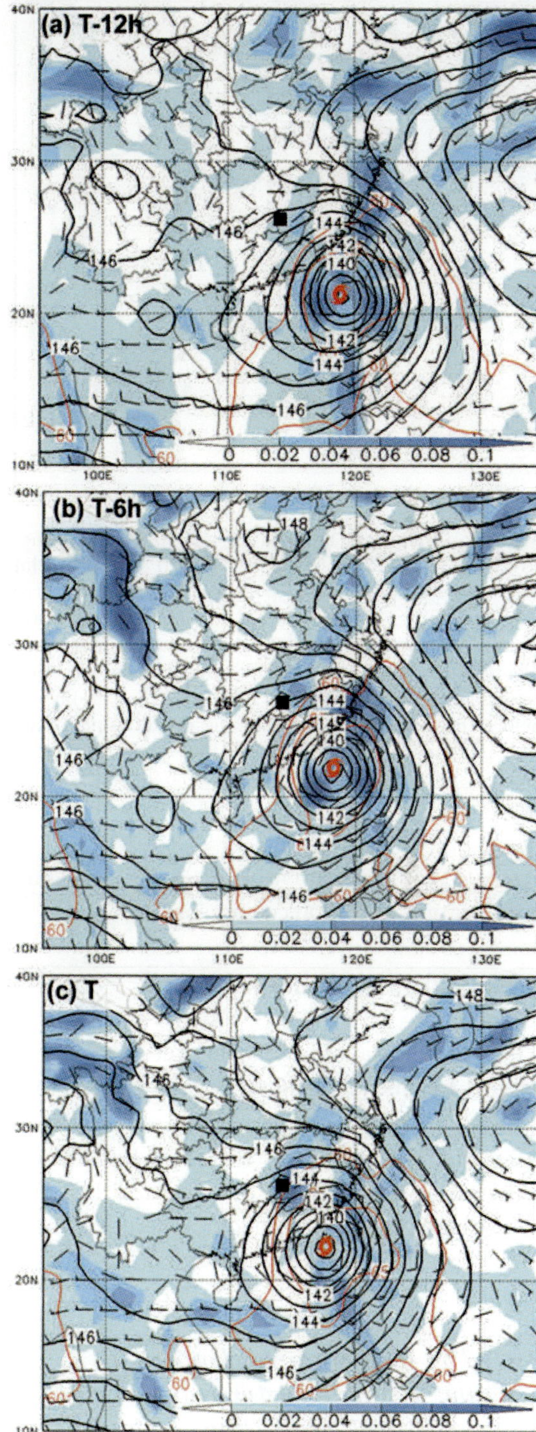

Fig. 4. Schematic diagram of the averaged features of pre-TC squall lines. The number in parentheses is the corresponding standard deviation. (Adapted from MZ12.)

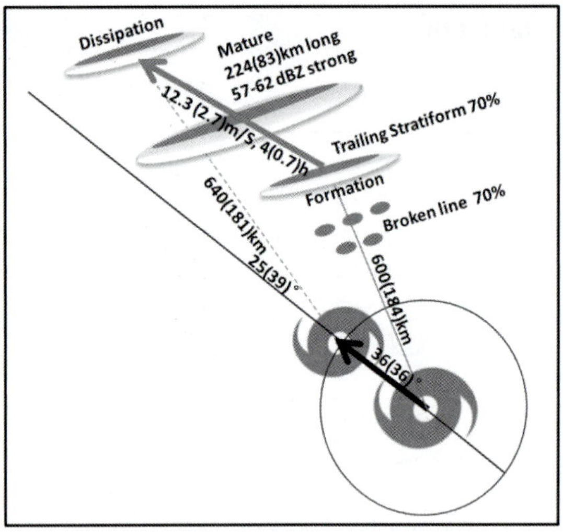

Fig. 5. Composite analysis of 925-hPa frontogenesis (102 K km^{-1} (3h)$^{-1}$; shaded), 850-hPa geopotential height (black contour every 10 gpm) and wind vector (full barb = 10 m s^{-1}) as well as column-integrated precipitable water larger than 60 kg m^{-2} (red contour every 5 kg m^{-2}) for the 9 cases that have westward motion component during 2007–09 at 12(a), 6(b), and 0(c) hours before the formation time T of the pre-TC squall line. The solid square denotes the median of the formation locations of pre-TC squall lines at time T. The red TC mark denotes the TC center. (Adapted from MZ12.)

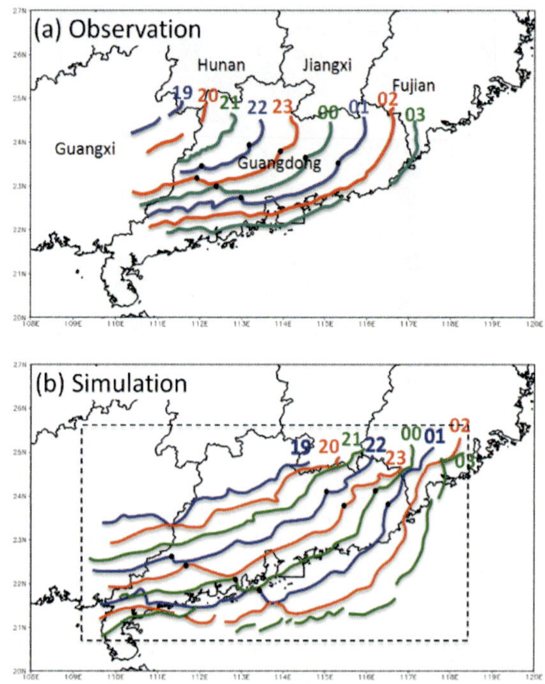

Fig. 6. Isochrones of the (a) observed and (b) simulated squall line. The numbers corresponding to each line are the time of the day (UTC). The segments between the two black dots are the big bow on which this work focused. (Adapted from Meng *et al.* 2012.)

strength could apparently affect the size and organization of the preceding squall line.

3. Squall Line Structure

The disasters caused by squall lines are closely related to their detailed structure. This section reviews two works that examined strong winds associated with detailed structure of squall lines: One was on a bow echo (Meng *et al.* 2012), and the other was on a tornadic supercell (Meng and Yao 2014) embedded in a squall line in east China.

3.1. *The formation mechanism of the bowing structure of a pre-trough squall line*

Bow echo is a structure of squall lines that is commonly associated with strong surface wind. The formation mechanism of bow echo has been examined through idealized numerical experiments (e.g. Weisman and Davis 1998) or observational studies (e.g. Grim *et al.* 2009). By performing a real-world numerical study, Meng *et al.* (2012) examined the formation mechanism of the bowing structure of a squall line in front of a westerly trough near Meiyu Front that occurred over South China in April 2007 (Fig. 6a).

The numerical experiment was performed using WRFV3.2 (Skamarock *et al.* 2008) with four domains and two-way nesting with grid-point intervals of 40.5, 13.5, 4.5 and 1.5 km respectively. The domains had 35 vertical layers with the model top at 10 hPa. The model was integrated for 24 h starting at 1200 UTC 23 April 2007 with the initial and boundary conditions provided by 6-hourly FNL/NCEP analysis of 1° by 1°. Physical parameterization schemes included the WRF single-moment (WSM) six-class microphysics scheme with graupel (Hong *et al.* 2010), the Yonsei State University (YSU) scheme (Noh *et al.* 2003) for planetary boundary layer processes, and the Grell-Devenyi cumulus scheme (Grell and Devenyi 2002) for the 40.5 and 13.5-km domains (no cumulus parameterization scheme was used for 4.5 and 1.5-km domains).

The observed bowing structure was successfully captured by the model simulation (Fig. 6b). It was found to be closely associated with the formation, expansion and consolidation of rear inflow. The environmental flow and horizontal pressure gradient force associated with the midlevel pressure deficit were not major contributors to the formation of the rear inflow. The storm-relative environmental flow at the back of the squall lines was mostly parallel to the orientation of the squall line (Fig. 7), and the rear inflow was maximized at the weakest pressure gradients associated with the low pressure deficit (not shown).

Results showed that the rear inflow and the resulted bow segment were mainly forced by several pairs of bookend vortices (Fig. 7). Three separated bow segments developed along the squall line at the beginning. Each of them was associated with a rear inflow and a pair of bookend vortices. Then the northern two pairs of the bookend vortices merged into one with the canceling of vorticity associated with the neighboring vortices of opposite sense. Finally, the remaining two pairs of bookend vortices merged into one again. The originally separated rear inflow associated with individual bookend pairs expanded, intensified and formed a coherent band of rear inflow, which resulted in the development of the large bowing structure (Fig. 7).

The formation mechanism of the bookend vortices was also found to be different from the idealized works (e.g. Weisman 1993; Weisman and Davis 1998). Weisman (1993)

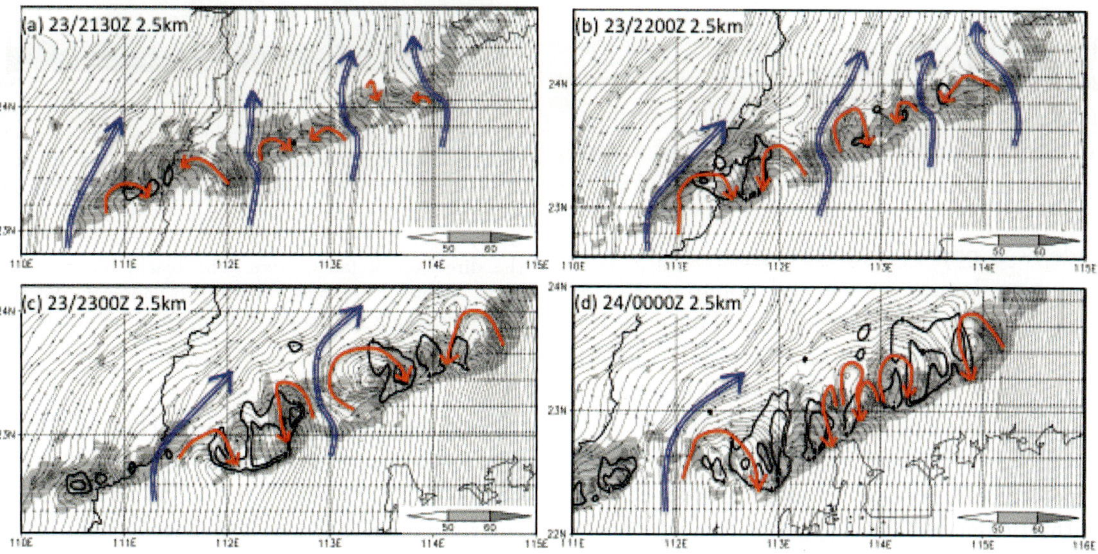

Fig. 7. The merging process of the three pairs of bookend vortices as shown by the evolution of line-relative streamline at 2.5 km together with composite radar reflectivity (shaded, dBZ) at (a) 2130 UTC 23 Apr, (b) 2200 UTC 23 Apr, (c) 2300 UTC 23 Apr, and (d) 0000 UTC 24 Apr 2007. The black contour denotes the rear inflow (every 4 m s^{-1}). The double blue arrows represent front-to-rear flow. The red curved arrows denote bookend vortices. (Adapted from Meng *et al.* 2012.)

proposed that the bookend vortices were produced by first upward tiliting of the environmental shear associated horzontal vorticity, which then experienced a downward tilting. Weisman and Davis (1998) proposed that the bookend vortices might have been formed by directly upward tilting of baroclinically generated horizontal vorticity of the cold pool instead of the environmental vertical shear. Meng *et al.* (2013) showed that though the bookend vortices formed mainly through tilting of horizontal vorticity, the sources of the horizontal vorticity for different pairs of the bookend vorticies were somewhat different. One pair was forced by the upward tilting of lower baroclinically and/or ambient shear forced horizontal vorticity, which was similar to Weisman and Davis (1998), whereas another pair formed through the downward tilting of upper level baroclinic horizontal vorticity.

3.2. *The formation of a tornadic supercell in a cold-vortex associated squall line*

This section reviews a work (Meng and Yao 2014) that examined the detailed radar characteristics of a tornadic supercell embedded in a squall line that formed in a cold vortex system on 21 July, 2012. Through detailed damage surveys, the occurrence of an EF-3 tornado was proved with the main evidence of a narrow damage swath and apparent storm-scale convergent surface debris. This was the first documented tornado in Beijing (Fig. 8). The Beijing tornado was embedded in a supercell that developed at the southern end of a squall line (Fig. 9). Detailed structures of the supercell were analyzed using the level-II data of the Beijing S-band Doppler radar, which was located ∼20 km west of the tornado.

Results showed that the intensification and expansion of the mesocyclone near the later stage of the tornado could be the reason for the zigzag tornado track likely because the tornado

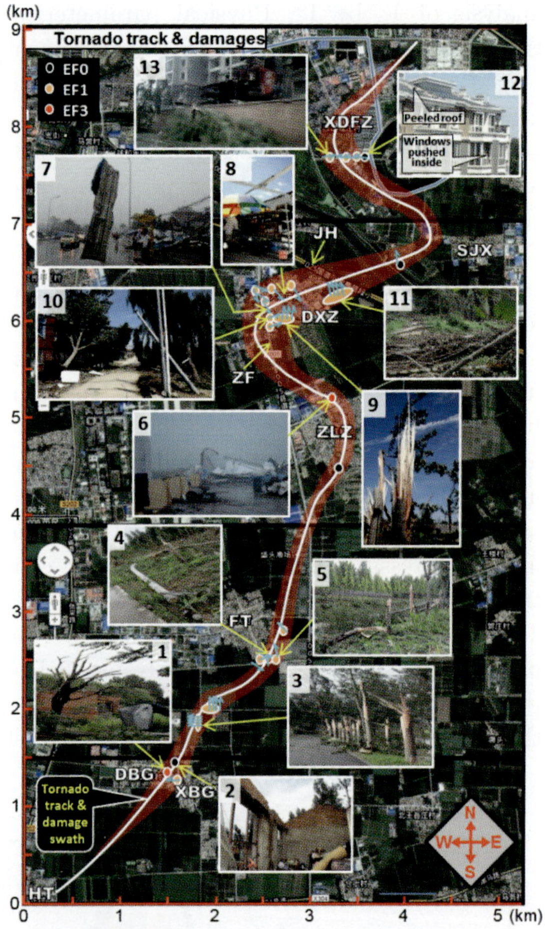

Fig. 8. The track of the tornado and its damage swath on a satellite map obtained from Google Earth. Pictures of 1–5 were provided by Bato.cn, 6 and 7 by 591hx.com, 8, 9, and 10 by Mingyuan Zhang, 12 by Sina.com.cn, 13 by Youku.com. The small light-blue arrows denote the direction of fallen trees. The black, blue and red dots denote EF scales of the tornado at different positions along the track. (Adapted from Meng and Yao 2014.)

was steered by the mesocyclone (Fig. 9f). Relative to the mesocyclone, the tornado vortex signature (TVS) showed a better coincidence with the wind damage in terms of their location, time, and intensity variation. The most severe damage was observed when the TVS was detected near the ground at the lowest scan.

The relation between the descending reflectivity core (DRC) and tornadogenesis has been

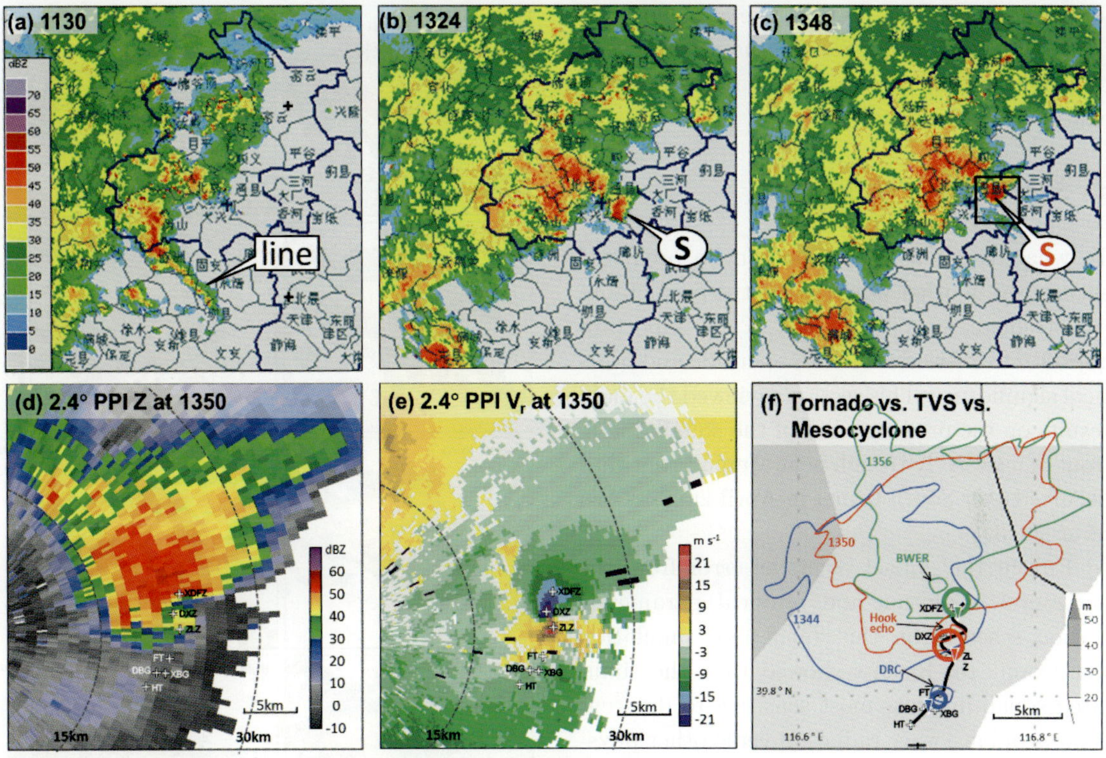

Fig. 9. Panel (a–c) demonstrate the evolution of radar reflectivity (ZH) at 0.5° PPI showing the squall line (denoted by "line") and the tornadic supercell (denoted by "S". "S" in red color represents the time when the tornado damage was occurring). The radar station of Beijing is marked by the black cross. The snapshots of 2.4° PPI ZH and VR at 1350 LST in the enlarged area denoted by the rectangle in (c) are shown in (d, dBZ) and (e, $\mathrm{m\,s^{-1}}$), respectively. The crosses in (d) and (e) denote the affected villages. Panel (f) shows the evolution of the supercell in terms of the contour of 40 dBZ ZH at 2.4° PPI between 1344–1356 LST every 6 min. Also plotted in (f) are the hook echo at 1350 LST, the bounded weak echo region (BWER) at 1356 LST, the descending reflectivity core (DRC) at 1344 LST, the tornado track (black heavy curve), the mesocyclone on 2.4° PPI (solid circles denoting the diameter of the mesocyclone), and the 0.5° PPI TVS locations (denoted by the bottom tip of the triangles) during the damage duration with different times in different colors. The topography is shaded progressively darker every 10 m. (Adapted from Meng and Yao 2014.)

an open question in tornado community (e.g. Rasmussen *et al.* 2006). In this case, two DRCs were detected respectively preceding the tornadogenesis and re-intensification of the wind damage. The first DRC developed about 15 min before the tornado touched the ground. The second DRC appeared right before the damage intensified after a short period of weakening. Besides, a tornado debris signature (TDS) (Ryzhkov *et al.* 2005) was detected in the later stage of the tornado likely due to the reflectivity by the lifted debris.

The formation environment of this tornadic supercell was then examined using a radiosonde that was released about 30 min before the tornadogensis about 20 km to the west of the tornado. Relative to U.S. climatology (Rasmussen and Blanchard 1998; Rasmussen 2003), this storm formed in an environment with a low instability, an extremely low convective inhibition (CIN), a close-to-median lifting condensation level (LCL), and high 0–3-km shear. An examination of multi-parameter indices such as 0–1-km energy-helicity index (EHI1) suggested less of a tornado risk in this case. The environment was quite favorable for the development of a nontornadic or weakly rather than significantly tornadic supercell.

4. Squall Line Predictability

To what extent a squall line can be predicted accurately, in other words, what the predictability of squall lines looks like, is an important question to answer with both scientific and practical significance. This section reviewed the main results on the predictability of the squall line in front of a westerly trough near Meiyu Front that occurred over South China in April 2007 by Wu *et al.* 2013.

Firstly, the impact of model error from the horizontal resolution and physical parameterization schemes on the squall line simulation was examined. Results showed that the simulated squall line was most sensitive to different choices of long-wave radiation verse other physical parameterization schemes such as cumulus, microphysics and PBL schemes. An interesting result was that the sensitivity to grid size depended on whether cumulus parameterization scheme was turned on or off.

When cumulus parameterization scheme was turned off, higher resolution produced better simulation (Fig. 10). However, turning on a cumulus parameterization scheme was found to degrade the simulation with a grid size from 20 to 5 km. More severe degradation was observed in higher resolution.

Secondly, the impact of uncertainty in the initial fields on the squall line simulation was examined using an ensemble forecast approach. It was found that about 15% of the ensemble successfully captured the evolution of the squall line. About 25% failed to capture the squall line, while the remaining 60% captured the squall line but with a large location error. Consistent with Zhang *et al.* (2012), the growth of the ensemble spread was found to be associated with moist convection development.

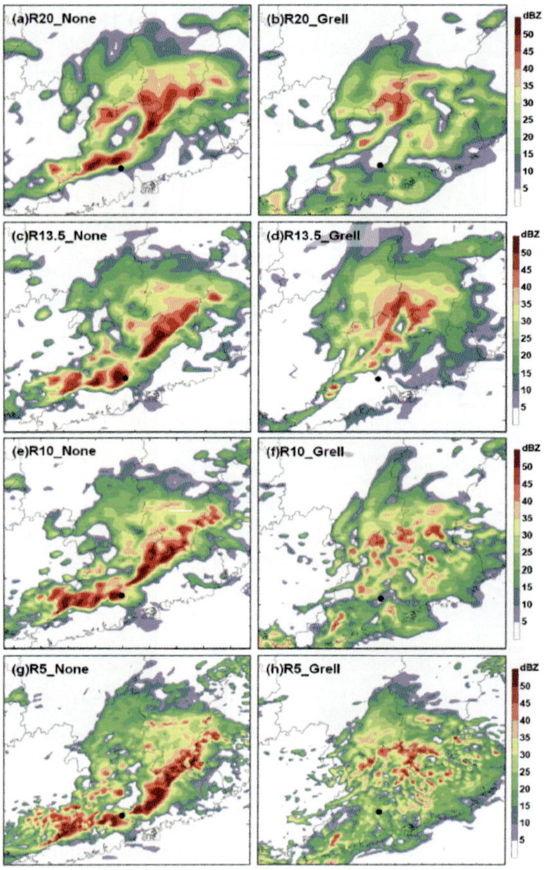

Fig. 10. The simulated composite radar reflectivity at 0000 UTC 24 April 2007 for experiments with a grid size of 20, 13.5, 10, and 5 km without using a cumulus parameterization (left column). The right column are as left one but using Grell cumulus scheme. (Adapted from Wu *et al.* 2013.)

Results also showed that linear improvement in the initial field led to linear improvement in the squall line simulation. This result was quite different from the nonlinear improvement feature obtained in a US squall line study (Melhauser and Zhang, 2012). Sensitivity tests demonstrated that the improvement in initial moisture field relative to other variables resulted in the largest improvement in the squall line simulation (Fig. 11).

Furthermore, the sensitivity of the squall line simulation to different lead times was examined. When the lead time was increased from the

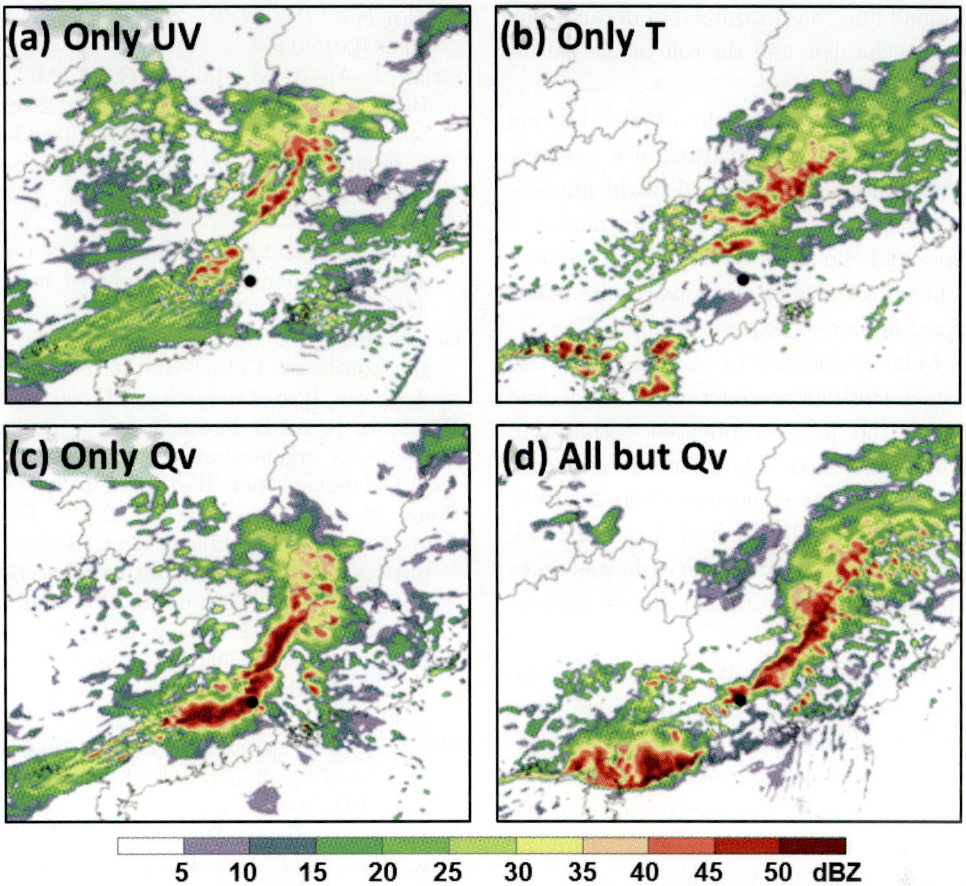

Fig. 11. The simulated composite radar reflectivity at 0000 UTC 24 April 2007 in experiments that only improved the initial U and V (a), T(b), Qv(c), and all other variables except for Qv(d). The black dot denotes the location of Guangzhou Radar. (Adapted from Wu *et al.* 2013.)

original 6 h to 12 or 18 h, the forecast skill apparently decreased. None of the ensemble members successfully captured the squall line with a correct time and location. This result indicated that the lead time of this squall line case with the current model configuration was only a few hours.

Finally, the way how to improve the forecast skill of squall line was investigated. It was found that using different combinations of physical parameterization schemes for different members apparently improved forecast capability for squall line. The percentage of the members that captured the squall line was increased by 10%.

5. Summary

This chapter reviewed recent progresses made on radar climatology, strong wind producing structures, and predictability of squall lines in east China.

Based on radar mosaic of 2008 and 2009, the geographical distribution, monthly and diurnal variations, basic radar traits such as length, orientation, organization, formation and dissipation mode as well as the basic synoptic flow patterns were obtained. The formation environment of squall lines in east China was found to be moister with a weaker low-level vertical shear than that in the U.S. A new type of squall line, pre-TC squall line, was examined in details. The formation mechanism and the role of the parent TC were investigated.

The formation mechanism of the bowing structure of a squall line in front of a westerly trough was studied using a real-world numerical simulation. New insights on the formation of bow echoes and the associated bookend vortices were obtained relative to previous results based on idealized numerical experiments.

The radar structure of a supercell that formed in a cold-vortex-associated squall line that caused the first-documented tornado in Beijing was examined. The tornadic supercell showed several radar signatures such as TVS, DRC and TDS. Results showed that the track and intensity variation of the tornado was quite consistent with the evolution of the mesocyclone and TVS.

Finally, the predictability of a pre-trough squall line was examined. It was found that the simulation of the squall line was very sensitive to the uncertainties of both the initial field and physical processes. The improvement in the initial moisture field played a key role in the improvement of squall line simulation.

Acknowledgments

This work was sponsored by the Ministry of Science and Technology of China with the Grants Nos. of 2013CB430104 and GYHY200906025 and the Natural Science Foundation of China with the Grants Nos. of 41375048, 41425018 and 4141101075.

References

Bluestein, H. B. and M. H. Jain, 1985: Formation of mesoscale lines of precipitation: Severe squall lines in Oklahoma during the spring. *J. Atmos. Sci.*, **42**, 1711–1732.

Grell, G. A. and D. Devenyi, 2002: A generalized approach to parameterizing convection combining ensemble and data assimilation techniques. *Geophys. Res. Lett.*, **29**, 1693. doi:10.1029/2002GL015311.

Grim, J. A., R. M. Rauber, G. M. McFarquhar, B. F. Jewett, and D. P. Jorgensen, 2009: Development and forcing of the rear inflow jet in a rapidly developing and decaying squall line during BAMEX. *Mon. Wea. Rev.*, **137**, 1206–1229.

Chen, G. T.-J. and H.-C. Chou, 1993: General characteristics of squall lines observed in TAMEX. *Mon. Wea. Rev.*, **121**, 726–733.

Geerts, B., 1998: Mesoscale convective systems in the southeast United States during 1994–95: A survey. *Wea. Forecasting*, **13**, 860–869.

Houze, Jr. R. A., B. F. Smull, and P. Dodge, 1990: Mesoscale organization of springtime rainstorms in Oklahoma. *Mon. Wea. Rev.*, **118**, 613–654.

Hong, S.-Y. and coauthors, 2010: Evaluation of the WRF doublemoment 6-class microphysics scheme for precipitating convection. *Adv. Meteor.*, 2010, 707253. doi:10.1155/2010/707253.

Melhauser, C. and F. Zhang, 2012: Practical and intrinsic predictability of severe and convective weather at the mesoscales. *J. Atmos. Sci.*, **69**, 3350–3371.

Meng, Z., D. Yan, and Y. Zhang, 2013: General features of squall lines in East China. *Mon. Wea. Rev.*, **141**, 1629–1647.

Meng, Z., F. Zhang, P. Markowski, D. Wu, and K. Zhao, 2012: A modeling study on the development of bowing structure and associated rear inflow within a squall line over South China. *J. Atmos. Sci.*, **69**, 1182–1207.

Meng, Z. and Y. Zhang, 2012: On the squall lines preceding landfalling tropical cyclones in China. *Mon. Wea. Rev.*, **140**, 445–470.

Meng, Z. and D. Yao, 2014: Damage survey, radar and environment analyses on the first-ever documented tornado in Beijing during the heavy rainfall event of 21 July 2012, *Wea. Forecasting*, accepted.

Noh, Y., W.-G. Cheon, S.-Y. Hong, and S. Raasch, 2003: Improvement of the K-profile model for the planetary boundary layer based on large eddy simulation data. *Bound.-Layer Meteor.*, **107**, 401–427.

Parker, M. D. and R. H. Johnson, 2000: Organizational modes of midlatitude mesoscale convective systems. *Mon. Wea. Rev.*, **128**, 3413–3436.

Rasmussen, E. N. and D. O. Blanchard, 1998: A baseline climatology of sounding derived supercell and tornado forecast parameters. *Wea. Forecasting*, **13**, 1148–1164.

Rasmussen, E. N., 2003: Refined supercell and tornado forecast parameters. *Wea. Forecasting*, **18**, 530–535.

Rasmussen, E. N., J. M. Straka, M. S. Gilmore, and R. Davies-Jones, 2006: A preliminary survey of rear-flank descending reflectivity cores in supercell storms. *Wea. Forecasting*, **21**, 923–938.

Ryzhkov, A., T. J. Schuur, D. W. Burgess, and D. S. Zrnic, 2005: Polarimetric tornado detection. *J. Appl. Meteor.*, **44**, 557–570.

Skamarock, W. C. and coauthors, 2008: A Description of the Advanced Research WRF Version 3. *NCAR Tech. Note TN-475_STR*, 113 pp.

Weisman, M. L., 1993: The genesis of severe, long-lived bow echoes. *J. Atmos. Sci.*, **50**, 645–670.

Weisman, M. L. and C. A. Davis, 1998: Mechanisms for the generation of mesoscale vortices within quasi-linear convective systems. *J. Atmos. Sci.*, **55**, 2603–2622.

Wu, D., Z. Meng, and D. Yan, 2013: The predictability of a squall line in South China on 23 April 2007. *Adv. Atmos. Sci.*, **30**, 485–502.

Wyss, J. and K. A. Emanuel, 1988: The pre-storm environment of midlatitude prefrontal squall lines. *Mon. Wea. Rev.*, **116**, 790–794.

Chapter 17

Advances in Understanding the Early-Summer Heavy Rainfall over South China

Yali Luo

State Key Laboratory of Severe Weather,
Chinese Academy of Meteorological Sciences,
Beijing 100081, China
yali@camscma.cn

In South China (SC), the annual precipitation commonly exceeds 2000 mm and about 40%–50% of the amount falls during April–June, which is called the pre-summer rainy season over SC. Frequent occurrences of extreme rainfall in the pre-summer rainy season often lead to severe flooding and inundations. In this paper, previous research and field programs aiming at better understanding of the heavy rainfall formation are reviewed. Current scientific issues of relevance to the heavy rainfall events are then presented followed by an introduction to SCMREX, which has been approved by World Meteorological Organization as a World Weather Research Program research and development project, including its background, scientific objectives, field experiment design, and preliminary scientific results.

1. Introduction

South China is one of the rainiest regions in the world, with an annual precipitation amount over 2000 mm. About half of the annual amount takes place during April to June, i.e. the pre-summer rainy season of South China, which is concurrent with the Meiyu season over Taiwan. There is a close relationship between the onset and advance of summer monsoon and the main rain belts in China (Tao and Chen 1988). The pre-rainy season in South China reaches its peak in terms of occurrence frequency and intensity of rainfall around and after the onset of summer monsoon over South China Sea (SCS) (Ding 1994; see Table 3.2 therein). This rainy season ends when the monsoonal rain belt moves northward to the Yangtze River Valley in central East China in mid-June, i.e. the Meiyu season starts over there. During the pre-summer rainy season, South China and vicinity often experiences most frequent occurrences of extreme rainfall, leading to severe flooding and inundations. So these storms endanger the safety of lives and cause marked property damages, often producing devastating economic losses.

The large-scale environmental characteristics that are favorable for persistent heavy rainfall occurrence over South China in the pre-summer rainy season have been studied since 1950's. The location and intensity of rain belt are determined by distributions of the middle-high-latitude troughs and ridges, the south Asia high pressure, the SCS monsoon surge, and the subtropical high over Western Pacific Ocean. Most heavy rainfall in the pre-summer rainy season results from the appropriate interaction and horizontal juxtaposition of cold air and moist air that are in close association with the westerlies and the SCS monsoon, respectively (Chen and Zhao 2000, 2004). The presence of front, shear line, frontogenesis zone, and convergence zone between the southerly and northerly winds in the lower troposphere provide a favorable environment, which produces

The Global Monsoon System: Research and Forecast (3rd Edition)
Edited by C. P. Chang et al.

the deep moist unstable layer and the upward motion branch of vertical circulation (e.g. Zhao *et al.* 2007).

Most precipitation during the pre-summer rainy season is of convective nature with mesoscale organizational characteristics. The precipitation distribution exhibits inhomogeneous feature with major rainfall centers being distributed over the coastal areas of Guangdong and Guangxi provinces, central Guangdong, and northern Guangxi. Total rainfall over these centers is contributed significantly (more than 50%) by the rates of larger than $50\,\text{mm}\,\text{day}^{-1}$. The rainfall centers mostly correspond well to most frequent occurrences of convection. The precipitation taking place hundreds of kilometers ahead of a cold/stationary front, namely, the warm-sector precipitation, is most significant for the pre-summer rainy season and normally has more intense rainfall than precipitation occurring within the frontal zone of a cold/stationary front or behind a cold/stationary front during the pre-summer rainy season (Huang *et al.* 1986; Ding 1994).

This paper will present an overview of the previous research and field programs (Sec. 2) and discussions on the current scientific issues that are of relevance to research and forecast of the pre-summer heavy rainfall over South China (Sec. 3), followed by an introduction to the Southern China Monsoon Rainfall Experiment (SCMREX) that has been approved by World Meteorology Organization (WMO) as a World Weather Research Project (WWRP) Research and Development Project (Sec. 4). Concluding remarks are given at the end.

2. Previous Research and Field Programs

2.1. *The program in 1977–1981*

The first research and field programming at better understanding the pre-summer rainy season heavy rainfall lasted five years (1977–1981) and conducted field experiments in Guangxi, Guangdong, Fujian, and the southern parts of Guizhou, Hunan and Jiangxi provinces during May–June 1977–1980. The field program consists of a pilot experiment in May–June 1977, formal experiments in May–June 1978–1979, and a boundary layer experiment in May–June 1980. This program is the very first one conducted in China aiming at better understanding heavy rainfall with participation of many researchers and operational people from 8 provinces of China. It significantly enhanced the knowledge about the heavy rain during the pre-summer rainy season in South China, mainly on the characteristics at the synoptic and intermediate scales. The research results revealed the importance of fronts, low-level jets (LLJs), processes in the planetary boundary layer (PBL), topography, and ocean-land contrast, and also suggested the roles played by mesoscale cyclones/anticyclones, mesoscale shearlines, and mesoscale fronts. Moreover, results indicated that the heavy rainfall often takes place to the south of a surface front, or without presence of a front from the Nanling mountain range to the northern SCS, i.e. South China is dominated by relatively warm and moist air. Therefore they are named "warm-sector heavy rainfall". This finding was an important improvement as occurrence of heavy rainfall in China had always been linked to influence of cold air. Research based on the data set collected during this field program and scientific findings obtained were summarized in the book by Huang *et al.* (1986).

2.2. *'98 Heavy Rainfall Experiment in South China (HUAMEX)*

The HUAMEX was conducted in May–June 1998 (Zhou *et al.* 2003), concurrent with an international field program, namely, the South China Sea Monsoon Experiment (SCSMEX; Lau *et al.* 2000). The major objective of HUAMEX is better understanding characteristics of mesoscale systems and their interaction with synoptic-scale systems. The mesoscale observing area of HUAMEX covers Guangdong, Hong Kong, and Fujian. During the HUAMEX experiment, observational data were collected for seven heavy rainfall events. In addition to the enhanced (6 hourly) radiosonde observations at operational upper air sounding stations and the land-based (*in situ*) operational observing network, multiple advanced observing facilities also participated in HUAMEX, including 4 Doppler weather radars, 2 wind profilers, and 3 GPS/MET stations. The conventional and non-conventional data collected were used to investigate the large-scale circulation characteristics, synoptic weather systems, and interactions between cold air from the north and the Asian monsoonal flow (e.g. Chen and Zhao 2004; Bei and Zhao 2005). The results emphasized the importance of conditions and processes in the PBL. Several studies conducted mesoscale analysis of the heavy rainfall events and revealed some mesoscale structures of MCSs and certain mechanisms by which they form; while topographical influence of Hainan island and Nanling mountain range on rainfall location/amount was also investigated (e.g. Zhao *et al.* 2007; Zhou *et al.* 2003). However, these studies mostly utilized numerical modeling with a horizontal grid spacing of about 15 km or larger, which could not explicitly represent the initiation and movement of convection.

Moreover, HUAMEX promoted development and test of algorithms to derive meteorological parameters from remote sensing observations. Two algorithms to derive 3D wind fields in rainy storms from Doppler radar measurements were tested; precipitable water was estimated from GPS-MET measurements and compared to radiosonde observations; surface properties such as vegetation type and albedo were estimated from satellite observations and applied to modeling heavy rainfall event (Wang *et al.* 2001); a scheme was developed to derive atmospheric motion vectors from satellite observations and compared to another method developed at the European Organization for the Exploitation of Meteorological Satellites (Xu *et al.* 2002).

2.3. *The Southern China Heavy Rainfall Experiments (SCHeREX) in 2008–2009*

The SCHeREX (Zhang *et al.* 2011; Ni *et al.* 2013) was staged during May–July 2008 and 2009 in four observing zones, one of them being located in South China (1 May to 10 June) and the other three in central East China (11 June–20 July). The SCHeREX experiments build upon the operational observing network of China Meteorological Administration which was enhanced significantly compared to '98 HUAMEX. The Doppler radar network at that time covered a large portion of South China. The measurements from the Doppler radars in Guangzhou and Shenzhen on 6 June 2008 were processed using the SOLOII and REORDER software packages developed at National Center for Atmospheric Research (NCAR) of the United States to investigate internal circulations of a quasi-linear-shaped mesoscale convective system (Zhao *et al.* 2013). For the first time in mainland China, two polarimetric Doppler radars operating at X-band and C-band, respectively, and one millimeter-wave cloud radar (MMCR) were placed in the South China observing zone. Efforts have been made to analyze the measurements from the polarimetric radars and MMCR after the SCHeREX experiments. For example, a data quality-control procedure for the dual-polarization radars was established and methods of quantitative precipitation estimate were developed and evaluated (Hu *et al.* 2012); the MMCR was able to show microphysical and dynamic structures of clouds associated with typhoon Nuri (Zhong *et al.* 2011). Moreover, high-resolution gridded analysis was produced by putting the collected multiple-source observational data into the Local Analysis and Prediction System (LAPS) (Albers *et al.* 1996) and used to investigate finer-scale structures and evolution of the rainy storm and the associated weather systems (Xu *et al.* 2013). Furthermore, the observations were applied to short-term numerical weather prediction (NWP) experiments with timely data updates during the field experiments and showed some promising results in predicting the heavy rainfall events (Wan *et al.* 2013).

3. Current Scientific Issues

There are still unknowns about the mechanisms governing formation and evolution of the heavy rainfall over South China during the pre-summer rainy season, especially those related to convective-scale processes and their interactions with larger-scale conditions and processes.

(1) The importance of environmental cold air, LLJs, and PBL processes has been noticed. However, knowledge on their specific roles in convective initiation/maintenance and in formation/evolution of MCSs is insufficient. The mesoscale structures of LLJs and PBL atmosphere also need to be revealed by observations.
(2) Influence of Nanling mountain range and Hainan island on the location and intensity of rainbands has been suggested by numerical modeling experiments. However, it is unclear how land-sea distribution, mountains of a range of scales, and cities over South China effect convective initiation and maintenance under a variety of environmental atmospheric conditions, i.e. different patterns of near surface wind fields and LLJs of different orientations. Relative importance of the land-sea variation and the multiple hills/mountains near the coastal regions needs more investigation.
(3) Current knowledge on the internal structures of MCSs are mostly from modeling studies with a horizontal grid spacing of 10–20 km that cannot explicitly represent the initiation and propagation of convection. Internal circulations of MCSs that were previously obtained from dual-Doppler radar wind retrieval are also associated with large uncertainties due to the too far distance (exceeding 100 km) between the radars. Acquirement and analysis of higher-quality observations of high resolutions are the basis for advancing knowledge on the MCSs' internal dynamic and microphysical structures as well as their interactions with mesoscale environments.
(4) Understanding predictability of the heavy rainfall events is of importance for both physical mechanism study and practical forecast. However, very little is known about this because of few studies on this topic. Simulation experiments could be designed to investigate the heavy rainfall's predictability and to explore impacts of initial conditions, model physical schemes, and grid spacing on the predictability.
(5) The quantitative precipitation forecast (QPF) skill for heavy rainfall especially the warm-sector extreme rainfall over South China during the pre-summer rainy season is very low. Approaches to improving the QPF skill include improving initial conditions for NWP by data assimilation, reducing uncertainties associated with physical schemes in models, and developing convection-permitting ensemble NWP systems.

4. Southern China Monsoon Rainfall Experiment (SCMREX)

4.1. *Background*

As a Research and Development Project of WMO/WWRP, SCMREX aims to expedite the understanding of processes key to the heavy rain formation and the efforts to improve simulation and prediction of the heavy rainfall during the pre-summer rainy season in South China and vicinity. SCMREX consists of four integrated components: field campaign, data collection/processing, NWP study, and physical mechanism study. SCMREX has its first (2013–2015) and second (2016–2018) stages (Luo *et al.* 2016).

The NWP study of SCMREX aims to improve quality of initial fields for NWP by data assimilation, evaluate and possibly improve cloud microphysics and PBL schemes in numerical models, and conduct/evaluate convection-permitting ensemble simulation experiments. The physical mechanism studies are mainly about the roles of underlying surface, PBL processes, LLJs, and mesoscale circulation in association with fronts on initiation and development of the rainy storms. A scientific database has been setup to share the data sets from the field campaign and NWP study. The data can be accessed through application at http://scmrex.cma.gov.cn where the Implementation Plan of SCMREX may also be available.

4.2. *Scientific objectives*

(1) Roles of underlying surface and boundary layer processes on the convective initiation and evolution

The topography of South China is very complicated which involves land-sea, mountain-valley, and city-rural contrasts. The nonlinear interactions between the underlying surface and larger-scale environmental flow may lead to the development of mesoscale disturbances such as the convergent lines and vortex to trigger convection initiation. Land-sea contrast may produce apparent convergence in coastal area during land breeze phase from midnight to early morning and apparent convergence in inner land area during sea breeze phase from afternoon to early evening. The multiple hills/mountains distributed near the coastal line could lift moist air from the adjacent ocean and trigger convection. The trumpet-shape mountain-valley distribution around Zhujiang River delta area is favorable for the convergence between westerly or southwesterly flow and southeasterly flow. Heating differences between city and rural areas in the central part of Guangdong province may trigger local mesoscale circulation and thus mesoscale disturbance. Mesoscale boundaries such as convergent lines may form due to the complicated interactions between the above mentioned various local mesoscale circulations forced by topography and PBL processes or between the local mesoscale circulations and background flow. These hypotheses need to be further evaluated. The dense surface observations and advanced radar capability in SCMREX will help investigate the roles of boundary layer processes and underlying surface on the initiation and evolution of the heavy-rain-producing convective systems.

(2) Roles of LLJs and mesoscale circulations in association with fronts on the convective initiation and development of the heavy-rain-producing storm

LLJs are recognized as one of the key factors that may trigger the heavy rainfall during the pre-summer rainy season in South China. It is hypothesized that the intensification and downward expansion of LLJs play important roles in triggering the heavy rains likely through providing moisture, heat, energy and disturbance and/or triggering inertial gravity wave. Mesoscale circulations in association with fronts are another important component that significantly influences the formation and evolution of storms and the associated heavy rain. SCMREX will provide time-continuous mesocale wind fields in the lower troposphere over Guangdong, Hainan, and coastal region of Guangxi, as well as information about PBL conditions over northern SCS. The comprehensive dataset enables us to analyze characteristics of the LLJs and mesoscale circulations in association with fronts, and also helps investigate their roles on the development and maintenance of the heavy-rain-producing convective systems.

(3) Microphysical processes of the heavy-rain-producing storms

Microphysical processes in heavy-rain-producing convective systems are a major factor controlling the intensity and efficiency of precipitation and characteristics of lightening. During the pre-summer rainy season in South China, the rain rates in the warm-sector hundreds kilometers ahead of the fronts are generally much larger than those near or behind the fronts, which is probably related to different microphysical properties and processes. Limited knowledge about the clouds and mesoscale environment and inadequate representation of microphysical processes in numerical models can be major contributors to the low skill for the pre-summer rainy season QPF. Observations on the microphysical properties of storms during the pre-summer rainy season were very limited in previous field experiments over mainland China. The dual-polarization radars, micro rain radar, millimeter-wave cloud radars, raindrop disdrometers in SCMREX, combined with available satellite-based long-term products such as those from TRMM (Simpson *et al.* 1996) and CloudSat (Stephens *et al.* 2002), will help understand the key microphysical processes and the distribution/phases of hydrometers in the heavy-rainfall-producing convective systems over South China, which provides necessary information for model evaluation.

(4) Improving QPF skill by better understanding the multi-scale precipitation processes, the assimilation of high-resolution observations into numerical models, the conduction and evaluation of convective-permitting ensemble simulation experiments

Because errors in the initial conditions of numerical prediction/simulations are present and the physical parameterization schemes in models are insufficiently representative of meso- and micro-scale physical processes in reality, QPF skill for the heavy rainfall during the pre-summer rainy season needs improvement, while the skill for the warm-sector rainfall is lower than that for the near-front rainfall. It has been demonstrated that numerical simulations of MCSs in mainland China during the warm season can be quite sensitive to the initial atmospheric conditions (Wu *et al.* 2013; Luo and Chen 2015) and the model physical parameterization schemes adopted (Luo *et al.* 2010). Moreover, while a single simulation or prediction cannot provide information on uncertainty, ensemble prediction serves to provide the statistical state of a collection of possible results (Toth and Kalnay 1993, Tracton and Kalnay 1993), which can help improve the forecast quality and extend valid prediction periods (Xue *et al.* 2007, 2011; Schwartz *et al.* 2010; Clark *et al.* 2012). The comprehensive dataset collected during SCMREX provides a unique opportunity to improve representation of the model initial conditions, evaluate/refine the microphysical and PBL schemes in NWP models, and better evaluate the ensemble NWP experiments.

4.3. *Experiment design*

The field campaign of SCMREX aims to obtain unique and composite observations of internal structures of the rainy storms and of their atmospheric environment during the pre-summer rainy season of South China. The field experiments are launched in a region mainly covering Guangdong, southeastern Guangxi, Hainan, Hong Kong, and the offshore of SCS during May to mid-June of 2013–2018. The observing network builds upon the meteorological operational observation network that consists of the sounding stations, Doppler weather radars, ground-based GPS water vapor stations, automatic weather stations, and satellites. During the intensive observation periods (IOPs) of 2014–2016, eight sounding stations in Guangdong, Hainan, eastern Guangxi, and Hong Kong launched two extra soundings per day (0600 UTC and 1800 UTC) in addition to the routine observations at 0000 UTC and 1200 UTC. Air-borne observations are collected over the offshore of SCS for

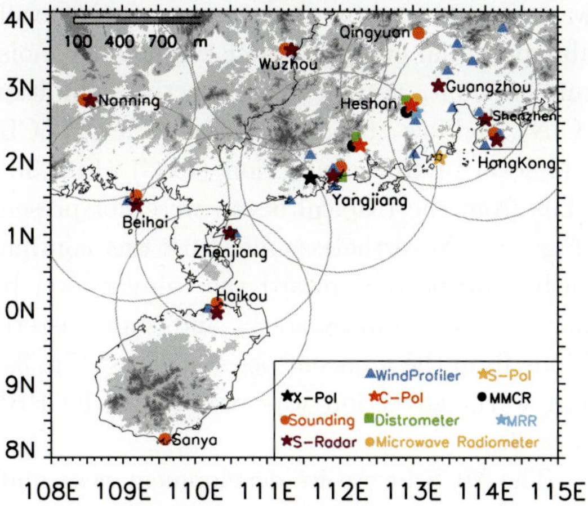

Fig. 1. Map of the distribution of major facilities deployed during the SCMREX-2014 field campaign. All the facilities are consistent with markers in the legend. Gray shadings represent topography. Luo *et al.* (2016) provide more information on the facilities.

the experiment. In the observing network, 15 fixed wind-profilers and 2 portable boundary layer wind-profilers are deployed, which helps better reveal mesoscale wind fields especially in the PBL and lower troposphere. A range of portable instruments, including polarimetric radar, frequency-modulated continuous-wave radar, millimeter-wave cloud radar, micro-rain radar, and raindrop distrometer, are placed over the western coastal region of Guangdong (Fig. 1). More information about the SCMREX field campaigns and preliminary results can be found in Luo *et al.* (2016).

4.4. *Preliminary results*

The pre-phase field experiment of SCMREX was successfully carried out from mid-April to end of May 2013. Four heavy rainfall events occurred during the IOPs (8–17 May and 24–28 May). An extreme-rain-producing MCS develops over western coastal region of Guangdong and drops about a half meter of rain over Yangjiang city on 10 May 2013 (Fig. 2). Observational analyses for this extreme rainfall event are summarized by Wang *et al.* (2014) and are concisely described here as an example of some preliminary results.

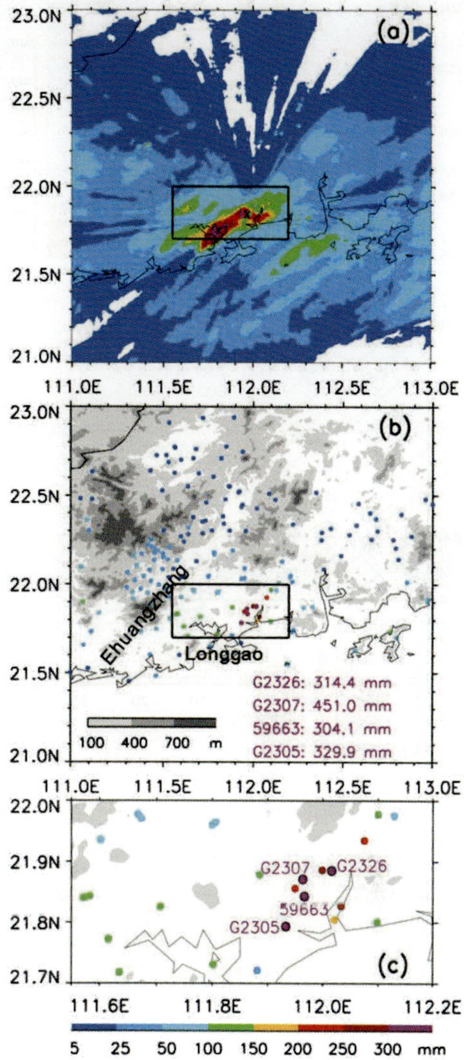

Fig. 2. Distribution of accumulated rainfall during 0000–1900 BST 10 May 2013 based on (a) rainfall retrieval from radar reflectivity data and (b, c) rain gauge observations. The black rectangle box in (a) and (b) denotes the control region; cross symbol in (a) represents location of the Yangjiang station (station number 59663). Names of the 4 stations receiving accumulated rain >300 mm are highlighted by black circles in (c) and the corresponding rain amounts are labeled in (b). Gray shadings in (b) and (c) denote topography. Thin black lines represent the coastal lines.

The extreme rainfall event occurred about 200–300 km to the south of a northeast-southwest-oriented shear line in the lower troposphere (Fig. 3a) and several hundred kilometers

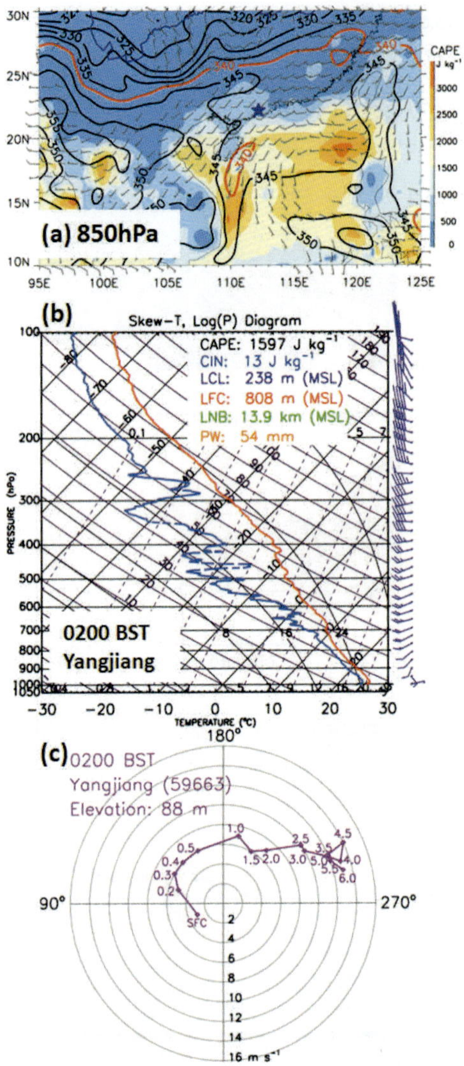

Fig. 3. (a) Equivalent potential temperature (θ_e, contoured at intervals of 5 K, with the 340 K contour in red), CAPE (J kg^{-1}, shaded), superimposed with horizontal wind barbs, at 850 hPa in the ERA-interim analysis at 0200 BST10 May 2013. The star denotes the location of Yangjiang sounding station (station number 59663). (b) Vertical profiles plotted on skew T-log P diagram of temperature (red lines) and dew point temperature (blue lines) over Yangjiang sounding station (59663) at 0200 BST 10 May 2013. (c) Hodograph from surface to 6.0 km (above MSL) over Yangjiang at 0200 BST 10 May 2013.

downstream of a trough in the mid-troposphere, i.e. the control region was dominated by relatively warmer and moister airs. The thermodynamic environment near the rainy storm was characterized by moderate convective available potential energy (CAPE), abundant moisture amount, very little convective inhibition (CIN), and low lifting condensation level (LCL) and level of free convection (LFC) (Fig. 3b). LLJs from the tropical ocean were not present (Fig. 3c). Nevertheless, moisture was continuously transported toward the rainy region by the low-level southwesterly and southeasterly winds from the adjacent oceanic areas (Fig. 3a) and warm advection was present in the PBL (Fig. 3c).

The MCS began its development near midnight (i.e. 2300 BST 9 May) from the initiation of convective cells near the east of Mt. Longgao about 30 km southwest from Yangjiang city, where higher-θ_e air, continuously fed by the southeasterly current in the lowest several hundred meters, was lifted (Fig. 4a). Convection was also initiated by orographic lifting near the east of Mt. Ehuangzhang starting near 0600 BST. The continuous convective initiation and their organization into two northeast-southwest-oriented rainbands (Fig. 4b) led to the production of a cold pool centered over Yangjiang city and northerly outflows in the early morning (i.e. about 0900 BST 10 May) (Fig. 5a). A boundary was consequently formed between the cold outflows and warmer air mass to the south where surface temperature increased substantially due to enhanced sensible heating (Fig. 5a). While the cold pool was maintained by persistent precipitation, the boundary propagated southward (Figs. 5b–d).

Intensity of the locally developed convection was enhanced considerably after 0600 BST. However, it was weaker in terms of convective intensity than the squall line that moved into the control region from the west and passed over in about 3 hours (1530–1830 BST) (Figs. 4e, f), suggested by analysis of maximum radar reflectivity, lightning flash intensity, and satellite-observed IR brightness temperature. Nevertheless, this squall line contributed only about 10%–15% to the total accumulative

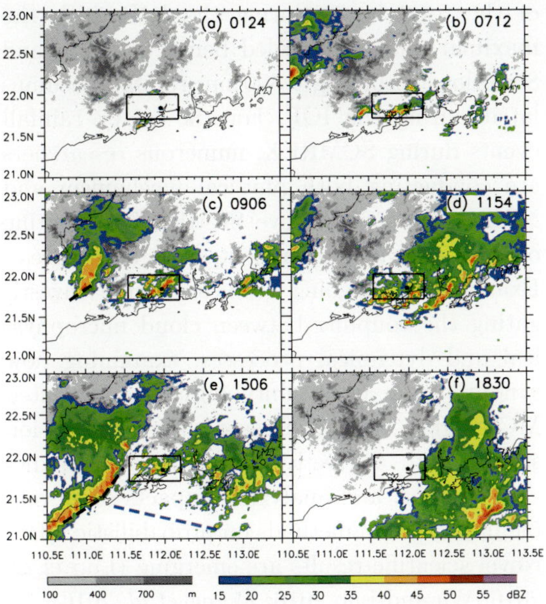

Fig. 4. Radar reflectivity (dBZ) at 3-km altitude above MSL derived from the SA radar and C-Pol radar at 6 selected times (a–f). Location of the SA radar is marked with black dot. The black rectangle box represents the control region. The dashed lines represent the approximate locations of the cold outflow boundary in association with the locally developed convection (dashed lines in blue) and the rainy system moving from the west (dashed lines in black), respectively. Gray shadings represent topography.

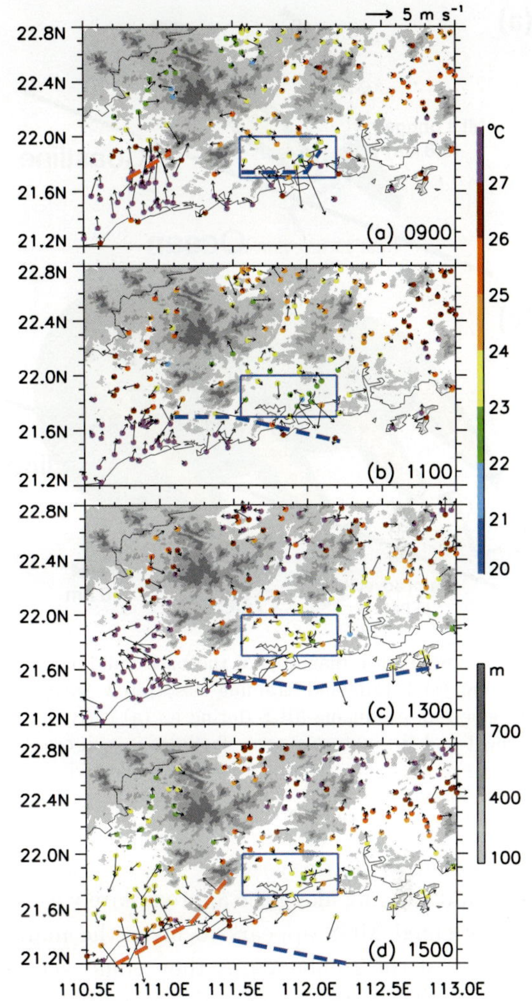

Fig. 5. Surface meso-analysis at 4 selected times (a–d). The locations of surface automatic weather stations are marked with dots, with colors representing the air temperature and black arrows representing the surface wind. The thick dashed lines represent the approximate locations of the cold outflow boundary in association with the locally developed convection (dashed lines in blue) and the system moving from the west (dashed lines in red), respectively. The blue rectangle box represents the control region. Gray shadings represent topography.

rainfall over the control region (Wang *et al.* 2014).

The influencing factors and convective organization taking place during the early development and the mature stages of the MCS are summarized in Fig. 6. The organization of the individual rainbands in the present case appears similar to that of the linear MCSs with parallel stratiform precipitation in the United States identified by Parker and Johnson (2000). The present case also resembles to a certain extent the backbuilding/quasi-stationary pattern described by Schumacher and Johnson (2005) for the extreme-rain-producing MCSs in the United States. However, the present MCS exhibits two scales of convective organization during its mature stage: One is the northeastward "echo training" of convective cells along individual rainbands, the other is the east- or southeastward "band training" of the rainbands along the MCS. The repeated convective back building and the subsequent echo training along the same path could account for the extreme rainfall production in the present case, whereas

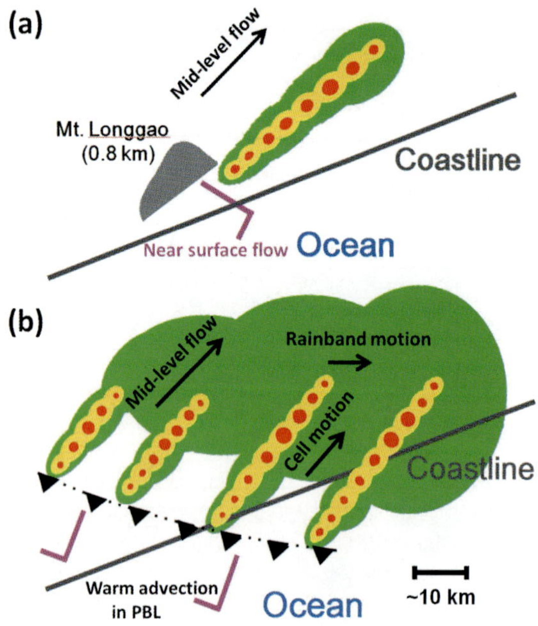

Fig. 6. Schematic diagrams of the back-building, echo training, and rainband training associated with the extreme-rain-producing MCS during its (a) early development and (b) mature stages. Shadings in red, orange, and green represent roughly the radar reflectivity values of 50, 35, 20 dBZ at 3-km above MSL, respectively. (From Wang *et al.* 2014.)

the band training along the approximately west-east-oriented MCS appears to favor the maintenance of the rainbands as they could access higher CAPE in the larger-scale southerly flows. The echo- and band trainings of convective elements found in this particular case differ from Parker and Johnson (2000) and Schumacher and Johnson (2005); but were recently noticed by Luo *et al.* (2014) in an extreme-rain-producing MCS over central east China during a Meiyu season.

5. Concluding Remarks

The formal field campaign of SCMREX has been successfully carried out with a continuous IOP from 0000 UTC 1 May to 1800 UTC 15 June 2014. Heavy rain $>50\,\mathrm{mm\,h^{-1}}$ ($20\,\mathrm{mm\,h^{-1}}$) fell over South China in 20% (67%) of the total hours during the IOP, with the maximal rainfall recorded being 1491 mm and 8 stations recording >800 mm rainfall accumulated during the IOP. For the heavy rainfall events during SCMREX, numerous researchers are conducting both physical mechanism and NWP studies to achieve SCMREX's scientific objectives, while data control is in progress. Examples of ongoing studies include investigating the coupling between cloud microphysical and dynamical processes, examining how synoptic-scale uncertainty influence simulated MCS errors, ensemble-based data assimilation studies with convective-scale ensemble prediction systems, and analyzing impacts of ensemble initialization methods on probabilistic QPF. Novel scientific results are emerging (Luo *et al.* 2016; Wu and Luo 2016; Zhang *et al.* 2016).

SCMREX provides unprecedented opportunity to study the detailed structures of the heavy-rain-producing MCSs and their interactions with convective-scale processes and larger-scale weather systems, and to improve QPF for the heavy rainfall events over South China. Future field experiments can be improved in several aspects, e.g. higher temporal resolution (three-hourly) sounding data, increased observations in data void regions such as ocean and complex terrain, and carefully designed polarimetric radar observations. With the better observations, important scientific issues related to convective initiation, possible role of land-sea contrast, and cloud-precipitation microphysical processes would be better addressed.

Acknowledgments

I would like to thank Professor Russ Schumacher (Colorado State University) for providing constructive comments and suggestions that helped improve the quality of the manuscript. I would also like to thank Dr. Hui Wang (Chinese Academy of Meteorological Sciences) for making the figures used in this paper. This research has been supported by the

Meteorology Public Welfare Scientific Research Projects (GYHY201406013, GYHY201306004), the Basic Research Funds of the Chinese Academy of Meteorological Sciences (2013Z003), and the National Natural Science Foundation of China (Projects 41175049 and 41221064).

References

Albers, S. C., J. A. McGinley, D. A. Birkenheuer, and J. R. Smart, 1996: The local analysis and prediction system (LAPS): Analysis of clouds, precipitation and temperature. *Wea. Forecast*, **11**, 273–287.

Bei, N. F. and S. X. Zhao, 2005: Diagnostic analysis of severe heavy rainfall in Taiwan area during Heavy Rainfall Experiment in South China. *Climatic and Environmental Research*, **10**(3), 474–491 (in Chinese with English abstract).

Chen, H. and S. X. Zhao, 2000: Heavy rainfall in South China and related circulation during First GARP Global Experiment Period. *Chinese J. Atmos. Sci.*, **24**(2), 238–252 (in Chinese with English abstract).

Chen, H. and S. X. Zhao, 2004: Heavy rainfalls in South China and related circulation during HUAMEX Period. *Chinese J. Atmos. Sci.*, **28**(1), 32–47 (in Chinese with English abstract).

Clark, A. J. *et al.*, 2012: An overview of the 2010 Hazardous Weather Testbed Experimental Forecast Program spring experiment. *Bull. Amer. Meteor. Soc.*, **93**, 55–74.

Ding, Y. H., 1994: Monsoons over China. Kluwer Academic Publishers, The Netherlands. 419 pp.

Hu, Z. Q., L. P. Liu, and L. R. Wang, 2012: A quality assurance procedure and evaluation of rainfall estimates for C-Band polarimetric radar. *Adv. Atmos. Sci.*, **29**(1), 144–156. doi: 10.1007/s00376-011-0172-y.

Huang, S. S. *et al.*, 1986: *Heavy Rainfall over Southern China in the Pre-summer Rainy Season.* Guangdong Science and Technology Press, 244 pp (in Chinese).

Lau, K. M., Y. H. Ding, J. T. Wang, *et al.*, 2000: A report of the field operations and early results of the South China Sea Monsoon Experiment (SCSMEX). *Bull. Amer. Meteor. Soc.*, **81**, 1261–1270.

Luo, Y. L., Y. Gong, and D.-L. Zhang, 2014: Initiation and organizational modes of an extreme-rain-producing mesoscale convective system along a Mei-Yu front in east China. *Mon. Wea. Rev.*, **142**, 203–221.

Luo, Y. L., Y. J. Wang, H. Y. Wang, Y. J. Zheng, and H. Morrison, 2010: Modeling convective-stratiform precipitation processes on a Mei-Yu front with the WRF model: Comparison with observations and sensitivity to cloud microphysics parameterizations. *J. Geophys. Res.*, **115**, D18117. doi:10.1029/2010JD013873.

Luo, Y. L. and Yangruixue Chen, 2015: Investigation of the predictability and physical mechanisms of an extreme-rainfall producing mesoscale convective system along the Meiyu front in East China: An ensemble approach. *J. Geophys. Res. Atmos.*, **120**, 10593–10618. doi:10.1002/2015JD023584.

Luo, Y. L., R. H. Zhang, Q. L. Wan, *et al.*, 2016: The Southern China Monsoon Rainfall Experiment (SCMREX). *Bull. Amer. Meteor. Soc.*, available online (http://journals.ametsoc.org/doi/abs/10.1175/BAMS-D-15-00235.1)

Ni, Y., R. Zhang, L. Liu, *et al.*, 2013: *The Southern China Heavy Rainfall Experiment (SCHeREX).* China Meteorological Press, Beijing, 283 pp (in Chinese).

Parker, M. D. and R. H. Johnson, 2000: Organizational imodes of midlatitude mesoscale convective systems. *Mon. Wea. Rev.*, **128**, 3413–3436.

Schumacher, R. S. and R. H. Johnson, 2005: Organization and environmental properties of extreme-rain-producing mesoscale convective systems. *Mon. Wea. Rev.*, **133**, 961–976.

Schwartz, C. S. *et al.*, 2010: Toward improved convection-allowing ensembles: Model physics sensitivities and optimizing probabilistic guidance with small ensemble membership. *Wea. Forecasting*, **25**, 263–280.

Simpson, J., C. Kummerow, W.-K. Tao, and R. F. Adler, 1996: On the Tropical Rainfall Measuring Mission (TRMM). *Meteor. Atmos. Phys.*, **60**, 19–36.

Stephens, G. L. *et al.*, 2002: The CloudSat mission and the A-Train. *Bull. Amer. Meteor. Soc.*, **83**, 1771–1790.

Tao, S. Y. and L. X. Chen, 1988: A review of recent research on the East Asian summer monsoon in China. *In Monsoon Meteorology*, eds. C. P. Chang and T. N. Krishnamurti, Oxford University Press, Oxford, UK, pp. 60–92.

Toth, Z. and E. Kalnay, 1993: Ensemble forecasting at NMC: The generation of perturbations. *Bull. Amer. Meteor. Soc.*, **74**, 2317–2330.

Tracton, M. Steven and Eugenia Kalnay, 1993: Operational ensemble prediction at the National Meteorological Center: Practical aspects. *Wea. Forecasting*, **8**, 379–398.

Wan, Q. L., Z. T. Chen, X. F. Wang, *et al.*, 2013: Application of multiple observations in regional numerical weather prediction experiments during SCHeREX. In: *The Southern China Heavy Rainfall Experiment (SCHeREX)* [Ni, Y., R. Zhang, L. Liu, *et al.*]. China Meteorological Press, Beijing, China, pp. 163–214 (in Chinese).

Wang, H., Y. L. Luo, and B. J.-D. Jou, 2014: Initiation, maintenance, and properties of convection in an extreme rainfall event during SCMREX: Observational analysis. *J. Geophys. Res.-Atmospheres*, accepted.

Wang, P., Q. Xiao, *et al.*, 2001: Land vegetation retrieving from satellite remote sensing and application test in mesoscale simulation for heavy rainfall of South China. *Quart. J. Appl. Meteor.*, **12**(3), 287–296 (in Chinese with English abstract).

Wu, M. W. and Y. L. Luo, 2016: Mesoscale observational analysis of lifting mechanism of a warm-sector convective system producing the maximal daily precipitation in China mainland during the 2015 pre-rainy season. *J. Meteor. Res.*, **30**(5), doi: 10.1007/s13351-016-6089-8.

Wu, D., Z. Meng, and D. Yan, 2013: The predictability of a squall line in South China on 23 April 2007. *Adv. Atmos. Sci.*, **30**(2), 485–502.

Xu, J., K. Holmlund, Q. Zhang, and J. Schmetz, 2002: Comparison of two schemes for derivation of atmospheric motion vectors. *J. Geophys. Res.*, **107**(D14), 4196, 10.1029/2001JD000744.

Xu, W. H., X. K. Wang, and Y. Q. Ni, 2013: Study on meso-β-scale convective vortices over Dabiemountain during SCHeREX using mesoscale reanalysis data. In: *The Southern China Heavy Rainfall Experiment (SCHeREX)* [Ni, Y., R. Zhang, L. Liu, and coauthors]. China Meteorological Press, Beijing, China, pp. 163–214 (in Chinese).

Xue, M. *et al.*, 2011: Realtime convection-permitting ensemble and convection-resolving deterministic forecasts of CAPS for the Hazardous Weather Testbed 2010 spring experiment. *Extended Abstracts, 24th Conference on Weather Analysis and Forecasting/20th Conference on Numerical Weather Prediction*, Seattle, WA, Amer. Meteor. Soc. [Available online at https://ams.confex.com/ams/91Annual/webprogram/Manuscript/Paper183227/Xue_CAPS_2011_Spring Experiment_24thWAF20thNWP_ExtendedAbstract.pdf].

Zhang, R. H., Y. Q. Ni, L. P. Liu, Y. L. Luo, and Y. H. Wang, 2011: South China Heavy Rainfall Experiments (SCHeREX). *J. Meteor. Soc. Japan*, **89A**, 153–166.

Zhang, X. B., Y. L. Luo, Q. L. Wan, *et al.*, 2016: Impact of assimilating wind profiling radar observations on convection-permitting quantitative precipitation forecasts during SCMREX. *Wea. Forecast.*, http://journals.ametsoc.org/doi/full/10.1175/WAF-D-15-0156.1

Zhao, S. X., N. F. Bei, and J. H. Sun, 2007: Mesoscale analysis of a heavy rainfall event over Hong Kong during a pre-rainy season in south China. *Adv. Atmos. Sci.*, **24**(4), 555–572.

Zhao, K., Z. Y. Meng, and D. C. Wu, 2013: Application of dual-Doppler observations and numerical modeling to investigating meso-β-scale convective system over South China. In: *The Southern China Heavy Rainfall Experiment (SCHeREX)* [Ni, Y., R. Zhang, L. Liu, and coauthors]. China Meteorological Press, Beijing, China, pp. 215–241 (in Chinese).

Zhong, L. Z., L. P. Liu, and S. Feng, 2011: A 35-GHz Polarimetric Doppler radar and its application in observing clouds associated with typhoon. *Adv. Atmos. Sci.*, **28**, 945–956.

Zhou, X. J. *et al.*, 2003: *Heavy Rainfall Experiment in South China during Pre-summer Rainy Season (HUAMEX)*, 1998. Beijing: China Meteorological Press, 220 pp (in Chinese).

Chapter 18

Recent Development of Heavy Rain Research Associated with Summer Monsoon in East Asia Excited by Advances in Weather Radar

Hiroshi Uyeda*, Ben Jong-Dao Jou† and Dong-In Lee‡

*HyARC, Nagoya University, Nagoya 464-8601, Japan
uyeda@hyarc.nagoya-u.ac.jp

†National Taiwan University, Taipei, Taiwan

‡Pukyong National University, Busan, Korea

Heavy rainfalls associated with summer monsoon in East Asia are closely related to the local frontal activities: Meiyu, Baiu and Changma. Recent studies utilizing Doppler radar data on heavy rain weather systems related to East Asian summer monsoon are reviewed. Field experiments on heavy rainfalls with Doppler radar in East Asia are highlighted.

1. Introduction

Recent concerns on precipitation associated with summer monsoon in East Asia are flash flood producer: extreme rainfall in localized areas. People are taking a lifestyle that is susceptible to localized heavy rainfalls with the development of transportation networks and urbanization. Popularization of mobile phone made immediate transmission of heavy rain disaster-related information possible, however the temporal and spatial resolution of the prediction information provided is low. Recent advancement of radar monitoring of heavy rainfall assures accurate estimation of rainfall intensity in high spatial resolution and provides further innovation of technology in short-term forecasting of heavy rainfalls. In order to meet this imperative, we need to improve our understanding on triggering mechanism and kinematic and precipitation structure of precipitation systems producing heavy rainfalls in localized area. Studies with radar on heavy rainfall events associated with summer monsoon in East Asia in recent 10 years are reviewed in this article.

Heavy rainfalls associated with summer monsoon in East Asia are closely related to the local frontal activities, in China it is called Meiyu front and in Japan Baiu front and in Korea Changma front, respectively. The Meiyu/Baiu/Changma front marches from South China Sea and Taiwan in mid-May and mid-June. In the period of mid-June to mid-July, the front moves northward over the Yangtze River Valley, East China Sea, and the western part of Japan, and finally Korean Peninsula and the northeastern part of Japan. Characteristics of the Meiyu/Baiu front have been studied previously by many Chinese and Japanese researchers (Chen and Chang 1980; Ninomiya and Muraki 1986; Tao and Chen 1987; Chen 2004). Development of Doppler radar networks in recent decades in China, Hong Kong, Japan, Korea and Taiwan, and major mesoscale field campaigns with new research radars, greatly improved the understanding of heavy rain weather systems. Studies on these heavy rain weather systems related to East Asia summer monsoon in recent 10 years are reviewed in this article.

The Global Monsoon System: Research and Forecast (3rd Edition)
Edited by C. P. Chang et al.

2. Monsoon Rain in East Asia

The East Asian summer monsoon is one important branch of the Asian monsoon system over East Asia and western North Pacific. The system is characterized by a quasi-stationary, southwest-northeast elongated rain band extending from Japan to China. This rain band is associated with a pronounced frontal activity called Meiyu in China, Baiu in Japan and Changma in Korea (Fig. 1a taken from Chen and Bordoni (2014)). From the view point of climatology (following Sampe and Xie 2010), the Meiyu-Baiu system is defined as 15 June–14 July and 15 May to 14 June as pre-Meiyu (or early summer rainy season in south China and also called Meiyu in Taiwan, the rainband is about 10 degree south of the square indicated in Fig. 1a) and 15 July to 13 August as post-Meiyu (or Changma in Korea, about 10 degree north). Embedded pronounced cold-cloud tops are frequently observed within the front cloud

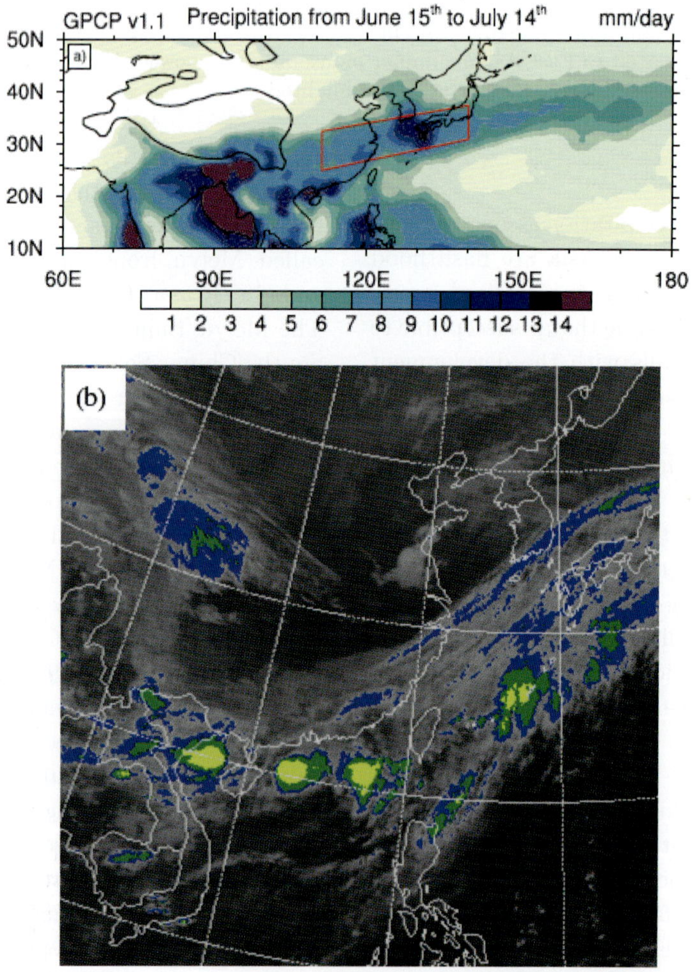

Fig. 1. (a) The climatology of Meiyu-Baiu rainfall distribution taken from Global Precipitation Climatology Project (GPCP) 1 degree Daily precipitation dataset (1DD), the climatology is from 1997–2008 (after Chen and Bordoni (2014)). (b) The enhanced satellite images for pre-Meiyu-Baiu frontal cloud system with MCSs embedded (cold-cloud tops with temperature well below −52°C).

band. These cold-cloud tops are associated with well-developed thunderstorm-type deep cumulonimbus clouds and are called as the Mesoscale Convective Systems (MCSs) in general (Fig. 1b). The MCSs are the heavy rain producers. Significant rainfall, sometimes with hails, strong winds and lightning activities, are the major characteristics of these severe weather systems. The favorable conditions for these severe storms to develop have been the major scientific interests of the scientists from China, Japan, and Korea (Chen and Chang 1980; Ninomiya 1984; Tao and Chen 1987; and Ninomiya and Murakami 1987; Ding 1992). However, the focus of these studies were mostly the large-scale environment and seldom touched the mesoscale features associated with the storms or the kinematic and thermodynamic structures of these systems. In this article, we intend to overview the efforts that have been taken in order to have a better understanding of the Meiyu-Baiu system with emphasis on the kinematic and thermodynamic structures of MCSs and its nearby mesoscale environments and associated mesoscale features.

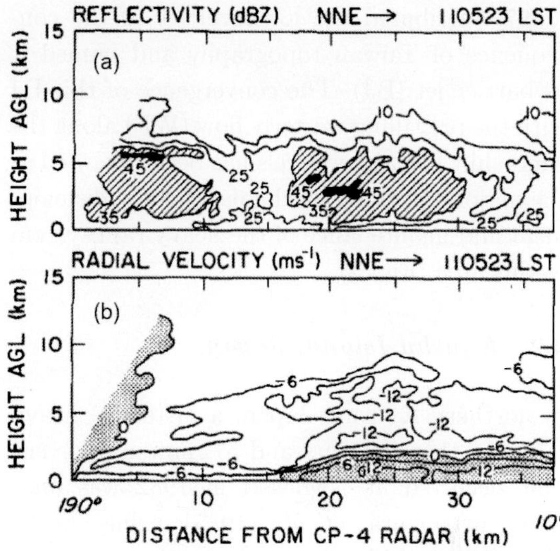

Fig. 2. Range-height indicator scans of (a) reflectivity and (b) radial velocity at 1108 LST (UTC + 8 hrs) 8 June 1987 at the NCAR CP-4 radar site (about 68 km southwest of Taoyuan International Airport: on the west coast of Taiwan). The azimuth angle is 10°. The hatched region in (a) indicates reflectivity greater than 35 dBZ, and the darkened regions are greater than 45 dBZ. The shaded regions in (b) are for radial velocity (northeasterly flow) moving toward the radar.

2.1. *TAMEX initiation*

The first analysis on rainfall systems with Doppler radar and aircraft data in this area was carried out in Taiwan Area Mesoscale Experiment (TAMEX; Kuo and Chen (1990)). Trier *et al.* (1989) have shown the not-too-tall convection with large reflectivity over ocean around Taiwan (Fig. 2). Ray *et al.* (1991) described the Meiyu frontal systems observed by Doppler radar over northwest of Taiwan. The front was shallow and the precipitation was widespread both ahead of and behind the front. The front was only 1.6-km deep over a distance of 100 km. Using velocity-azimuth display (VAD) data, a portion of the frontogenetic function was computed during the times the front was in the vicinity of the radar. The increase in both convergence and deformation contributed to large values of the frontogenetic function. Wang *et al.* (1990) and Lin *et al.* (1990) studied the fast propagating subtropical squall line (IOP 2 of TAMEX) and found the similarity with the tropical squall lines.

Jorgensen *et al.* (1991), Jou and Yu (1993), and Yu *et al.* (1999) applied NOAA P3 aircraft data to study the MCS over the SE ocean of Taiwan. The MCS was associated with an oceanic MCV with a diameter of about 70 km and the development of vorticity was related to stretching mechanism. Jou and Deng (1992 and 1998) used dual-Doppler wind synthesized technique to study the prevailing low-level jet over northwest of Taiwan and found the jet was lifted from 925 hPa to 850 hPa or higher when encountered the deep convective precipitation systems along the Meiyu front. Shallow frontal rainband was triggered by the speed convergence of the low-level jet and the deep convective systems were related to the pre-frontal convergence line formed by the LLJ and westerly flow. Li *et al.*

(1997) attributed the low-level jet as the consequence of Taiwan topography and named it as barrier jet (BJ). The convergence of this BJ with the prevailing westerly flow (WF) along the wind-shift line at low levels has been regarded as the major triggering mechanism for the development and maintenance of the heavy rain system in this area (Fig. 3).

2.2. *Kyushu Island, Japan*

In northern Kyushu, Japan, a historical heavy rainfall (313 mm/3 hrs and 572 mm/day) event that occurred at Nagasaki in 1982 was analyzed by Ogura *et al.* (1985) using Japan Meteorological Agency radar data to reconstruct the detailed mechanism. They discovered that convective echoes remained stationary over Nagasaki as a mesoscale cloud cluster during the heavy rainfall event. In addition, a new echo cell which formed to the west of Nagasaki moved into the cloud cluster. Formation mechanism of the linear rain band was explained by Seko (2001) through introducing a back and side building type. Along the west coast of Kyushu, orographic heavy rainfalls are associated with continuous rainfalls in the downstreams of linear peninsula or islands parallel to the moist southwesterly. For this area, Kanada *et al.* (2000) analyzed a rainfall enhancement of band-shaped convective cloud systems in the downwind side of an isolated island. Yoshizaki *et al.* (2000) revealed a feature of orographic rainband observed in western Kyushu by analytical and numerical studies.

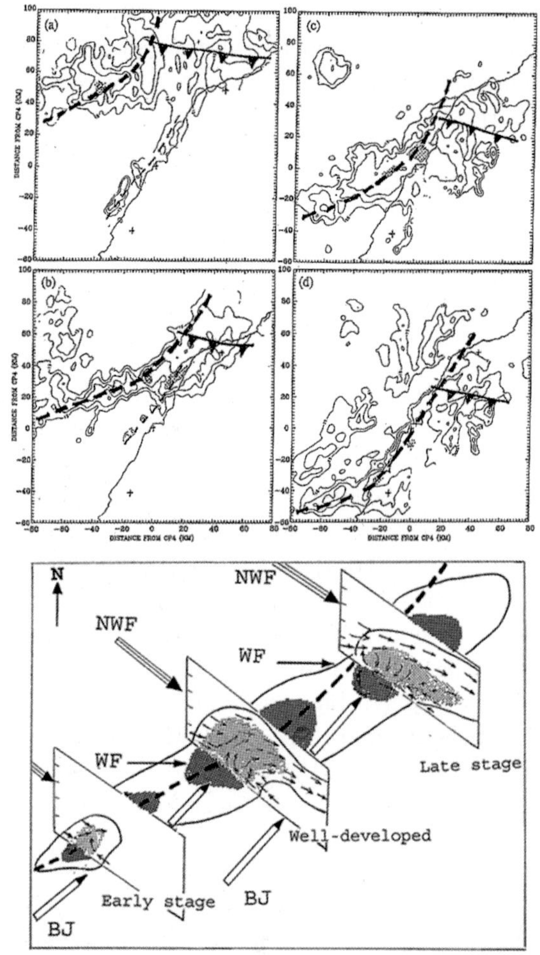

Fig. 3. Horizontal distributions of reflectivity (dBZ) obtained from CP4 radar for (a) 0120 LST (UTC+8 hrs), (b) 0320 LST, (c) 0520 LST, and (d) 0730 LST 25 June at the 2.5 km level (upper panel), and Schematic diagram for the structure of the rainband (bottom panel). The barrier jet (BJ) converges with the westerly flow (WF) along the wind-shift line (heavy, dashed line) in low levels. The northwesterly flow (NWF) dominates in the upper troposphere. The vertical cross sections show the long-lived reflectivity maxima embedded in the rainband at the early, well developed, and late stage, respectively. The solid lines represent radar echo boundary with reflectivity core areas shaded. The thin arrows on the cross sections indicate the relative airflow to the rainband system (from Li *et al.* (1997)).

Based on fundamental and local studies, the first Doppler radar observation in northwest Kyushu was performed in 1988. In this experiment, mesoscale precipitation systems with heavy rainfall have demonstrated the important role of dry air intrusion from the north in enhancing convergence along the Baiu front and convective systems (Ishihara *et al.* 1995; Takahashi *et al.* 1996). After this experiment, Doppler radar observations on the Baiu frontal heavy rainfalls were made in several places in Kyushu.

Intensive field experiments in the eastern part of East China Sea and western Kyushu

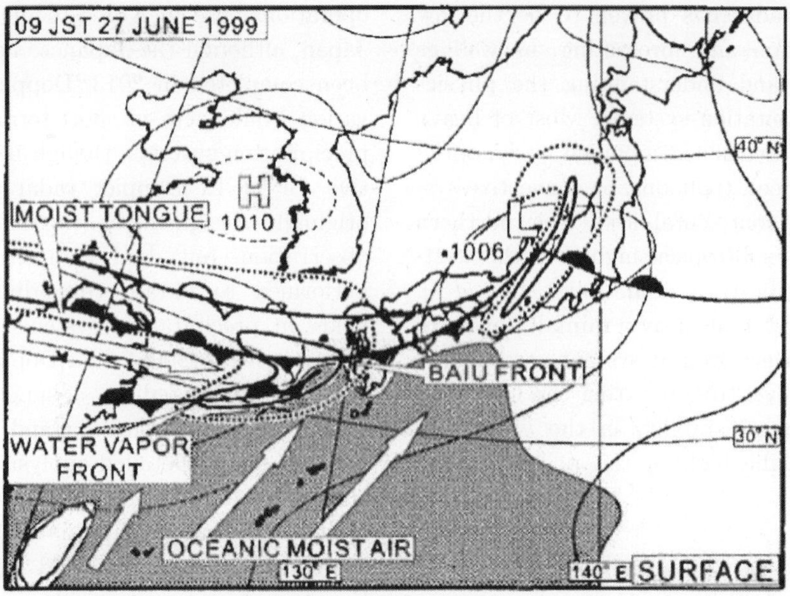

Fig. 4. Schematic diagram of synoptic-scale structure of the Baiu frontal region based on the surface weather map at 09 JST (UTC + 9 hrs) 27 June 1999. The dark and light shaded areas indicate distribution of the oceanic moist air at the surface and the moist tongue at 700 hPa, respectively. Areas of weak and convective rainfall are surrounded by dotted and solid lines, respectively. The symbol of a stationary front in the south of the Baiu front over the East China Sea denotes the "water vapor front", a convergence line with a large water vapor gradient distinct from the Baiu front about 100 km south of the Baiu front. The open symbol of a stationary front to the north of the Baiu front indicates the cyclonic wind-shear line at 700 hPa (from Moteki *et al.* (2004b)).

were conducted from 1998 to 2002 (Yoshizaki *et al.* 2002) with Doppler radars, wind profilers, an airplane, and a research vessel with on-board radar. One interesting result was the finding of a water vapor front (Moteki *et al.* 2004a, b), i.e. a convergence line with a large water vapor gradient distinct from the Baiu front about 100 km south of the Baiu front as shown in Fig. 4.

2.3. *Yangtze River Basin*

The first intensive field experiment on Meiyu frontal precipitation systems with Doppler radar in the upper air sounding network was conducted in 1998 and 1999 during Global Energy and Water Cycle Experiment (GEWEX) Asian Monsoon Experiment/Huaihe River Basin Experiment (GAME/HUBEX) (Fujiyoshi and Ding 2006). Maesaka (2003) showed two types of precipitation systems identified with IOP radar data in 1998: one without a temperature gradient in the subtropical air mass and the other with a temperature gradient resulting from merging with a cold front. In cases without strong temperature gradients on 29 June 1998, two dominant precipitation systems were recognized. One was a pair consisting of a convective precipitation system on the convergence line of the Meiyu front near the ground and a stratiform precipitation system. The other was a linear convective precipitation system south of the convergence line near the ground.

2.4. *Korean Peninsula*

In the Korean Peninsula, the heavy rainfall of the summer monsoon brings sudden and excessive amounts of precipitation locally causing losses of lives and extensive property damages.

The weather radar has proven to be the key tool in detection and forecasting, as well as in identifying and understanding the physics of heavy precipitation systems. Most of heavy rainfall over Korea is accompanied by synoptic-scale disturbances, typhoons, or convective systems between continental lows over northern China and the subtropical high over the western North Pacific (Lee *et al.* 2008). Lee *et al.* (1998) reported that heavy rainfall tends to occur in a surface frontal system accompanying an upper-level trough during the Changma period or strong instability in the vicinity of the North Pacific high in the post-Changma period.

3. Development of Radar in East Asia

The initiation of Doppler radar observation in TAMEX in 1987 propelled the observation of heavy rainfalls with Doppler radar in East Asia. Field experiments on heavy rainfalls with Doppler radar have been carried out in 1990's and 2000's over Japan, China and Korea. Doppler radar network completed in China, Taiwan, Korea and Japan recently are used for understanding heavy rainfall systems even in the area without research radar. Field experiments with polarimetric radar as were done in SoWMEX/TIMREX are proving the benefit of the equipment for quantitative precipitation estimation (QPE).

In the last three decades, weather radar has been developed rapidly, from conventional weather and Doppler radar to polarimetric (multi-parameter) radar. Conventional weather radar detects precipitation and estimates precipitation intensity by measuring radar reflectivity. Doppler weather radar has been widely used in research worldwide since the 1980s, and the Next-Generation Weather Radar (NEXRAD) has had full-coverage Doppler radar networks over the United States since the early 1990s. Doppler radar sets have been installed for operational use in Korea, China, Taiwan, and Japan, although the Japanese network has just been completed in 2013. Doppler radar observation would help in short-term forecasting of precipitation systems, though it is not so effective for QPE. Doppler radar networks were originally designed for single Doppler radar observation, but dual Doppler radar can be performed to obtain three-dimensional wind fields in precipitation systems. As shown by Shimizu *et al.* (2008) dual Doppler radar analysis was conducted on a supercell-like storm near Tokyo using Haneda and Narita airport Doppler radar. Detailed analysis and numerical simulation showed that the supercell-like storm might form based on the balance between a weak downdraft in the storm and weak inflow into the storm in a moist environment. This study emphasized the importance of dual Doppler radar observation of precipitation systems in research and operational use.

So far polarimetric radar has been used only for research purposes in East Asia. Polarimetric radar would be very useful in QPE using parameters of specific differential phase (KDP), correlation coefficient $\rho_{hv}(0)$ and differential reflectivity factor (ZDR). Recent textbooks on radar meteorology explain the merits of polarimetric radar for the measurement of rainfall intensity by using polarimetric factors (Doviak and Zrnic, 1993; Fukao and Hamazu, 2013). Ministry of Land, Infrastructure, Transport and Tourism (MLIT), Japan, has been installing X-band polarimetric radar (multi-parameter radar) network for large cities since July 2010. Ministry of Land, Infrastructure and Transport (MoLIT), Korea, has been installing S-band polarimetric radar network for forecasting flood since 2009. Korea Meteorological Administration (KMA) has started to replace Doppler radar to S-band polarimetric radar for improving weather forecast since 2014. The extension of polarimetric radar network in East Asia would make an effective quantitative precipitation forecasting (QPF) in near future.

4. Recent Studies on Heavy Rainfalls in East Asia

Recent field experiments with new radar, detailed analyses with operational radar data and numerical simulations with cloud resolving model are revealing characteristics of heavy rainfalls in each areas of East Asia.

4.1. *Korea*

Prominent researches on heavy rainfalls have been made recently in Korea.

4.1.1. *Heavy rainfall in central Korean Peninsula*

Understanding on heavy rainfall of the summer monsoon in Korean Peninsula and surrounding area has been improved by analyzing radar observation data. Jeong *et al.* (2012) pointed out that heavy rainfall is related to a low-level jet (LLJ) transporting warm and moist air from southern China. Sun and Lee (2002) suggested that MCSs tended to form between the upper-level jet (ULJ) to the north and LLJ to the south. In addition, the northeastward moving typhoon enhances moisture advection toward the central Korean Peninsula along the LLJ (Choi *et al.* 2011). Hence, heavy rainfall of summer monsoon is characterized by strong baroclinicity even though the atmosphere over Korea is generally thermodynamically neutral. This contrasts with the large convective available potential energy in the central U.S. (Hong 2004).

In order to elucidate the characteristics of the inner structure and flow in convective systems, high-resolution observational dataset are required. For instance, Kim and Lee (2006) investigated the characteristics of the convective systems occurring over the central Korean Peninsula using single Doppler radar data and showed the evolution process of the precipitation systems.

4.1.2. *Heavy rainfall in southern Korea*

Observational research of the Changma frontal region has been actively conducted at the intensive observation period (IOP) of the Global Research Laboratory Pukyong-HyARC Nagoya University Observation Network in the East China Sea (GRL PHONE) which is designed to examine multi-scale aspects of frontal precipitation system near the East China Sea (Lee *et al.* 2011). During the GRL PHONE period, Jeong *et al.* (2012 and 2013) revealed that a strong equivalent potential temperature associated with warm and moist air producing convectively unstable atmosphere was detected over the core of the LLJ. You *et al.* (2010) also proved that deep warm-air advection (WAA) supports the maintenance of a convective system for a longer time and results in greater rain intensity, producing drops of larger size (Fig. 5). Based on dual-Doppler radar analysis, a rear-to-front jet corresponding to a low-level jet was concurrent with a high equivalent potential temperature environment inflow over the warm front, which suggests that the rear-to-front jet with moist air induced destabilization.

4.1.3. *Orographic effect of rainfall enhancement*

Jeju Island, the southern part of Korea has frequently suffered from flooding and landslides due to orographically-intensified rainfall systems during the rainy season (June to mid-July). The island is an isolated terrain with an elliptical-shaped mountain extending east-to-west (width 35 km, length 78 km, height 1.95 km; Fig. 6a). During the rainy season when a stationary front is located off the northern shore of the island, a moist environment and ambient southwesterly winds are well predominant around the island. Lee *et al.* (2010, 2012 and 2014) revealed the effects of the isolated elliptic terrain on the flow modification and the related rainfall enhancement in a moist environment by dual Doppler

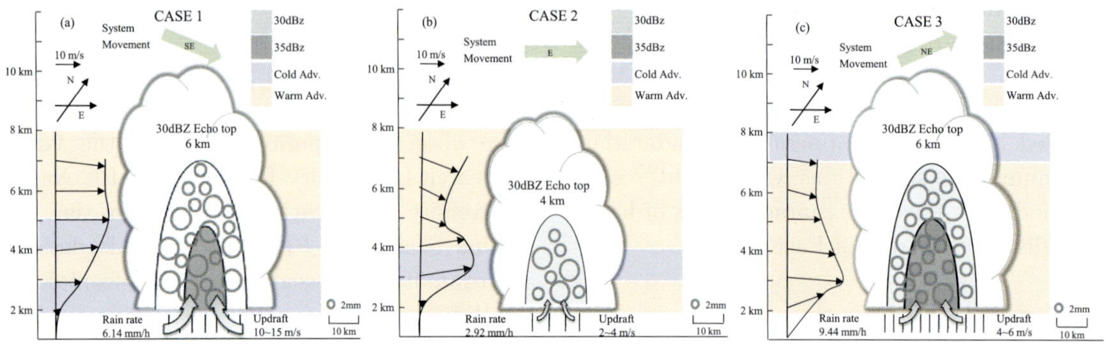

Fig. 5. Schematic diagram of three identified rainfall systems of the observed Changma front, (a) case 1, (b) case 2, and (c) case 3, respectively (from You *et al.* (2010)).

radar analyses and a modeling study. According to their result, when an eastward pre-existing rainfall system travels off northwestern shore, the system develops by accelerated southwesterly wind blowing parallel to the coastline, in the narrowed space between the system and terrain (Fig. 6b), as similar result showed in the northwestern offshore of Taiwan (Wang *et al.* 2005). On the terrain, the moist southwesterly wind, which is orographically-redirected to blow surrounding the terrain, generates an arc-shaped localized moist updraft zone (grey hatched area, Fig. 6a) on the northern side of the terrain; thus lateral side of the island becomes favorable for rainfall enhancement of pre-existing rainfall system (Fig. 6c). These results suggest that the low-level southwesterly wind in a moist environment could be a useful meteorological parameter to improve rainfall forecast over Jeju Island.

4.2. *Japan*

Recent studies are confirming the structure of linear rainband explained by Seko (2001). Band-shaped convective cloud systems in the moist southwesterly to the south of the Baiu front are not quite deep. Their characteristics were shown by Doppler radar observation (Shinoda *et al.* 2009) and polarimetric radar observation (Shusse *et al.* 2009, Oue *et al.* 2010) in Okinawa region.

Kato and Goda (2001) analyzed the formation factors of heavy rainfalls in Niigata-Fukushima, Japan. Kato (2005) discussed the problems in prediction using a cloud-resolving model. A heavy rainfall from a band-shaped precipitation system, associated with a cold front over northern Kyushu, Japan, has been analyzed by Kato (2006). Low-level humid air from the southwest and mid-level dry air from the west continuously moved into the precipitation system. Low-level humid air initiated the MCSs and the middle-level dry air enhanced the convective intensity (Fig. 7).

Morotomi *et al.* (2012) revealed the enhancement of rainfall by orographic effect of valley facing to the southerly moist air flow. A valley facing to the south played a role in enhancing convections by making large convergence of moist airflow in low altitude from the south.

4.3. *China*

4.3.1. *Around the Meiyu front*

To study precipitation in the main Meiyu regions, intensive field experiments in the downstream of the Yangtze River were conducted with the Doppler radar and wind profiler in 2001 and 2002 within the upper-air sounding network (Yamada *et al.* 2003; Geng *et al.* 2004). This experiment was conducted by the Frontier

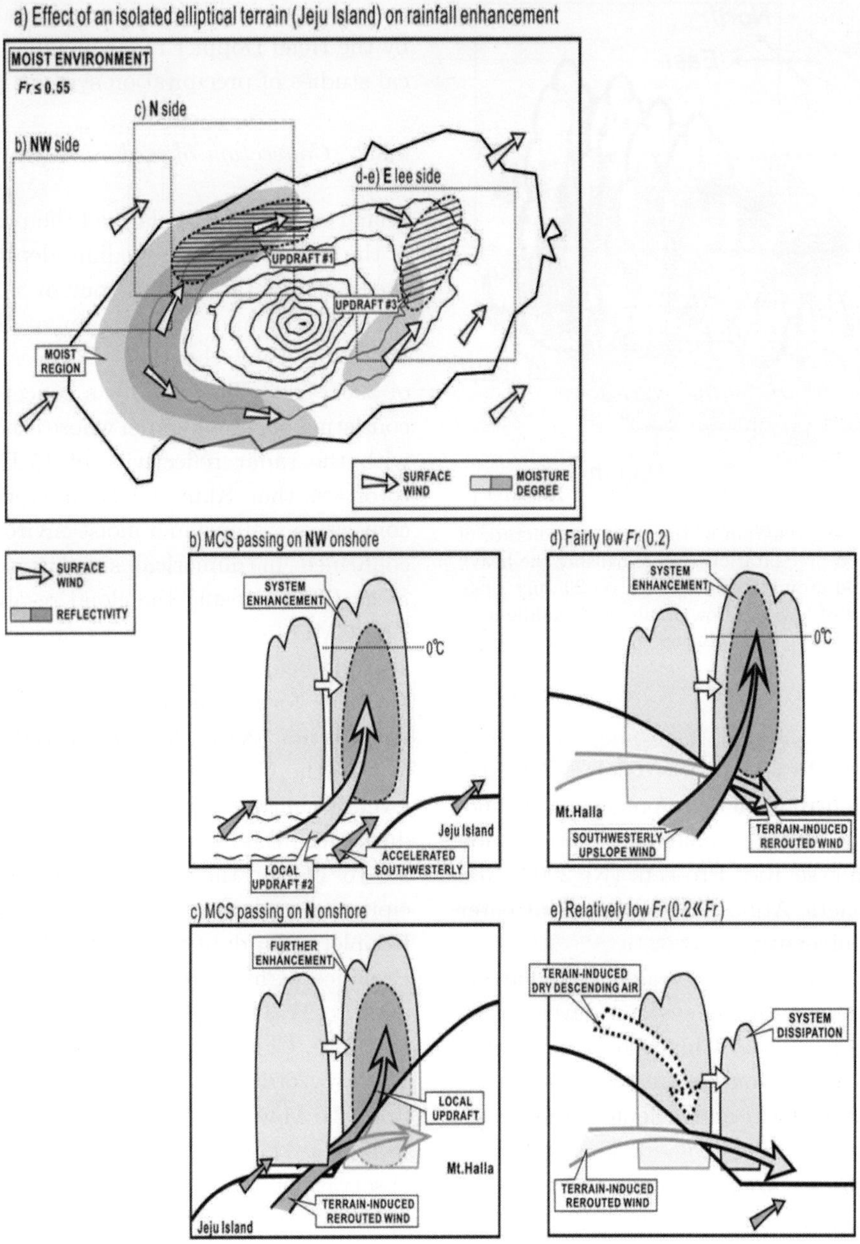

Fig. 6. Schematic diagram of effect of an isolated elliptical shaped terrain on the distribution of surface moisture contents and evolution of rainfall system in the vicinities of Jeju Island: (a) surface wind (white arrows), moisture contents (shading), and updraft (hatched); and terrain–induced (b) rainfall enhancement off the northwestern shore, (c) rainfall enhancement on the northern lateral side, (d) rainfall enhancement on the lee side, and (e) rainfall dissipation on the lee side of the terrain. Thick solid line in each panel indicates topography of Jeju Island. The grey arrows in (b)–(e) indicate the low–level wind (from Lee *et al.* (2014)).

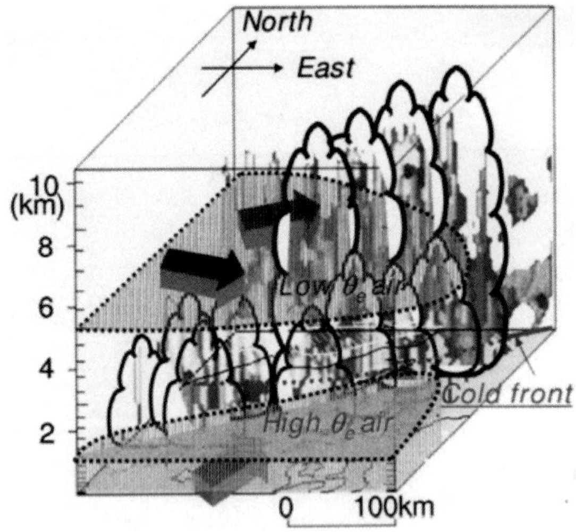

Fig. 7. Schematic diagram of the vertical structure of the band-shaped precipitation system causing the heavy rainfall observed around Fukuoka area on 29 July 1999, and the inflows of low-level low θ_e air, and middle-level high θ_e air into the precipitation system (from Kato (2006))

Observational Research for Global Change, JAMSTEC (currently Japan Agency for Marine-Earth Science and Technology) in cooperation with the Chinese Key Projects (Ni 2002) and the Hydrospheric Atmospheric Research Center of Nagoya University.

Yamada *et al.* (2003) revealed a mechanism of mesoscale convective system development along the east coast of China, i.e. convergence of a moist, strong, southwesterly inflow to the Meiyu frontal system and a low-altitude easterly flow from the ocean (Fig. 8). The convective systems west of the coast had precipitation cores at around 2 km in height. The mesoscale convective system developed higher near the coast and moved to the Japan Sea the next day, causing heavy rainfall in southern Kyushu and Shikoku, Japan.

Geng *et al.* (2004), Zhang *et al.* (2006a) and Yamada *et al.* (2007) others are clarified a variety of precipitation systems around the Meiyu front using the observational data from 2001 to 2003 in the downstream of Yangtze River. This experiment yielded continuous data acquisition by the Hefei Doppler radar, making the statistical studies of precipitation systems possible.

4.3.2. *Convection of medium depth*

Zhang *et al.* (2006a) clarified the predominance of the convection with medium-depth (CMD) in the south of the Meiyu front over the Huaihe River Basin and downstream of the Yangtze River by using the Hefei Doppler radar data of 2001–2003. The CMD is a concept for MCSs consisting of cumulonimbi whose echo top height with the radar reflectivity of 15 dBZ is equal to or less than 8 km. The characteristics of the convective systems in a moist environment were confirmed in numerical simulation by Zhang *et al.* (2006b) using the cloud resolving model. The CMD formed raindrops rapidly in the low altitude below 3 km in height and precipitation particles formed mostly below the 0°C level. Kato *et al.* (2007) gave a theoretical explanation to the CMD by statistical analyses on the level of neutral buoyancy (LNB) around Japan during the Baiu season.

To clarify the statistical features of precipitation cells during the Meiyu/Baiu period, Doppler radar data obtained at Shouxian, Anhui Province, China (continental area) from 17 June to 17 July 1998, at Zhouzhuang, Jiangsu Province, China (coastal area) from 10 June to 13 July 2001, and at Okinawa (oceanic area) from 27 May to 11 June 2004 were analyzed (Shinoda *et al.* 2007). From the radiosonde observations, moist layers were found to exist in the lower (less than 2 km) and middle (from 2 to 5 km) troposphere.

4.3.3. *Statistical study of squall line system*

After the completion of Doppler radar network over China, statistical analyses on squall lines based on mosaics of radar reflectivity were made by Meng *et al.* (2013). The study revealed the

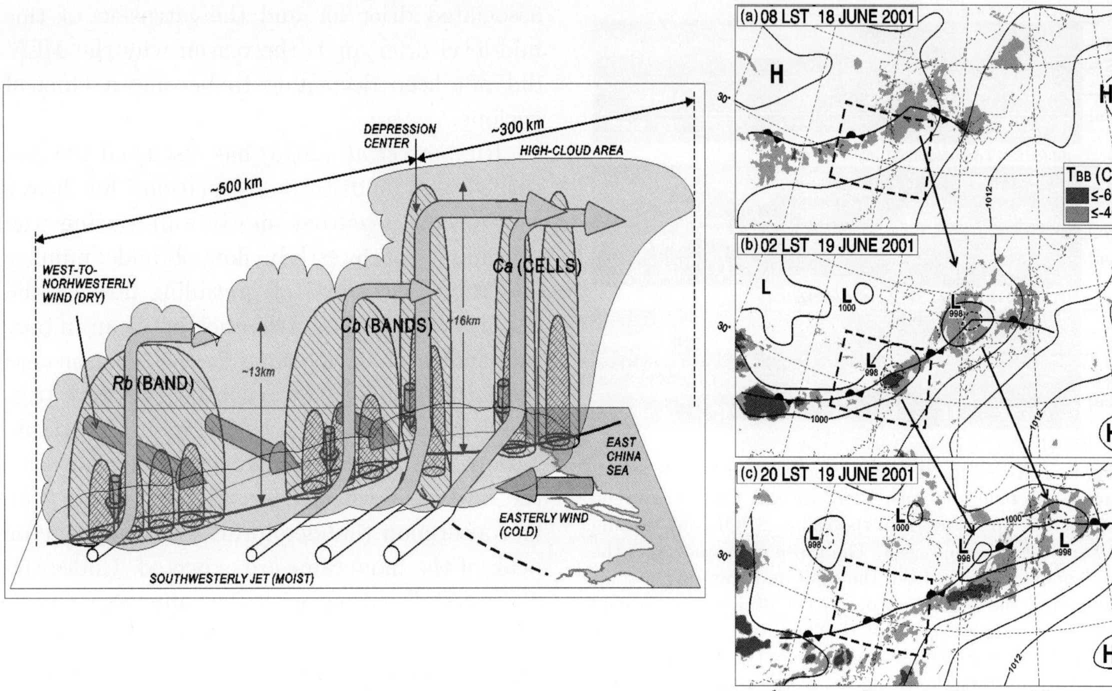

Fig. 8. Schematic illustration of the three-dimensional structure of the mesoscale convective systems developed in the downstream of the Yangtze River near the coast (left panel). Series of surface weather map at 18 hour intervals from 08 LST 18 June to 20 LST 19 June 2001 (right panel). The distribution of cloud top black-body temperature (*TBB*) is superimposed on each panel. (Figure taken from Yamada *et al.* (2003).)

squall lines in midlatitude east China tend to form in a moister environment with comparable background instability, and weaker vertical shear relative to their U.S. counterparts. Meng *et al.* (2012) simulated bowing structure of a case associated with a squall line in April. Zheng *et al.* (2013) classified mesoscale convective systems (MCs) over central China into seven morphologies using composite reflectivity Doppler radar data from June to September of 2007–2010. These studies are suggesting a new direction of the field experiment in the southeast coast of China.

4.4. *Taiwan*

The most recent and systematic field campaign has been performed in southwest Taiwan: the Southwest Monsoon Experiment (SoWMEX)/Terrain-influenced Monsoon Rainfall Experiment (TiMREX). The experiment SoWMEX/TiMREX is a cooperative field observational program conducted jointly by the scientists of the U.S. and Taiwan to study the mesoscale environment and the microphysical characteristics of the heavy rain weather systems (Jou *et al.* 2011). During SoWMEX/TiMREX (May and June 2008), there were two major heavy rain events observed by the intensive field campaign. One is in late May and early June and was studied by Davis and Lee (2012), Lai *et al.* (2011), and Ruppert *et al.* (2013), and the other case is in middle of June and was studied extensively by Xu *et al.* (2012) and Jung *et al.* (2012), respectively (Fig. 9).

Davis and Lee (2012) showed that when the Meiyu front approached to the island, shallow secondary fronts formed along the coast

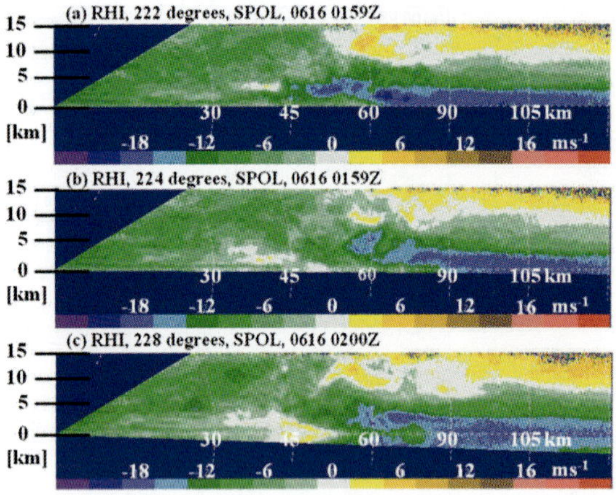

Fig. 9. RHI scans of radial velocity from NCAR SPOL radar, the blue and purple are toward radar the green and yellow are away from the radar. SPOL was looking toward west to the ocean. The green color indicates the cold pool signature from the land and the purple color indicating the flow from ocean colliding with the cold pool flow and getting uplifted. The lifting of the oceanic warm moist unstable air was far from the coastal line. It seems to indicate the extension of the island into ocean as suggested by Xu *et al.* (2012). (Figure taken from Xu *et al.* (2012).)

area and these shallow coastal fronts formed closely associated with the earlier precipitation. Using enhanced dropsonde data over the Taiwan Strait, the shallow frontogenetic processes can be delineated well. These shallow coastal fronts played a significant role in triggering and enhancing heavy rainfall. Lai *et al.* (2012) studied the mesoscale convective vortex embedded within the Meiyu frontal system using dropsonde and Doppler radar.

Their results indicate that the MCV was associated with strong equivalent potential temperature gradient (majorly the moisture gradient) and the heavy rain was associated with the strong southwesterly flows over the south and the east quadrants. The enhanced low level convergence associated with the low-level jet at the east and southeast of the system was the mechanism to maintain the convection and concentrated the vorticity associated with the MCV. It was noted that the northwest quadrant was associated drier air and the intrusion of this mid-level drier air is the reason why the MCV did not keep deepening to become a tropical cyclone.

Ruppert *et al.* (2013) has discussed the disturbed and undisturbed situations for heavy rain events occurred in Taiwan during the prevailing southwesterly flow. Froude number (U/Nh, U the speed of prevailing flow, N the stability, and h the terrain height) is an important indicator in determining the location and the mechanisms how the convections were triggered and maintained. Under undisturbed situation, the Froude number is usually low and the flows are deflected around the island. The late afternoon thunderstorms over the slope or peak of the mountains are expected. Under the disturbed situation, with the influence of mid-latitude frontal system, the Froude number is usually larger and there is high possibility the flow can be lifted to a higher altitude. During the disturbed period, the precipitation is more widespread, upslope, intense, and can be elevated to a higher altitudes.

Xu *et al.* (2012) studied the case of June 14–16, 2008 heavy rain case over Kaohsiung during SoWMEX/TiMREX. Their results suggested that the orography of Taiwan was extended to the upstream ocean and had an indirect effects on the long-duration of the heavy rain event in SE of Taiwan. It was shown that the warm, moist, unstable low level jet (LLJ) was only found over the upstream ocean while the island of Taiwan was under the control of a weak cold pool. The LLJ was lifted upward at the boundary between the cold pool and the LLJ. Most convective clusters supporting the long-lived rainy mesoscale system were initiated and developed along the boundary. It is suggested by Xu *et al.* (2012) that the initiation and maintenance of the system is a back-building quasi-stationary process.

A better understanding on the microphysics of heavy rainfall systems during monsoon period is one of the major scientific objectives of

SoWMEX/TiMREX. Jung *et al.* (2012) studied the precipitation characteristics of the MCS on June 2, 2008. S-Pol, Verti-X, and disdrometer data were used to identify the microphysical features. The MCS was a leading convection and trailing stratiform type squall line system. For convenience, Jung *et al.* has separated the system into several different sectors, convective line with center, leading edge, trailing edge, stratiform, and reflectivity trough, respectively. It was shown that the convective center has smaller number concentration and mass-weighted mean drop diameter than observations from other maritime storms, for example, Darwin Australia and TOGA-COARE. The drop size distribution also showed the Taiwan squall line has a smaller value and narrower range of shape, slope, and intercept compared with the maritime storms, but with a larger value and broader spread of median volume drop diameter. Lightning data is also important to heavy rain study. Tong (2009) studied the environmental conditions and the precipitation characteristics of 40 significant continuous rain periods (SCRPs) identified by rain gauge and radar data during SoWMEX/TIMREX. The 40 SCRPs were divided into three different categories according to their origin location and further development, i.e. the land-type SCRP, the ocean-type SCRP, and the mixed type SCRP, respectively. It is shown that these SCRPs have strong diurnal variations and are associated with lightning.

Around Taiwan, the heavy rainfall area and situation shown in Fig. 1 are not quite clear, though heavy rainfall in Typhoon Megi (2010) and Typhoon Morakot (August 2009) where a valley in the land and orography play important roles. In the very moist environment under less baloclinic situation, microphysical studies on formation of precipitation systems will be essential. Observation of raindrop size distribution and identification of solid precipitation particles in precipitating cloud systems should be further investigated.

5. Future Perspective

Structure and characteristics of heavy rainfalls associated with the Meiyu/Baiu front have been revealed by field experiments using Doppler radars. In recent 10 years, the development processes of heavy rainfalls occurring in East Asia, in a moist environment at the western edge of the subtropical high (Pacific high) have been clarified for the island and peninsula surrounded by warm ocean. These research developments are based on the completion of operational Doppler radar network in each area and investigation using new research radar as polarimetric radar.

Collaborative field experiments and studies broadened the perspective on precipitation systems around the Meiyu, Baiu and Changma front. Further collaboration and cooperation among researchers on heavy rainfalls associated with summer monsoon in East Asia is expected. Further advancement of operational weather radar network and utilization of high resolution numerical model for assimilation would assure a better understanding of the precipitation systems and the forecasting of heavy rainfalls.

Acknowledgments

The authors express their thanks to Dr. K.-O. Lee, Nagoya University, and Dr. Jeong, Pukyong National University, for their provision of figures and materials for this review. The authors express their thanks to Professors H.-C. Kuo, National Taiwan University, and C.-C. Wang, National Taiwan Normal University, for their suggestions on heavy rainfalls in East Asia. This work is supported by the Korea Meteorological Industry Promotion Agency under Grant KMIPA 2015-5060.

References

Chen, G. T. J., 2004: Research on the phenomena of Meiyu during the past quarter century: An overview. *East Asian Monsoon*, Chap. 10, World Scientific Series on Meteorology of East Asia, Vol. 2, Eds. C. P. Chang *et al.*, World Scientific, pp. 357–403.

Chen, G. T. J. and C. P. Chang, 1980: The structure and vorticity budget of an early summer monsoon trough (Mei-Yu) over southeastern China and Japan. *Mon. Wea. Rev.*, **108**, 942–953.

Chen, J. and S. Bordoni, 2014: Orographic effects of the Tibetan Plateau on the East Asian summer monsoon: An energetic perspective. *J. Climate*, **27**, 3052–3072.

Choi, H. Y., J. H. Ha, D. K. Lee, and Y. H. Kuo, 2011: Analysis and simulation of mesoscale convective system accompanying heavy rainfall: The Goyang case. *Asia-Pac. J. Atmos. Sci.*, **47**, 265–279.

Davis, C. A. and W. C. Lee, 2012: Mesoscale analysis of heavy rainfall episodes from SoWMEX/TiMREX. *J. Atmos. Sci.*, **69**, 521–537.

Ding, Y.-H., 1992: Summer monsoon rainfalls in China. *J. Meteor. Soc. Japan*, **70**, 373–396.

Doviak, R. J. and D. S. Zrnic, 1993: Doppler radar and weather observation, *Academic Press*, 562 pp.

Fujiyoshi, Y. and Y. H. Ding, 2006: Final report of GAME/HUBEX: GEWEX Asian Monsoon Experiment/Huaihe river Basin Experiment. GAME Pub. 43, GAME/HUBEX Project Office, 610 pp.

Fukao, S. and K. Hamazu, 2013: Radar for meteorological and atmospheric observations, Springer, 537 pp.

Geng, B., H. Yamada, K. K. Reddy, H. Uyeda, and Y. Fujiyoshi, 2004: An observational study of the development of a rainband on a Meiyu front causing heavy rainfall in the downstream region of the Yangtze River. *J. Meteor. Soc. Japan*, **82**, 1095–1115.

Hong, S. Y., 2004: Comparison of heavy rainfall mechanisms in Korea and the central US. *J. Meteor. Soc. Japan*, **82**, 1469–1479.

Ishihara M., Y. Fujiyoshi, A. Tabata, H. Sakakibara, K. Akaeda, and H. Okumura, 1995: Dual Doppler radar analysis of an intense mesoscale rainband generated along the Baiu front in 1988: Its kinematical structure and maintenance process. *J. Meteor. Soc. Japan*, **73**: 139–163.

Jeong, J. H., D. I. Lee, C. C. Wang, S. M. Jang, C. H. You, and M. Jang, 2012: Environment and morphology of mesoscale convective systems associated with the Changma front during 9–10 July 2007. *Annales Geophysicae*, **30**, 1235–1248.

Jeong, J. H., D. I. Lee, C. C. Wang, S. M. Jang, S.-H. Park, and S. A. Jung, 2013: Structure and evolution of line-shaped convective systems associated with Changma front during GRL PHONE-09: 6 July 2009 case. *Meteorological Applications*, **21**, 786–794. (online version released).

Jorgensen, D. P., M. L. LeMone, and B. J. D. Jou, 1991: Precipitation and kinematic structure of an oceanic mesoscale convective system. Part I: Convective line structure. *Mon. Wea. Rev.*, **119**, 2608–2637.

Jou, B. J. D. and S. M. Deng, 1992: Structure of a low-level jet and its role in triggering and organizing the moist convection over Taiwan: A TAMEX case study. *TAO*, **3**, 39–58.

Jou, B. J. D. and S. M. Deng, 1998: The organization of convection in a Mei-Yu frontal rainband. *TAO*, **9**, 553–572.

Jou, B. J. D., W. C. Lee, and R. H. Johnson, 2011: An overview of SoWMEX/TiMREX operation. In *Global Monsoon System*, Eds. C. P. Chang *et al.*, World Scientific Publishing, 214–232.

Jou, B. J. D. and C. K. Yu, 1993: Airborne measurement of an oceanic mesoscale convective system during TAMEX. In *Proc. 2nd International Conference on East Asia and Western Pacific Meteorology and Climate.* World Scientific, pp. 453–462.

Jung, S. A., D. I. Lee, B. J. D. Jou, and H. Uyeda, 2012: Microphysical properties of maritime squall line observed on 2 June 2008 in Taiwan. *J. Meteor. Soc. Japan*, **90**(5), 833–850.

Kanada, S., H. Minda, B. Geng, and T. Takeda, 2000: Rainfall enhancement of band-shaped convective cloud system in the downwind side of an isolated island. *J. Meteor. Soc. Japan*, **78**, 47–67.

Kato, T., 2005: Formation factors of 2004 Niigata-Fukushima and Fukui heavy rainfalls and problems in predictions using a cloud-resolving model. *SOLA*, **1**, 1–4.

Kato, T., 2006: Structure of the band-shaped precipitation systems inducing the heavy rainfall observed over Northern Kyushu, Japan on 29 June 1999. *J. Meteor. Soc. Japan*, **84**, 129–153.

Kato, T. and H. Goda, 2001: Formation and maintenance processes of a stationary band-shaped heavy rainfall observed in Niigata on 4 August 1998. *J. Meteor. Soc. Japan*, **79**, 899–924.

Kato, T., S. Hayashi, and M. Yoshizaki, 2007: Statistical study on cloud top heights of cumulonimbi thermodynamically estimated from objective analysis data during the Baiu season. *J. Meteor. Soc. Japan*, **84**, 529–557.

Kim, H. W. and D. K. Lee, 2006: An observational study of mesoscale convective systems with heavy rainfall over the Korean Peninsula. *Wea. Forecast.*, **21**, 125–148.

Kuo, Y. H. and G. T. J. Chen, 1990: The Taiwan Area Mesoscale Experiment (TAMEX): An overview. *Bull. Amer. Meteor. Soc.* **71**, 488–503.

Lai, H. W., C. A. Davis, and B. J.-D. Jou, 2011: A subtropical oceanic mesoscale convective vortex observed during SoWMEX/TiMREX. *Mon. Wea. Rev.*, **139**, 2367–2385.

Lee, D. I., H. Uyeda, B. J. D. Jou, M. Maki, and Y. Iwasaka, 2011: Introduction of Korean GRL activities for international cooperation field experiments between Korea, Japan and Taiwan. Japan Geoscience Union Meeting 2011.

Lee, D. K., H. R. Kim, and S. Y. Hong, 1998: Heavy rainfall over Korea during 1980–1990. *Kor. J. Atmos. Soc.*, **1**, 32–50.

Lee, D. K., J. G. Park, and J. W. Kim, 2008: Heavy rainfall events lasting 18 days from July 31 to August 17, 1998, over Korea, *J. Meteor. Soc. Japan*, **86**, 313–333.

Lee, K. O., S. Shimizu, M. Maki, C. H. You, H. Uyeda, and D. I. Lee, 2010: Enhancement mechanism of the 30 June 2006 precipitation system observed over the northwestern slope of Mt. Halla, Jeju Island, Korea. *Atmos. Res.*, **97**, 343–358.

Lee, K. O., H. Uyeda, and D. I. Lee, 2014: Effect of an isolated elliptical terrain (Jeju Island) on rainfall enhancement in a moist environment. *Tellus A*, **66**, 20484, http://dx.doi.org/10.3402/tellusa.v66.20484.

Lee, K. O., H. Uyeda, S. Shimizu, and D. I. Lee, 2012: Dual-Doppler radar analysis of the enhancement of a precipitation system on the northern side of Mt. Halla, Jeju Island, Korea on 6 July 2007. *Atmos. Res.*, **118**, 133–152.

Li, J., Y. L. Chen, and W.-C., Lee, 1997: Analysis of a heavy rainfall event during TAMEX. *Mon. Wea. Rev.*, **125**, 1060–1082.

Lin, Y.-J., H. Shen, T.-C. C. Wang, Z.-S. Deng, and R. W. Pasken, 1990: Characteristics of a subtropical squall line determined from TAMEX dual-Doppler data. Part II: Dynamic and thermodynamic structures and momentum budgets. *J. Atmos. Sci.*, **47**, 2382–2399.

Maesaka, T., 2003: Study on the Formation Process of Precipitation Systems in the Meiyu Front on the China Continent. Doctoral dissertation, Hokkaido University.

Meng, Z., D. Yan, and Y. Zhang, 2013: General Features of Squall 1 lines in East China. *Mon. Wea. Rev.*, **141**, 1629–1647.

Meng, Z., F. Zhang, P. Markowski, D. Wu, and K. Zhao, 2012: A modeling study on the development of a bowing structure and associated rear inflow within a squall line over South China. *J. Atmos. Sci.*, **69**, 1182–1207.

Morotomi, K., T. Shinoda, Y. Shusse, T. Kouketsu, T. Ohigashi, K. Tsuboki, H. Uyeda, and I. Tamagawa, 2012: Maintenance mechanisms of a precipitation band formed along the Ibuki-Suzuka Mountains on September 2–3, 2008. *J. Meteor. Soc. Japan*, **90**, 737–753.

Moteki, Q., H. Uyeda, T. Maesaka, T. Shinoda, M. Yoshizaki, and T. Kato, 2004a: Structure and development of two merged rainbands observed over the East China Sea during X-BAIU-99 Part I: Meso-β-scale structure and development processes. *J. Meteor. Soc. Japan*, **82**, 19–44.

Moteki, Q., H. Uyeda, T. Maesaka, T. Shinoda, M. Yoshizaki, and T. Kato, 2004b: Structure and development of two merged rainbands observed over the East China Sea during X-BAIU-99 Part II: Meso-α-scale structure and build-up processes of convergence in the Baiu frontal region. *J. Meteor. Soc. Japan*, **82**, 45–65.

Ni, Y., 2002: Experiment on heavy rainfall in Meiyu front in the stream of the Yangtze River. Int'l Conf. on Mesoscale Convective Systems and Heavy Rainfall/Snowfall in East Asia. Tokyo, Japan, 86–88.

Ninomiya, K., 1984: Characteristics of Baiu front as a predominant subtropical front in the summer northern hemisphere. *J. Meteor. Soc. Japan*, **62**, 880–894.

Ninomiya, K. and H. Muraki, 1986: Large-scale circulations over East Asia during Baiu period of 1979. *J. Meteor. Soc. Japan*, **64**, 409–429.

Ogura, Y., T. Asai, and K. Dohi, 1985: A case study of a heavy precipitation event along the Baiu front in northern Kyushu, 23 July 1982: Nagasaki heavy rainfall. *J. Meteor. Soc. Japan*, **63**, 883–900.

Oue, M., H. Uyeda, and Y. Shusse, 2010: Two types of precipitation particle distribution in convective cells accompanying a Baiu frontal rainband around Okinawa Island, Japan. *J. Geophys. Res.*, **115**, D02201. doi:10.1029/2009JD011957, 2010.1.19.

Ray, P. S., A. Robinson, and Y. Lin, 1991: Radar analysis of a TAMEX frontal system. *Mon. Wea. Rev.*, **119**, 2519–2539.

Ruppert, Jr. J. H., R. D. Johnson, and A. K. Rowe, 2013: Diurnal circulations and rainfall in Taiwan during SoWMEX/TiMREX (2008). *Mon. Wea. Rev.*, **141**, 3851–3872.

Sampe, T. and S.-P. Xie, 2010: Large-scale dynamics of the Meiyu-Baiu rainband: Environmental forcing by the westerly jet. *J. Climate*, **23**, 113–134.

Seko, H., 2001: Study of the Shapes and Maintenance Mechanisms of Meso-scale Line-shaped Precipitation Systems in the Middle-Latitudes. Doctoral dissertation, University of Tokyo.

Shimizu, S., H. Uyeda, Q. Moteki, M. Yoshizaki, T. Maesaka, Y. Takaya, T. Kato, and K. Akaeda, 2008: Structure and formation mechanism on the 24 May 2000 supercell-like storm developing in a moist environment over the Kanto Plain, Japan. *Mon. Wea. Rev.*, **136**, 2389–2407.

Shinoda T., T. Amano, H. Uyeda, and K. Tsuboki, 2009: Structure of line-shaped convective systems obliquely training to the Baiu front observed around the southwest islands of Japan. *J. Meteor. Soc. Japan*, **87**(4), 739–745.

Shinoda, T., R. Maki, S. Shimizu, K. Tsuboki, and H. Uyeda, 2007: Statistical features of precipitation cells observed in East Asia during the Meiyu/Baiu period. In *Proc. of the ICMCS-VI.*

Shusse, Y., K. Nakagawa, N. Takahashi, S. Satoh, and T. Iguchi, 2009: Characteristics of polarimetric radar variables in three types of rainfalls in a Baiu front event over the East China Sea. *J. Meteor. Soc. Japan*, **87**, 865–875.

Sun, J. and T. Y. Lee, 2002: A numerical study of an intense quasistationary convection band over the Korean Peninsula. *J. Meteor. Soc. Japan*, **80**, 1221–1245.

Takahashi, N., H. Uyeda, K. Kikuchi, and K. Iwanami, 1996: Mesoscale and convective scale features of heavy rainfall events in late period of the Baiu season in July 1988, Nagasaki prefecture. *J. Meteor. Soc. Japan*, **74**, 539–561.

Tao, S. Y. and L. Chen, 1987: A review of recent research on the East Asian summer monsoon in China. In *Monsoon Meteorology*, Chap. 3. Eds. C.-P. Chang and T. N. Krishnamurti, Oxford Univ. Press, pp. 60–92.

Tong, C. C., 2009: Precipitation Characteristics and Associated Environmental Regimes during SoWMEX/TiMREX. Master's thesis, National Taiwan University, 90 pages.

Trier, S. B., D. B. Parsons, and T. J. Matejka, 1989: An observational and numerical study of a subtropical cold front during TAMEX. Preprints, 24th Conf. on Radar Meteorology. Tallahassee, FL. Amer. Meteor. Soc., 561–564.

Wang, C. C., G. T.-J. Chen, T. C. Chen, and K. Tsuboki, 2005: A numerical study on the effects of Taiwan topography on a convective line during the Mei-Yu season. *Mon. Wea. Rev.*, **133**, 3217–3242.

Wang, C. T. C., Y. J. Lin, H. Shen, and R. W. Pasken, 1990: Characteristics of a subtropical squall line determined from TAMEX dual-Doppler data. Part I: Kinematic structure. *J. Atmos. Sci.*, **20**, 2357–2381.

Xu, W., E. J. Zipser, Y. L. Chen, C. Liu, Y. C. Liou, W. C. Lee, and B. J. D. Jou, 2012: An orography-associated extreme rainfall event during TiMREX: Initiation, storm evolution, and maintenance. *Mon. Wea. Rev.*, **140**, 2555–2574.

Yamada, H., B. Geng, K. K. Reddy, H. Uyeda, and Y. Fujiyoshi, 2003: Three-dimensional structure of a mesoscale convective system in a Baiu-frontal depression generated in the downstream region of the Yangtze River. *J. Meteor. Soc. Japan*, **81**, 1243–1271.

Yamada, H., B. Geng, H. Uyeda, and K. Tsuboki, 2007: Thermodynamic impact of the heated landmass on the nocturnal evolution of a cloud cluster over a Meiyu-Baiu front. *J. Meteor. Soc. Japan*, **85**, 663–685.

Yoshizaki, M., T. Kato, C. Muroi, H. Eito, S. Hayashi, and CREST Observation Group, 2002: Recent activities of field observations on mesoscale convective systems (MCSs) over East China Sea and Kyushu in the Baiu season and over the Japan Sea in winter. Int'l Conf. on Mesoscale Convective Systems and Heavy Rainfall/Snowfall in East Asia, Tokyo, Japan, 80–85.

Yoshizaki, M., T. Kato, Y. Tanaka, H. Takayama, Y. Shoji, H. Seko, K. Arao, K. Manabe, and M. of X-BAIU-98 observation, 2000: Analytical and numerical study of the 26 June 1998 orographic rainband observed in Western Kyushu, Japan. *J. Meteor. Soc. Japan*, **78**, 835–856.

You, C. H., D. I. Lee, S.-M. Jang, M. Jang, H. Uyeda, and T. Shinoda, 2010: Characteristics of rainfall systems accompanied with Changma front at Chujado in Korea. *Asia-Pac. J. Atmos. Sci.*, **46**, 41–51.

Yu, C. K., B. J. D. Jou, and B. F. Smull, 1999: Formative stage of a long-lived mesoscale vortex observed by airborne Doppler radar. *Mon. Wea. Rev.*, **127**, 838–857.

Zhang, C. Z., H. Uyeda, H. Yamada, B. Geng, and Y. Ni, 2006a: Characteristics of mesoscale convective systems over the east part of continental China during the Meiyu from 2001 to 2003. *J. Meteor. Soc. Japan*, **84**, 763–782.

Zhang, C.-Z., H. Uyeda, H. Yamada, and B. Geng 2006b: Characteristics of convections of medium depth to south of the Meiyu front analyzed by using numerical simulation. *SOLA*, **2**, 160–163.

Zheng, L., J. Sun, X. Zhang, and C. Liu, 2013: Organizational modes of mesoscale convective systems over central east China. *Wea. Forecast.*, **28**, 1081–1098

Chapter 19

Heat Waves over East Asia: Model Simulation of Synoptic Characteristics and Projections for the 21st Century

Ngar-Cheung Lau

Institute of Environment, Energy and Sustainability and
Department of Geography and Resource Management,
The Chinese University of Hong Kong, Shatin, N.T., Hong Kong
gabriel.lau@cuhk.edu.hk
www.cuhk.edu.hk

The synoptic behavior of present-day heat waves (HW) over East Asia is studied using a global high-resolution atmospheric model (HiRAM) with 50-km grid-spacing. The simulated HW characteristics are compared with those derived from Climate Forecast System Reanalysis products. Additional runs of HiRAM are conducted for the "time slice" of 2086–2095 in the climate scenario corresponding to Representative Concentration Pathway (RCP) 4.5. By the end of the 21st century, the averaged duration and frequency of HW in selected East Asian sites are projected to increase by a factor of 1.4–2.1 and 2.0–2.3, respectively, from the present-day values. The output from a continuous integration of a coupled general circulation model with a lower resolution through the 1901–2100 period indicates a notable increase in severity, duration and HW days during the 21st century.

1. Introduction

Episodes of extremely high temperature lasting for more than several days occur in many parts of the world, including various monsoon regions in China (Ding *et al.* 2010), India (De *et al.* 2005), Indo-Pacific (Caesar *et al.* 2011) and South America (Marengo *et al.* 2010). These extreme weather events, hereafter referred as "heat waves" (HW), pose a severe threat to public health and economic activities, and are directly linked to substantial human casualties as well as financial losses (IPCC 2011). Improved understanding of the physical causes of the HW, as well as the impacts of climate change on the future behavior of this phenomenon, are of strong interest to the meteorological community.

Much effort has been devoted to the study of HW in midlatitudes using observational data and model output. Recently, Lau and Nath (2012; hereafter referred as LN1) examined the synoptic characteristics of HW simulated in various regions of North America by a global general circulation model (GCM) with a grid resolution of about 50 km. LN1 have also noted the projected increase in the severity, duration and frequency of North American HW from the 20th to the 21st century. Lau and Nath (2014, hereafter referred as LN2) used similar tools and methodology to investigate the behavior of HW occurring in the European sector.

In the present study, the diagnostic procedure employed in LN1 and LN2 is applied to the output from a suite of model runs contributing to the Fifth Coupled Model Intercomparison

The Global Monsoon System: Research and Forecast (3rd Edition)
Edited by C. P. Chang et al.

Project (CMIP5), with the focus being shifted to HW in East Asia.

2. Model Tools and Experimental Design

The primary model tools for this study are developed at the Geophysical Fluid Dynamics Laboratory (GFDL), and consist of the following GCMs:

- A fully coupled climate model version 3 (CM3). Details of the atmospheric component of this model (AM3) are described by Donner *et al.* (2011). The model atmosphere has a horizontal resolution of approximately 200 km. Vertical variations are represented in 48 layers. The oceanic component of CM3 is based on the Modular Ocean Model (MOM4p1) code of Griffies (2009). The horizontal resolution of the model ocean is 1° in latitude and longitude poleward of 30°. Vertical variations are represented in 50 layers.
- A high-resolution atmosphere-only model (HiRAM). The formulation of this model is described by Zhao *et al.* (2009). The horizontal resolution is approximately 50 km. There are 32 atmospheric layers in the vertical direction.

The primary data base for the present study is generated in various experiments with the above models. These model runs include:

- Integration of CM3 through the sequence of climate scenarios in the 1860–2100 period. In the historical era (1860–2005), the emission rates or concentrations of various trace atmospheric constituents are based on observational estimates. From 2006 onwards, the radiative forcings are assumed to follow the Representative Concentration Pathway 4.5 (RCP4.5) issued by the Intergovernmental Panel for Climate Change.
- "Time-slice" runs with the HiRAM model for simulating the current climate in the historical period of 1979–2008 (hereafter referred to as the H experiment), and for projecting the future climate scenario in the period of 2086–2095 (referred to as the S2 experiment). In the H experiment, the lower boundary conditions at oceanic grid points are prescribed using intermonthly variations of observed SST and sea ice. The model atmosphere is also subjected to temporally varying radiative forcings based on observational data. In the S2 experiment, radiative forcings corresponding to the RCP4.5 scenario are prescribed, and the SST condition and sea ice concentration are computed using the projection by the CM3 experiment.

Some of the model findings are checked against the corresponding observational results based on the new Climate Forecast System Reanalyses (CFSR) produced by the U.S. National Centers for Environmental Prediction (Saha *et al.* 2010). The horizontal resolution of this dataset is 38 km. The global reanalysis fields cover the 1979–2010 period.

3. Methodology for Identifying Heat Waves

Spatially coherent regions with high-amplitude fluctuations in the daily maximum temperature (T_{max}) field are identified by applying a rotated empirical orthogonal function (REOF) analysis (Horel 1981) to model and reanalysis data. The results of this analysis are displayed in Fig. 1, which shows the patterns of the regression coefficients of T_{max} versus the normalized temporal coefficients of selected leading REOF modes. The T_{max} field in all model and observational datasets for the present-day climate exhibit pronounced variability in three specific regions: northeastern China centered near 35°N 110°–115°E (Region 1; see panels a, b, c), southwestern China near 30°N 105°E (Region 2; see panels d, e, f) and northern China near 40°N 105°E (Region 3; see panels g, h, i). For a given

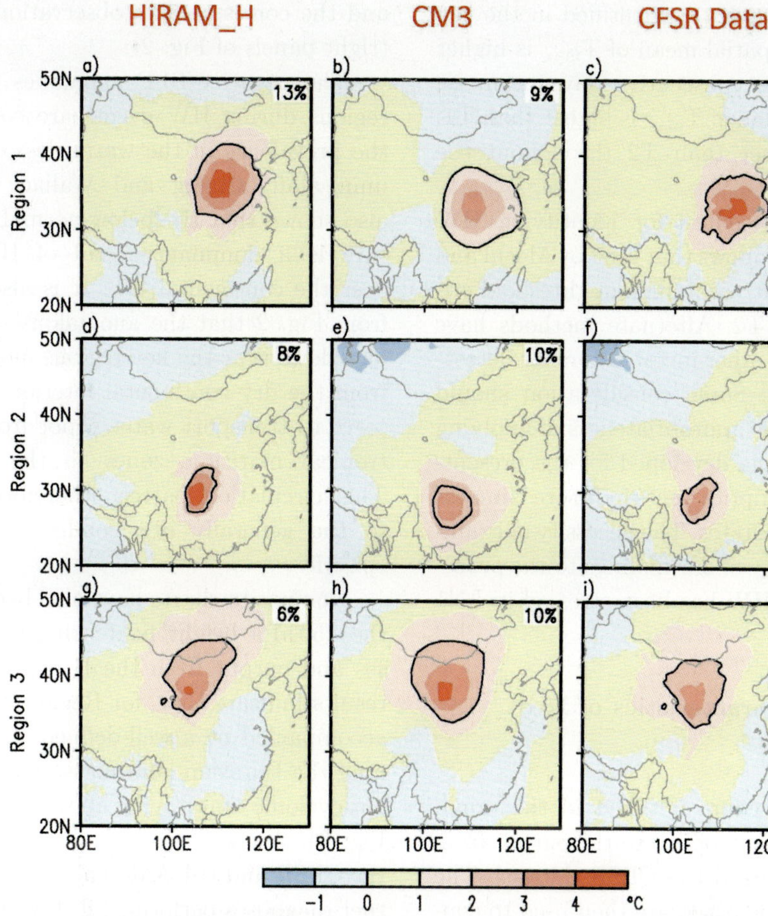

Fig. 1. Regression charts of daily maximum surface air temperature (T_{max}) in the May-June-July-August-September season versus the normalized temporal coefficients of selected leading REOF modes of Tmax in the 20°–40°N 80°–130°E domain, as computed using output from the (left column) H experiment, (middle column) CM3 experiment for the 1976–2005 period, and (right column) CFSR data. The values shown in these charts represent the amplitude of local Tmax fluctuations corresponding to one standard deviation of the REOF temporal coefficients. The fraction of variance of the respective dataset explained by a given mode is indicated at the upper right corner of each panel. Each of the rows depicts enhanced temporal variability in a "key region" of East Asia. The labels assigned to these regions (1, 2 and 3) are shown on the left side of each row of panels. The spatial extent of the key regions is indicated by bold contours, which correspond to selected isopleths in the regression patterns.

dataset, each of the key regions is defined as the collection of grid points in the corresponding regression chart with coefficients larger than a certain threshold value. The boundary of the key regions as determined by this procedure is indicated in various panels of Fig. 1 using bold contours. The locations of the three key regions, as inferred from output of the HiRAM (left column of Fig. 1) and CM3 (middle column), are in general agreement with observations (right column).

In a given dataset, the search for HW events in a key region is conducted on the basis of the daily time series of the spatial average of T_{max} over that key region. The 90th and 75th percentile values (hereafter referred as T1 and T2, respectively) of this population of daily data for the summer season in all available years are noted. A HW event is identified in the key region when the spatial mean of T_{max} is higher than T1 for three consecutive days or more; event-averaged value of T_{max} is higher than T1; and T_{max} is higher than T2 throughout the event.

The above procedure for identifying HW events essentially follows that used by Meehl and Tebaldi (2004), but with different threshold criteria for T1 and T2. Alternate methods have been employed by other investigators for detecting HW episodes. Some consideration should also be given to the appropriateness of applying the selection criteria developed for the present-day climate to synoptic events in future climates with altered mean states. The necessity of examining the sensitivity of our findings to the procedure for defining HW has been stressed in LN1 and LN2.

4. Synoptic Characteristics of Heat Wave Events

The analysis procedure outlined in the previous section has been applied to the output from the H experiment and the CFSR dataset. The HW events thus identified are then used to construct the composite patterns in Fig. 2. These charts show the distributions of the anomalous sea level pressure (SLP; contours), near-surface horizontal wind vector (arrows) and precipitation (shading) fields for HW occurring in each of the three key regions.

The HW in all three key regions are characterized by below-normal SLP, cyclonic circulation near the surface, and negative precipitation anomaly. Enhanced southerly flows prevail to the east of the anomalous low SLP center, thus contributing to the high temperatures at the HW sites. Region 2 is located in the vicinity of the Sichuan Basin. The composite charts for HW occurring in this region (panels c, d) are indicative of strong orographic influences. There exists a general agreement between the model-generated synoptic pattern (left panels of Fig. 2) and the corresponding observational composite (right panels of Fig. 2).

The negative SLP anomalies near the key regions during HW events are consistent with the prevalence of the warm, less dense air column aloft. Chang and Wallace (1987) have also shown that the below-normal surface pressure is a dominant signal of HW occurring over the continental U.S. It is also noteworthy from Fig. 2 that the anomalous southerly surface flows near the key regions mostly originate from the dry continental interior. They do not serve to transport water vapor from the moist, tropical maritime zones to the key regions. This circulation pattern therefore contributes to the generally dry conditions during HW episodes.

Composite charts have also been plotted for the 250 hPa height field using CFSR reanalyses and output from the H experiment. These results indicate that, for Region 3, the HW are accompanied by a well-defined wavetrain spanning the Eurasian land mass, with a prominent anticyclonic center lying above the HW site (see Fig. 3). A good agreement is evident between the CFSR and HiRAM-simulated patterns. Further diagnoses performed in LN2 illustrate that the temporal evolution of the individual nodes of such wavetrains is characterized by successive downstream development. This behavior is suggestive of Rossby wave dispersion. Analogous wavelike patterns are also obtained by compositing 250 hPa height data for HW events in Region 1. The typical vertical structure of the pressure anomalies over the HW sites, with a low center near the surface (Fig. 2) and a high center in the upper troposphere (Fig. 3), is indicative of increased thickness of the atmospheric column above the HW region.

As noted in LN1 and LN2, the upper level anticyclone over the HW site acts to block the passage of high-frequency transient disturbances across the region affected by the HW. The reduced level of transient activity is conducive to a stagnant circulation pattern, which

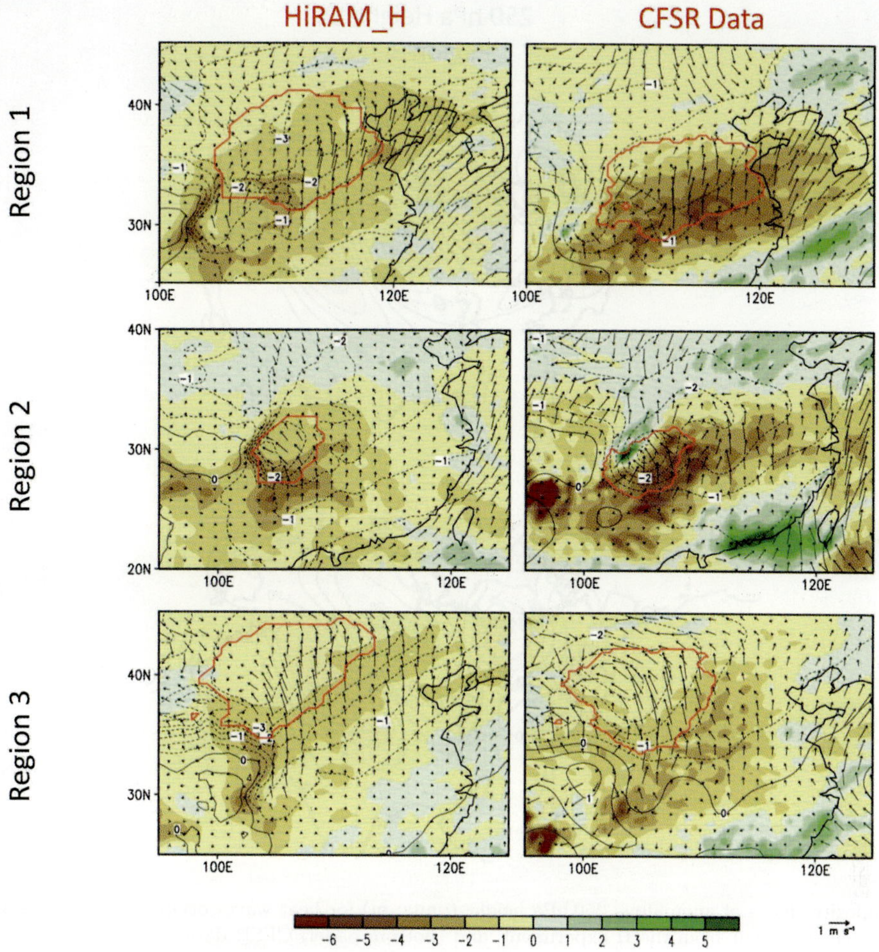

Fig. 2. Composite charts of anomalous sea level pressure (contours; with dashed contours indicating negative values; units: hPa), surface wind vectors (arrows) and precipitation (shading; units: $mm\,d^{-1}$) for heat waves occurring in (top row) Region 1, (middle row) Region 2 and (bottom row) Region 3. Results are based (left column) on output from the H experiment (consisting of three parallel runs), and (right column) CFSR data. The boundaries of the key regions, as determined from the REOF patterns in Fig. 1, are indicated by bold contours.

serves to lengthen the persistence of the HW episodes.

5. Model Projection of Heat Wave Measures for the 21st Century

The HW events in the S2 experiment are identified by following the same procedure as that applied to the H simulation (see Sec. 3), and using the same T1, T2 criteria as determined from the H dataset. As in LN1 and LN2, the severity of a given HW is defined as the average of T_{max} over the lifetime (referred hereafter as the duration) of that event. The mean frequency is evaluated by dividing the total population of events by the number of available years, and the averaged number of HW days per year is obtained by multiplying the mean duration and frequency. All of these HW metrics are computed separately for each key region in the H and S2 runs. The ratios of these metrics in the H runs versus those in the S2 runs are

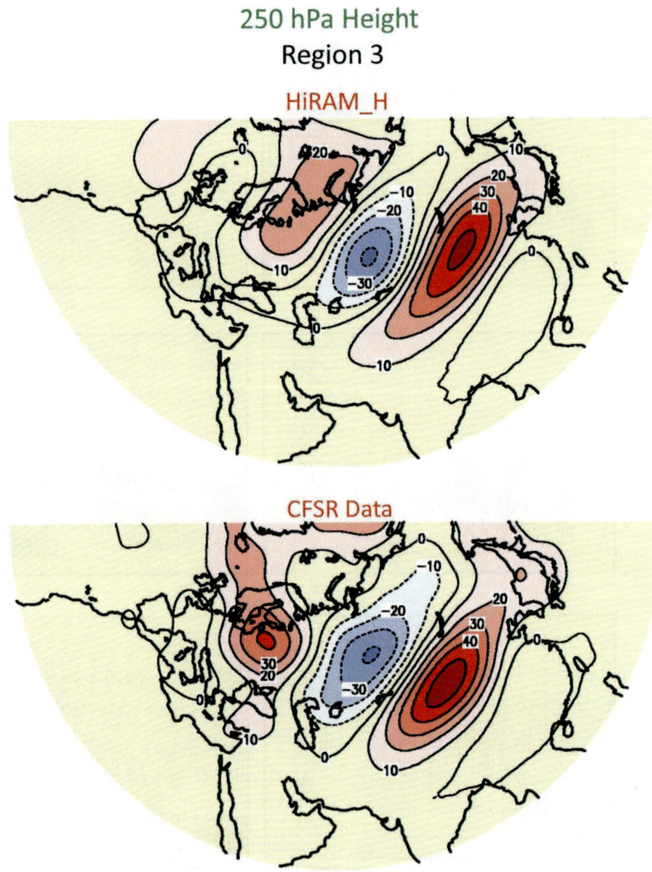

Fig. 3. Composite charts of anomalous 250 hPa height (units: m) for heat waves occurring in Region 3. Results are based on (top panel) output from the H experiment, and (bottom panel) CFSR data.

Table 1. Ratio of various HW metrics (duration, number of events per year, and number of HW days per year) in the HiRAM_S2 experiment (2086–2095) vs. those in the HiRAM_H experiment (1979–2008), for (first row) Region 1, (second row) Region 2 and (third row) Region 3.

	Duration	# HW Events per year	# HW days per year
Region 1	1.9	2.0	3.8
Region 2	1.4	2.3	3.3
Region 3	2.1	2.3	4.8

displayed in Table 1 for each of the three regions considered here. These results indicate that the HiRAM projects a marked increase in duration, frequency and number of HW days per year from the 1979–2008 period to the 2086–2095 period.

Probability distribution functions (PDFs) of the daily T_{max} output from the HiRAM and CM3 models have been computed for each key region. The results based on the present era are contrasted with those for the end of the 21st century. These PDFs are displayed in Fig. 4. The peak in the PDFs for the future climate, as projected by both the HiRAM and CM3 models, exhibits a distinct shift from the peak in corresponding PDFs for the present climate. However, there is little change in the shape of the PDFs for the two periods. In particular, the degree of "spread" of the T_{max} fluctuations, and the population of extreme outliers relative to

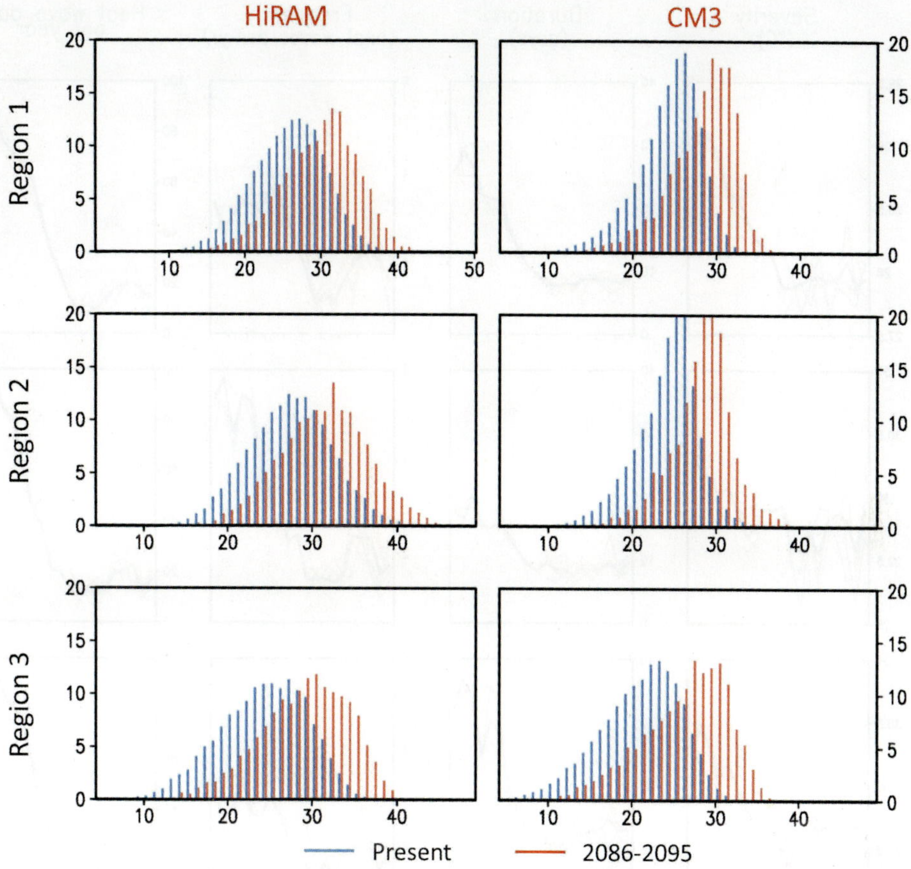

Fig. 4. Probability distribution functions of daily maximum surface air temperature (units: °C) output from (left column) HiRAM and (right column) CM3 experiments, for (first row) Region 1, (second row) Region 2 and (third row) Region 3. Results based on data for the present and 2086–2095 periods are displayed using blue and red vertical stripes respectively.

the peak of the PDFs, are not altered noticeably from the 20th to the 21st century. The relationships between the PDFs in present and future climates, as illustrated in Fig. 4, suggest that projected changes in the metrics of HW events (see Table 1) can be reproduced by adding the mean shift between the present and future climatological temperatures to the daily fluctuations in the present-day simulation. This implication is supported by the evidence presented by LN1 and LN2 for the North American and European sectors.

A more detailed depiction of the temporal evolution of the HW characteristics has been made using the output from the CM3 experiment for the 1901–2100 period. In each decade of the 20th and 21st century, the averaged severity, duration, frequency and number of HW days are computed for HW identified in that decade. The time series of various decadal HW measures for each of the three key regions are shown in Fig. 5.

Examination of the ensemble-mean results (bold curves in Fig. 5) reveals that, in the course of the 21st century, the ensemble-mean severity increases by up to 1°C, and the duration is prolonged by a factor of 2 to 3. The time series for mean frequency

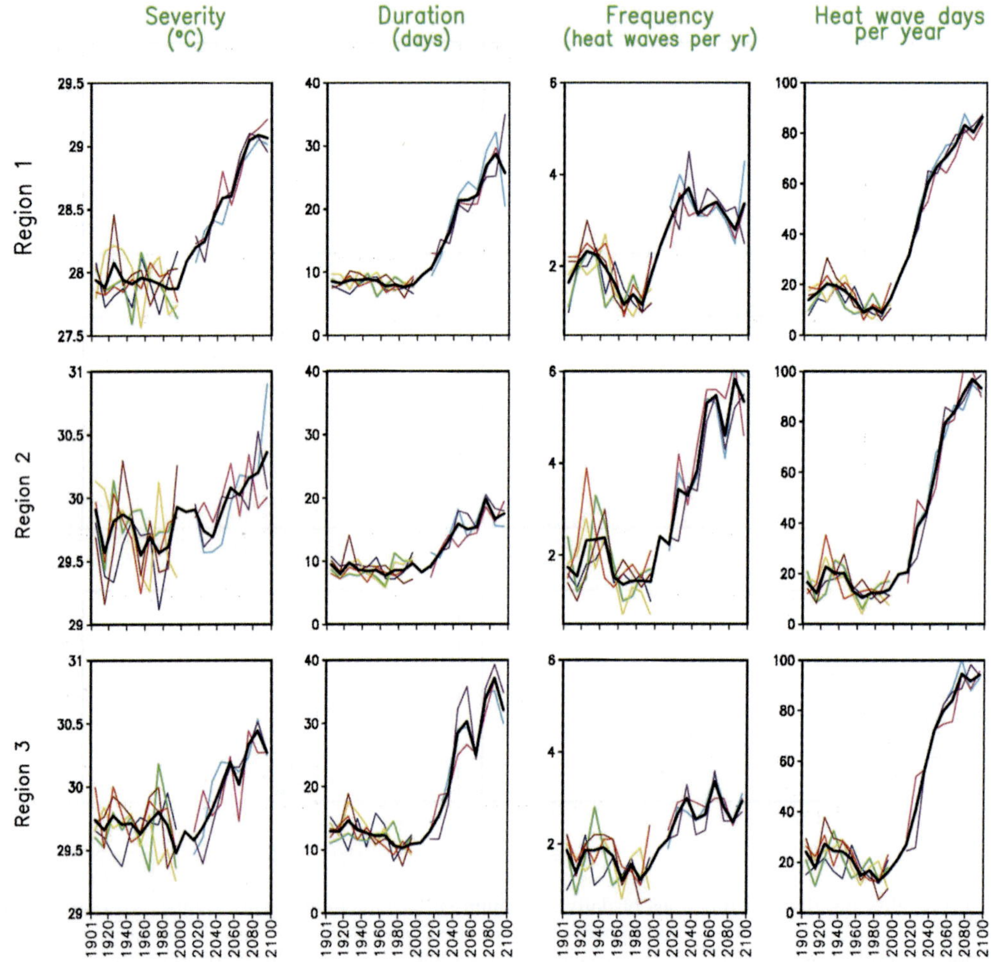

Fig. 5. Time series of the (first column from left) severity, (second column) duration, (third column) frequency of occurrence, and (fourth column) number of heat wave days per year, as computed for the May-June-July-August-September season in individual decades of the 20th and 21st centuries of the CM3 experiment. Results are shown for heat waves occurring in (top row) Region 1, (second row) Region 2 and (third row) Region 3. Time series for each of the five (three) parallel model runs in the 1901–2000 (2011–2100) periods are displayed using thin curves. Ensemble-mean HW measures based on data from all available model runs are depicted using bold curves.

exhibit considerable multi-decadal fluctuations throughout the 200-year period. A prominent increase in mean frequency (by a factor of almost 3 in some regions) is discernible within the 1980–2060 period. The ensemble-averaged number of HW days exhibits only minor changes during much of the 20th century, but rises steeply and almost monotonically beyond 2000, with a total increase by a factor of 3–4 above the present-day level.

The HW severity and frequency, as estimated using data for individual model runs (see thin curves Fig. 5), exhibit notable excursions from the ensemble-mean values (bold curves). The scatter of the individual realizations about the mean response is relatively lower in the results for duration and number of HW days per year. For all four HW measures considered here, the magnitude of projected ensemble-mean changes through the 21st century generally exceeds the level of variability among the individual samples.

6. Discussion

The results presented in this study are generally in support of the key findings reported in LN1 and LN2 on the basis of HW in North America and Europe, respectively. In particular, the synoptic flow pattern accompanying HW in North America, Europe and East Asia is dominated by a low pressure anomaly and enhanced southerly flows near the surface, and an anticyclone at upper levels. Moreover, the projected increases in duration and frequency of East Asian HW during the 21st century are similar in magnitude to the increases in these measures for the North American and European HW.

More analysis remains to be conducted on the relationships between HW occurrences in East Asia and various phases of monsoon development in that region during the summer season. In particular, it would be of interest to differentiate the characteristics and primary driving mechanisms for HWs in the period prior to monsoon onset from those occurring after the onset. The contrasts between HW behavior in East Asia and that in monsoon regions in other parts of the globe, such as South Asia and the Americas, also require further study.

Many studies (e.g. Gong *et al.* 2004; You *et al.* 2010) have related the occurrence of HW events in East Asia to the variability of the circulation associated with the western North Pacific subtropical high (WNPSH). The role of this feature in the HW episodes examined in the present work requires further clarification. Composite synoptic charts similar to those in Fig. 2, but based on data in the lower and middle troposphere (e.g. 850–500 hPa), where the WNPSH signal attains maximum strength, would be helpful for identifying relationships between the WNPSH and HW phenomena in this region.

The robustness of the findings of this study should be assessed by performing similar analyses of the output from other models, and by testing the sensitivity of our results to the identification procedure for HW episodes. The common set of criteria used in our current series of studies for selecting HW events (see Sec. 3) might not be universally applicable to all regions of the globe. It might be necessary to develop different methodologies for specific regions of interest.

Our investigation has primarily focused on the atmospheric circulation patterns associated with HW events. It should be stressed that other processes could also play significant roles in the formation, maintenance and termination of HW spells. These mechanisms include interactions between the atmosphere and ground hydrology, modulation of cloud cover by the atmospheric flow field, remote forcing of sea surface temperature anomalies, and orographic effects. The contributions of these individual mechanisms to HW occurrences remain to be diagnosed in more detail using observational and model datasets.

Acknowledgments

The author thanks the GFDL staff for developing the model tools that provide the data bases for this study and Mary Jo Nath at GFDL for performing the computations. The author's appointment at the Chinese University of Hong Kong is partially supported by the AXA Research Fund.

References

Caesar, J. and coauthors, 2011: Changes in temperature and precipitation extremes over the Indo-Pacific region from 1971 to 2005. *Int. J. Climatol.*, **31**, 791–801.

Chang, F.-C. and J. M. Wallace, 1987: Meteorological conditions during heat waves and droughts in the United States Great Plains. *Mon. Wea. Rev.*, **115**, 1253–1269.

De, U. S., R. K. Dube, and G. S. Prakasa Rao, 2005: Extreme weather events over India in the last 100 years. *J. Ind. Geophys. Union*, **9**, 173–187.

Ding, T., W. Qian, and Z. Yan, 2010: Changes in hot days and heat waves in China during 1961–2007. *Int. J. Climatol.*, **30**, 1452–1462.

Donner, L. J. and coauthors, 2011: The dynamical core, physical parameterizations, and basic simulation characteristics of the atmospheric component AM3 of the GFDL Global Coupled Model CM3. *J. Climate*, **24**, 3484–3519.

Gong, D., Y. Pan, and J. Wang, 2004: Changes in extreme daily mean temperatures in summer in eastern China during 1955–2000. *Theor. Appl. Climatol.*, **77**, 25–37.

Griffies, S. M., 2009: Elements of MOM4p1. *GFDL Ocean Group Tech. Rep. 6*, 377 pp. [Available online at http://data1.gfdl.noaa.gov/~arl/pubrel/o/old/doc/mom4p1_guide.pdf.]

Horel, J. D., 1981: A rotated principal component analysis of the interannual variability of the Northern Hemisphere 500 mb height field. *Mon. Wea. Rev.*, **109**, 2080–2092.

IPCC, 2011: Special Report on Managing the Risks of Extreme Events and Disasters to Advance Climate Change Adaptation, Eds. C. B. Fields *et al.*, Cambridge University Press.

Lau, N.-C. and M. J. Nath, 2012: A model study of heat waves over North America: Meteorological aspects and projections for the twenty-first century. *J. Climate*, **25**, 4761–4784.

Lau, N.-C. and M. J. Nath, 2014: Model simulation and projection of European heat waves in present-day and future climates. *J. Climate*, **27**, 3713–3730.

Marengo, J. A., M. Rusticucci, O. Penalba, and M. Renom, 2010: An intercomparison of observed and simulated extreme rainfall and temperature events during the last half of the twentieth century: Part 2: historical trends. *Climatic Change*, **98**, 509–529.

Meehl, G. A. and C. Tebaldi, 2004: More intense, more frequent, and longer lasting heat waves in the 21st century. *Science*, **305**, doi: 10.1126/science.1098704.

Saha, S. and coauthors, 2010a: The NCEP Climate Forecast System Reanalysis. *Bull. Amer. Meteor. Soc.*, **91**, 1015–1057.

You, Q., S. Kang, E. Aguilar, *et al.*, 2010: Changes in daily climate extremes in China and their connection to the large scale atmospheric circulation during 1961–2003. *Clim. Dyn.*, doi:0.1007/s00382-009-735-0.

Zhao, M., I. M. Held, S.-J. Lin, and G. A. Vecchi, 2009: Simulations of global hurricane climatology, interannual variability, and response to global warming using a 50-km resolution GCM. *J. Climate*, **22**, 6653–6678.

Chapter 20

Large Increasing Trend of Tropical Cyclone Rainfall in Taiwan and the Roles of Terrain and Southwest Monsoon

Hung-Chi Kuo*, Chih-Pei Chang*,†,¶, Yi-Ting Yang*, Yu-Han Chen*, Shih-Hao Su‡ and Lee-Yaw Lin§

*National Taiwan University, Taipei, Taiwan

†Naval Postgraduate School, Monterey, California, U.S.A.

‡Chinese Culture University, Taipei, Taiwan

§National Center for Disaster Reduction, Taipei, Taiwan

¶cpchang@nps.edu

Since the beginning of the 21st century Taiwan has experienced a dramatic increase in typhoon-related rainfall. Some investigators suggested that they are the manifestation of global warming effects. However, an analysis of typhoon rainfall intensity with respect to typhoon tracks in different landfall phases relative to the Central Mountain Range indicates this is unlikely the cause. Rather, most of the recently observed large increase in typhoon rainfall in the pre-landing and overland phases is the result of slower moving tropical cyclones (TCs) and their tracks relative to the high mountains. A positive feedback mechanism in which the convective heating pattern forced by topography acts to slow down the TC motion, which is most efficient for the slowly-moving northern-track storms. Another factor contributing to increased TC-related rainfall is the interaction between the typhoon circulation and southwest monsoon wind surges. This factor is most important after the typhoon center exits Taiwan and led to the increase of both typhoon rainfall intensity and rainfall amount in the new century. Both the slower TC motion and the increased monsoon interaction are consistent with the recently observed multidecadal trend of intensifying subtropical monsoon and tropical circulations, which is contrary to some theoretical and model projections of global warming. There is also no evidence of a positive feedback between global warming-related water vapor supply and TC intensity in the typhoon rainfall increase, as the number of strong landfalling TCs has decreased significantly since 1960 and the recent heavy rainfall typhoons are all of weak to medium intensity.

1. Introduction

Taiwan, which is situated in one of the main paths of western North Pacific tropical cyclones (TCs), has experienced a series of TCs with extraordinary amount of rainfall since the late 1990s. As of 2012, nine of the top 12 typhoons in total rainfall since hourly rainfall observations started in 1960 have occurred in the 21st century (Table 1). The most extreme case is the record-breaking Typhoon Morakot in 2009 (Ge *et al.* 2010; Lee *et al.* 2011; Chien and Kuo 2011, Wu *et al.* 2011) that caused huge economic and life losses, and became the first natural disaster in Taiwan that directly triggered a change of government. Whether the heavy rainfall of these typhoons is an impact of global warming is the subject of intense debate (Hsu *et al.* 2011; Chang *et al.* 2012). Here we review recent analyses that show it would be a mistake to attribute the apparently large increasing trends of TC rainfall in Taiwan in the recent decades to anthropogenic global warming.

The Global Monsoon System: Research and Forecast (3rd Edition)
Edited by C. P. Chang et al.

Table 1. The twelve typhoons in 1960–2011 with total rainfall over Taiwan exceeding 3500 mm during the three phases. The eight since 2004 are highlighted in boldface. (Adapted from Chang *et al.* 2013.)

Rank	Year	Typhoon Name	Rainfall (mm)	Central Weather Bureau Track type
1	**2009**	**Morakot**	**8996**	**CWB 3 (C)**
2	2001	Nari	8108	CWB Special
3	**2008**	**Sinlaku**	**8105**	**CWB 2 (N)**
4	**2005**	**Haitang**	**5589**	**CWB 3 (C)**
5	1996	Herb	4836	CWB 2 (N)
6	1989	Sarah	4655	CWB 3 (C)
7	1960	Shirley	4637	CWB 2 (N)
8	**2007**	**Krosa**	**3936**	**CWB 2 (N)**
9	**2004**	**Mindulle**	**3856**	**CWB 6**
10	**2008**	**Jangmi**	**3800**	**CWB 2 (N)**
11	**2008**	**Kalmaegi**	**3763**	**CWB 2 (N)**
12	**2005**	**Talim**	**3526**	**CWB 3 (C)**

2. The Roles of Terrain and Southwest Monsoon

Chang *et al.* (2013) examined hourly rainfall averaged over 21 stations in Taiwan (Fig. 1a) for the 84 typhoons that made landfall during 1960–2011. They focused on the three main leading track types that directly cross Taiwan whose maximum width is 144 km: Northern type (N, 26 cases, Fig. 1b); Central type (C, 23 cases, Fig. 1c); and Southern type (S, 14 cases, Fig. 1d). The average separation between tracks of the North and Central types and that of the Central and South types are both about 100 km, which is about one half of the length of the Central Mountain Range (CMR). The remaining 21 cases are spread over six less frequent track categories. Examples of the 925 hPa streamlines for each of the three main types are shown in Fig. 1e.

A most important factor on the amount of typhoon rainfall is the terrain effect of the CMR. From Fig. 1, it is clear the circulations of the three track types give rise to different degrees of wind-terrain interactions on the western slope of the CMR. The terrain effect is strongest for the N type and weakest for the S type. The track of each landfall typhoon can be separated into three different phases of its life history: the pre-landing (PR) phase that starts when the typhoon center moves to within 100 km from the nearest coastline point; followed by the overland phase (OL) that starts when the center makes landfall; and the exit phase (EX) that starts when the center exits the coast and ends when the center reaches 100 km away from the nearest coastline.

The first question to be answered is whether the large increase in TC rainfall implies a significant trend of rainfall intensity in the past decades. This can be examining a scatter plot of the accumulated rain amount versus the duration of that typhoon for the different track types during each phase.

Figure 2a shows the plots for the pre-landing case for the N, C, and S track types separately, and for the All track type (all nine track types with a total of 84 cases). When all typhoons are considered, the variation of rainfall intensity is far more scattered with a coefficient of determination $R^2 = 0.31$ than those of the individual N ($R^2 = 0.64$), C ($R^2 = 0.57$), and S ($R^2 = 0.72$) types. This large variation is mainly due to the large differences in the mean rainfall intensities among the three types (Table 2), which is highest for the N type (1960 mm/12 h), followed by the C type (1305 mm/12 h) and the S type (1033 mm/12 h). Prior to landfall, a typhoon situated to the north will have the most exposure of westerly and southwesterly winds against the terrain therefore its rainfall will be the largest, but this exposure will decrease if the typhoon is farther south. These results indicate that when the typhoons are stratified according to these track types, the rainfall intensity during the pre-landing phase changes over a far smaller range over the 52 years than might be inferred from the long-term rainfall time series of all typhoons. So the impression of a significant trend from combining all TCs could be misleading.

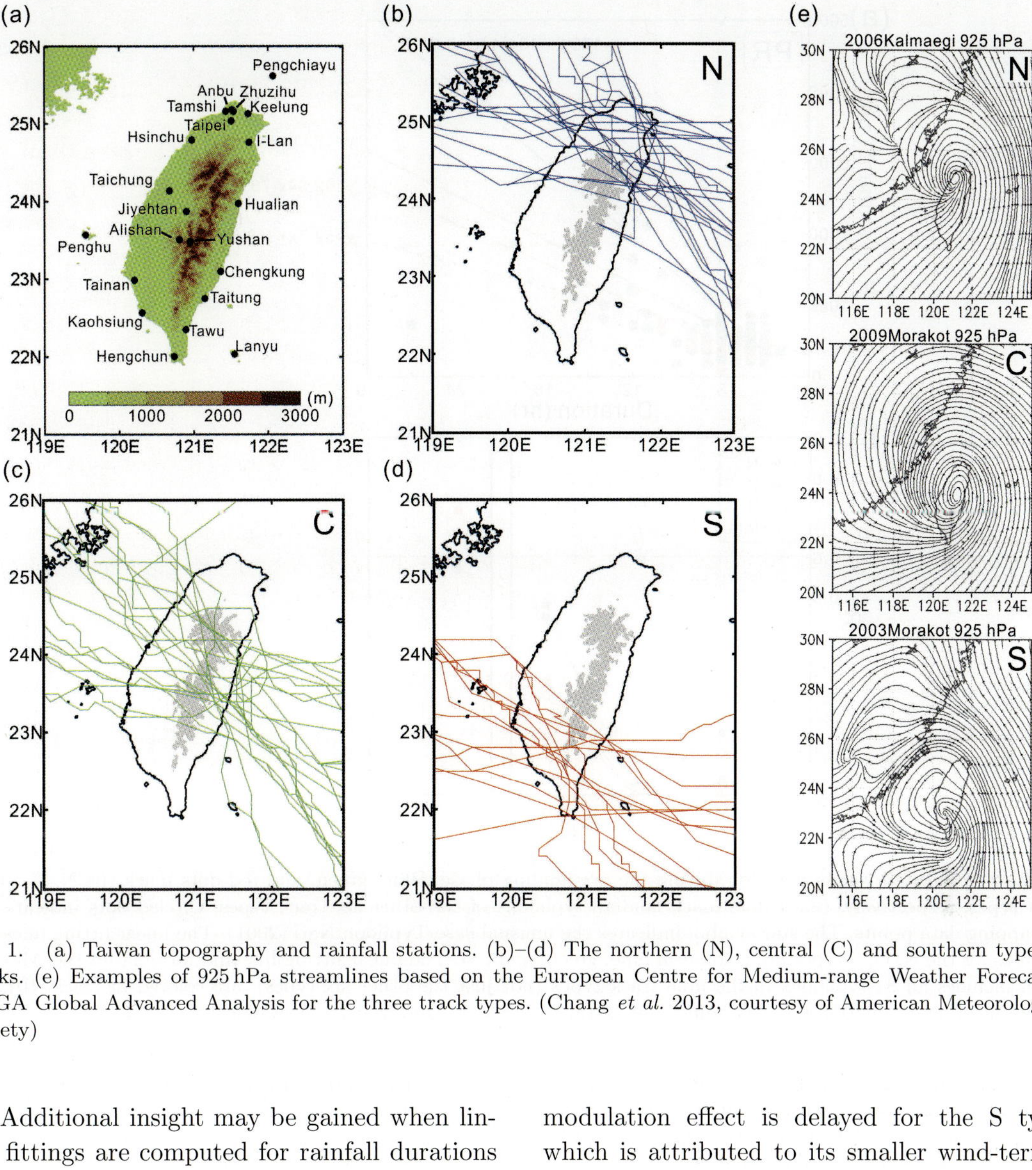

Fig. 1. (a) Taiwan topography and rainfall stations. (b)–(d) The northern (N), central (C) and southern type (S) tracks. (e) Examples of 925 hPa streamlines based on the European Centre for Medium-range Weather Forecasts–TOGA Global Advanced Analysis for the three track types. (Chang *et al.* 2013, courtesy of American Meteorological Society)

Additional insight may be gained when linear fittings are computed for rainfall durations up to six, nine, and 12 hours. During the pre-landing phase the coefficients of determination (Fig. 2b) show that clustering of rainfall intensity over 12 h becomes evident earlier for the N and C types ($R^2 = 0.47$ and 0.49 at 12 h, respectively) than for the S ($R^2 = 0.29$) and All ($R^2 = 0.25$) types. These results suggest that starting from 12 hours after a typhoon is within 100 km of the coast, the wind-terrain interaction has a modulating effect on the rainfall intensity that constrains its variations. However, this modulation effect is delayed for the S type, which is attributed to its smaller wind-terrain interaction. The decrease of the terrain effect to the south is consistent with the decrease of rain intensity from the N type to the S type in Table 2.

During the overland phase, the rainfall intensity (Table 2) again is highest for the N type, followed by the C type, and lowest for the S type. The N and C types have higher intensities than the corresponding rainfall intensities during the pre-landing phase because the maximum wind-terrain interaction occurs while the center is

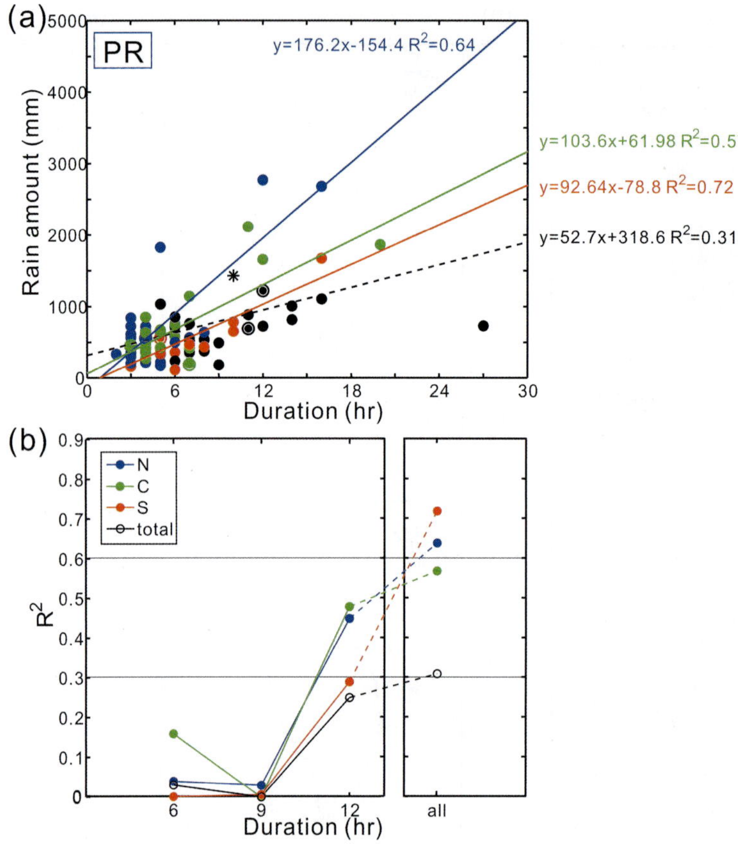

Fig. 2. (a) Rainfall versus duration during the pre-landing phase. Blue, green, and red dots mark the N, C, and S track types, respectively. Black dots mark landfall typhoons of the other six track types. Circled dots indicate two overlapping data points. The star symbol indicates the unusual case Typhoon Nari (2001). The linear fitting lines and formulas for the N, C, S types are colored in blue, green, and red. The linear fitting line and formula for the All type, which includes all 84 typhoons in the nine track types (not just the other six types), are colored in black. (b) The coefficient of determination (R^2). (Chang *et al.* 2013, courtesy of American Meteorological Society)

Table 2. The 12-hour rainfall intensity (mm) for the three track types during the pre-landing (PR), overland (OL) and exit (EX) phases (Chang *et al.* 2013, courtesy of American Meteorological Society.)

Phase track type	PR	OL	EX
N	1960	2132	1458
C	1305	1558	1467
S	1033	812	572

over land, even though the TC intensity typically weakens over land. Figure 1e shows westerly and southwesterly winds interacting with the terrain over the entire western coast of Taiwan for the N type and over the southern half of the coast for the C type. On the other hand, the S type has lower rainfall intensity than for the pre-landing phase due to a smaller wind-terrain interaction. Rain amounts for the All type are substantially distorted by inclusion of the Typhoon Nari (2001), which has a rare southward track over nearly the entire length of the island and a record-breaking rainfall of 6282 mm over 51 hours.

The strong terrain control on rain intensity is also reflected in the fact that all three types as well as the All track type have high R^2 values throughout the overland phase, and the values are the highest among the three phases

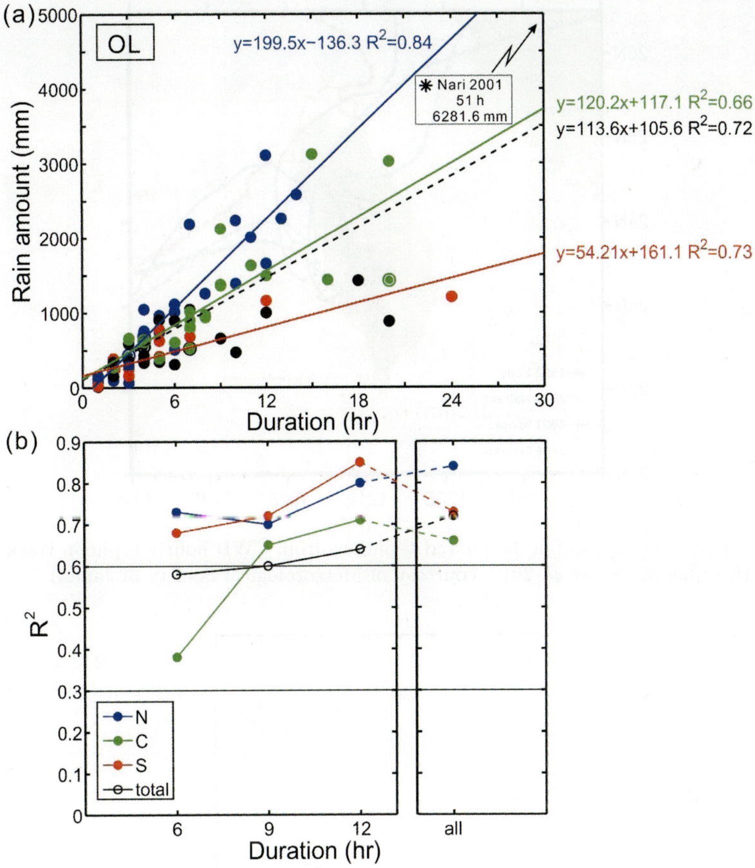

Fig. 3. Same as Fig. 2 except for the overland phase. (Chang *et al.* 2013, courtesy of American Meteorological Society)

for all track types (Fig. 3). It appears that the strong terrain effect and the abundant moisture supply from the South China Sea may have caused the TC rain intensity to approach the maximum limit for each type, so the rain intensity during the overland phase for each type has even less variations than the pre-landing phase.

There are some exception cases where a typhoon has more rainfall during the pre-landing phase than the overland phase (Su *et al.* 2012). Figure 4 shows these PR-rainfall-dominated tracks, which all possess a looping pattern. The looping tracks prolong the duration of TCs and thus produce large rainfall. These tracks tend to occur in the north side of the east coast, suggesting that the topographic stretching effect of flow across the mountain (Kuo *et al.* 2001) or the mountain channeling effect (Jian and Wu 2008) may be important.

For the exit phase (Fig. 5), the modulating effect of the terrain on the rainfall becomes smaller compared to the previous two phases. The largest decrease in the coefficient of determination is for the S type ($R^2 = 0.42$ versus 0.72 for PR and 0.73 for OL), which have the smallest terrain modulation effect on rainfall intensity. These TCs are moving farther away into the South China Sea during the exit phase (Fig. 1d) and thus the terrain effect is even more diminished. As indicated in Table 2, the N and the S types also have significantly lower rainfall intensity in the exit phase than during the other two

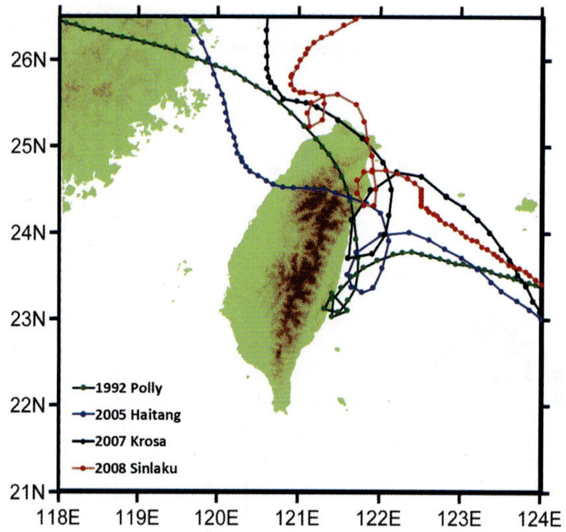

Fig. 4. Tracks of four pre-landing rainfall dominated typhoons from CWB hourly typhoon track data. The dots are the positions every three hours. (Su *et al.* 2012, courtesy of Meteorological Society of Japan)

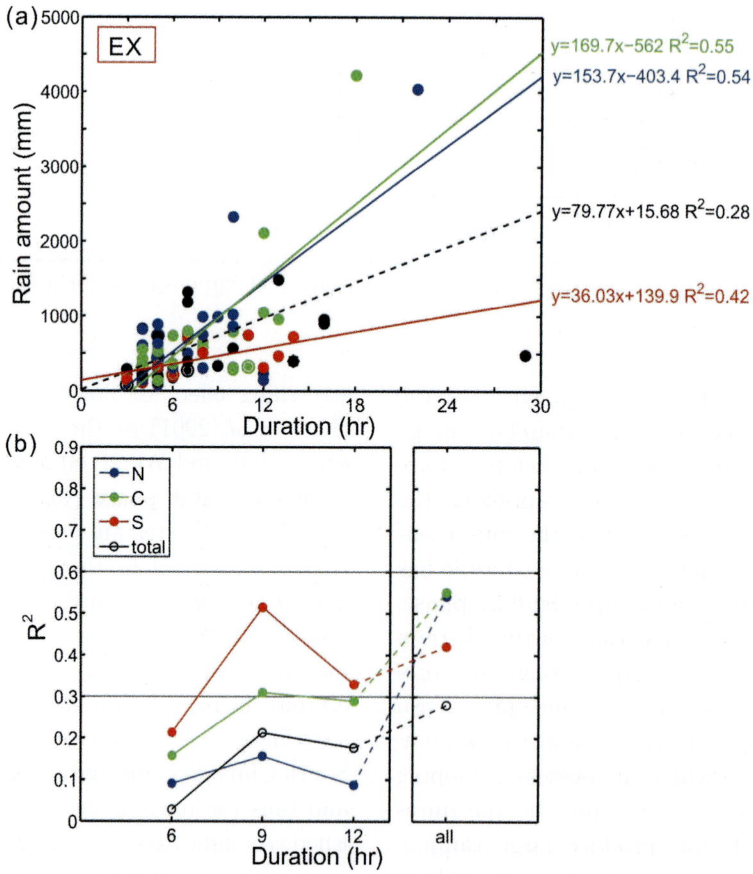

Fig. 5. Same as Fig. 2 except for the exit phase. (Chang *et al.* 2013, courtesy of American Meteorological Society)

phases. On the other hand, the rainfall intensity for the C type during the exit phase is only slightly reduced from the peak value achieved during the overland phase, and it matches the rainfall intensity of the N type. This continued high rainfall intensity for the central track is due to the typhoon center being in the middle of the Taiwan Strait during the exit phase (Fig. 1c). As a result rainfall in the C type typhoons is supported by two mechanisms. On the one hand their circulation still has a robust wind-terrain interaction. On the other hand there is a stronger interaction with the moist southwest monsoon winds originating from the South China Sea that contributes to enhanced wind-terrain interaction. The N type storms receive less such contribution because it moves northwestward away from the South China Sea (Fig. 1a).

A most dramatic example of the effect of the interaction of the southwest monsoon flow with the TC circulation and the terrain is Typhoon Morakot (2009), which led to torrential rainfall in southern Taiwan during the exit phase. Figure 6 shows the 925 hPa water vapor flux and the wind vectors for Typhoon Morakot (2009) during the exit phase. A convergent zone in the Taiwan Strait is caused by the typhoon northwesterly flow and the monsoon southwesterly flow, the origin of the latter can be traced back to the Bay of Bengal.

Table 3 shows the correlations of the typhoon rainfall amount (upper panel), and the rainfall intensity (lower panel), with two moisture flux indexes of the exit phase during the southwest monsoon season (JJA) of 1960–2012. The correlations with the duration of the exit phase are also shown. The indexes, computed from the Japanese 55-year Reanalysis data, are the southwest monsoon water vapor flux (SWF) and the total monsoon water vapor flux (TSWF) at 925 hPa in the region 16.25°N–22.5°N, 110°E–120°E, the domain used by as Pan *et al.* (2013) to cover the upstream inflow of southwest monsoon adequately. The SWF is the average moisture flux over the region during the exit phase and the TSWF is the product of SWF and the exit phase duration. Because of the very moist air over the warm sea surface, the SWF is mainly determined by the wind speed and is not sensitive to either the moisture content or the choice of the depth of the moisture layer within 1000–700 hPa.

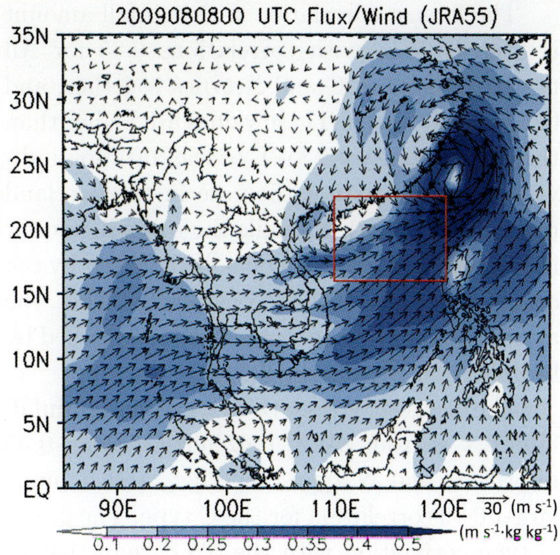

Fig. 6. Typhoon Morakot (2009) 925 hPa average wind vector (vector, m s^{-1}) and average water vapor flux (shaded, m s^{-1}) during exit period. The small rectangular (16.25–22.5° N, 110.0–120.0° E) is the region for computing the southwest monsoon water vapor flux (SWF).

Table 3. The correlations of the typhoon rainfall amount (upper panel), and the rainfall intensity (lower panel), with the southwest monsoon water vapor flux (SWF), the total monsoon water vapor flux (TSWF), and duration, during the exit phase.

	TSWF	SWF	Duration
Typhoon rainfall (mm)			
All	0.85	0.58	0.57
N+C	0.91	0.69	0.66
N	0.62	0.50	0.14
C	0.97	0.91	0.85
S	0.75	0.74	0.70
Typhoon rainfall intensity (mm hr^{-1})			
All	0.47	0.62	0.00
N+C	0.55	0.66	0.14
N	0.42	0.59	0.30
C	0.77	0.79	0.60
S	0.00	0.57	0.26

The largest correlation with rainfall amount is from TSWF, with correlations 0.85 for All type, 0.91 for the C and N types together, and 0.97 for the C type. These values are higher than the correlations with SWF or the duration individually, which is not the case in the overland phase (not shown). So while the rainfall amount in the overland phase is controlled mainly by the duration because of the strong terrain effect, the rainfall amount in the exit phase is controlled by both the duration and SWF.

The largest correlation with the rainfall intensity is from SWF, 0.62 for All type, 0.55 for the C and N types together, and a much larger 0.79 correlation for the C type. The larger C type correlation with the SWF than that of the N type is expected as the C type tracks are closer to the South China Sea. The winds in its southwest sector converge with the monsoon wind to produce a strengthened interaction with the CMR. The low SWF correlation for the S type may be due to the fact that being further away from Taiwan the circulation of the S type TCs is less likely to converge with the southwest monsoon wind and move together towards the CMR. Furthermore, the S type is the only track type that does not have duration longer than 14 hours in the exit phase (Fig. 5a).

3. Decadal Changes

Excluding Typhoon Nari (2001), the other eight heavy rainfall typhoons between 2000–2011 all occurred in or after 2004 (boldfaced in Table 1). Table 4 compares the frequencies and durations of typhoons of the three track types before and since 2004. When a year has more than one typhoon of the same type, the average duration is used. Overall, the averaged total number of TCs of all track types making landfall in Taiwan each year increased during recent decades. But this increasing trend in Taiwan does not reflect global or basin scale changes, as it is opposite to the decreasing trends of both the western North Pacific TCs (Lau and Zhou 2012) and TCs affecting the East Asian summer monsoon region (Chang *et al.* 2012) since middle-20th century, and also the century-trend of decreasing number of landfalling typhoons on Taiwan (Lee and Chia 2008).

Table 4. The frequency and duration of the N, C, and S typhoons before and after 2004, and the frequency of the three tracks typhoons before and after 1986. (Adapted from Chang *et al.* 2013.)

Frequency	1960–2003	2004–2011
N	21	5
C	15	8
S	11	3
Duration (hour)		
N	15.9	24.0
C	23.9	23.3
S	21.7	20.0

The bulk of the increase of rainfall since 2004 was due to Types N and C. During the 44 years before 2004, N type typhoons occurred about once every two years, C type typhoons occurred about once every three years, and S type typhoons occurred about once every four years. For all three types, the number of typhoons per year increased after 2004, but the increase is minimum for the N and S types each of which has one extra typhoon over the eight-year period. On the other hand, the C type had a drastic increase from averaging one every three years to one every year, a 300% increase. While the frequency for the N type did not increase much, its average duration increased 51%. Since the N type has the highest rainfall intensity and the C type has the second highest rainfall intensity (and highest intensity during the exit phase), the longer-duration N type typhoons and the more frequent C type typhoons since 2004 are the key reasons for the large increase in rainfall and rainfall intensity. The S type, on the other hand, has the lowest rainfall intensity, and its average duration has decreased by 10%.

The increase in the duration of the N type TCs may be related to a change of the large-scale easterly steering flow. Chu *et al.* (2012) reported a slowing trend between 1959 and 2009 of the steering flow in the subtropical western North Pacific east of Taiwan and the South China Sea. Although this would be consistent with the increase of average duration of typhoons, most of the slowing down in this study was for the N type typhoons among the three leading tracks. The explanation may lie in the location of the maximum terrain interaction, which is in southern Taiwan behind the N type track in which the typhoons move in a prevailing northwestward direction. Hsu *et al.* (2013) used mesoscale model simulations to demonstrate that the topographically locked convection acts to slow down (speed up) the northern (southern) landfall typhoons. A slower steering flow will slow down all three track types, but the effect is largest on the N type due to this topographic locking effect during the pre-landing and overland phases. A slower-moving northern typhoon will have more time to experience a positive feedback mechanism, in the sense more prolonged precipitation and latent heat release in the windward side of the southern Taiwan mountains will result in further speed reductions.

Table 5 shows that although the rainfall amount per typhoon or per year increases substantially from the first half to the second half of the 52 years for all three phases, the average rainfall intensity did not change much during the pre-landing and the overland phases. These results are a strong indication that the effect of the wind-terrain interaction may constrain potential climate changes of typhoon rainfall intensity in Taiwan. That is, long-term trends are small in the pre-landing and overland phases when the typhoon-terrain interaction is strong and important in contributing to the heavy rainfall. However, during the exit phase the rainfall intensity has an increasing trend. This increase cannot be explained by the terrain effect associated with track and duration changes. It also cannot be attributed to an increase in the TC intensity because the number of strong typhoons actually decreased over the 52 years. Su *et al.* (2012) have also shown that Taiwan the amount of rainfall has poor correlation with the TC intensity. The possible implication of this multidecadal increase in rainfall intensity during the exit phase will be discussed in the next section.

4. Discussion and Concluding Remarks

The analysis of the 52-year data shows that typhoon rainfall intensity in Taiwan is strongly regulated by the terrain effects, particularly as the typhoon approaches Taiwan from the east and during the period its center is over land. Because of the strong terrain effect during these two phases, the long-term trend of rainfall intensity is small. On the other hand, there is increase in the typhoon rainfall amount for all phases, and this may be caused by two factors. The first is a change in the TC tracks. The slowdown of the northern track storms and the increase of the frequency of the central track storm both contribute to an increase in rainfall. The second is the strengthening of the southwest monsoon so that the TC circulation-monsoon-terrain interaction effect leads to more rainfall. This

Table 5. The average rainfall amount per TC, rainfall amount per year and rainfall intensity of the three phases during the first (1960–1985) and second half (1986–2011) of the 52-yr period.

	1960–1985 (41 TCs)			1986–2011 (43 TCs)		
	mm TC^{-1}	mm yr^{-1}	mm hr^{-1}	mm TC^{-1}	mm yr^{-1}	mm hr^{-1}
PR	563.1	888.0	112.2	831.9	1375.8	112.1
OL	632.2	1003.2	128.8	1131	1870.5	130.4
EX	496.8	783.4	75.9	749.1	1238.9	92.0

factor contributes significantly to an increasing trend of rainfall intensity during the exit phase. Table 5 also shows a very small effect on the overland phase. An updated analysis including data up to 2015 shows that the increasing trend for the over-land phase (not shown) has continued, suggesting that the strengthening of the southwest monsoon has began to contribute to the increasing rainfall intensity in the over-land phase as well.

Tu *et al.* (2009), Chou *et al.* (2010) and Wang *et al.* (2011) have suggested that changes in western Pacific SST associated with global warming may have caused more typhoons to move northward toward midlatitude East Asia rather than westward to the South China Sea. However, changes in the frequencies among the three track types in this study are unlikely related to this process, because the separation distances between the tracks are only around 100 km. Furthermore, the shift of tracks has been from northern and southern types to the central type, rather than have been biased toward the north, and these three types are all within the East Asia domain in these previous studies.

The largest long-term trend in typhoon rainfall intensity detected was during the exit phase, when the TC circulation interacts strongly with the monsoon southwesterlies while experiencing less terrain effects. Wang *et al.* (2013) found from ERA-40 reanalysis wind data that the Northern Hemisphere subtropical monsoon and the Hadley and Walker circulations have a coherent increasing trend since the late 1970s. Over East Asia the subtropical westerlies penetrates more eastward into the western part of the western North Pacific, about 20 degrees longitude past Taiwan. An increasing subtropical monsoon trend offers a plausible explanation for the increasing trend of exit phase typhoon rainfall, since the typhoon-monsoon interaction will be enhanced. Increased subtropical westerlies over East Asia would also be consistent with the slowing down of the easterly steering flow and typhoon movement during the recent decades.

These results suggest that the main reasons for the increase in typhoon rainfall in Taiwan cannot be attributed to global warming. Using coupled-model experiments, Wang *et al.*'s (2013) attributed the main cause of the strengthening of the monsoon and tropical circulations to natural variations associated with multi-decadal variations of the Mega-El Nino Southern Oscillation (ENSO) and the Atlantic Multidecadal Oscillation (AMO). They concluded that the impacts of the Mega-ENSO and AMO are so strong that they would be larger than the anthropogenic effects during the recent decades that have been predicted in the global warming model studies. The increase in typhoon-related rainfall in Taiwan also cannot be attributed to a positive feedback between global warming-related water vapor supply and typhoon intensity, because on the one hand the rainfall intensity and typhoon intensity is poorly correlated, and also the frequency of strong typhoons affecting Taiwan has actually decreased over the last half century.

References

Chang, C.-P., Y. Lei, C.-H. Sui, X. Lin, and F. Ren, 2012: Tropical cyclone and extreme rainfall trends in East Asian summer monsoon since mid-20th century. *Geophys. Res. Lett.*, **39**, L18702. doi:10.1029/2012GL052945.

Chang, C.-P., Y. T. Yang, and H. C. Kuo, 2013: Large increasing trend of tropical cyclone rainfall in Taiwan and the roles of terrain. *J. Climate*, **26**, 4138–4147. doi:10.1175/JCLI-D-12-00463.1.

Chien, F.-C. and H.-C. Kuo, 2011: On the extreme rainfall of Typhoon Morakot (2009). *J. Geophys. Res.*, **116**, D05104. doi:10.1029/2010JD015092.

Chou, C., J.-Y. Tu, and P.-S. Chu, 2010: Possible impacts of global warming on typhoon activity in the vicinity of Taiwan, in *Climate Change and Variability*, Suzanne Simard (Ed.), pp. 79–93.

Chu, P.-S., J.-H. Kim, and Y. R. Chen, 2012: Have steering flows in the western North Pacific and the South China Sea changed over the last 50 years? *Geophys. Res. Lett.*, **39**, L10704. doi:10.1029/2012GL051709.

Ge, X., T. Li, S. Zhang, and M. S. Peng, 2010: What causes the extremely heavy rainfall in Taiwan during Typhoon Morakot (2009)? *Atmos. Sci. Lett.*, **11**, 46–50.

Hsu, H.-H. and Coauthors, 2011: Science Report of Climate Change in Taiwan 2011 (in Chinese). National Science Council, 362 pp.

Hsu, L.-H., H.-C. Kuo, and R. G. Fovell, 2013: On the geographic asymmetry of typhoon translation speed across the mountainous island of Taiwan. *J. Atmos. Sci.*, **70**, 1006–1022.

Jian, G.-J. and C.-C. Wu, 2008: A numerical study of the track deflection of supertyphoon Haitang (2005) prior to its landfall in Taiwan. *Mon. Wea. Rev.*, **136**, 598–615.

Kuo, H.-C., R. T. Williams, J.-H. Chen, and Y.-L. Chen, 2001: Topographic effects on barotropic vortex motion: No mean flow. *J. Atmos. Sci.*, **58**, 1310–1327.

Lau, W. K. M. and Y. P. Zhou, 2012: Observed recent trends in tropical cyclone rainfall over the North Atlantic and the North Pacific. *J. Geophys. Res.* **117**, D03104, 13 pp. doi:10.1029/2011JD016510.

Lee, C.-T., and H. H. Chia, 2008: Variation of landfalling typhoons in Taiwan during the recent 100 years. Report presented at the 2008 Taiwan Climate Change Workshop, 25–26 August 2008, Central Weather Bureau, Taipei, Taiwan.

Lee, C.-S., C.-C. Wu, T.-C. C. Wang, and R. L. Elsberry, 2011: Advances in understanding the "perfect monsoon-influenced typhoon": Summary from International Conference on Typhoon Morakot (2009). *Asia-Pac. J. Atmos. Sci.*, **47**, 213–222.

Pan, T.-Y., Y.-T. Yang, H.-C. Kuo, Y.-C. Tan, J.-S. Lai, T.-J. Chang, C.-S. Lee, and K. Hsu, 2013: Improvement of watershed flood forecasting by typhoon rainfall climate model with an ANN-based southwest monsoon rainfall enhancement. *J. Hydrol.*, **506**, 90–100.

Su.-S. H., H.-C. Kuo, L.-H. Hsu, and Y.-T. Yang, 2012: Temporal and spatial characteristics of typhoon extreme rainfall in Taiwan. *J. Meteor. Soc. Japan*, **90**, 721–736.

Tu, J.-Y., C. Chou, and P.-S. Chu, 2009: The abrupt shift of typhoon activity in the vicinity of Taiwan and its association with western North Pacific–East Asian climate change. *J. Climate*, **22**, 3617–3628.

Wang, R., L. Wu, and C. Wang, 2011: Typhoon track changes associated with global warming. *J. Climate*, **24**, 3748–3752.

Wang, B., J. Liu, H. J. Kim, P. J. Webster, S. Y. Yim, and B. Xiang, 2013: Northern Hemisphere summer monsoon intensified by mega-El Nino/southern oscillation and Atlantic multidecadal oscillation. *Proc. Natl. Acad. Sci.* **110**, 5347–5352.

Wu, L., J. Liang, and C.-C. Wu, 2011: Monsoonal influence on Typhoon Morakot (2009). Part I: Observational analysis. *J. Atmos. Sci.*, **68**, 2208–2221.

Chapter 21

Sudden Track Change of Typhoon Megi (2010) and Its Remote Effect on Rainfall over Taiwan — Evaluation of Variability Based on Ensemble Simulations

Chun-Chieh Wu and Ting-Chen Chen

Department of Atmospheric Sciences, National Taiwan University, Taipei, Taiwan

cwu@typhoon.as.ntu.edu.tw

http://typhoon.as.ntu.edu.tw

In this study, the uncertainties of (1) the sudden recurvature of Typhoon Megi (2010), and (2) precipitation over Taiwan under the joint influence of the outer circulation of Megi and the northeasterly monsoon are examined with ensemble simulations based on an ensemble Kalman filter data assimilation system and the Weather Research and Forecasting (WRF) Model. Our analysis shows that the overall movement of Megi well aligned with the direction of steering flow associated with the monsoon trough. After Megi headed north, the approaching upper-level trough helped enhance Megi's northward movement.

In the second part of the study, the results show that the low-level moisture convergence of northeasterly monsoon and the outer circulation of TC provides a favorable environment for the potential occurrence of such heavy rainfall events. As the prevailing wind helps advect the moisture from the adjacent seas inland, the steep slopes of the windward mountains creates a strong orographic lifting effect that further enhances the rainfall.

This study shows that ensemble simulations can provide new insights into the uncertainty and predictability in understanding tracks and the typhoon-induced remote impact on rainfall.

1. Introduction

One of the major threats of a tropical cyclone (TC) is the accompanying excessive torrential rainfall, which often leads to devastating mudslides and floods, resulting in enormous economic loss and potential threats to human life. The quantitative precipitation forecast (QPF) associated with TCs remains a major challenge for numerical models, since the precipitation amount is affected by multiple factors, such as large-scale moisture distribution, cloud microphysics, the internal structure of the TC, track variation and the complicated interaction between the TC circulation and topography (e.g., Wu and Kuo 1999; Lin *et al.* 2001; Wu *et al.* 2002; Zhang *et al.* 2010; Wu *et al.* 2013). In particular, the accuracy of the TC track is often highlighted as a deciding factor for accurate forecast of typhoon-related rainfall, especially for landfalling cases (e.g., Wu and Kuo 1999; Wang *et al.* 2009).

In general, TC-related rainfall can be categorized into two types: direct and indirect. The former suggests the precipitation is mainly caused by the TC circulation itself and the latter indicates the rainfall event is associated with strong interactions between the TC and other synoptic systems, such as an upper-level trough or cold frontal systems. With the latter case, as the phenomenon may occur at a distance of hundreds of kilometers from the TC center, it is also known as the remote effect of a TC on precipitation (Wang *et al.* 2009). Although remote rainfall events can occur frequently in Taiwan due to the existence of a monsoon system, relatively

The Global Monsoon System: Research and Forecast (3rd Edition)
Edited by C. P. Chang et al.

less attention has been placed on such events. During autumn or early winter, the warm, moist air advected northward by the outer circulation of TC can interact with the cold, dry winter northeasterly monsoon, creating a frontal-type system that brings even more rainfall to certain regions in Taiwan (Wu *et al.* 2009). Previous studies show that recurving TCs are more likely to cause a remote rainfall event positioned in a distance northward to the main TC circulation (Wu *et al.* 2009; Galarneau *et al.* 2010). In the United States, such a remote rainfall event often occurs beneath the equatorward jet-entrance region of an upper-level jet streak and is referred to as a predecessor rain event (PRE), in which a long-lived, quasi-stationary mesoscale convective system is produced by the frontogenetical forcing with the arrival of tropical moisture from a TC, and thus resulting in heavy rainfall (Galarneau *et al.* 2010).

To address the predictabilities and uncertainties of a single deterministic forecast for the emergency management decision makers and the general public, ensemble simulations have recently been applied to TC-related rainfall issues (e.g., Zhang *et al.* 2010; Wu *et al.* 2013). Wu *et al.* (2013) investigated the impact of TCs with different track bias on the rainfall amounts and distribution in the ensemble simulations during Typhoon Sinlaku (2008), concluding that the uncertainties in rainfall patterns and amounts can be assessed from ensemble track variations.

Since TC-related precipitation events that have been studied are mostly landfalling cases, understanding in the uncertainty of TC tracks and associated remote rainfall patterns are relatively limited. In order to further address the issue, a case of Typhoon Megi (2010) with a sudden northward track change and accompanying remote rainfall over the northeastern part of Taiwan is studied. Qian *et al.* (2013) examined the THORPEX Interactive Grand Global Ensemble data provided by ECMWF, and found that compared to the mean of members without northward recurvature, the mean of members with accurate track forecasts shows slower TC movement, larger TC size, having a stronger eastward moving midlatitude trough, and the Pacific subtropical high is weaker and located more eastward.

Our objectives are twofold. One is to evaluate the uncertainty of the sudden TC track change and to quantitatively identify the dominant factors leading to Megi's sharp recurvature. The other is to investigate the predictability and detailed mechanism of this heavy rainfall event under the influence of the complicated TC-monsoon-terrain interaction. It is also of interest to know if improvement in TC track predictions could guarantee better precipitation forecasts in this kind of remote rainfall events. The details of the second portion of this study on precipitation mechanism have been shown in Chen and Wu (2016).

2. An Overview of Typhoon Megi

Typhoon Megi formed as a tropical storm on 13 October 2010 to the west-southwest of Guam, reaching its peak intensity with a minimum central sea level pressure of 885 hPa at 1800 UTC 17 October 2010 when it moved west-northwestward across the west Pacific ocean at around $4.5\,\mathrm{m\,s^{-1}}$ and then made landfall at Palanan Bay at northeastern Philippines on 18 October. After Megi left the Philippines, its translation speed slowed to around $2.5\,\mathrm{m\,s^{-1}}$ and made a sharp poleward turn at 0000 UTC 20 October, heading for Fujian province in China. During Megi's sharp recurvature over the South China Sea, located 800 km away from northeastern Taiwan, heavy precipitation was observed not only along the TC track but also northeastern Taiwan and the adjacent seas during 0000 UTC 19 October to 0000 UTC 23 October (Fig. 1a). The maximum amount of the 4-day accumulated rainfall was recorded at 1,186 mm in Yilan (Fig. 1b).

As shown in Fig. 2, Megi was forecasted to move westward by all of the operational centers two days before its recurvature. Just 18 hours before Megi's sudden northward turn, the large

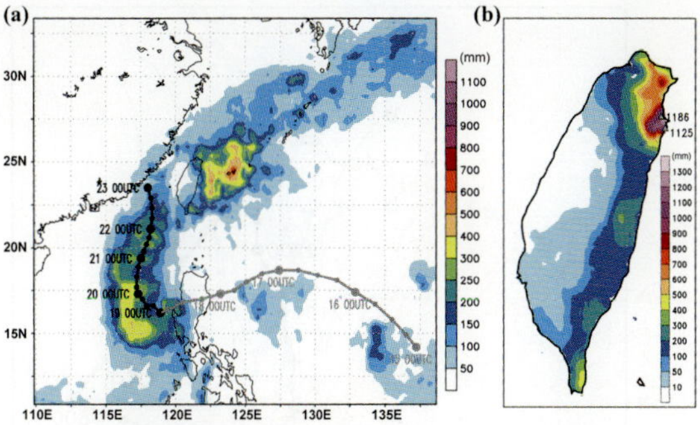

Fig. 1. The 4-day accumulated rainfall (mm) during 0000 UTC 19–23 Oct 2010 from (a) the TRMM 3B42 V7 3-hourly rainfall products and (b) the rain gauge stations in Taiwan. The observation track of Typhoon Megi from JMA is shown in (a) (from Chen and Wu 2016).

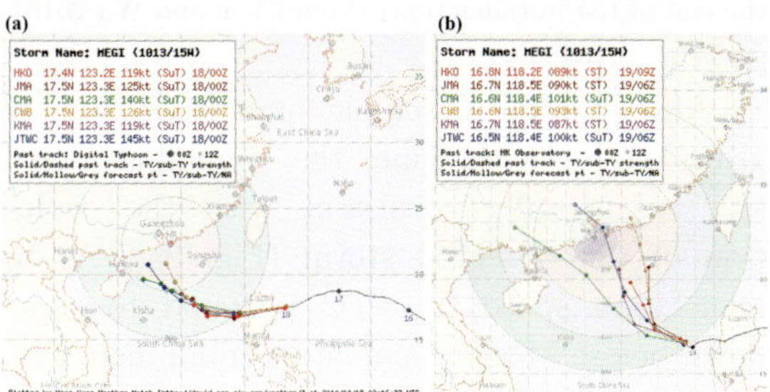

Fig. 2. Official track forecasts of Typhoon Megi from Hong Kong Observatory (HKO), Japan Meteorological Agency (JMA), China Meteorological Administration (CMA), Central Weather Bureau (CWB, Taiwan), Korea Meteorological Administration (KMA) and Joint Typhoon Warning Center (JTWC) at (a) 00Z 18 and (b) 06Z (09Z for HKO) 19 October 2010. (Derived from http://david.samsiu.hk/weather/2010/review/review201015w.html).

spread of simulations among different operational centers indicates that the uncertainty of Megi's recurvature was quite large during the time.

3. Methodology

3.1. *Model configuration*

The Advanced Research WRF (ARW) (version 3.2.1) based on an EnKF data assimilation system is used in this study. The sizes of the four domains are 142×100, 166×136, 124×133 and 139×223, each with a grid spacing of 54, 18, 6, and 2 km, respectively. The physics parameterization schemes used in this study include the WSM 6 microphysics scheme, the Rapid Radiative Transfer Model (RRTM) scheme for longwave radiation, the simple shortwave scheme for shortwave radiation, and the YSU planetary boundary layer scheme. Taking the National Centers for Environmental Prediction (NCEP) Final Analysis (FNL; $1^\circ \times 1^\circ$) at 0000 UTC 17 October 2010 as the initial condition, 36 ensemble members are generated by randomly perturbing the mean analysis in a transformed stream function field as described in Zhang *et al.* (2006), and a 24-hr-initialization is followed. During the initialization, with valuable aircraft observations collected during the field campaign of "Impact of Typhoons on the Ocean in the Pacific (ITOP)" (Fig. 3), the EnKF update cycle is conducted every 30 min to assimilate not only the state variables but also the TC center position, storm motion vector and axisymmetric surface wind profile for better initial TC track, motion and structure (Wu *et al.* 2010). A 5-day ensemble simulation is then carried out from 0000 UTC 18 October to 0000 UTC 23 October 2010. A terrain-removed numerical experiment

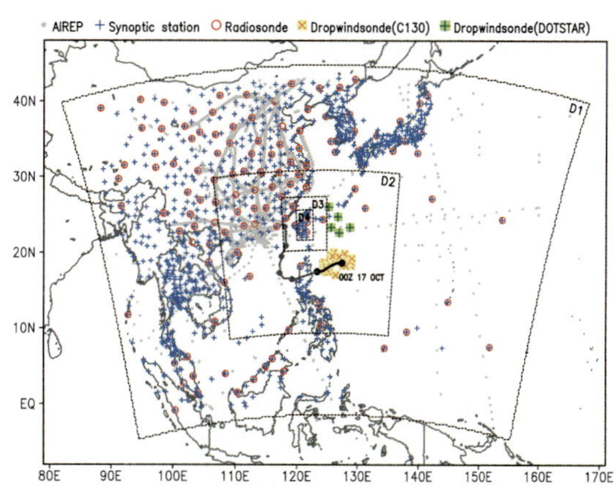

Fig. 3. The aircraft reports, conventional data and dropwindsonde data assimilated during initialization (from Chen and Wu 2016).

is performed to examine the role of terrain in this remote rainfall event.

3.2. *PV diagnosis*

The basic concept of PV invertibility is that, given a distribution of PV and prescribed balance and boundary conditions, the balanced mass and wind fields (nondivergent wind) can be recovered (Wu *et al.* 2003). Taking the advantage of piecewise inversion, when the flow field is appropriately divided into the mean and perturbation components, the balanced fields associated with each individual PV perturbation can be rederived at a given time (Davis 1992). Then the steering flow, defined as the deep-layer-mean (975–300 hPa) wind vector, $\overline{V_{SDLM}}$, averaged over the inner 3 degree latitude around the storm center associated with each dynamical system is evaluated:

$$\overline{V_{SDLM}} = \frac{\int_{925\,\mathrm{hPa}}^{300\,\mathrm{hPa}} \overline{V_s}(p)\,dp}{\int_{925\,\mathrm{hPa}}^{300\,\mathrm{hPa}} dp} \qquad (1)$$

where $\overline{V_s}(p) = \frac{\int_0^{3^\circ}\int_0^{2\pi} \overline{V} r\,dr\,d\theta}{\int_0^{3^\circ}\int_0^{2\pi} r dr\,d\theta}$ is the steering flow at each level. The detailed algorithm can be found in Wu *et al.* (2003).

4. Results

The ensemble members associated with different tracks are classified into three categories: the northward (NG), westward (WG) and remaining groups (RG) based on the northward-turning angle relative to the best track (Fig. 4). The northward-turning angle, α, is defined as

$$\alpha = \tan^{-1}\left(\frac{\mathrm{TC\,lat}_{2218} - \mathrm{TC\,lat}_{1800}}{\mathrm{TC\,lon}_{1800} - \mathrm{TC\,lon}_{2218}}\right) \qquad (2)$$

where $\mathrm{TC\,lat}_{\mathrm{ddhh}}$ and $\mathrm{TC\,lon}_{\mathrm{ddhh}}$ represent the location of TC center at the time noted in the suffix "ddhh". Note that the members in NG (9 members) is fewer than those in WG (24 members).

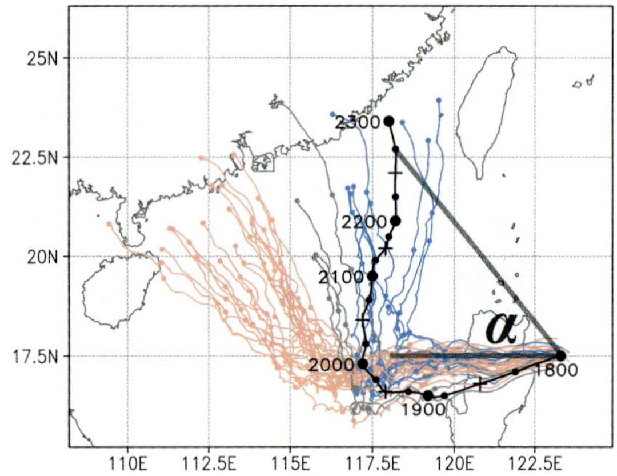

Fig. 4. The best track from JMA (black) and the simulated tracks. Lines colored in pink, blue and gray are WG, NG and RG, respectively. Red line is the ensemble mean run (initialized with the ensemble mean state at the end of the initialization) (from Chen and Wu 2016).

Although the ensemble spread in intensity grows larger with time, the simulated intensity is generally consistent with the evolution of the observed storm (Fig. 5a). As to translation speed, the TC is found to slowdown before Megi's recurvature in all members, which is in agreement with the best track data (Fig. 5b).

4.1. *Uncertainty in simulated TC tracks*

Since there are no significant systematic differences in TC intensity and translation speed between NG and WG, the TC size, measured by the radius of $17\,\mathrm{m\,s^{-1}}$ azimuthal surface wind, is further examined. Prior to the recurvature of Megi, the TC size of the NG ensemble mean has become smaller than that of WG [which is not consistent with part of the results form Qian *et al.* (2013)], yet the relationship between the simulated TC size and track variation remains unclear in our study (Fig. 5c). We attribute the uncertainty of the TC track to the difference in large-scale dynamical systems, such as the subtropical high, monsoon trough, upper-level trough and so on (Fig. 6).

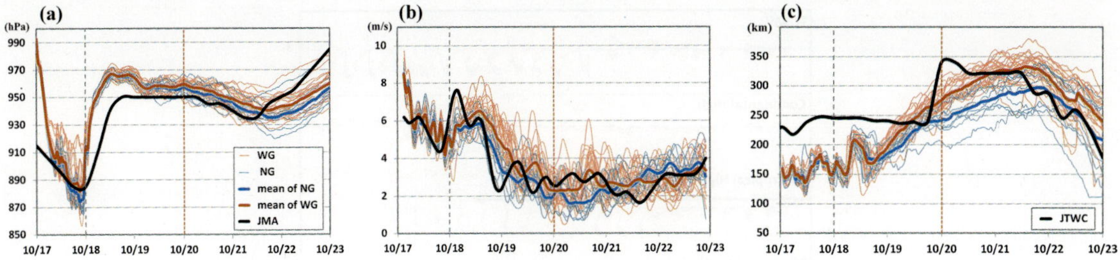

Fig. 5. (a) The time evolution of the TC intensity (hPa), (b) translation speed ($m\,s^{-1}$) and (c) TC size (km) of 36 ensemble members. Pink thin lines belong to westward group (WG), the blue ones are for the northward group (NG) and the thick lines in red and blue represent the ensemble means of WG and NG, respectively. The thick gray (black) line represents the ensemble mean run (JMA best track data). The vertical gray (red) dash lines indicate the time when initialization ends and when Megi is observed to make a sudden northward turn (from Chen and Wu 2016).

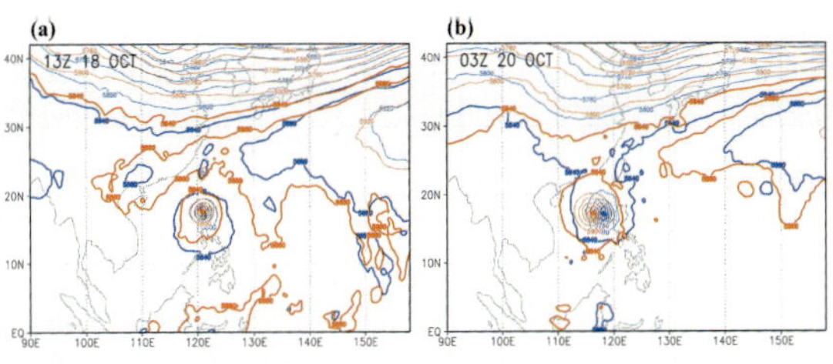

Fig. 6. 500 hPa geopotential heights of the member with the largest northward turning angle (m028, blue) and the one with the smallest northward turning angle (m007, red) at (a) 13Z 18 and (b) 03Z 20 October 2010.

To better understand the possible factors leading to Megi's sharp recurvature, m028 and m007 are selected due to its largest and smallest northward-turning angle, respectively. Piecewise potential vorticity diagnosis is employed to quantitatively evaluate the relative contribution of individual dynamical systems to the TC's steering flow.

The results show that before the recurvature occurs, the TC's deceleration is mainly caused by the following three steering flows: (1) the weaker southeasterly/easterly steering flow associated with the eastward retreat of SH, (2) the stronger westerly steering flow associated with the approaching upper-level trough, and (3) the weaker easterly steering flow associated with monsoon trough (Figs. 6 and 7). After Megi (m028) turns north, the overall movement of Megi well aligns with the direction of the steering flow associated with the MS trough and an anticyclone that located east to the TC. The sudden northward change of steering flow associated with MS trough occurs when the associated positive PV perturbations embedded in the MS trough move from the south to the west of Megi (m028; not shown), thus inducing an easterly to southeasterly steering flow. Moreover, two major differences between m007 and m028 are the steering flows associated with the upper-level trough and the anticyclone located to the east of the TC, both of which are stronger in m028 than in m007 and thus accelerate Megi (m028)'s northward movement significantly (Fig. 7).

4.2. *Uncertainty in simulated rainfall over Taiwan*

Figure 8 shows the means and standard deviations of 4-day accumulated rainfall distribution of NG and WG, respectively. Most of the members can well capture the heavy precipitation over Yilan, no matter what their track category is. However, variability still exists over the eastern part of Taiwan and some members still fail to simulate such an event even with relative smaller track errors (not shown). The quantitative evaluation based on the equitable threat score (ETS; Schaefer 1990) indicates that,

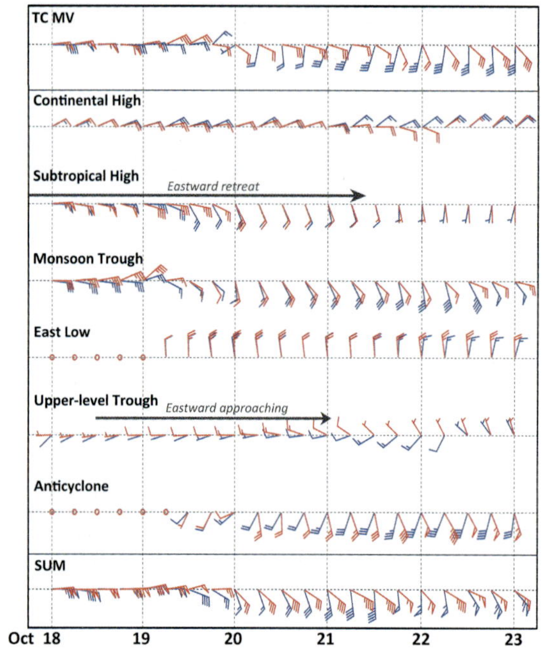

Fig. 7. The deep-layer-mean steering flow associated with the PV perturbation of the continntal high, the subtropical high, the monsoon trough, low pressure system east to the TC (East Low), the upper-level trough, the anticyclone east to the TC and the summation of the above five systems (SUM). The movement of Megi is indicated by TCMV. One full wind barb represents $1\,\mathrm{m\,s^{-1}}$. The red and blue arrows indicate the analyses for m007 and m028, respectively.

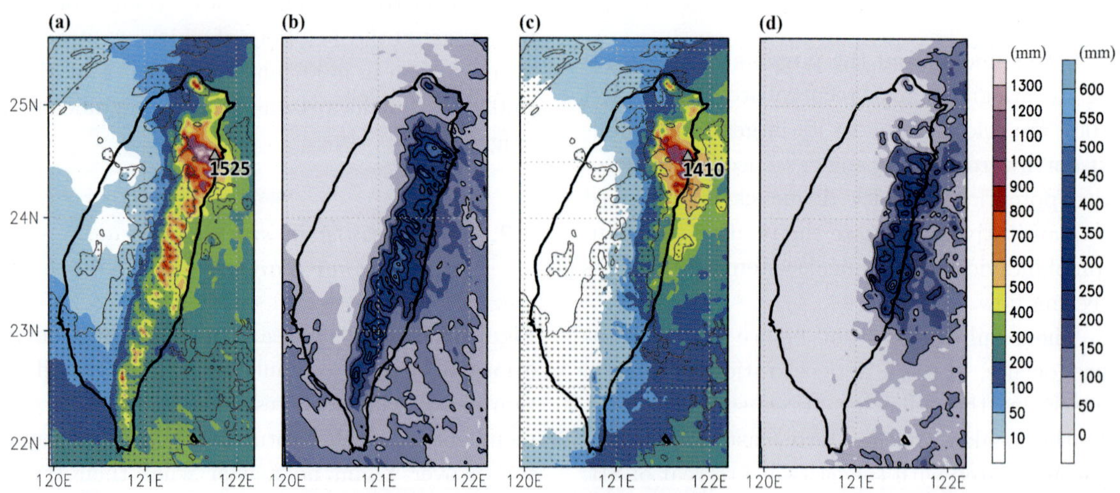

Fig. 8. 4-day accumulated rainfall (mm) from 0000 UTC 19 to 23 Oct 2010 in the ensemble mean of (a) NG and (c) WG, respectively. Dotted areas indicate where the NG-WG difference is statistically significant at the 95% confidence level. (b) and (d) are the standard deviation (std) of the four-day accumulated rainfall in NG and WG, respectively. (From Chen and Wu 2016).

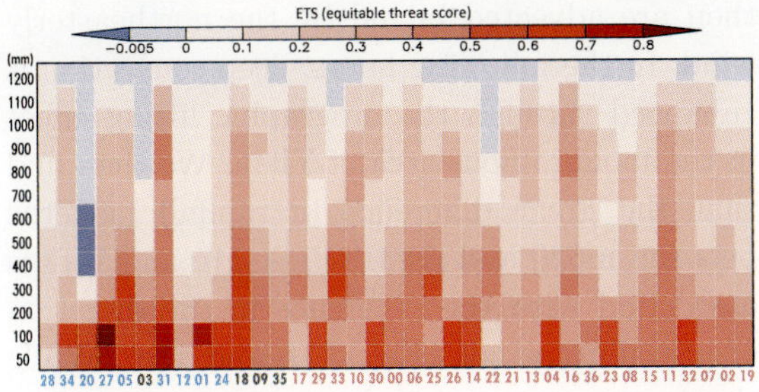

Fig. 9. ETS of the 4-day accumulated rainfall forecasts from 19 to 23 October 2010 calculated over the land area in Taiwan. Ensemble members (colored in blue, pink and gray are NG, WG and RG, respectively) are arranged in increasing order of track errors from left to right along the abscissa. Vertical axis represents the selected rainfall thresholds (mm) (from Chen and Wu 2016).

with one or two exceptions, members with tracks close to the best track produce more consistent rainfall distribution (i.e., with ETS closer to 1) at smaller rainfall threshold (50–400 mm) (Fig. 9). However, the relation between accuracy of track forecasts and that of precipitation forecasts is unclear at heavy rainfall thresholds, indicating that the rainfall distribution in this remote rainfall event is not dominated by TC track, and that the uncertainty could be larger than that in the rainfall event directly caused by the TC. It is of interest to see why some members with relative smaller track errors (NG) fail to predict the heavy rainfall, while several representative members are selected to see what other factors cause the difference in rainfall patterns.

4.2.1. *The role of TC circulation*

The low-level-averaged water vapor mixing ratio and wind fields show that deep tropical moisture associated with the TC is advected from the TC into the coherent area of rainfall by the TC's outer circulation, providing a humid environment (Fig. 10). Similar patterns can be observed in most of the members except the one with extremely-westward-biased track and relative smaller TC sizes. In such members, without deep moisture advected by the southeasterly

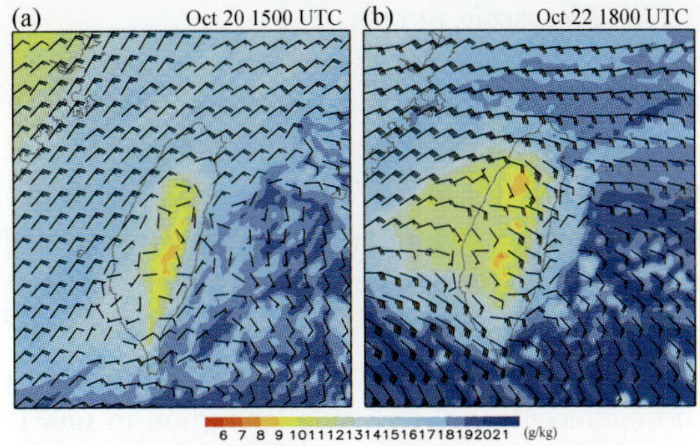

Fig. 10. Low-level-averaged water vapor mixing ratio (shaded, $\mathrm{g\,kg^{-1}}$) and wind ($\mathrm{m\,s^{-1}}$) fields in a selected member (m031) at (a) 15Z 20 and (b) 18Z 22 Oct. 2010.

wind associated with the outer circulation of TC, the accumulated rainfall over northeastern Taiwan, solely produced by the northeasterly monsoon, peaks at a location similar to the observation while the maximum rainfall is largely reduced (by more than 50%, figure not shown).

4.2.2. *Interaction between TC circulation and the northeasterly monsoon*

The south or southeasterly wind associated with the TC circulation is likely to cause a low-level water vapor flux convergence with the northeasterly monsoon over the outer seas off the coast of northeastern Taiwan. However, since the low-level wind speed is greatly reduced over land due to surface friction, the influence of water vapor flux convergence is relative small over land compared to when it is over sea. In other words, the abundant moisture is mainly accumulated over the outer seas instead of in the heavy rainfall area. Note that the location of moisture convergence is determined by the TC size and the area covered by the northeasterly monsoon (the strength of the monsoon system). An examination of the ensemble members shows that those with northeasterly wind extending to northeastern Taiwan is in favor of the occurrence of heavy rainfall in Yilan.

4.2.3. *Mesoscale forcing*

Comparing two members, m031 and m020, both in NG but the former with high ETS and the latter with low ETS (Fig. 9), we found that extensive water vapor flux convergence is present in both members during a selected 24-hour period (Figs. 11a and 11d), although the amount of moisture convergence is larger and is more concentrated in a smaller region in m031. Further examination of m031 shows that convections form initially over the outer seas, and then are advected inland by the northeasterly wind. In the following hours, deep convection is enhanced through the orographic lifting effect in the mountainous area of Yilan. We then find that the great difference in rainfall patterns between m020 and m031 (Figs. 11c and 11f) is associated with the difference in the direction of prevailing winds (Figs. 11b and 11e). While the northeasterly wind dominates in m031, the southeasterly wind dominating in m020 advects moisture to mountains with gentler slopes, where the lifting effect is not as strong as that in m031.

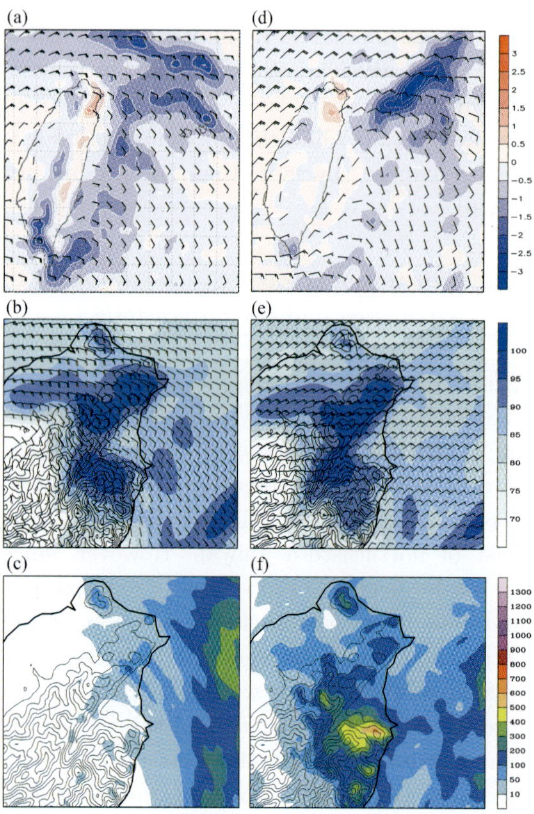

Fig. 11. (a) 24-hr, low-level averaged water vapor flux divergence (shaded, $10^{-6}\,\mathrm{kg\,m^{-2}\,s^{-1}}$), (b) 24-hr, low-level averaged water vapor mixing ratio (shaded, $\mathrm{g\,kg^{-1}}$) and 24-hr averaged wind fields ($\mathrm{m\,s^{-1}}$ at 0.75 km height) and (c) 24-hr-accumulated rainfall during 0000 UTC 20 to 21 October 2010 and terrain heights (contour) in m020. (d), (e) and (f) are same as in (a), (b) and (c), respectively, but they are for m031.

4.2.4. *The role of topography*

To verify the role of topography in this heavy rainfall event, a terrain removal experiment is carried out on m031 (m031_TR), in which the land of Taiwan is set to be ocean. The simulated track has a slight westward bias after the terrain of Taiwan is removed (not shown), but the intensity has no significant changes. The simulated maximum accumulated rainfall has a 80% reduction in m031_TR. Figure 12 shows that after the terrain is removed, the rainfall caused by the low level convergence near the outer seas can still be observed, but the heavy rainfall over Yilan is greatly reduced, indicating that the orographic lifting effect contributes significantly to this remote heavy rainfall event.

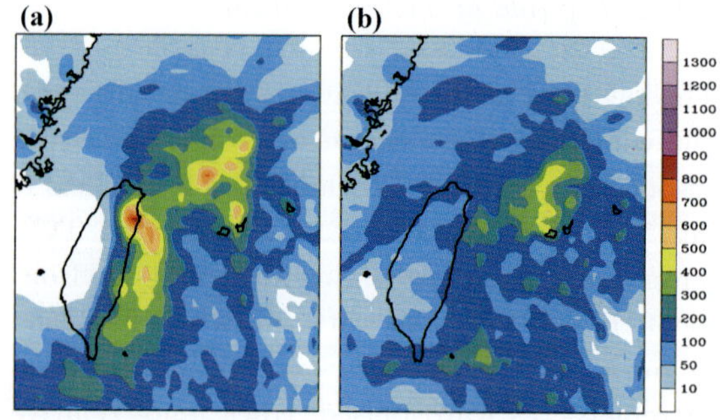

Fig. 12. 4-day accumulated rainfall in (a) m031 and (b) in the terrain-removed run (m031_TR).

5. Summary and Discussion

Quantitative precipitation forecast of the typhoon-induced rainfall over Taiwan has always been a challenging task, especially for cases with higher uncertainty in track forecasts.

In this study, we first quantitatively evaluate the relative contribution of individual dynamical systems to the TC's steering flow, and identify the possible factors leading to Megi's sudden track change based on piecewise PV diagnosis. Our analysis shows that the overall movement of Megi is in good agreement to the direction of steering flow associated with the monsoon trough. And the northward change in steering flow occurs when positive PV perturbations embedded in the monsoon trough move from the south to the west of TC. The upper-level trough plays a role in enhancing Megi's northward movement at a later time. The analysis also attributes the initial deceleration of Megi mainly to the retreat of subtropical high and the approach of upper-level trough.

In the second part of our study, the quantitative evaluation of precipitation forecast based on the equitable threat score (ETS) shows that the TC track is not the only dominant factor affecting the performance of rainfall forecast in this remote rainfall event, especially at heavy rainfall thresholds. When Megi is about to recurve over the South China Sea, the outer circulation of the TC helps transport moisture northward, causing a low-level moisture convergence with the northeasterly monsoon. However, the deep convection produced by the low-level convergence is not the main factor leading to this heavy rainfall event. It is the prevailing wind that further advects the moisture inland, and the northeasterly wind is especially favorable for heavy rainfall because it can advect the moisture to steeper mountain areas in Yilan, where the stronger orographic lifting effect can result in greater precipitation. Overall, in this study we highlight that the uncertainty of remote rainfall simulation is quite large, not strongly dependent on storm track as in most of the landfalling cases and can be affected by the TC size, strength of monsoon system, the direction of prevailing wind, the orientation and slopes of windward mountains, and the effects of orographic lifting.

This study shows that ensemble simulations can provide new insights into the uncertainty and predictability for both understanding and forecasting tracks and the typhoon-induced remote impact on rainfall. A more detailed discussion on the remote precipitation mechanism shown in this article has been published in Chen and Wu (2016).

Acknowledgments

This work is supported by the National Science Council of Taiwan through Grant MOST 103-2628-M-002004 and the Office of Naval Research through Grant N62909-13-1-N073.

References

Chen, T.-C. and C.-C. Wu, 2016: The remote effect of Typhoon Megi (2010) on the heavy rainfall over northeastern Taiwan. *Mon. Wea. Rev.*, **144**, 3109–3131.

Davis, C. A., 1992: Piecewise potential vorticity inversion. *J. Atmos. Sci.*, **49**, 1397–1411.

Galarneau, T. J., L. F. Bosart, and R. S. Schumacher, 2010: Predecessor rain events ahead of tropical cyclones. *Mon. Wea. Rev.*, **138**, 3272–3297.

Lin, Y.-L., S. Chiao, T.-A. Wang, M. L. Kaplan, and R. P. Weglarz, 2001: Some common ingredients for heavy orographic rainfall. *Wea. Forecast.*, **16**, 633–660.

Qian, C., F. Zhang, B. W. Green, J. Zhang, and X. Zhou, 2013: Probabilistic evaluation of the dynamics and prediction of super Typhoon Megi (2010). *Wea. Forecast.*, **28**, 1562–1577.

Schaefer, J. T., 1990: The critical success index as an indicator of warning skill. *Wea. Forecast.*, **5**, 570–575.

Wang, Y., Y. Wang, and H. Fudeyasu, 2009: The role of Typhoon Songda (2004) in producing distantly located heavy rainfall in Japan. *Mon. Wea. Rev.*, **137**, 3699–3716.

Wu, C.-C. and Y.-H. Kuo, 1999: Typhoons affecting Taiwan: Current understanding and future challenges. *Bull. Amer. Meteor. Soc.*, **80**, 67–80.

______, T.-H. Yen, Y.-H. Kuo, and W. Wang, 2002: Rainfall simulation associated with Typhoon Herb (1996) near Taiwan. Part I: The topographic effect. *Wea. Forecast.*, **17**, 1001–1015.

______, T.-S. Huang, W.-P. Huang, and K.-H. Chou, 2003: A new look at the binary interaction: Potential vorticity diagnosis of the unusual southward movement of Typhoon Bopha (2000) and its interaction with Typhoon Saomai (2000). *Mon. Wea. Rev.*, **131**, 1289–1300.

______, K. K. W. Cheung, and Y.-Y. Lo, 2009: Numerical study of the rainfall event due to the interaction of Typhoon Babs (1998) and the northeasterly monsoon. *Mon. Wea. Rev.*, **137**, 2049–2064.

______, G.-Y. Lien, J.-H. Chen, and F. Zhang, 2010: Assimilation of tropical cyclone track and structure based on the Ensemble Kalman Filter (EnKF). *J. Atmos. Sci.*, **67**, 3806–3822.

______, S.-G. Chen, S.-C. Lin, T.-H. Yen, and T.-C. Chen, 2013: Uncertainty and predictability of tropical cyclone rainfall based on ensemble simulations of Typhoon Sinlaku (2008). *Mon. Wea. Rev.*, **141**, 3517–3538.

Zhang, F., Z. Meng, and A. Aksoy, 2006: Test of an ensemble Kalman filter for mesoscale and regional-scale data assimilation. Part I: Perfect model experiments. *Mon. Wea. Rev.*, **134**, 722–736.

______, Y. Weng, Y.-H. Kuo, J. S. Whitaker, and B. Xie, 2010: Predicting Typhoon Morakot's catastrophic rainfall with a convection-permitting mesoscale ensemble system. *Wea. Forecast.*, **25**, 1816–1825.

Part IV
Climate Change and Decadal Variability

Chapter 22

Global Monsoon: Interannual-to-Centennial Variability and Future Changes

Bin Wang

Department of Meteorology and International Pacific Research Center, University of Hawaii, Honolulu, HI 96822, USA

wangbin@hawaii.edu

The global monsoon (GM) is an evolving concept that has not yet been fully recognized, especially whether its interannual-multidecadal climate variability can be viewed as one of the major modes of the global climate system. In this chapter we focus on reviewing the common variability and linkages among all regional monsoons. We begin with a brief review of the concept of the GM and elaborate what the essence of the GM is and why it is a defining feature of the Earth's climatology. The GM variability and change are driven by two sets of factors. One is external forcing, both natural (solar orbital change, volcanic aerosols) and man-made (greenhouse gases concentration, anthropogenic aerosols, land use, etc.). The other is internal feedback processes within the coupled Earth's climate system. Evidence is reviewed to show that on the interannual-to-multidecadal time scale, the monsoon variations are not merely of isolated regional characters; rather, the monsoon can be driven by common internal feedback processes on global scale such as El Nino-Southern Oscillation (ENSO), mega-ENSO, and Atlantic Multidecadal Oscillation (AMO). The changes in solar radiation flux and volcanic forcing is shown to generate a coherent variability of all regional monsoons; thus the GM represents a leading mode of variability on centennial or longer time scales in response to changing solar forcing. In addition, we review how anthropogenic forcing affects future change of GM precipitation. It is important to understand the difference in GM responses to anthropogenic and natural forcings. In summary, it is argued that the GM is not only a defining feature of the Earth's climatology but also a major mode of Earth's climate variability across time scales ranging from interannual to centennial.

1. Concept of the Global Monsoon

Monsoon climate is characterized by both the annual reversal of surface winds and the contrast between rainy summer and dry winter (Webster 1987). The classical monsoon definition was solely based on annual reversal of surface winds (Ramage 1971), which confines monsoons to Asia, Australia, and tropical Africa and Indian Ocean, all in the eastern hemisphere. Wang (1994) first attempted to delineate monsoon domains based on synthesis of characteristics of seasonal distribution of deep convection (a surrogate to rainfall); the resulting monsoon domains extend from eastern to western hemispheres, including North and South American and southern African monsoons.

Figure 1 shows the monsoon precipitation domains defined by (a) the local summer-minus-winter precipitation exceeds 300 mm and (b) the local summer precipitation exceeds 55% of the annual total (Wang and Ding 2008). Here the local summer denotes May through September for the northern hemisphere (NH) and November through March for the southern hemisphere (SH). The defined monsoon domains include all major monsoon regions: Asia, Australia, Northern Africa, Southern Africa, North America, and South America.

The global monsoon (GM) areas occupies about 19% of the total surface area of the

The Global Monsoon System: Research and Forecast (3rd Edition)
Edited by C. P. Chang et al.

Earth, while the GM rainfall accounts for about 31% of the total amount of global precipitation (Wang and Ding 2008). Note that the monsoons entail substantial oceanic regions, including the marginal seas in South and East Asia, and the tropical western North Pacific (WNP), southwest Indian Ocean, and tropical eastern North Pacific Ocean. The oceanic monsoons are integral parts of the monsoon systems. Note as well that dry regions are generally located to the west and poleward side of the monsoons due to the descent resulting from the interaction between monsoon heating-induced Rossby waves and the mean westerly flow on their poleward side (Hoskins 1996, Rodwell and Hoskins 1996). The monsoon and desert together occupy majority of the global tropics and subtropics.

The GM has intimate relationship with annual migration of the inter-tropical convergence zone (ITCZ) (Gadgil *et al.* 1988). However, the GM should also be distinct from the ITCZ. The bulk (about three quarters) of the ITCZ (from Africa to Australia and over America) is embedded within the monsoon regions and exhibits prominent seasonal migration (Fig. 1). The portion of ITCZ in the monsoon regions has large annual march across the equator and is also called monsoon trough. Conversely, in the central Pacific Ocean, where the ITCZ's annual displacement is less than a few degrees of latitude, the zonal winds (trade winds) on both sides of the ITCZ do not change directions between summer and winter, and therefore monsoon is absent. This portion of the ITCZ can be referred to as "trade wind convergence zone" (Wang 1994). Monsoon trough occupies more than third quarters of the ITCZ, thus global monsoon largely drives ITCZ movement. Since the ITCZ represents the ascent and descent branch of the Hadley circulation, the Hadley circulation is also largely driven by the GM.

Regional monsoons are bonded by the global divergent circulation. Trenberth *et al.* (2000) depicted GM as the global-scale seasonally varying overturning circulation throughout the tropics. Wang and Ding (2008) demonstrate that GM can be quantitatively defined by the dominant mode of the annual variation of precipitation and circulation in the global tropics and subtropics. In physical essence, GM is a forced response of the coupled climate system to the annual cycle of insolation. Land-sea thermal contrast, moisture processes, topography and Earth's rotation are all critical in determining monsoon rainfall-circulationpatterns and vigor.

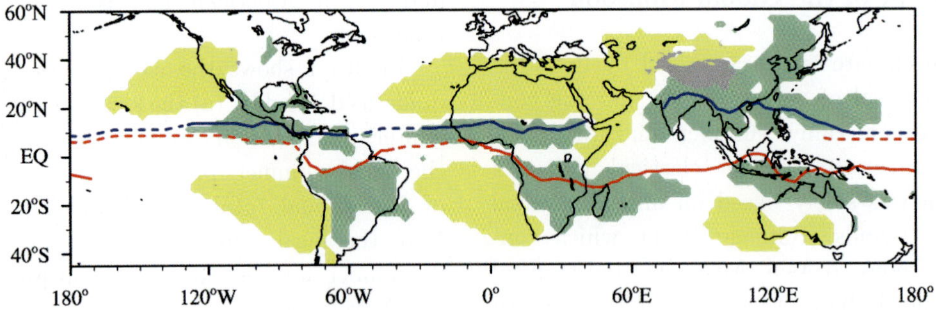

Fig. 1. The GM precipitation domain defined by local summer-minus-winter precipitation exceeds 300 mm and the local summer precipitation exceeds 55% of the annual total (in green). Summer denotes May through September for the northern hemisphere and November through March for the southern hemisphere. The dry regions, where the local summer precipitation is less than 150 mm are shown (yellow). The blue lines indicate ITCZ position for August; the red lines indicate ITCZ position for February (solid for monsoon trough and dashed for trade wind convergence). In February, double ITCZ exist between 140E and 170W. The 3000 m height contour surrounding Tibetan Plateau is shaded. The merged Global Precipitation Climatology Project/Climate Prediction Center Merged Analysis of Precipitation data and ERA interim data were used for 1979–2012.

Regional monsoons exhibit distinct features due to their indigenous land-sea configuration and particular atmosphere-ocean-land interaction involved. This is why the monsoons have been studied intensively on regional scales. The Indian monsoon or South Asian monsoon has the strongest and most typical monsoon climate on Earth. The East Asian monsoon is a subtropical monsoon which distinguishes from tropical monsoons and is uniquely driven by east-west land-ocean thermal contrast. There is linkage among some regional monsoon systems. For instance the interannual variations of the Indian summer monsoon are positively correlated with summer monsoon in northern China.

In summary, the GM is a planetary scale circulation system with a seasonal reversal of three-dimensional monsoon circulation that is accompanied by migration of the monsoon rainfall zones. However, it remains a debated issue as to what extent and at which time scales GM can be viewed as a major mode of climate variability (Wang *et al.* 2014).

2. Future Projection: How Does Anthropogenic Forcing Affect GM Precipitation?

The observed records contain both internal modes of variability and externally-forced change. To better understand the interannual-to-multideacadal variability of the GM, it is advantageous to first identify the signals of anthropogenic change.

How much GM precipitation would increase for a given temperature increase due to global warming and how would the spatial distribution of precipitation change? To assess current fidelity and future conditions, 20 coupled models that participated in the phase five of Coupled Model Intercomparison Project (CMIP5) are examined, mainly their historical runs for 1850–2005 and their Representative Concentration Pathway (RCP) 4.5 runs (assuming that radiative forcing will stabilize at about 4.5 Wm^{-2} after 2100) for 2006–2100 (Lee and Wang 2014). To select only those projections from models that best depict monsoons in the current climate, only best four CMIP5 models are selected based on the models' performance on a metrics of climatology fields (refers to Lee and Wang 2014 for details). The four best models' multi-model ensemble projects an overall intensification of summer monsoon rainfall, signifying an amplifying annual variation of the hydrological cycle. Meanwhile, the precipitation in the global subtropical desert and trade wind regions will decrease due to a monsoon-desert coupling mechanism mentioned in the previous section. The annual mean GM precipitation shows about 1.9% increase per Kelvin global mean temperature rise with a huge difference between the NH and SH. While the NH monsoon precipitation shows a 3.1% increase per Kelvin global mean temperature increase, there is a slight decreasing trend in the future projection of the SH monsoon precipitation (Fig. 2). The significant increase of the NH monsoon precipitation, is mainly attributed to the increase of the temperature difference between the NH and SH (warm NH-cold SH), enhancement of the Hadley circulation, and atmospheric moistening, against the stabilization of the lower as well as upper troposphere (Lee and Wang 2014). The warm NH-cold SH tends to reduce the SH summer monsoon, which accounts for the large difference between the NH and SH monsoon changes. There will also be a prominent east-west asymmetry in future changes with moistening of the eastern hemisphere monsoons and drying of the western hemisphere monsoon, especially in the NH. The reasons need further investigated, but may have to do with the land distribution between the northern eastern and western hemisphere. The land monsoon domain over Asia will expand westward by about 10%. The NH monsoon onset will be advanced and withdrawal will be delayed. However, the uncertain role of aerosols in general and carbon aerosols in particular, complicates

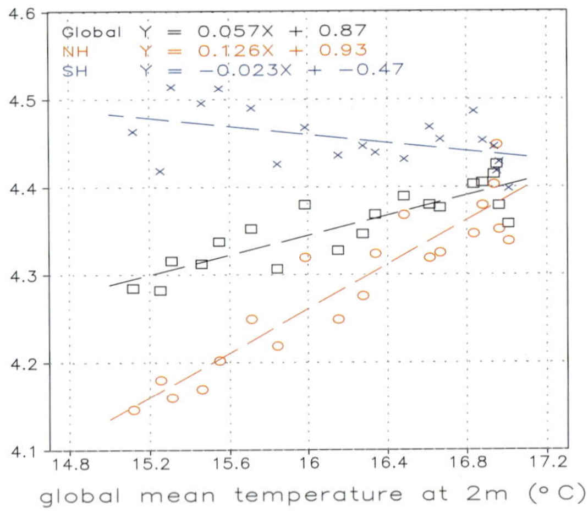

Fig. 2. Scatter diagram showing area mean and monsoon precipitation rates as functions of global mean 2 m-air temperature for the period of 2000–2100 in the CMIP5 best 4 models' multi-model ensemble over the globe (square and black line), northern hemisphere (circle and red line), and southern hemisphere (cross and blue line), respectively. 5-year average was applied. The difference between the NH and SH slopes is significant at 95% confidence level for both area mean and monsoon precipitation.

future projections of monsoon precipitation over land, particularly for Asia. Further understanding of the driving mechanisms behind monsoon changes holds a key for their faithful prediction.

3. Interannual to Multidecadal Variations Driven by Internal Feedback Processes

The monsoon variability on interannual to multidecadal time scales arise primarily from dynamic and thermodynamic feedback processes within the components (e.g. atmosphere, ocean, land, and Sea ice) of the Earth climate system. Monsoon interannual-multidecadal variations have been studied primarily on regional scales due to their indigenous characteristics associated with specific land-ocean configuration and differing feedback processes. These include South Asia (e.g. Webster *et al.* 1998), East Asia (e.g. Tao and Chen 1987), Australia (e.g. McBride 1987), Africa (e.g. Nicholson and Kim 1997), North America (e.g. Higgins *et al.* 2003), and South America (e.g. Zhou and Lau 1998).

To what extent can the interannual variations of the GM precipitation be driven by internal feedback processes in the climate system? Wang *et al.* (2012) has shown that from one monsoon year (May to the next April) to the next, most continental monsoon regions, separated by vast areas of arid trade winds and deserts vary in a cohesive manner driven by El Niño-Southern Oscillation (ENSO). The ENSO has tighter regulation on the northern hemisphere summer monsoon (NHSM) than on the southern hemisphere summer monsoon (SHSM). ENSO also tends to more strongly influence continental monsoon rainfall than the oceanic monsoons. As a result, the total amount of the NHSM land rainfall is highly related to ENSO with correlation coefficient $r = 0.73$ for 1979–2008. Thus, regional monsoons are coordinated not only by external (e.g. orbital) forcing but also by internal feedback processes, such as ENSO.

On interdecadal timescales, numerous studies have investigated the linkage between regional monsoons and other major modes of climate variability. For instance, Indian summer monsoon precipitation has been shown to exhibit a correlation with the North Atlantic Oscillation (Goswami *et al.* 2006) and multidecadal variations of the Pacific SST (Meehl and Hu 2006), northern China's rainfall is correlated with the Pacific Decadal Oscillation (Qian and Zhou 2013), East Asian summer monsoon is related to Atlantic multidecadal Oscillation (Lu *et al.* 2006), west African and North American monsoon variability is related to the Atlantic Multidecadal Oscillation (AMO) (Sutton and Hodson 2005; Zhang and Delworth 2006; Gaetani and Mohino 2013), and Australian summer monsoon was linked to Pacific decadal oscillation (Power *et al.* 1999a). A variety of decadal and interdecadal variations of regional monsoons has been identified, with differing periodicity and phase change points (Yim *et al.* 2013). However, a coherent global structure and the underlying causes of GM interdecadal variability have yet to be widely studied.

The total amount of global land monsoon rainfall during the period of 1948–2003 seems to exhibit an interdecadal fluctuation with a decreasing trend mainly due to weakening West African and South Asian monsoons (Wang and Ding 2006). These variations were found reproducible in global climate models when observed global SST anomalies are specified (Zhou *et al.* 2008). These studies of the trends in relatively short precipitation records have not been able to nail down whether these trends are part of longer-period fluctuation.

The global precipitation observation over both land and ocean are available only after the late 1970's. Using this limited dataset, Wang *et al.* (2013) has shown that during this recent period, a coherent decadal change of precipitation and circulation emerges in the entirety of NHSM system. The NHSM, as well as Hadley and Walker circulations have all shown substantial intensification during boreal summer. Using the CMIP 5 projected anthropogenic forced change, they estimated that only 1/3 of the trend during 1979–2013 is likely related to global warming. The other portion is a part of multidecadal variation as elaborated in the next paragraph.

Using a circulation index for the NHSM, it is possible to investigate the decadal-multidecadal variability of the GM as reanalysis data has become available for the past century (e.g. monthly circulation data taken from 20th century reanalysis for 1871–2010, Compo *et al.* 2011). Wang *et al.* (2013) used the NHSM circulation index defined by the vertical shear of zonal winds between 850 hPa and 200 hPa averaged in (0°–20°N, 120°W–120°E) (Fig. 3a). The NHSM circulation index is highly correlated with the NHSM rainfall intensity over the modern record ($r = 0.85$ for 1979–2011) and also positively correlated with Walker circulation strength. It was demonstrated that the NHSM circulation has experienced large-amplitude multidecadal fluctuations since 1871, which is associated with "Mega-ENSO" defined by the SST anomaly difference between the western Pacific K-shape area and eastern Pacific triangle (Wang *et al.* 2013) (Fig. 3b). What does mega-ENSO index mean? The mega-ENSO index is defined by using the total anomaly field, rather than EOF decomposition (such as PDO or IPO). The mega-ENSO is an integrated measure of the year-to-year variability of the tropical Pacific (the ENSO signal), the Pacific Decadal Oscillation (PDO, Mantua and Hare 2002), and the Interdecadal Pacific Oscillation (IPO, Power *et al.* 1999b). For more details of the Mega-ENSO the reader is referred to Wang *et al.* (2013).

On the decadal time scale, as measured by the 3-y running mean time series, the long-term variation of the NHSM circulation index is strongly correlated to the Mega-ENSO index ($r = 0.77$, for 1958–2010) using ERA-40 reanalysis data (Uppsala *et al.* 2005) and $r = 0.62$ for the period 1871–2010 by using the 20th century reanalysis dataset (Fig. 3c). Physically, the eastern Pacific cooling and the western Pacific warming are consistent with a strengthening of the Pacific Subtropical Highs in both the North and South Pacific and their associated trade winds, causing moisture to converge into the Asian and African monsoon regions and contribute to the intensification of NHSM rainfall. It was also found that the NHSM strength is significantly linked to Atlantic multidecadal oscillation (AMO) (Enfield *et al.* 2001).

3.1. *Centennial variation over the past millennium: Effects of solar-volcanic forcing*

Is the GM precipitation a useful concept for examining response of the Earth's climate to external forcing? A set of millennial simulations

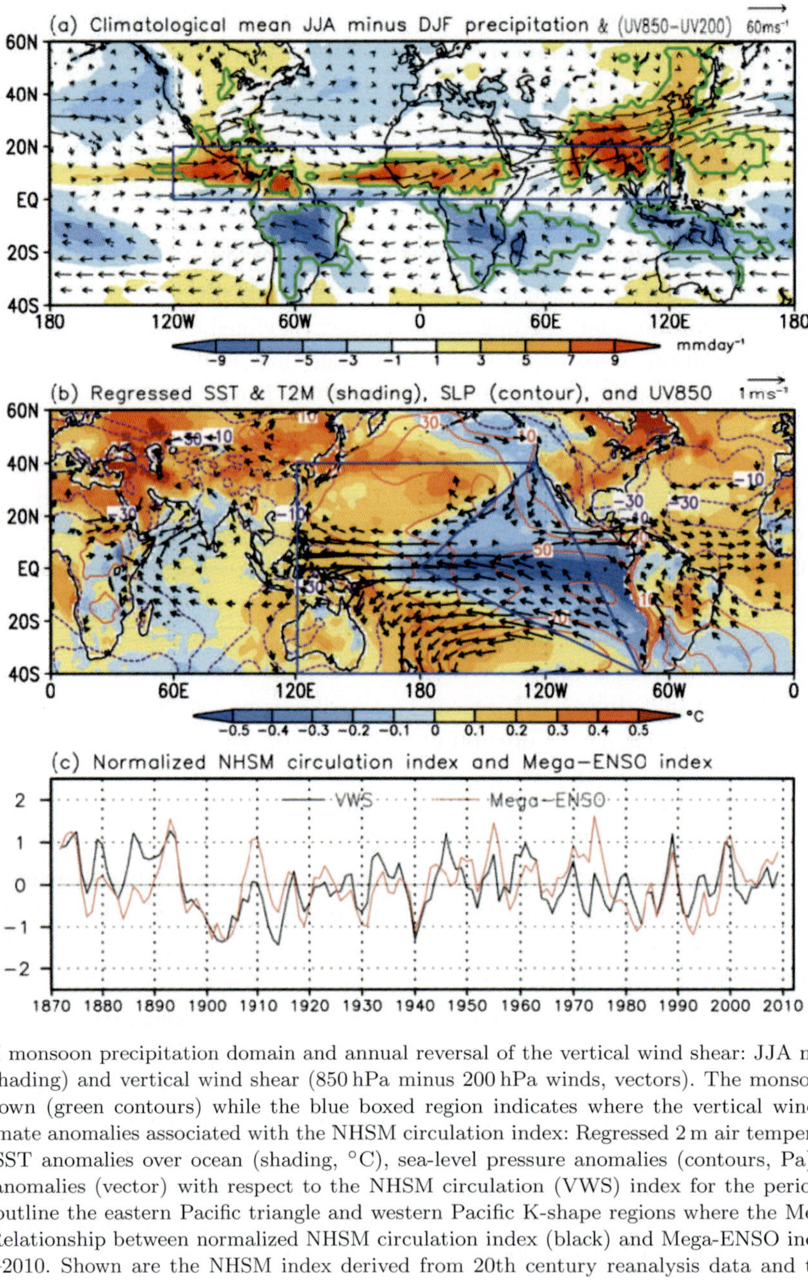

Fig. 3. (a) NH monsoon precipitation domain and annual reversal of the vertical wind shear: JJA minus DJF mean rainfall (color shading) and vertical wind shear (850 hPa minus 200 hPa winds, vectors). The monsoon precipitation domains are shown (green contours) while the blue boxed region indicates where the vertical wind shear index is defined. (b) Climate anomalies associated with the NHSM circulation index: Regressed 2 m air temperature anomalies over land and SST anomalies over ocean (shading, °C), sea-level pressure anomalies (contours, Pa) and significant 850 hPa wind anomalies (vector) with respect to the NHSM circulation (VWS) index for the period of 1979–2010. The blue lines outline the eastern Pacific triangle and western Pacific K-shape regions where the Mega-ENSO index is defined. (c) Relationship between normalized NHSM circulation index (black) and Mega-ENSO index (red) for the period of 1871–2010. Shown are the NHSM index derived from 20th century reanalysis data and the Mega-ENSO index derived from the HadISST dataset. All data used are 3-year running means.

with the coupled climate model ECHO-G forced by changes in solar radiation, volcanic aerosols and greenhouse gas (GHG) concentration (Legutke and Voss 1999) provides useful insight into GM rainfall variability on centennial-submillennial time scale. The ECHO-G model-simulated global precipitation climatology (annual mean, annual cycle, and GMP domain etc.) is comparable to the assimilated climatology in modern reanalysis and comparable

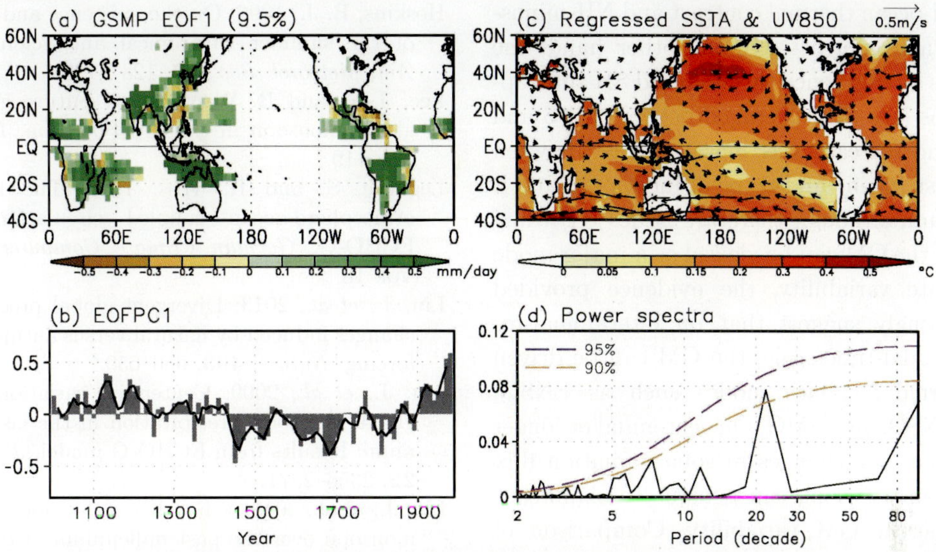

Fig. 4. The leading EOF mode of the global summer monsoon precipitation (GSMP) variation over the past millennium: (a) spatial structure, (b) the corresponding principal component (PC); the smoothed curve is 3-decade running mean, (c) regressed SST and 850 hPa wind anomalies, shown are only wind vectors that are significant above 90% confidence level; and (d) the corresponding PC spectra. The fraction on the top of panel (a) denotes the fractional variance. The data were decadal mean time series derived from ECHO-G forced (ERIK) simulation. The correlation matrix was used for EOF analysis.

to the top three climate models that participated in the CMIP5 (Liu *et al.* 2013).

Figure 4 shows that the simulated leading pattern of centennial variability of the monsoon precipitation in global monsoon domain is characterized by a nearly uniform increase of precipitation across all regional monsoons. The corresponding principal component (PC1) shows a slow variation on centennial-millennial time scale with enhanced global summer monsoon precipitation generally occurring in the medieval warming period (MWP) from 1000–1400 AD and the 20th century (1900–1990 AD), along with a suppressed monsoon rainfall occurring in the Little Ice Age (LIA) from 1450–1850 AD. The total amount of global summer monsoon precipitation is highly related to the global mean temperature change, indicating a close relationship between the global warming and change of the GM or the annual cycle of the climate system. Spectral analysis reveals a bicentennial peak significant at 90% significance level. The sub-millennial variation and the bicentennial peak are primarily a response to the effective solar forcing because the solar-volcanic forcing shows a similar sub-millennial variation and a bi-centennial peak (Liu *et al.* 2009).

Of note is that the increase of global summer monsoon precipitation during the 20th century is unprecedented, resulting in a maximum that is significantly stronger than that during the MWP. Since the effective solar forcing in the MWP is about the same as in the 20th century, the excessive warming in the 20th century reflects the impact of the GHG forcing. Note also that this mode is totally absent in the forcing-free run, indicating that it is externally forced by effective solar radiation (Solar-Volcanic) and GHG forcing. In addition, Liu *et al.* (2012) found that the NHSM precipitation responds to GHG forcing more sensitively, while the SHSM precipitation responds to the Solar-volcanic radiative forcing more sensitively. The NHSM is enhanced by increased

NH land-ocean thermal contrast and NH-minus-SH thermal contrast. On the other hand, the SHSM is strengthened by enhanced SH subtropical highs (the SLP averaged between 20°S and 40°S) and the east-west mass contrast between Southeast Pacific and tropical Indian Ocean.

In sum, although there are issues regarding whether the GM can be viewed as a major mode of climate variability, the evidence provided here strongly suggest that on interannual-to-multidecadal time scale, the GM can be driven by internal feedback modes such as ENSO, mega-ENSO, and AMO. On centennial or longer time scales the changes in solar radiation flux and volcanic forcing can generate a significant and coherent GM variability. Comparison of the differences and commonality among different regional monsoons will deepen our understanding of the fundamental processes that drives monsoon variability.

References

Cheng, Q. and T. Zhou, 2013: Multidecadal variability of North China aridity and its relationship to PDO during 1900–2010. *J. Climate*, **27**, 1210–1222.

Compo, G. P., *et al.*, 2011: The twentieth century reanalysis project. *Quart. J. Roy. Meteor. Soc.*, **137**, 1–28.

Enfield, D. B., A. M. Mestas-Nuñez, and P. J. Trimble, 2001: The Atlantic Multidecadal Oscillation and its relation to rainfall and river flows in the continental U.S. *Geophys. Res. Lett.*, **28**, 2077–2080.

Gadgil, S., 1988: Recent advances in monsoon research with particular reference to the Indian monsoon. *Aust. Met. Mag.*, **36**, 193–204.

Gaetani, M. and E. Mohino, 2013: Decadal prediction of the Sahelian precipitation in CMIP5 simulations. *J. Climate*, **26**, 7708–7719.

Goswami, B. N., *et al.*, 2006: A physical mechanism for North Atlantic SST influence on the Indian summer monsoon. *Geophys. Res. Lett.*, **33**, L02706.

Higgins, R. W., A. Douglas, and A. Hanmann, 2003: Progress in Pan American CLIVAR research: The North American monsoon system. *Atmosfera*, **16**, 29–65.

Hoskins, B. J., 1996: On the existence and strength of the summer subtropical anticyclones. *Bull. Am. Meteorol. Soc.*, **77**, 1287–1291.

Lee, J.-Y. and B. Wang, 2014: Future change of global monsoon in the CMIP5. *Clim. Dyn.*, **42**, 101–119.

Legutke, S. and R. Voss, 1999: The Hamburg atmosphere-ocean coupled circulation model ECHO-G, *German Climate Computer Center*, Vol. 18

Liu, J., *et al.*, 2013: Divergent global precipitation changes induced by natural versus anthropogenic forcing. *Nature*, **493**, 656–659.

Liu, J., *et al.*, 2009: Centennial variations of the global monsoon precipitation in the last millennium: Results from ECHO-G model. *J. Climate*, **22**, 2356–2371.

Liu, J., *et al.*, 2012: What drives the global summer monsoon over the past millennium? *Clim. Dyn.*, **39**, 1063–1072.

Lu, R., B. Dong, and H. Ding, 2006: Impact of the Atlantic Multidecadal Oscillation on the Asian summer monsoon. *Geophys. Res. Lett.* **33**, L24701.

Mantua, N. and S. Hare, 2002: The Pacific decadal oscillation. *J. Oceanogr.*, **58**, 35–44.

McBride, J. L., 1987: The Australian summer monsoon. In: C. P. Chang and T. N. Krishnamurty (eds.), *Monsoon Meteorology*. Oxford University Press, Oxford, pp. 203–231.

Meehl, G. A. and A. Hu, 2006: Megadroughts in the Indian Monsoon Region and Southwest North America and a mechanism for associated Multi-decadal Pacific Sea Surface Temperature anomalies. *J. Climate*, **19**, 1605–1623.

Nicholson, S. E. and J. Kim, 1997: The relationship of the El Nino Southern oscillation to African rainfall. *Int. J. Climatol.*, **17**: 117–135.

Power, S., *et al.*, 1999a: Decadal climate variability in Australia during the twentieth century. *Int. J. Climatol.*, **19**, 169–184.

Power, S., *et al.*, 1999b: Inter-decadal modulation of the impact of ENSO on Australia. *Clim. Dyn.*, **15**, 319–324.

Ramage, C. S., 1971: *Monsoon Meteorology* (International Geophysics), **15**, Academic Press, San Diego.

Rodwell, M. J. and B. J. Hoskins, 1996: Monsoons and the dynamics of deserts. *Quart. J. Roy. Meteor. Soc.*, **122**, 1385–1404.

Sutton, R. T. and D. L. R. Hodson, 2005: Atlantic Ocean Forcing of North American and European Summer Climate. *Science*, **309**, 115–118.

Tao, S. and L. Chen, 1987: A review of recent research on the East Asian summer monsoon in China. In: C. P. Chang and T. N. Krishnamurti (eds.), *Monsoon Meteorology*. Oxford University Press, Oxford, pp. 60–92.

Trenberth, K. E., D. P. Stepaniak, and J. M. Caron, 2000: The global monsoon as seen through the divergent atmospheric circulation. *J. Climate*, **13**, 3969–3993.

Uppala, S. M., *et al.*, 2005: The ERA-40 re-analysis. *Quart. J. Roy. Meteor. Soc.*, **131**, 2961–3012.

Wang, B., 1994: Climatic regimes of tropical convection and rainfall. *J. Climate*, **7**, 1109–1118.

Wang, B. and Q. Ding, 2006: Changes in global monsoon precipitation over the past 56 years. *Geophys. Res. Lett.*, **33**, L06711.

Wang, B. and Q. Ding, 2008: Global monsoon: Dominant mode of annual variation in the tropics. *Dynam. Atmos. Oceans*, **44**, 165–183.

Wang, B., *et al.*, 2012: Recent change of the global monsoon precipitation, 1979–2008. *Clim. Dyn.*, **39**, 1123–1135.

Wang, B., *et al.*, 2013: Northern hemisphere summer monsoon intensified by mega-El Niño/southern oscillation and Atlantic multidecadal oscillation. *Proc. Natl. Acad. Sci.*, **110**, 5347–5352.

Wang, P. X., B. Wang, H. Cheng, J. Fasullo, Z. T. Guo, T. Kiefer, and Z. Y. Liu, 2014: The global monsoon across time scales: Is there coherent variability of regional monsoons? *Clim. Past*, **10**, 2007–2052, 2014, www.clim-past.net/10/2007/2014/doi:10.5194/cp-10-2007-2014.

Webster, P. J., 1987: The elementary monsoon. In: J. S. Fein and P. L. Stephens (Eds.), *Monsoons*. John Wiley & Sons, New York, pp. 3–32.

Webster, P. J., *et al.*, 1998: Monsoons: Processes, predictability, and the prospects for prediction. *J. Geophys. Res.*, **103**, 14451–14510.

Yim, S.-Y., B, Wang, and M. Kwon, 2014: Interdecadal change of the controlling mechanisms for East Asian early summer rainfall variation around the mid-1990s. *Clim. Dyn.*, **42**, 1325–1333.

Zhou, J. and K. M. Lau, 1998: Does a monsoon climate exist over South America? *J. Climate*, **11**, 1020–1040.

Zhou, T., *et al.*, 2008: Ocean forcing to changes in global monsoon precipitation over the recent half-century. *J. Climate*, **21**, 3833–3852.

Chapter 23

Interdecadal Variability of the Asian Summer Monsoons

A. G. Turner
NCAS-Climate/Department of Meteorology, University of Reading, Reading, Berkshire RG6 6BB, UK
a.g.turner@reading.ac.uk
www.met.reading.ac.uk

B. Wang
International Pacific Research Center, School of Ocean and Earth Science and Technology, University of Hawaii, Honolulu, Hawaii 96822, USA
wangbin@hawaii.edu
iprc.soest.hawaii.edu

The Asian monsoon represents one of the major modes of the seasonal cycle on the planet and undergoes variations on intraseasonal through interannual time scales that affect the water supply for more than half the global population. But the monsoon also undergoes appreciable variability on decadal time scales, noted both in the Asian monsoon and also as part of more coherent variations in the global monsoon. In the observed data, large variations on multi-decadal time scales are present but are confounded by external drivers including anthropogenic aerosols and greenhouse gas emissions. This review examines some of the mechanisms linking internal multi-decadal ocean modes with Asian monsoon decadal variability, including from the Pacific and Atlantic. Since much focus is still being given to seasonal monsoon prediction, we also examine perturbations caused by long time-scale oceanic modes to the monsoon-ENSO teleconnection, affecting seasonal monsoon predictability. Finally, we examine new prospects for decadal monsoon prediction in research considering the CMIP5 decadal hindcasts and forecasts, which are particularly encouraging for Atlantic multi-decadal variation modes. If future work can establish links between synoptic or intraseasonal monsoon variability and longer-term variations in the mean state, then there is potential for a cascade of predictability, improving our forecasts of the monsoon.

1. Introduction

The Asian monsoon represents one of the dominant annual modes of climate variability on the planet and affects the lives of more than half of the global population, its seasonal variation in prevailing wind direction bringing significant rainfall during the summer months. The Asian monsoon spans around 90° of longitude and features within it numerous interactions from synoptic-scale monsoon depressions through intraseasonal variability (with northward and eastward propagating components). Together with the time-mean state, these can be driven on interannual or longer time scales by slowly-varying coupled ocean-atmosphere modes of variability such as ENSO.

How secular decadal variability, such as that arising in the large ocean basins, may affect the mean state Asian monsoon and aspects of its variability, potentially affecting monsoon predictability on subseasonal to seasonal time scales, is a topic of increasing study. We examine recent evidence for the effect of long time-scale

The Global Monsoon System: Research and Forecast (3rd Edition)
Edited by C. P. Chang et al.

ocean variability on the Asian monsoon and its variability, in addition to decadal variability forced from the extratropics or high latitudes.

But we first must understand the observed system in the context of human-induced external long-term variations in the monsoon. Studies using observed data are complicated both by interactions with and confusion arising from signals of anthropogenic change, the relatively short observational record, and other non-oscillating drivers such as volcanic forcing or localized and time-varying anthropogenic aerosol emissions. So we must also understand anthropogenic changes in the mean monsoon.

The global monsoon may also serve as a useful concept for understanding coordinated changes in all the planet's monsoon regions, using common drivers such as oceanic modes of variability. However there are important differences between regions and within the Asian monsoon belt.

A further portion of this review is devoted to the effect of decadal variability on the major monsoon teleconnections, including ENSO and snow cover, thereby acting to modulate seasonal monsoon predictability.

In order to translate the improved understanding of decadal monsoon variability into better impacts for society, we need to be able to make better predictions. So finally, we examine some of the new opportunities and challenges emerging in decadal monsoon prediction, spurred on by the 5th Coupled Model Intercomparison Project (CMIP5) decadal hindcast experiments.

This review will consider the above issues from the recent literature and of outcomes of the International Workshop on Interdecadal Variability of the Global Monsoons held in September 2012.[a] Section 2 will examine decadal monsoon change in the context of anthropogenic change while Sec. 3 will describe in detail the decadal variability in the mean monsoon and its drivers. In Sec. 4 we will set the decadal variability of the Asian monsoon in the context of the global monsoons. In Sec. 5 we discuss the impact of decadal variability on monsoon-ENSO teleconnections, thereby modulating the seasonal predictability of the monsoon. Finally, in Sec. 6 we discuss future prospects and challenges for decadal prediction. We will not review the extensive literature on variability in Asian palaeo-monsoon records.

2. Decadal Change in the Context of Anthropogenic Greenhouse Gas and Aerosol Forcing

In the observed system with competing internal modes of variability and externally-forced change, the interdecadal variability of the Asian monsoon can only be understood in the context of signals related to anthropogenic change.

Examination of future projections of Asian monsoon rainfall under idealised enhanced greenhouse gas forcing generally suggest increases in time-mean monsoon rainfall (e.g. as reviewed in Turner and Annamalai 2012), relating to increased moisture convergence over the warmer oceans. This signal is illustrated in Fig. 1, which shows the summer-mean (June to September) projection of Asian monsoon precipitation at the time of doubled CO_2 concentrations relative to pre-industrial conditions in transient integrations from 31 of the CMIP5 models.

Recent works have also shown that similar signals emerge by the end of the 21st century under more realistic projections based on emissions scenarios (representative concentration pathways; RCPs), for example, Kitoh *et al.* (2013). The Asian monsoon rainfall domain will expand over land by about 10%

[a]A workshop organised by the WCRP CLIVAR Asian-Australian Monsoon Panel, held at NUIST, Nanjing, China, 10–12 September 2012, http://www.clivar.org/organization/aamp/activities/AAMP_decadal_workshop

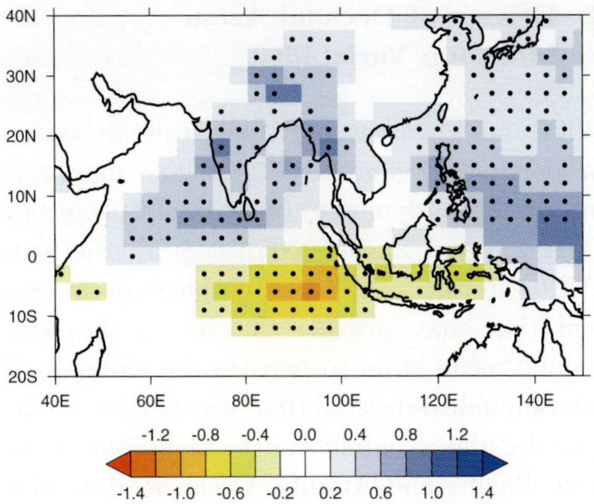

Fig. 1. Future projection of summer (JJAS) Asian monsoon precipitation computed as the increase in rainfall from pre-industrial control integrations to the time of CO_2 doubling in transient 1pctCO_2 CMIP5 integrations. The multi-model mean from 31 models is used; stippling indicates where more than 20 models agree on the sign of change. Units are mm/day.

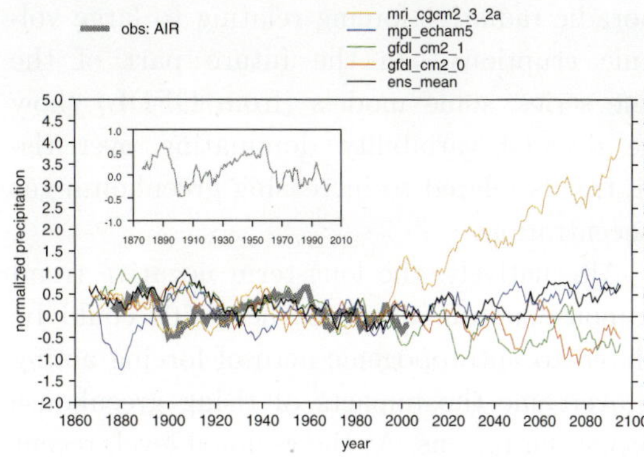

Fig. 2. Time series of summer (JJAS) monsoon rainfall over India in observations ("All-India Rainfall"; thick grey curve, also shown in the inset for clarity) and four CMIP3 models judged to reasonably simulate the South Asian monsoon (coloured lines). Curves are first normalized by their standard deviation before smoothing by an 11-year moving window. The multi-model mean is also shown (black). The historical and SRES-A1B model experiments are used. [Adapted from Turner and Annamalai (2012).]

(Lee and Wang 2014), and the total South Asian (East Asian) summer monsoon precipitation will increase by around 5.0%/°C (6.4%/°C) of global warming in the multi-model ensemble mean (Wang *et al.* 2014). Indices of heavy precipitation are also projected to increase much more than those for mean precipitation, especially over the Asian monsoon domain, indicating the strong sensitivity to global warming (Kitoh *et al.* 2013).

Comparing large signals in future projections is more straightforward than an analysis of the past, however. Given the consistency of future projections dominated by greenhouse gas forcing, the relative stability of Asian monsoon rainfall in recent observations has been puzzling (Goswami *et al.* 2006a). However, such analysis is strongly dependent on the region chosen and the availability of data. Further study of the observed rainfall record in several datasets reveals declining trends for South Asia (Bollasina *et al.* 2011). Figure 2 shows long-term monsoon rainfall changes (using South Asia as an example) in pre-industrial to future climates of four CMIP3 models judged to reasonably simulate the monsoon (Annamalai *et al.* 2007) and the observed record. Decadal-filtered rainfall in observations also reveals robust decadal variability, with several extended periods of strong and weak monsoons. The recent period since the 1950s shows a generally declining trend over the global land monsoon regions mainly from 1950 to 1980; after 1980 there is slight upward trend (Wang and Ding 2006). There is a broad upward trend from the beginning of the 20th century.

In the models shown in Fig. 2, strong decadal variability is present, although here we do not ascribe these variations to internal modes perhaps originating in the ocean or external factors. If the origin of these variations is internal modes of long-time scale variability in the oceans, the lack of phase coherence with the observed record reflects the experimental design in which the oceans are not initialized from the observed state. (Note that we could still expect coherent signals amongst the models and with observations relating to sporadic radiative forcing relating to large volcanic eruptions.) In the future part of the time series, some models (from GFDL) show the decadal variability dominating over rising trends related to increasing greenhouse gas concentrations.

Alternatively, the long-term negative trend in monsoon rainfall since the 1950s could be related to anthropogenic aerosol forcing acting to overcome the impacts of rising greenhouse gas concentrations. At the regional level, recent observations suggest that localized aerosol forcing, either from sulphur dioxide or black carbon emissions (resulting from rapidly increasing population and industrialization in South and East Asia) may be acting to weaken the monsoon over South Asia in particular (Ramanathan *et al.* 2005; Bollasina *et al.* 2011). This relates to negative radiative forcing acting to cool the surface and the inter-hemispheric temperature gradient, perturbing the large-scale driving mechanism of the monsoon. As we shall discuss later, aerosol forcing may also be implicated in driving multi-decadal oceanic modes of variability that then go on to modulate the monsoon. No such consensus signals emerge over East Asia.

These factors mean that disentangling secular modes of interdecadal variability in the observed Asian monsoon is a difficult task, pointing to the role of controlled experiments in modelling work to establish mechanisms involved in the absence of other factors. While assessments can be made of monsoon simulation and variability in contemporary periods of CMIP-type modelling (e.g. Sperber *et al.* 2013), it is more challenging to do so on decadal time scales owing to the difficulty in making direct comparisons to observations.

We therefore cannot ignore anthropogenic drivers when measuring decadal variability of the Asian monsoon. We should also bear in mind the uncertainties in the models' future projection. The uncertainty in precipitation is particularly large due to model cumulus parameterization schemes and aerosol effects.

3. Drivers of Decadal Asian Monsoon Variability

While on intraseasonal time scales a certain proportion of monsoon variability may arise from internal dynamics of the atmosphere (e.g. Goswami and Xavier 2005), it is internal variability in the coupled atmosphere-ocean system that may provide changes in the background mean state to perturb the monsoon on interannual-to-decadal time scales. Here we discuss decadal variability in the monsoons arising from Pacific and Atlantic Ocean modes, while we defer predictability arising from the Indian Ocean until later.

3.1. *The Pacific Ocean*

The El Niño-Southern Oscillation (ENSO) represents one of the best understood variations in lower boundary conditions of the atmosphere lending seasonal predictability to the tropics (Charney and Shukla 1981). The analog on decadal time scales is the Pacific Decadal Oscillation (PDO), which features a broadly similar warming pattern in the eastern tropical Pacific (although extending further poleward from the equator than El Niño) surrounded by a horseshoe-shaped cold anomaly (Mantua and Hare 2002).

Krishnan and Sugi (2003) studied the impact of the PDO (defined as 8-year smoothed SST measured over 45°N–45°S in the Pacific basin) on Indian rainfall since the 1870s. By compositing Indian rainfall station data on warm minus cold periods of the PDO, they showed typical decadal-mean declines of around 10%, or around 30% in the dry northwestern part of India.

Similarly in the East Asian monsoon, Cheng and Zhou (2014) found that the Palmer drought severity index (PDSI) in north China is significantly correlated with the PDO index. Approximately 70% of the drying trend in north China during 1960–90 originates from 50–70-yr multidecadal variability related to PDO phase changes.

An understanding of decadal variability in the Pacific Ocean can therefore yield significant signal in Asian monsoon rainfall.

3.2. *The Atlantic Ocean*

Despite the proximity of the Indo-Pacific basins to Asia, much research has focused on the Atlantic as a modulator of the monsoon on decadal time scales. Zhang and Delworth (2006) first noted that an idealised dumping of freshwater into the North Atlantic in a coupled model experiment could simulate a weakened thermohaline circulation (THC) that resulted in a redistribution of ocean heat content and a southward shift of the ITCZ. They argued that the resulting El Niño-like changes in the Pacific SST distribution and to the Walker Circulation led to a Rossby wave response that weakened both the circulation and precipitation of the Asian monsoon.

In the observed system, Goswami *et al.* (2006b) looked in more detail at the Atlantic multi-decadal oscillation (AMO) and determined an in-phase relationship between low-pass filtered AMO SST and Asian monsoon rainfall. (They also noted an inverse relationship on these time scales with Niño-3 region SST, emphasizing the interaction between ocean basins.) Goswami *et al.* (2006b) proposed that as the AMO weakens (cooling of the North Atlantic relative to the global mean surface temperature), it cools Eurasia thereby weakening the meridional tropospheric temperature gradient known to drive the Asian monsoon (Li and Yanai 1996; Xavier *et al.* 2007). The weakening of the meridional contrast in tropospheric temperature was particularly apparent in late summer, hastening the withdrawal of the monsoon.

The AMO link was explored further by Lu *et al.* (2006) who imposed AMO-related SST anomalies in the HadCM3 coupled model. Their positive AMO (warm North Atlantic, cold South Atlantic) led to stronger Southeast Asian and East Asian monsoons and an extension of the Indian monsoon, consistent with Goswami *et al.* (2006b). The tropospheric temperature gradient mechanism was particularly important for India while changes in other components of the Asian monsoon were attributed to coupled ocean-atmosphere feedbacks in the Indian and West Pacific Oceans.

3.3. *Tibetan Plateau forcing*

In addition to internal dynamical modes in ocean basins, long-time scale changes in the land surface may drive decadal variability of the Asian summer monsoon. This is particularly relevant over the Tibetan Plateau, marking the northern pole of the meridional tropospheric temperature contrast, where variations in snow cover have long been known to influence the monsoon through changes to this gradient (Blanford 1884).

Zhang *et al.* (2004) found that during spring (March–April), a sharp increase in Tibetan Plateau snow depth after the late 1970s was accompanied by increasing summer rainfall over the Yangtze River valley and a decrease on the southeast coast of China and over the Indochina peninsula. The excessive snowmelt results in surface cooling over and around the plateau and high pressure anomalies, possibly causing a northwestward extension of the western Pacific subtropical high in the subsequent summer. Additionally, the increased surface moisture supply provides more energy for the development of the eastward-migrating low-level vortex over the eastern flank of the Tibetan Plateau. Both factors lead to a wetter summer in the vicinity of the Yangtze River valley.

The cause of the snow depth change over the plateau was linked to extensive changes of circulation over Eurasia and the Indian Ocean that are associated with a climate shift in the North

Pacific after the mid-1970s (Trenberth and Hurrell 1994), suggesting that Pacific decadal change can also exert significant influence on the East Asian summer monsoon through changing snowfall over the Tibetan Plateau. On longer time scales however, anthropogenic warming may lead to more rapid melting of Himalayan and Tibetan Plateau snow cover during the seasonal cycle, providing another pathway mechanism for long time-scale monsoon variability.

4. Coherent Drivers of the Global Monsoon

So far numerous studies have investigated the linkage between regional monsoons and other major modes of SST decadal variability as reviewed in the previous sections. A variety of decadal and interdecadal variations of regional monsoons has been identified, with differing periodicity and phase change points (Yim *et al.* 2013). Liu and Chiang (2012) have suggested coordinated abrupt weakening of West African and Asian monsoon rainfall together since the 1960s relating to cooling in the extratropical North Atlantic, following similar mechanisms to those discussed earlier. Tropical Atlantic variability noted by Kucharski *et al.* (2009a) has also been determined to affect the African and Indian monsoons together. However, it remains an issue for debate as to what extent the regional monsoons vary in a cohesive manner.

Since the multi-decadal variability normally also involves larger three-dimensional spatial scales, its study for the monsoon will be more meaningful if we look at coherence on the global scale. Much has been achieved in documenting and understanding decadal variations of monsoon precipitation on individual monsoon regions but development of a coherent global picture of monsoon decadal variations helps to reveal shared interregional features and the common mechanisms that produce such variability. Considerable recent work has discussed the concept of the global monsoon (or its hemispheric components), generally defined as the leading solsticial mode of tropical precipitation variability (Wang and Ding, 2008), which corresponds to local summer minus winter mean precipitation. Since some of the decadal drivers outlined above may affect monsoons elsewhere than Asia, much can be learnt about the mechanisms involved using this approach as local small-scale features are smoothed out.

The total amount of global land monsoon rainfall during the period of 1948–2003 seems to exhibit an interdecadal fluctuation with a decreasing trend mainly attributable to the aforementioned weakening West African and South Asian monsoons (Wang and Ding, 2006). These variations were found to be reproducible in GCMs when forced by observed global SSTs over the same period (Zhou *et al.* 2008). However these studies attempting to reproduce observed precipitation trends over relatively short periods have not been able to nail down whether these signals form part of long-period oscillations.

Analysis of the precipitation over both land and ocean indicates that since 1979, northern hemisphere-average monsoon precipitation and circulation as well as Hadley and Walker circulations have all shown substantial and coherent intensification (Wang *et al.* 2013). Using the CMIP5 historical integrations, which include variations in anthropogenic forcing terms, they estimated that only a third of the trend during 1979–2013 is likely related to global warming.

Wang *et al.* (2013) devised a unified circulation index for the northern hemisphere summer monsoons, based on the vertical wind shear over a large region of the northern deep tropics and an extension of the Webster and Yang (1992) dynamical monsoon index for the Asian monsoon. Since the monsoon circulation features a strong shear in the vertical (first baroclinic mode), the shear corresponds to the large-scale diabatic heating within the monsoon domain. Using ERA-40 reanalysis data (Uppala *et al.* 2005), the 3-year smoothed northern hemisphere monsoon circulation index has experienced large-amplitude multidecadal fluctuations over 1958–2010, which are associated with the "Mega-ENSO" index with a correlation of 0.77. This multi-decadal variability is reproducible over the entire 1871–2010 period using the 20th Century Reanalysis dataset (Compo *et al.* 2011) with a correlation of 0.62. The mega-ENSO index is an integrated measure of ENSO, the decadal ENSO (Zhang *et al.* 1996), the PDO, and Interdecadal Pacific Oscillation (IPO, Power *et al.* 1999). It is defined by the SST anomaly difference between the western Pacific K-shape area and eastern Pacific triangle located between 40°S and 40°N.

It was also found that the NHSM strength has a significant linkage with AMO (reflecting the mechanisms described earlier) and a measure of the hemispheric thermal contrast. The hemispheric circulation index may itself possess components of aerosol forcing (dependent upon local emissions changes, e.g. Bollasina *et al.* 2011) and on changes in land-sea contrast (itself a result of warming since there is a biased distribution of land towards the northern hemisphere) that project onto the hemispheric mode. In this vein, Wang *et al.* (2014) compared future precipitation projections for the Australian and Asian monsoons and found that the Australian monsoon would increase only moderately with warming, at 2.6%/°C. The greater relative warming of the northern compared to the southern hemisphere induces an SLP difference that favours enhanced increases in the Asian monsoon as shown earlier.

While the large-scale northern hemisphere monsoon circulation index is positively correlated with recent upward trends in all three identified driving factors, relationships with long-term precipitation change in the regional monsoons are more complex. This makes the usefulness of any enhanced predictability in global monsoon indices more difficult to realize. Nevertheless, global monsoon indices allow the effects of various internal and external driving factors on the monsoon to be unified via common mechanisms.

5. Modulation of Teleconnections

In addition to the potential predictability offered by decadal modes of the climate system to the time-mean Asian monsoon, these modes can also act to modulate teleconnections with which forecasts of the monsoon can be made at the seasonal scale. This is potentially of greater benefit, since society is more concerned with variability at the shorter time scales rather than in the long-term mean. The monsoon-ENSO teleconnection, for example, features multi-decadal variations. Apparent weakening of the Indian monsoon-ENSO relation in the late 1970s (Krishna Kumar *et al.* 1999) has caused frequent failure of seasonal rainfall prediction. Figure 3 shows the recent observed weakening in the South Asian monsoon-ENSO teleconnection over the 20th century.

At the same time, a notable change in the relationship between the East Asian monsoon

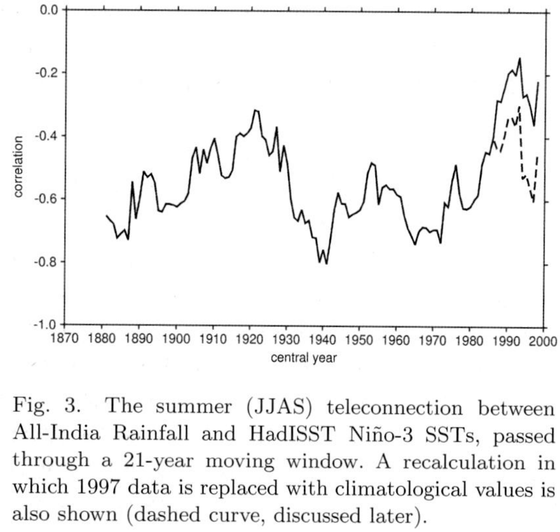

Fig. 3. The summer (JJAS) teleconnection between All-India Rainfall and HadISST Niño-3 SSTs, passed through a 21-year moving window. A recalculation in which 1997 data is replaced with climatological values is also shown (dashed curve, discussed later).

in northern China and Japan and ENSO has emerged (Wu and Wang 2002), and the negative correlation between Indonesian monsoon rainfall and ENSO has become enhanced (Chang *et al.* 2004). In further examination of the interannual variability of the entire Asian-Australian monsoon system with season-reliant EOF analysis, Wang *et al.* (2008) found that since the late 1970s the relationships between ENSO and the western North Pacific, East Asian and Indonesian monsoons have all enhanced during the developing, mature and decaying phases of ENSO, overriding the weakening of the Indian monsoon-ENSO anticorrelation during the developing phase.

Long-term variability of the monsoon-ENSO teleconnection also occurs in climate model simulations, both under control conditions with fixed CO_2 concentrations (Turner *et al.* 2007a) and in 20th century experiments (Annamalai *et al.* 2007). In such experiments, phase changes in this decadal-scale variability are out of phase with those in the observed record, suggesting: (i) that recent weakening is not necessarily related to anthropogenic warming and (ii) that variations are internal to the climate system.

Another feature revealed from Fig. 3 is that of sampling. A rather normal South Asian monsoon rainfall season occurred in the run up to the unusually strong East Pacific El Niño warming of 1997, against expectations. However, if the 1997 SSTs are removed from the dataset (replaced with the long-term mean climatology) then the weakening tendency in the last few decades is much reduced. In this vein, Krishna Kumar *et al.* (2006) compared El Niño events coinciding with monsoon droughts to those with normal rainfall. They showed that the strength of the teleconnection was strongly dependent upon the type (position) of the El Niño anomaly, with central Pacific ("Modoki") El Niño events having much greater influence. Other work has also shown very different behaviour for the East Asian monsoon during central Pacific events (Weng *et al.* 2007).

5.1. *Modulation of teleconnections via the Pacific*

Since we know that ENSO can undergo large regime shifts in terms of frequency and amplitude in long control model integrations (with fixed external forcing; Wittenberg *et al.* 2009), then its impact on the monsoon-ENSO teleconnection must be considered. For example, Turner *et al.* (2007b) have shown that multi-decade-length changes in ENSO regime have a strong impact on the monsoon-ENSO relationship. Likewise, in a 1000-year model integration, Chen *et al.* (2010) have demonstrated an in-phase relationship between the strength of the monsoon-ENSO teleconnection and ENSO power on decadal time scales. Meanwhile, Wang *et al.* (2008) have attributed coordinated interdecadal changes in ENSO teleconnections with different regions of the Asian monsoon to increased magnitude and periodicity of ENSO and strengthened monsoon–ocean interactions.

The mechanism for modulation of the Asian monsoon-ENSO teleconnection to PDO-like changes in the Pacific was envisaged by Krishnamurthy and Goswami (2000). They suggested that the PDO can alter the teleconnection, such that during its warm (cold) phase it acts to enhance the impact of El Niño (La Niña) on the monsoon, via a superposition of interannual Walker Circulation anomalies onto a mean background state perturbed by the PDO.

As we shall see next, the Atlantic Ocean can also perturb ENSO teleconnections to the monsoon via altering the Walker Circulation.

5.2. *Modulation of teleconnections via the Atlantic*

In addition to the effects on the mean state, the weakening of the THC via a freshwater hosing experiment, which cools the North Atlantic, in Zhang and Delworth (2006) was found to enhance ENSO variance and strengthen the monsoon-ENSO teleconnection.

Lu *et al.* (2008) and Chen *et al.* (2010) used long model integrations to show that a positive phase of the AMO (warm North Atlantic) led to induced coupled feedbacks in the tropical Pacific, relaxing the trade winds and deepening the thermocline in the east. This dampens ENSO variance and hence the teleconnection to the monsoon. Goswami *et al.* (2006b) also showed that the AMO could modulate the effect of the North Atlantic Oscillation on the monsoon on interannual time scales, thereby providing a competing driving signal to ENSO.

The *tropical* Atlantic also plays a role in modulating the monsoon-ENSO teleconnection. Kucharski *et al.* (2007) used a modelling study to demonstrate that changes to tropical Atlantic SSTs in boreal summer (SST cooling in the Benguela region) seem to weaken the monsoon-ENSO teleconnection. Indeed cooling in the Gulf of Guinea in the 1980s and 1990s excited Rossby waves thought to project onto the Indian Ocean, weakening the dominance of ENSO over the region. In this way, east/west shifts in heating centers can also perturb the monsoon, in addition to changes in the north/south temperature gradient forced by changes in the Atlantic Meridional Overturning Circulation.

5.3. *Tibetan Plateau forcing*

Finally, Wu *et al.* (2012) have shown that snow cover over the Tibetan Plateau can modulate ENSO teleconnections with the monsoon, by demonstrating that ENSO significantly correlates with the East Asian summer monsoon only in summers where snow cover is reduced, via a perturbation to the ENSO Rossby wave response. Some sort of decadal variability perturbing snow cover or long term warming trends leading to snowmelt may therefore drive this change.

6. Effects on Synoptic-to-Intraseasonal Variability

Since there are known relationships between the mean state and interannual variability (e.g. Turner *et al.* 2005), and between modes of intraseasonal variability and low pressure systems within the monsoon (e.g. Krishnamurthy and Shukla 2007), then if effects of long-term climate changes or decadal variability in the mean state can be passed down to shorter time scales, we may benefit from a *cascade of predictability*. However there has been little research in this area. An example is Kajikawa *et al.* (2009), who found a decadal change of the boreal summer ISV around the mid-1990s in the western Pacific-South China Sea. During 1979–1993, ISV over the South China Sea (a centre of strong ISV variance during boreal summer) has a spectral peak around 64 days, which is longer than the recent 1994–2007 period (around 42 days). ISV during 1979–1993 involves a merger of northward- and westward-propagating convection anomalies over the western North Pacific. ISV during 1994–2007 has no such combination but exhibits a tilted band-structure extending from the northern Indian Ocean to the SCS, which is strongly connected to the equatorial eastward-propagating Madden-Julian Oscillation. The change of ISV periodicity is a result of a change in the propagation route of the ISV, however the cause of this change remains unknown.

7. New Opportunities and Challenges in Decadal Prediction

It is still a challenge to make predictions of climate beyond a few seasons ahead (Mehta *et al.* 2011). However, as Kucharski *et al.* (2009b) note in their multi-model analysis forcing AGCMs with 20th century SSTs, observed long-term variations in the Asian monsoon appear more reproducible than those at interannual time scales. New results are also encouraging, particularly those emerging from the CMIP5 decadal hindcasts and forecasts, in which the ocean models are initialized as closely as possible to the observed temperature and salinity state, potentially allowing for the correct phase of internal modes of ocean decadal variability to be captured. Kim *et al.* (2012) examined decadal hindcasts from 7 CMIP5 coupled models and showed that all had high prediction skill for the Indian Ocean, West Pacific and North Atlantic regions. In these regions they suggest the domination of external forcing and low-frequency climate variability. Into the closest ocean basin to the Asian monsoon, Guemas *et al.* (2013) studied multi-year temperature forecasting skill in a 5-model CMIP5 decadal hindcast ensemble and a larger ensemble of decadal forecasts in the HadCM3 model. They noted that extended range prediction of 2–9 years for the Indian Ocean outperformed any other region. However, comparison between the initialized hindcasts and uninitialised historical integrations of the 20th century revealed little difference in skill (aside from small regions in the southeast equatorial Indian Ocean). This suggests that much of the skill there originates from radiative forcing (the climate change signal arising from increasing greenhouse gas concentrations, or potentially from volcanic eruptions), rather than the internal dynamical modes of the ocean described earlier.

Given the aforementioned studies describing the usefulness of North Atlantic for modulating the Asian monsoon and its teleconnections, multi-year prediction there is likely to be of benefit. Indeed Kim *et al.* (2012) showed 95%-significant skill at AMO prediction out to 4–7 year lead times, as shown in Fig. 4. As in many other measures of skill, the multi-model mean outperforms any individual model. On the other hand, extended range prediction of the PDO was found to be poor (not shown).

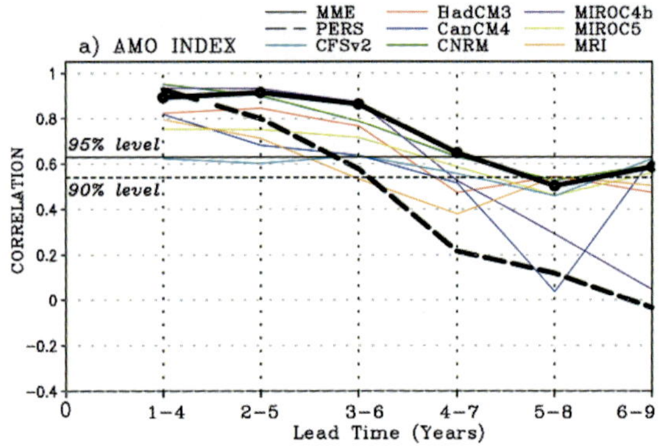

Fig. 4. Decadal prediction skill as a function of lead time for various CMIP5 model hindcast sets. The multi-model mean and persistence are also shown. [Taken from Kim *et al.* (2012).]

While the decadal forecasts and hindcasts from CMIP5 offer new opportunities, many challenges remain, not least in raising the societal usefulness of forecasts (Mehta *et al.* 2011). Particularly large uncertainties remain in ocean initialization, model uncertainty and biases that develop rapidly on seasonal and shorter time scales, especially in the face of large model biases in the Asian monsoon region (Sperber *et al.* 2013). There are new prospects however. For example new research suggests that abrupt changes in oceanic modes that lend predictability to the monsoon could themselves be forced by anthropogenic factors, for example aerosol-forced changes in the AMO (Booth *et al.* 2012), in which changes in anthropogenic aerosols and volcanic eruptions are found to explain the majority of variance in 20th century North Atlantic SST. Beyond changes in snow cover of the Tibetan Plateau discussed herein, there may also be largely untapped predictability in the monsoon arising from the land surface, perhaps through anthropogenic land-use/land-cover change. Some of the CMIP 5 models are now capable of simulating the multi-decadal mode of South Asian monsoon reasonably well (e.g. GFDL-CM3, HadGEM2-AOC). The question of decadal predictability and prediction skill should be revisited with such models.

Acknowledgments

AG Turner was funded by a Natural Environment Research Council (NERC) Fellowship number NE/H015665/1 and also acknowledges the generous support of the World Climate Research Programme in attending the IWM-V meeting in Macau. B Wang acknowledges the support by the National Research Foundation of Korea (NRF) through a Global Research Laboratory (GRL) grant (MEST 2011-0021927).

References

Annamalai, H., K. Hamilton, and K. R. Sperber, 2007: South Asian Summer Monsoon and its relationship with ENSO in the IPCC AR4 simulations. *J. Climate*, **20**, 1071–1092.

Blanford, H. F., 1884: On the connexion of the Himalaya snowfall with dry winds and seasons of drought India. *Proc. Roy. Soc. London*, **37**, 3–22.

Bollasina, M., Y. Ming, and V. Ramaswamy, 2011: Anthropogenic aerosols and the weakening of the South Asian Summer Monsoon. *Science*, **334** (6055), 502–505.

Booth, B. B. B., N. J. Dunstone, P. R. Halloran, T. Andrews, and N. Bellouin, 2012: Aerosols implicated as a prime driver of twentieth-century North Atlantic climate variability. *Nature*, **484**, 228–232.

Chang, C.-P., Z. Wang, J. Ju, and T. Li, 2004: On the relationship between Western Maritime Continent monsoon rainfall and ENSO during northern winter. *J. Climate*, **17**, 665–672.

Charney, J. G. and J. Shukla, 1981: Predictability of monsoons. In *Monsoon Dynamics*, (eds. J. Lighthill and R. P. Pearce), Cambridge University Press. pp. 99–109.

Chen, W., R. Lu, and B.-W. Dong, 2010: Impact of the Atlantic Ocean on the multidecadal fluctuation of El Niño-Southern Oscillation-South Asian monsoon relationship in a coupled general circulation model. *J. Geophys. Res. Atmos.*, **115**, D17109.

Cheng, Q. and T. Zhou, 2014: Multidecadal variability of North China aridity and its relationship to PDO during 1900–2010. *J. Climate*, **27**(3), 1210–1222.

Compo, G. P. *et al.* 2011: The twentieth century reanalysis project. *Quart. J. Roy. Meteorol. Soc.*, **137**(654), 1–28.

Goswami, B. N. and P. K. Xavier, 2005: Dynamics of 'internal' interannual variability of Indian Summer Monsoon in a GCM. *J. Geophys. Res.*, **110**, D24104.

Goswami, B. N., V. Venugopal, D. Sengupta, M. S. Madhusoodanan, and P. K. Xavier, 2006a: Increasing trend of extreme rain events over India in a warming environment. *Science*, **314**, 1442–1445.

Goswami B. N., M. S. Madhusoodanan, C. P. Neema, and D. Sengupta, 2006b: A physical mechanism for North Atlantic SST influence on the Indian summer monsoon. *Geophys. Res. Letts.*, **33**: L02706.

Guemas, V., S. Corti, J. Garcia-Serrano, F. J. Doblas-Reyes, M. Balmaseda, and L. Magnusson, 2013: The Indian Ocean: The region of highest skill worldwide in decadal climate prediction. *J. Climate*, **26**, 726–739.

Kajikawa, Y., T. Yasunari, and B. Wang, 2009: Decadal change in intraseasonal variability over the South China Sea. *Geophys. Res. Letts.*, **36**, L06810.

Kim, H.-M., P. J. Webster, and J. A. Curry, 2012: Evaluation of short-term climate change prediction in multi-model CMIP5 decadal hindcasts. *Geophys. Res. Letts.*, **39**, L10701.

Kitoh, A., H. Endo, K. K. Kumar, I. F. A. Cavalcanti, P. Goswami, and T. Zhou, 2013: Monsoons in a changing world: A regional perspective in a global context. *J. Geophys. Res. Atmos.*, **118**(8), 3053–3065.

Krishna Kumar, K., B. Rajagopalan, and M. A. Cane, 1999: On the weakening relationship between the Indian monsoon and ENSO. *Science*, **284**, 2156–2159.

Krishna Kumar, K., B. Rajagopalan, M. Hoerling, G. Bates, and M. Cane, 2006: Unraveling the mystery of Indian monsoon failure during El Niño. *Science*, **314**, 115–119.

Krishnamurthy, V. and B. N. Goswami, 2000: Indian monsoon-ENSO relationship on interdecadal timescale. *J. Climate*, **13**, 579–595.

Krishnamurthy, V. and J. Shukla, 2007: Intraseasonal and seasonally persistent patterns of Indian monsoon rainfall. *J. Climate*, **20**, 3–20.

Krishnan, R. and M. Sugi, 2003: Pacific decadal oscillation and variability of the Indian summer monsoon rainfall. *Clim. Dyn.*, **21**, 233–242.

Kucharski, F., A. Bracco, J. H. Yoo, and F. Molteni, 2007: Low-frequency variability of the Indian monsoon-ENSO relationship and the Tropical Atlantic: The "Weakening" of the 1980s and 1990s. *J. Climate*, **20**, 4255–4266.

Kucharski, F., A. Bracco, J. H. Yoo, A. Tompkins, L. Feudale, P. Ruti, and A. Dell'Aquila, 2009a: A Gill-Matsuno-type mechanism explains the Tropical Atlantic influence on African and Indian monsoon rainfall. *Quart. J. Roy. Meteorol. Soc.*, **135**, 569–579.

Kucharski, F., A. A. Scaife, J. H. Yoo, C. K. Folland, J. Kinter, J. Knight, D. Fereday, A. M. Fischer, E. K. Jin, and J. Kröger, 2009b: The CLIVAR C20C project: Skill of simulating Indian monsoon rainfall on interannual to decadal timescales. Does GHG forcing play a role? *Clim. Dyn.*, **33**, 615–627.

Lee J.-Y. and B. Wang, 2014: Future change of global monsoon in the CMIP5. *Clim. Dyn.*, **42**, 101–119.

Li, C. and M. Yanai, 1996: The onset and international variability of the Asian summer monsoon in relation to land-sea thermal contrast. *J. Climate*, **9**, 358–375.

Liu, Y. and J. C. H. Chang, 2012: Coordinated abrupt weakening of the Eurasian and North African monsoons in the 1960s and links to extra-tropical North Atlantic Cooling. *J. Climate*, **25**, 3532–3548.

Lu, R., B. Dong, and H. Ding, 2006: Impact of the Atlantic Multidecadal Oscillation on the Asian summer monsoon. *Geophys. Res. Letts.*, **33**, L24701.

Lu, R., W. Chen, and B.-W. Dong, 2008: How does a weakened Atlantic thermohaline circulation lead to an intensification of the ENSO-south Asian summer monsoon interaction? *Geophys. Res. Letts.*, **35**(8), L08706.

Mantua N. J. and S. R. Hare, 2002: The Pacific Decadal Oscillation. *Journal of Oceanography*, **58**: 35–44.

Mehta, V., G. Meehl, L. Goddard, J. Knight, A. Kumar, M. Latif, T. Lee, A. Rosati, and D. Stammer, 2011: Decadal climate predictability and prediction: Where are we? *Bull. Amer. Meteorol. Soc.*, **92**, 637–640.

Power, S., T. Casey, C. Folland, A. Colman, and V. Mehta, 1999: Inter-decadal modulation of the impact of ENSO on Australia. *Clim. Dyn.*, **15**, 319–324.

Ramanathan, V., C. Chung, D. Kim, T. Bettge, L. Buja, J. T. Kiehl, W. M. Washington, Q. Fu, D. R. Sikka, and M. Wild, 2005: Atmospheric brown clouds: Impacts on South Asian climate and hydrological cycle. *Proc. Nat. Acad. Sci.*, **102**(15), 5326–5333.

Sperber, K. R., H. Annamalai, I.-S. Kang, A. Kitoh, A. Moise, A. Turner, B. Wang, and T. Zhou, 2013: The Asian Summer Monsoon: An intercomparison of CMIP5 vs. CMIP3 simulations of the late 20th century. *Clim. Dyn.*, **41**, 2711–2744.

Trenberth, K. E. and J. W. Hurrell, 1994: Decadal atmosphere-ocean variations in the Pacific. *Clim. Dyn.*, **9**, 303–319.

Turner, A. G., P. M. Inness, and J. M. Slingo, 2005: The role of the basic state in the ENSO-Monsoon relationship and implications for predictability. *Quart. J. Roy. Meteorol. Soc.*, **131**, 781–804.

Turner, A. G., P. M. Inness, and J. M. Slingo, 2007a: The effect of doubled CO_2 and model basic state biases on the monsoon-ENSO system. I: Mean response and interannual variability. *Quart. J. Roy. Meteorol. Soc.*, **133**, 1143–1157.

Turner, A. G., P. M. Inness, and J. M. Slingo, 2007b: The effect of doubled CO2 and model basic state biases on the monsoon-ENSO system. II: Changing ENSO regimes. *Quart. J. Roy. Meteorol. Soc.*, **133**, 1159–1173.

Turner, A. G. and H. Annamalai, 2012: Climate change and the South Asian summer monsoon. *Nature Clim. Change*, **2**, 587–595.

Uppala, S. M. *et al.* 2006: The ERA-40 re-analysis. *Quart. J. Roy. Meteorol. Soc.*, **131**(612), 2961–3012.

Wang, B. and Q. Ding, 2006: Changes in global monsoon precipitation over the past 56 years. *Geophys. Res. Letts.*, **33**: L06711.

Wang B. and Q. Ding, 2008: Global monsoon: Dominant mode of annual variation in the tropics. *Dynam. Atmos. Oceans*, **44**, 165–183.

Wang, B., J. Liu, H.-J. Kim, P. J. Webster, S.-Y. Yim, and B. Xiang, 2013: Northern Hemisphere summer monsoon intensified by mega-El Niño/southern oscillation and Atlantic multidecadal oscillation. *Proc. Nat. Acad. Sci.*, **110**, 5347–5352.

Wang, B., J. Yang, T. Zhou, and B. Wang, 2008: Interdecadal changes in the major modes of Asian-Australian monsoon variability: Strengthening relationship with ENSO since the late 1970s. *J. Climate*, **21**, 1771–1789.

Wang, B., S.-Y. Yim, J.-Y. Lee, J. Liu, and K.-J. Ha, 2014: Future change of Asian-Australian monsoon under RCP 4.5 anthropogenic warming scenario. *Clim. Dyn.*, **52**, 83–100.

Webster, P. J. and S. Yang, 1992: Monsoon and ENSO: Selectively interactive systems. *Quart. J. Roy. Meteorol. Soc.*, **118**, 877–926.

Weng, H., K. Ashok, S. K. Behera, S. A. Rao, and T. Yamagata, 2007: Impacts of recent El Niño Modoki on dry/wet conditions in the Pacific rim during boreal summer. *Clim. Dyn.*, **29**, 113–129.

Wittenberg, A. T., 2009: Are historical records sufficient to constrain ENSO simulations? *Geophys. Res. Letts.*, **36**, L12702.

Wu, R. and B. Wang, 2002: A contrast of the East Asian summer monsoon-ENSO relationship between 1962–1977 and 1978–1993. *J. Climate*, **15**, 3266–3278.

Wu, Z., J.-P. Li, Z.-H Jiang, and T.-T. Ma, 2012: Modulation of the Tibetan Plateau snow cover on the ENSO teleconnections: From the East Asian Summer Monsoon perspective. *J. Climate*, **25**, 2481–2489.

Xavier, P. K., C. Marzin, and B. N. Goswami, 2007: An objective definition of the Indian summer monsoon season and a new perspective on the ENSO-monsoon relationship. *Quart. J. Roy. Meteorol. Soc.*, **133**, 749–764.

Yim, S.-Y., B. Wang, and M. Kwon, 2014: Interdecadal change of the controlling mechanisms for East Asian early summer rainfall variation around the mid-1990s. *Clim Dyn.*, **42**, 1325–1333.

Zhang, Y., J. M. Wallace, and D. S. Battisti, 1997: ENSO-like interdecadal variability: 1900–93. *J. Climate*, **10**, 1004–1020.

Zhang, Y., T. Li, and B. Wang, 2004: Decadal change of the spring snow depth over the Tibetan Plateau: The associated circulation and influence on the East Asian summer monsoon. *J. Climate*, **17**, 2780–2793.

Zhang, R. and T. L. Delworth, 2006: Impact of Atlantic multidecadal oscillations on India/Sahel rainfall and Atlantic hurricanes. *Geophys. Res. Letts.*, **33**, L17712.

Zhou, T., R. Yu, H. Li, and B. Wang, 2008: Ocean forcing to changes in global monsoon precipitation over the recent half-century. *J. Climate*, **21**, 3833–3852.

Chapter 24

Monsoon Precipitation in a Future Warmer World

Akio Kitoh

Faculty of Life and Environmental Sciences,
University of Tsukuba, Tsukuba, Ibaraki 305-8572, Japan
kito.akio.ff@u.tsukuba.ac.jp

Hirokazu Endo

Climate Research Department, Meteorological Research Institute,
Tsukuba, Ibaraki 305-0052, Japan

We provide a latest view of the current understanding of future changes in global as well as regional monsoon precipitation as projected by the state-of-the-art CMIP5 model simulations under RCP2.6 and RCP8.5 scenarios. The amount and intensity of the Asian summer monsoon rainfall is projected to increase under global warming, with the rate of increase being higher than that in other monsoon regions.

1. Introduction

Monsoons are the dominant modes of the annual variation of the tropical precipitation and circulation, which characterize the seasonality of Earth's climate (Wang and Ding 2008). Monsoons are responsible for the majority of summer rainfall within the tropics, where billions of people depend on the monsoon rainfall. Thus, reliable future projection for monsoon rainfall, especially on a regional-scale, has been strongly demanded. The monsoon region is distributed over all the tropical continents, and over the tropical oceans (Wang and Ding 2006). While different indices have been used to identify the life cycle of monsoons, the definition of the monsoon in terms of the precipitation characteristics (Wang and LinHo 2002; Wang and Ding 2006) rather than the wind reversal, permits to use similar monsoon indices for all the monsoon regions.

Substantial information on climate change projections is available from the Coupled Model Intercomparison Project phase 3 and 5 (CMIP3, CMIP5). Models show that global precipitation will increase with increased global mean surface temperature, however, changes in average precipitation in a warmer world will exhibit substantial spatial variation (Collins *et al.* 2014). Over tropical oceans, precipitation changes depend on the spatial pattern of sea surface temperature (SST) warming. Due to this SST pattern effect, the seasonal mean rainfall is projected to increase on the equatorward flank of tropical convergence zones, following a "warmer-get-wetter" pattern, increasing rainfall where warming of SST exceeds the tropical mean and vice versa (Xie *et al.* 2010; Chadwick *et al.* 2013; Christensen *et al.* 2014).

Skill of climate models in reproducing climatological features of the global monsoon is improving, but models still have biases in the representation of the monsoon domain and intensity (Sperber *et al.* 2013; Flato *et al.* 2014). The global monsoon precipitation is projected to strengthen in the 21st century with increases in its area and intensity, while the monsoon circulation weakens, due to enhanced moisture

The Global Monsoon System: Research and Forecast (3rd Edition)
Edited by C. P. Chang et al.

convergence (Christensen *et al.* 2014). However, the responses of the regional monsoons to a changing climate vary from region to region due to various forcing employed and complex feedbacks in the climate system. Thus, there is low confidence in projections of changes in large-scale patterns of natural climate variability, including changes in monsoons (Seneviratne *et al.* 2012; Christensen *et al.* 2014).

Future projections of the global monsoon as well as regional monsoons are essential. In the former edition of "The Global Monsoon System: Research and Forecast" (Chang *et al.* 2011), Turner *et al.* (2011) and Kitoh (2011) reviewed future behavior of monsoons. Several studies are available for future monsoon precipitation changes in the future projected by CMIP5 models (Hsu *et al.* 2013; Kitoh *et al.* 2013; Endo and Kitoh 2014; Lee and Wang 2014; Wang *et al.* 2014). In this paper, we review the projected changes of the global monsoon as well as regional monsoons as revealed by the CMIP5 models.

2. Global Monsoon: Present

There are evidences that the CMIP5 models are better than the CMIP3 models in simulating the Asian-Australian monsoon system. Recent works (e.g. Li *et al.* 2012; Sperber *et al.* 2013; Ogata *et al.* 2014) report that the CMIP5 multi-model mean is more skillful than the CMIP3 multi-model mean; however, most climate models still have deficiencies in the simulation of the spatial or intra-seasonal variation of monsoon precipitation. See Sperber *et al.* (in this book) for the review on the present monsoon simulated by current climate models.

Kim *et al.* (2008) point out that models with higher resolution generally reproduced a better spatial pattern of the global monsoon precipitation compared to lower-resolution models. Wang *et al.* (2011) also showed superiority of high-resolution models in simulating global monsoonal rainfall patterns.

CMIP5 models skill in simulating observed global monsoon domains against observations is investigated by Hsu *et al.* (2013) and Kitoh *et al.* (2013). Figure 1 shows the observed and simulated global monsoon precipitation domains for the present climate (Kitoh *et al.* 2013). Global monsoon area (GMA) is defined as the regions where the annual range of precipitation exceeds 2.5 mm day^{-1}, based on the definition of Wang *et al.* (2011). The annual range

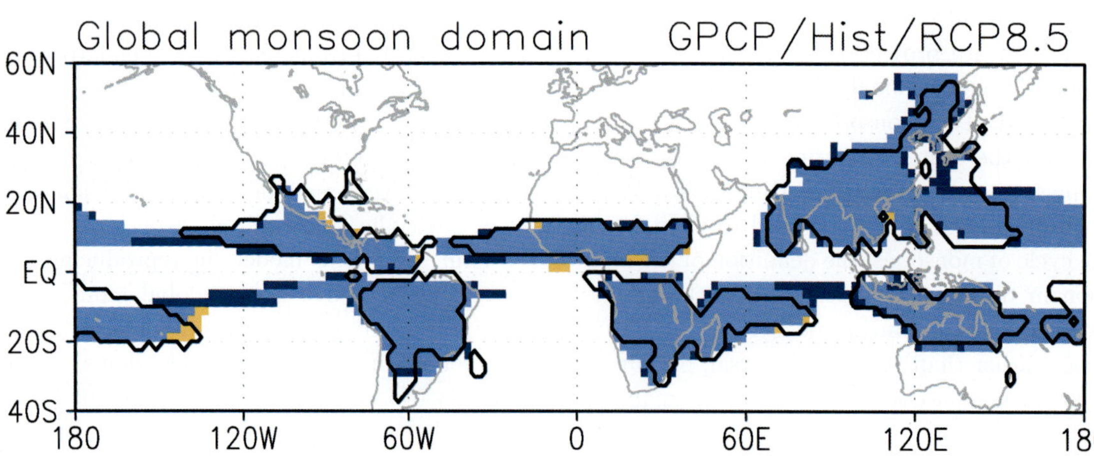

Fig. 1. Observed (thick contour) and simulated (shading) global monsoon domain where the seasonal range (between May–September and November–March) of precipitation exceeds 2.5 mm day^{-1} (Wang *et al.* 2011). The observation is based on GPCP data, and the simulations are based on 29 CMIP5 multi-model mean precipitation with a common 2.5 by 2.5 degree grid in the present-day (1986–2005) and the future (2080–2099; RCP8.5 scenario). Warm yellow (dark blue) shading: monsoon domain only in the present-day (future). Blue shading: monsoon domain in the both periods (from Kitoh *et al.* 2013).

is defined as the difference between the May–September (MJJAS) mean and the November–March (NDJFM) mean. Hsu *et al.* (2013) and Kitoh *et al.* (2013) use slightly different definition, but both show that the ensemble mean precipitation of CMIP5 models reproduces the observed GMA, global monsoon precipitation (GMP), and global monsoon intensity (GMI), although the spread of individual models is large. Here GMP is defined as the mean of summer rainfall in the monsoon area, and GMI is defined as the global monsoon precipitation amount per unit area.

Figure 1 shows that the multi-model ensemble mean of 29 CMIP5 models generally reproduces the observed global monsoon domain. However, models have difficulty in reproducing monsoon rainfall in East Asia and the tropical Pacific. The underestimation of monsoon domain in East Asia is pointed out by Kim *et al.* (2008) for the CMIP3 models, and is not improved in CMIP5 models. It is noted that even the best CMIP5 models cannot remedy this deficiency (Lee and Wang 2014). Hsu *et al.* (2011) and Kitoh *et al.* (2013) assessed GMA and GMI of each model against observations. Note that observations (GPCP and CMAP) have about 10% uncertainty in estimating GMA. They show that models have large variability both in reproducing GMA and GMI, but the multimodel ensemble closely matches the observations although with a slight overestimation.

3. Global Monsoon: Future

Observed data indicate that the global monsoon precipitation had an increasing trend over the past three decades (Hsu *et al.* 2011; Wang *et al.* 2012). Whether or not this increasing trend will continue in the 21st century is an issue. Hsu *et al.* (2012) investigated global monsoon area and precipitation in the late 21st century using CMIP3 and CMIP5 models as well as three high-resolution atmospheric models that were forced by different future SST warming patterns. They find consistent increase in the global monsoon area, precipitation and intensity among the model projections. Using the recent CMIP5 model simulations, Hsu *et al.* (2013), Kitoh *et al.* (2013) and Lee and Wang (2014) all find future increases of global monsoon area and precipitation intensity in the late 21st century.

Changes in the global monsoon domain at the end of the 21st century under RCP8.5 scenario are demarcated in Fig. 1 with yellow and dark blue shadings, where yellow shading denotes shrinkage and dark blue shading denotes expansion of monsoon domain, respectively. The CMIP5 model projections show that GMA will expand mainly over the central to eastern tropical Pacific, the southern Indian Ocean, and eastern Asia. It is notable that the southwestern Japan and some part of Korean peninsula, where the historical experiments underestimate precipitation amount, become a monsoon domain due to increased precipitation in the RCP8.5 scenario.

The future projections indicate that GMA, GMI and GMP all show an increase in the RCP2.6 scenario and more so in the RCP8.5 scenario. The change rates of GMA at the end of the 21st century compared with the present are +1.5, +3.6, +7.1% (+5.2, +9.4, +14.5%) for RCP2.6 (RCP8.5) at 10th, 50th and 90th percentile, respectively. The change rates of GMI are −0.4, +1.7, +3.2% (−0.4, +5.2, +8.3%) and those for GMP are +3.0, +5.0, +9.3% (+4.9, +16.6, +22.5%) for RCP2.6 (RCP8.5). Therefore, monsoon-related precipitation is projected to significantly increase in a warmer climate.

Figure 2 shows the time-series of precipitation, evaporation, lower-troposphere water vapor flux convergence, and lower-troposphere wind convergence averaged over global monsoon domain over land by CMIP5 historical simulation and RCP2.6/8.5 scenarios. Previous results show that the observed global monsoon precipitation over land experienced a slight increasing trend from 1900 through early 1940s, then an overall decreasing trend from 1940s to 1980s,

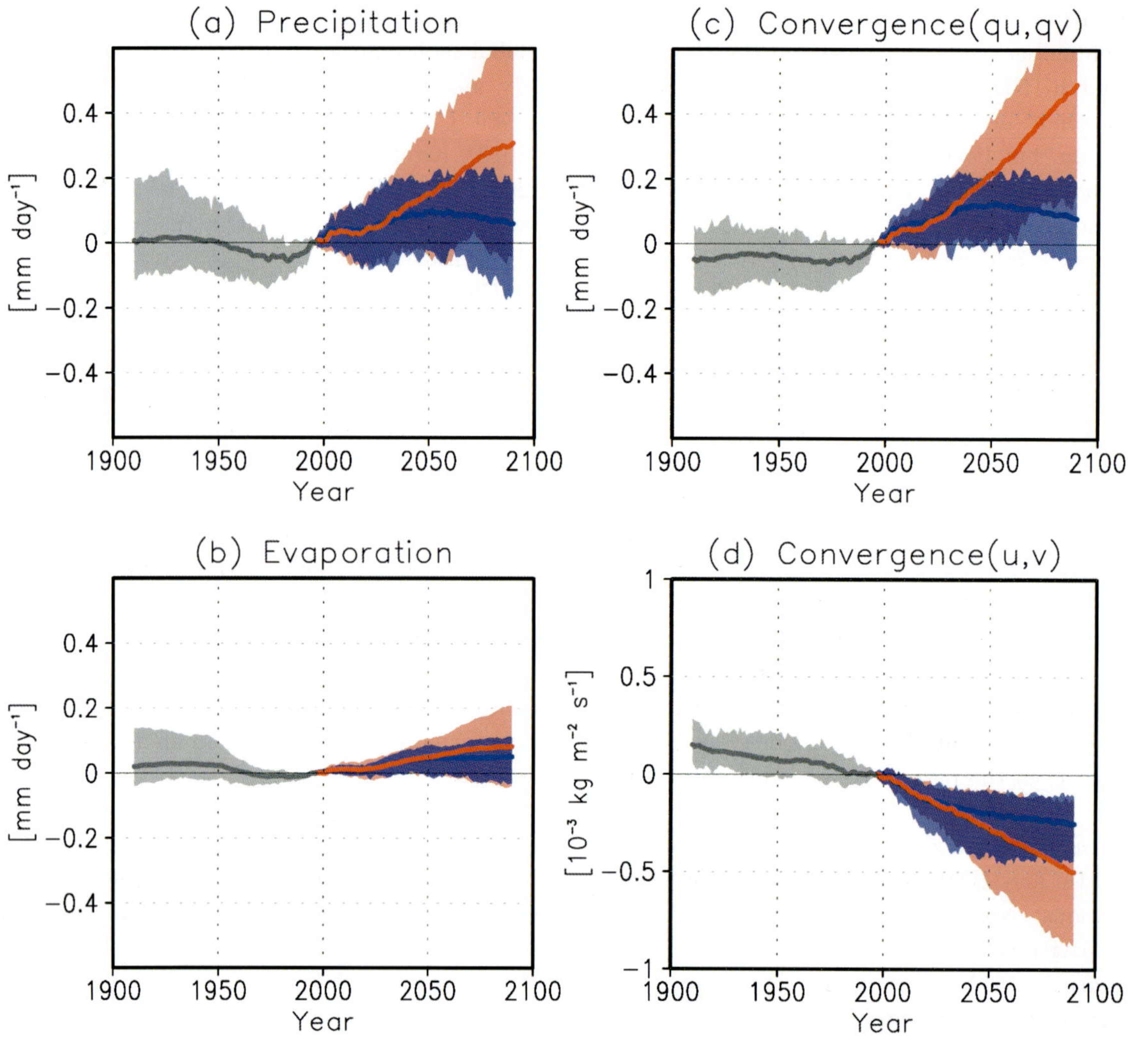

Fig. 2. Time series of simulated anomalies over global land monsoon domain for (a) precipitation (mm day^{-1}), (b) evaporation (mm day^{-1}), (c) water vapor flux convergence in the lower (below 500 hPa) troposphere (mm day^{-1}) and (d) wind convergence in the lower troposphere (10^{-3} kg m^{-2} s^{-1}), relative to the base period average (1986–2005), based on CMIP5 multi-model monthly outputs. Historical (grey; 29 models), RCP2.6 (dark blue; 20 models), and RCP8.5 (red; 24 models) simulations are shown in the 10th and 90th percentile (shading), and in all model averages (thick lines).

followed by a recovering trend until present (e.g. Zhang and Zhou 2011; Hsu *et al.* 2011). The CMIP5 models are not able to reproduce the mid-20th century maximum, but reproduce a gradual decline from the early 20th century until 1980s, and project an increase throughout the 21st century. The CMIP5 multi-model mean shows a large increase of global land monsoon precipitation in RCP8.5 scenario toward the end of the 21st century. In RCP2.6 scenario, global land monsoon precipitation reaches a maximum around 2050 and shows a slight decline thereafter, which coincides with global temperature changes in this scenario. As indicated by the 10th and 90th percentile lines, model spread is quite large (and larger in RCP8.5 than in

RCP2.6). It is noted that there are significant differences in the trends of rainfall between RCP2.6 and RCP8.5, reflecting remarkable temperature differences between the two scenarios.

There are various drivers of monsoons. Role of aerosols on monsoon is one of major research target. Aerosols have a net cooling effect on the global climate system, but show various roles on regional climate changes. Model simulations show that the direct effect of aerosols reduces the magnitude of Asian monsoon precipitation change (Bollasina *et al.* 2011; Polson *et al.* 2014; Song *et al.* 2014). Large uncertainty exists on indirect influence of aerosols on climate by modifying clouds.

In the CMIP5 experiments, in addition to "Historical" run that includes all forcing, additional experiments are performed for climate change detection and attribution studies. The "HistoricalNat" experiment uses only natural forcing, i.e. volcanoes and solar variability. The "HistoricalGHG" experiment uses greenhouse gas forcing only. The response to aerosols can be estimated as a residual of these three experiments.

Using the 14 CMIP5 model data, it is revealed that neither experiment with GHG forcing nor natural forcing reproduce the decreasing trend that was observed and simulated by the all forcing. This implies that the aerosol forcing is the essential agent in the decreasing monsoon precipitation in the latter half of the 20th century. Song *et al.* (2014) show that the aerosol forcing, i.e. cooling over continental East Asia, plays a primary role in driving weakened low-level monsoon circulation in this region in the late 20th century.

From Fig. 2d, it is clear that the CMIP5 models show a decreasing trend of lower-troposphere wind convergence (dynamical factor) throughout the 20th and 21st centuries. When moisture build-up effect is considered, moisture flux convergence shows a different story (Fig. 2c). In the 20th century, moisture convergence over the global land monsoon domain is almost constant until around 1980, and then turns into an increasing trend into the 21st century. This implies that the moisture increase due to temperature rise overwhelms the weakening of the large-scale circulation, thus resulting in increased moisture flux and its convergence around 1980, which is responsible for the intensification of mean monsoon rainfall. Surface evaporation showed a decreasing trend in the late 20th century probably due to aerosol increase (Turner and Annamalai 2012), but it turned to increase around 1980s due to warmer surface temperatures (Fig. 2b). Therefore the increase of the global land monsoon precipitation in the future can be attributed to the increase of moisture convergence due to increased water vapor in the air column and surface evaporation, offset to a certain extent by the weakening of the monsoon circulation.

4. Regional Monsoon Precipitation Changes

At regional scales, generally there is little consensus in GCM projections regarding the sign of future change in precipitation. We show future precipitation changes in global and 7 regional monsoon regions (Kitoh *et al.* 2013). Figure 3 shows change ratios (%) in averaged precipitation (Pav), simple precipitation daily intensity index (SDII), seasonal maximum 5-day precipitation total (R5d), and seasonal maximum consecutive dry days (CDD) during local summer season over each of the regional land monsoon domains.

In the North American monsoon region (NAM), Pav is projected to decrease in RCP8.5. The projected SDII change is around zero with half models showing different sign in changes, but R5d is projected to increase. In the South American monsoon region (SAM), although the median Pav change is small, the majority of models indicate increase of precipitation in this region in RCP8.5. Many models project an increase in the precipitation extreme indices

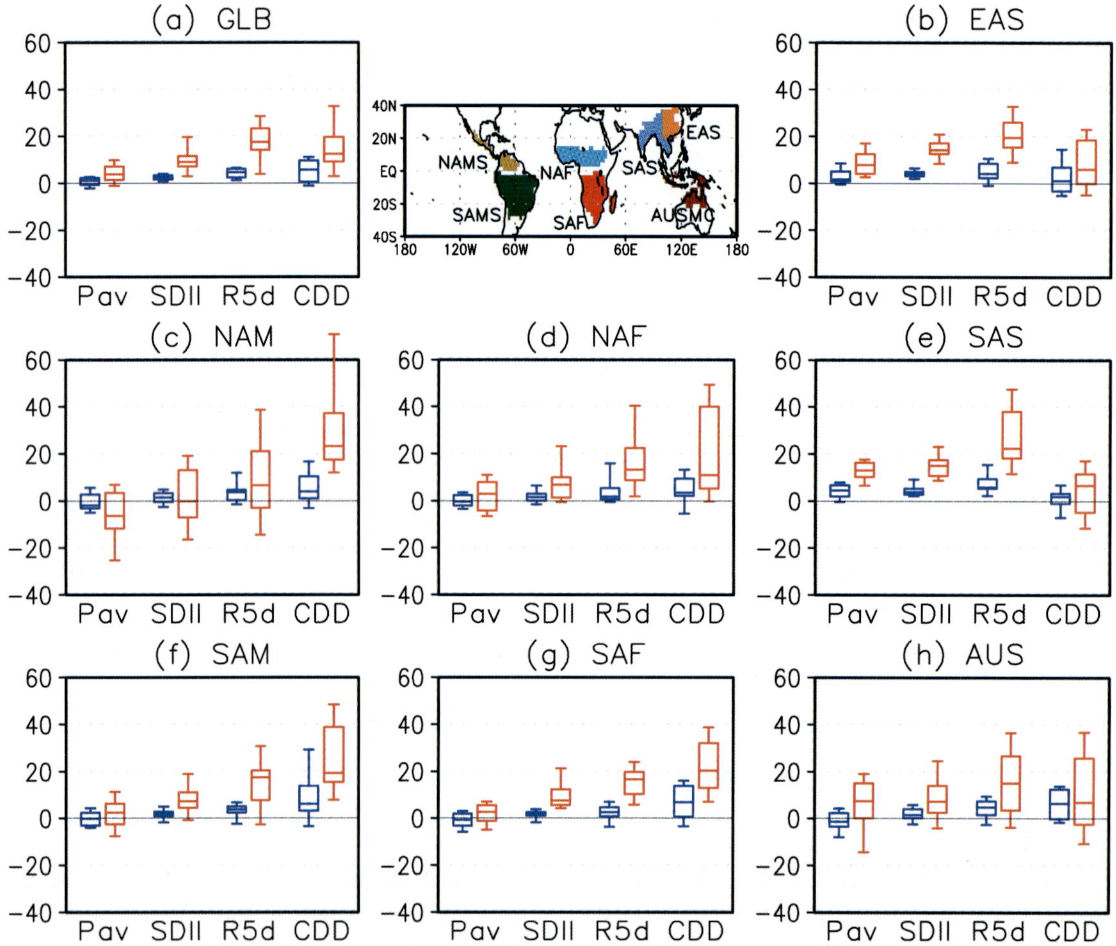

Fig. 3. Future change ratios (%) in averaged precipitation (Pav), simple precipitation daily intensity index (SDII), seasonal maximum 5-day precipitation total (R5d), and seasonal maximum consecutive dry days (CDD) for the summer season over (a) global land monsoon domain and (b)–(h) each regional land monsoon domain based on CMIP5 multi-model daily outputs. Box-whisker plots show the 10th, 25th, 50th, 75th, and 90th percentiles under RCP2.6 (dark blue; 18 models) and RCP8.5 (red; 26 models) scenarios. In the upper-middle, the regional land monsoon domains based on 26 multi-model mean precipitation is shown. Note that the precipitation statistics are calculated over the land monsoon domains determined by each model at the present-day.

(SDII and R5d) in both RCP2.6 and RCP8.5 scenarios, larger in the latter. The number of consecutive dry days is projected to increase both in NAM and SAM.

In Africa (NAF and SAF), changes in Pav are insignificant. SDII and R5d will increase in RCP8.5 scenario for both African monsoon regions. CDD will significantly increase.

Over the Asian monsoon domains (EAS and SAS), the median increase rate for precipitation is larger than that over other monsoon domains, indicating that the sensitivity of Asian monsoon to global warming is stronger than that of other monsoons. Over the East Asian monsoon region (EAS), more than 85% of models show an increase in Pav in RCP2.6, and more than 95% of models project an increase in SDII and R5d in RCP8.5. Also many models project an increase in CDD. Increase in the precipitation extremes in the South Asian monsoon region (SAS) is projected by all models in both scenarios.

Over the Australian monsoon region (AUS), many models agree in projecting an increase of Pav and SDII in RCP8.5 scenario. R5d will increase in both scenarios. CDD will also increase in the future but model scatter becomes larger in RCP8.5.

5. Thermodynamic versus Dynamic Effects: Regional Differences

In the previous section, we have shown that over the Asian monsoon domain, projected changes in extreme precipitation indices are larger than over other monsoon domains, indicating the strong sensitivity of Asian monsoon to global warming. Here we show that relative role of thermodynamic and dynamic effects are different between the Asian summer monsoon domain and other monsoon domains.

Endo and Kitoh (2014) separate the rainfall changes into a component related to atmospheric moisture content changes (thermodynamic component) and a component related to atmospheric mean circulation changes (dynamic component) following the method of Seager *et al.* (2010). Atmospheric moisture content is projected to increase everywhere in the tropics with a larger increment over the equatorial Pacific, giving a positive effect to the precipitation change. On the other hand, atmospheric mean ascending motion (dynamic component) generally weakens over tropical ascending areas in the present-day climatology, including the monsoon regions, largely canceling out the positive thermodynamic effect. Endo and Kitoh (2014) note that in the Asian monsoon domain, the slowdown of the ascending motion is clearly less than in the other monsoon domains.

Figures 4a and 4b show future changes in the spatial distributions of thermodynamic term

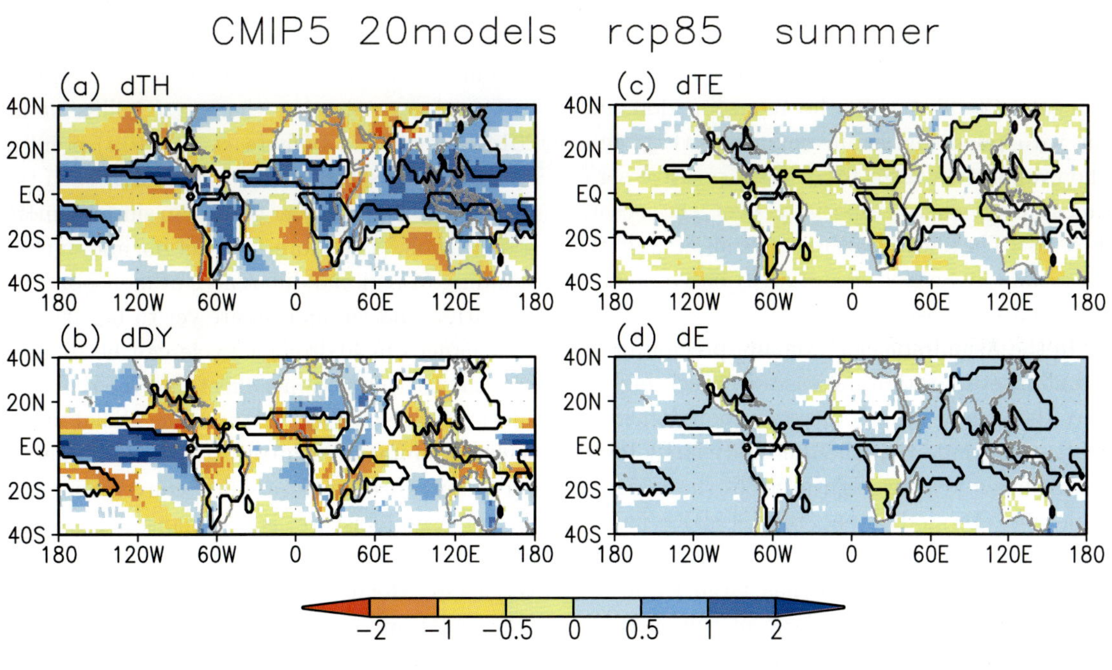

Fig. 4. Future changes in the moisture budget based on the 20 CMIP5 multi-model projections under the RCP8.5 scenario for the summer season. Shown are (a) thermodynamic term, (b) dynamic term, (c) transient eddy term, and (d) evaporation term. Units are mm d^{-1}. Only changes with the same sign in more than two thirds of the models are shaded. The monsoon domains are indicated by thick contours (from Endo and Kitoh 2014).

and dynamic term, respectively, in the moisture equation for the summer season. The thermodynamic term (δTH) is positive everywhere in the monsoon region, while it is negative in the subtropics outside of the monsoon region (Fig. 4a). The sign of δTH is reflected by the present-day climatological wind convergence/divergence pattern in the lower troposphere, because atmospheric moisture content is concentrated into the lower troposphere and enhanced everywhere in a warmer climate. The term δTH corresponds to the "wet-get-wetter" effect (Held and Soden 2006; Chou *et al.* 2009). Thus, there are strong correlations between the patterns of δTH and those of climatological precipitation field. However, the dynamic term (δDY) modifies the wet-get-wetter pattern through circulation changes partly following the "warmer-get-wetter" mechanism (Xie *et al.* 2010; Ma and Xie 2013), and it largely cancels the thermodynamic term because the tropical overturning circulation slowdowns (Chadwick *et al.* 2013). It is noted that the effect of δDY tends to be smaller over the Asian monsoon domain than over the other monsoon domains. The transient eddy term (δTE) shows a negative tendency in the monsoon domain, but with relatively small magnitude (Fig. 4c). The evaporation term (δE) has a positive sign almost everywhere, especially over the ocean (Fig. 4d). The above features are consistent with the results by Seager *et al.* (2010) based on the CMIP3 model ensemble.

In the Asian monsoon domain, precipitation as well as evaporation increases at much higher rates than those in the other monsoon domains. It is clearly found that the thermodynamic term (δTH) has a positive sign in all of the regions, while the thermodynamic term (δDY) has a negative sign. It is interesting that the magnitude of the dynamic term has large variation across the regions, while those of the thermodynamic term are fairly uniform. Smaller reduction of δDY in the Asian monsoon region is in contrast to a larger reduction in other monsoon regions, particularly that in NAM. In short, the thermodynamic effect is mostly compensated by the dynamic effect, except for the Asian monsoon region, where the dynamic effect is smaller, resulting in much larger increases in rainfall.

6. Concluding Remarks

The CMIP5 models project that the global monsoon precipitation is likely to strengthen in the 21st century with increase in its area as well as its intensity, while monsoon circulation weakens. Heavy precipitation indices are projected to increase much more than the mean precipitation over all regional monsoon domains. It is noted that as the atmosphere becomes more stable, frequency of precipitating events becomes fewer, but intensity can be stronger once an event occurs because more moisture is available due to warmer atmosphere. Over the Asian monsoon regions, projected changes in extreme precipitation indices are larger than the other monsoon regions.

Both the CMIP3 and CMIP5 models show an overall consistent view of future climate projections in terms of continental-scale temperature and precipitation changes (Collins *et al.* 2014). However, as Ogata *et al.* (2014) pointed out, there are some systematic differences in the circulation changes over the Asian summer monsoon region. Therefore, projection and understanding of regional differences of monsoon precipitation and circulation are yet to be clear.

Changes in hydrologic cycle can be divided into the thermodynamic effect and the dynamic effect (Seager *et al.* 2010). For global monsoon precipitation changes, the thermodynamic effect (enhanced moisture flux convergence due to increased moisture) is dominant, partly compensated by the dynamic effect (less wind convergence). Relative role of the two effects is different from region to region. In the most regional monsoons, the dynamic effect mostly compensates the thermodynamic effect. However in the Asian monsoon region, the dynamic effect is smaller than other monsoon regions, resulting in large increase in monsoon precipitation (Endo and Kitoh 2014). Why the response of the Asian summer monsoon is different from other monsoons is an interesting topic to be pursued. Although the tropical circulation weakens in its entity (Vecchi and Soden 2007), it may not affect largely the entire Asian monsoon circulation because the Asian monsoon region is located far from the equator, and is affected by a huge landmass of the Tibetan Plateau.

Considering the improvement of CMIP5 models against CMIP3 models (Sperber *et al.* 2012), we have now more confidence in the results. Changes in climate phenomena such as El Niño-Southern Oscillation (ENSO) can modulate regional monsoons. Currently, atmospheric aerosol loading is influencing regional climate in various ways, such as a black carbon-induced enhancement and a sulfate-induced suppression of regional monsoon precipitation (Turner and Annamalai 2012). Most CMIP5 models adopted in this study include both the direct and indirect effect of aerosols, but their level of understanding is limited, leaving uncertainties in precisely assessing the future state of the monsoon.

Acknowledgments

A.K. was supported by the SOUSEI Program of the Ministry of Education, Culture, Sports, Science and Technology (MEXT) of Japan. H.E. was supported by the Environment Research and Technology Development Fund (2-1503) of the Ministry of the Environment of Japan.

References

Annamalai, H., K. Hamilton, and K. R. Sperber, 2007: South Asian summer monsoon and its relationship with ENSO in the IPCC AR4 simulations. *J. Climate*, **20**, 1071–1083.

Bollasina, M. A., Y. Ming, and V. Ramaswamy, 2011: Anthropogenic aerosols and the weakening of the South Asian summer monsoon. *Science*, **334**, 502–505.

Chadwick, R., I. Boutle, and G. Martin, 2013: Spatial patterns of precipitation change in CMIP5: Why the rich don't get richer in the tropics. *J. Climate*, **26**, 3803–3822.

Chang, C.-P., Y. H. Ding, R. H. Johnson, G. N.-C. Lau, B. Wang, and T. Yasunari, eds., 2011: *The Global Monsoon System: Research and Forecast, 2ndEdition.* World Scientific, 608 pp.

Chou, C., J. D. Neelin, C.-A. Chen, and J.-Y. Tu, 2009: Evaluating the "rich-get-richer" mechanism in tropical precipitation change under global warming, *J. Climate*, **22**, 1982–2005.

Christensen, J. H., K. Krishna Kumar, E. Aldrian, S.-I. An, I. F. A. Cavalcanti, M. de Castro, W. Dong, P. Goswami, A. Hall, J. K. Kanyanga, A. Kitoh, J. Kossin, N.-C. Lau, J. Renwick, D. Stephenson, S.-P. Xie, and T. Zhou, 2014: Climate Phenomena and their Relevance for Future Regional Climate Change. In: *Climate Change 2013: The Physical Science Basis. Contribution of Working Group I to the Fifth Assessment Report of the Intergovernmental Panel on Climate Change* [Stocker, T. F., D. Qin, G.-K. Plattner, M. Tignor, S. K. Allen, J. Boschung, A. Nauels, Y. Xia, V. Bex and P. M. Midgley (eds.)]. Cambridge University Press, pp. 1217–1308.

Collins, M., R. Knutti, J. M. Arblaster, J.-L. Dufresne, T. Fichefet, P. Friedlingstein, X. Gao, W. J. Gutowski, T. Johns, G. Krinner, M. Shongwe, C. Tebaldi, A. J. Weaver, and M. Wehner, 2014: Long-term Climate Change: Projections, Commitments and Irreversibility. In: *Climate Change 2013: The Physical Science Basis. Contribution of Working Group I to the Fifth Assessment Report of the Intergovernmental Panel on Climate Change* [Stocker, T. F., D. Qin, G.-K. Plattner, M. Tignor, S. K. Allen, J. Boschung, A. Nauels, Y. Xia, V. Bex and P. M. Midgley (eds.)]. Cambridge University Press, pp. 1029–1136.

Endo, H. and A. Kitoh, 2014: Thermodynamic and dynamic effects on regional monsoon rainfall changes in a warmer climate. *Geophys. Res. Lett.*, **41**, doi:10.1002/2013GL059158.

Flato, G., J. Marotzke, B. Abiodun, P. Braconnot, S. C. Chou, W. Collins, P. Cox, F. Driouech, S. Emori, V. Eyring, C. Forest, P. Gleckler, E. Guilyardi, C. Jakob, V. Kattsov, C. Reason, and M. Rummukainen, 2014: Evaluation of Climate Models. In: *Climate Change 2013: The Physical Science Basis. Contribution of Working Group I to the Fifth Assessment Report of the Intergovernmental Panel on Climate Change* [Stocker, T. F., D. Qin, G.-K. Plattner, M. Tignor, S. K. Allen, J. Boschung, A. Nauels, Y. Xia, V. Bex and P. M. Midgley (eds.)]. Cambridge University Press, pp. 741–866.

Held, I. M. and B. J. Soden, 2006: Robust responses of the hydrological cycle to global warming, *J. Climate*, **19**, 5686–5699.

Hsu, P.-C., T. Li, and B. Wang, 2011: Trends in global monsoon area and precipitation in the past 30 years. *Geophys. Res. Lett.*, **38**, L08701.

Hsu, P.-C., T. Li, J.-J. Luo, H. Murakami, A. Kitoh, and M. Zhao, 2012: Increase of global monsoon area and precipitation under global warming: A robust signal? *Geophys. Res. Lett.*, **39**, L06701.

Hsu, P.-C., T. Li, H. Murakami, and A. Kitoh, 2013: Future change of the global monsoon revealed from 19 CMIP5 models. *J. Geophys. Res.*, **118**, doi.10.1002/jgrd.50145.

Kim, H.-J., B. Wang, and Q. Ding, 2008: The global monsoon variability simulated by CMIP3 coupled climate models. *J. Climate*, **21**, 5271–5294.

Kitoh, A., 2011: Impact of climate change on Asian monsoon characteristics. In: *The Global Monsoon System: Research and Forecast, 2nd Edition.* (eds. Chang, C.-P. *et al.*). *World Scientific*, pp. 557–568.

Kitoh, A., H. Endo, K. Krishna Kumar, I. F. A. Cavalcanti, P. Goswami, and T. Zhou, 2013: Monsoons in a changing world: A regional perspective in a global context. *J. Geophys. Res.*, **118**, 3053–3065. doi.10.1002/jgrd.50258.

Lee, J.-Y. and B. Wang, 2014: Future change of global monsoon in the CMIP5. *Clim. Dyn.*, **42**, 101–119.

Li, Y., N. C. Jourdain, A. S. Taschetto, C. C. Ummenhofer, K. Ashok, and A. Sen Gupta, 2012: Evaluation of monsoon seasonality and the tropospheric biennial oscillation transitions in the CMIP models. *Geophys. Res. Lett.*, **39**, L20713 doi:10.1029/2012GL053322.

Ma, J. and S.-P. Xie, 2013: Regional patterns of sea surface temperature change: A source of uncertainty in future projections of precipitation and atmospheric circulation. *J. Climate*, **26**, 2482–2501.

Ogata, T., H. Ueda, T. Inoue, M. Hayasaki, A. Yoshida, S. Watanabe, M. Kira, M. Ooshiro, and A. Kumai, 2014: Projected future changes in the Asian monsoon: A comparison of CMIP3 and CMIP5 model results. *J. Meteor. Soc. Japan*, **92**, 207–225.

Polson, D., M. Bollasina, G. C. Hegerl, and L. J. Wilcox, 2014: Decreased monsoon precipitation in the Northern Hemisphere due to anthropogenic aerosols. *Geophys. Res. Lett.*, **41**, doi: 10.1002/2014GL060811.

Seager, R., N. Naik, and G. A. Vecchi, 2010: Thermodynamic and dynamic mechanisms for large-scale changes in the hydrologic cycle in response to global warming. *J. Climate*, **23**, 4651–4668.

Seneviratne, S. L. and coauthors, 2012: Changes in climate extremes and their impacts on the natural physical environment. In: *Managing the Risks of Extreme Events and Disasters to Advance Climate Change Adaptation. A Special Report of Working Groups I and II of the Intergovernmental Panel on Climate Change* (eds. Field, C. B. *et al.*). Cambridge University Press, pp. 109–230.

Song, F., T. Zhou, and Y. Qian, 2014: Responses of East Asian summer monsoon to natural and anthropogenic forcings in the 17 latest CMIP5 models. *Geophys. Res. Lett.*, **41**, doi:10.1002/2013GL058705.

Sperber, K. R., H. Annamalai, I.-S. Kang, A. Kitoh, A. Moise, A. G. Turner, B. Wang, and T. Zhou, 2013: The Asian summer monsoon: An intercomparison of CMIP5 vs. CMIP3 simulations of the late 20th century. *Clim. Dyn.*, **41**, 2711–2744.

Turner, A. G. and A. Annamalai, 2012: Climate change and the South Asian summer monsoon. *Nature Climate Change*, **2**, doi:10.1038/NCLIMATE1495.

Turner, A., K. R. Sperber, J. Slingo, G. Meehl, C. R. Mechoso, M. Kimoto, and A. Giannini, 2011: Modelling monsoons: understanding and predicting current and future behavior. In: *The Global Monsoon System: Research and Forecast, 2nd Edition.* (eds. Chang, C.-P. *et al.*). World Scientific, pp. 421–454.

Vecchi, G. A. and B. J. Soden, 2007: Global warming and the weakening of the tropical circulation. *J. Climate*, **20**, 4316–4340.

Wang, B. and LinHo, 2002: Rainy season of the Asian-Pacific summer monsoon. *J. Climate*, **15**, 386–398.

Wang, B. and Q. Ding, 2006: Changes in global monsoon precipitation over the past 56 years. *Geophys. Res. Lett.*, **33**, L06711.

Wang, B. and Q. Ding, 2008: Global monsoon: Dominant mode of annual variation in the tropics. *Dynam. Atmos. Oceans*, **44**, 165–183.

Wang, B., H.-J. Kim, K. Kikuchi, and A. Kitoh, 2011: Diagnostic metrics for evaluation of annual and diurnal cycles. *Clim. Dyn.*, **37**, 941–955.

Wang, B., Y. Liu, H.-J. Kim, P. J. Webster, and S.-Y. Yim, 2012: Recent change of the global monsoon precipitation (1979–2008). *Clim. Dyn.*, **39**, 1123–1135.

Wang, B., S.-Y. Yim, J.-Y. Lee, J. Liu, and K.-J. Ha, 2014: Future change of Asian-Australian monsoon under RCP 4.5 anthropogenic warming scenario. *Clim. Dyn.*, **42**, 83–100.

Xie, S. P., C. Deser, G. A. Vecchi, J. Ma, H. Teng, and A. T. Wittenberg, 2010: Global warming pattern formation: Sea surface temperature and rainfall. *J. Climate*, **23**, 966–986.

Zhang, L. and T. Zhou, 2011: An assessment of monsoon precipitation changes during 1901–2001. *Clim. Dyn.*, **37**, 279–296.

Chapter 25

Review on Studies of Air Pollution and Climate Change Interactions in Monsoon Asia

Congbin Fu* and Aijun Ding

Institute for Climate and Global Change Research and School of Atmospheric Sciences, Nanjing University, Nanjing, 210023, China

Jian Wu

School of Resource Environment and Earth Science, Yunnan University, Kunming, 650091, China

fcb@nju.edu.cn

The relationship between air pollution and climate change in Monsoon Asia has long been studied, but most of these studies were conducted for a one-directional focus on the impact of air pollution on climate or the influence of Asian monsoon on air pollution. In this review, we give a summary on both directions with focus on their two-way interactions in Monsoon Asia. The main characteristics of air pollution and monsoon climate in Asia and the basic processes of air pollution-monsoon climate interaction are introduced. From the perspectives of observation analysis and numerical modeling, the main studies and findings on the impact of air pollution, mainly aerosols, on monsoon climate and the impact of monsoon climate change on air pollution at different temporal scales are reviewed. Typical cases are also illustrated to explain the main mechanisms of two-way interactions in regions with intensive and complex pollution, and their implications are also discussed. This review highlights the importance of long-term observations and two-way coupled numerical modeling on improving the current understanding of air pollution and climate change issues in the Monsoon Asia region.

1. Introduction

Air pollution and climate change are two of the most important issues related to social-economic development and global sustainability. Both topics have been intensively studied in the past decades. Air pollution studies mainly focus on understanding the emission, transport, chemical and deposition processes as well as their impacts and control measures, while climate researches aim to understand climatology, variability and change of main meteorological parameters, such as air temperature and rainfall etc., over a period of time. Both of them are mainly located in two separate traditional directions in the field of atmospheric sciences.

However, it has been known that air quality and climate change closely linked with each other (e.g. Leung and Gustafson 2005; Ramanathan and Feng 2009; Jacob and Winner 2009). On one hand, air pollutants, particularly those radiation important aerosol species, could change the climate by their direct effect on solar radiation transfer and by their indirectly influence via modification of clouds and precipitation. The direct effect mainly includes two parts: one is the absorption aerosols like black carbon and mineral dust, which absorb solar radiation and heat the atmosphere, and the other is scattering aerosols like sulfate, nitrate, and organic carbons etc., which can partly block the solar radiation consequently causing a net

The Global Monsoon System: Research and Forecast (3rd Edition)
Edited by C. P. Chang et al.

cooling effect of earth surface. The indirect effect of aerosols is mainly due to their roles as cloud condensation nuclei (IPCC 2013).

On the other hand, climate change also influences air quality because meteorological conditions always play important roles on local and regional air quality (Jacob and Winner 2009). For example, changes in air temperature, circulations and precipitation can substantially influence main processes related to air pollution, such as emission, transport and dispersion, chemical transform, and dry/wet deposition. Climate change can influence frequency or distribution of lightning activities, which produces nitrogen oxides (NO_x) in the atmosphere and also influences burning activities of forests and grasses. The change of air temperature and solar radiation can also influence emission rate of biogenic trace gases, such as volatile organic compounds (VOCs) from vegetation (Jacob and Winner 2009). The change of circulation patterns and atmospheric vertical structure will influence long-range transport and dispersion of air pollutants (Ding *et al.* 2004; Ding *et al.* 2013a; Zhou *et al.* 2013; Ding *et al.* 2016; Zhang *et al.* 2016). The change of air temperature, water vapor, and radiation can influence the chemical reaction rate and may cause a significant change of secondary pollutants (Squizzato *et al.* 2013; Zhang *et al.* 2015). In addition, the change of rainfall can directly influence wet deposition of air pollutants (Xiao and Liu 2002; Gratz *et al.* 2009), especially for those water-soluble species; also the change of surface characters (e.g. soil moisture and land-cover etc.) may influence dry deposition of air pollutants (Wesely and Hicks 2000).

The Monsoon Asia is a unique area to study air pollution, monsoon climate change and their interactions. First, Asia is still suffering from heavy air pollution in some regions. The latest Intergovernmental Panel on Climate Change (IPCC) assessment (IPCC 2013) pointed out that "*The emissions, properties and concentrations of aerosols or aerosol precursors could respond significantly to climate change, but there is little consistency across studies in the magnitude or sign of this response*" and that "*models simulate relatively small feedback parameters (i.e. within $\pm 0.2\ Wm^{-2}K^{-1}$) with very low confidence, however regional effects on the aerosol may be important.*" As one of the main regions with intensive aerosol pollution, Asia is a focus region from the perspective of aerosol's climate effects. At the same time, this region has unique climate with the strongest monsoon and also the largest inter-annual and decadal variability in monsoon precipitation (Fu and Wen 1999; Li and Zeng 2005), and also the monsoon circulation may has been found to strongly influence air pollution transport in this region (Ding *et al.* 2013a; Zhou *et al.* 2013).

Although large numbers of studies have been conducted to understand the relationships between air pollution and climate in this region in the past decades, the existing reviews mainly focused on the impact of aerosols on climate change, such as that done by Lau and Kim (2011) for the South Asia region, and the latest systemic review done in the IPCC AR5 report (IPCC 2013). There is still a lack of comprehensive review on their two-way interactions in Monsoon Asia, especially in the East Asia region with the strongest emission of air pollutants in the world. Therefore, this work will try to make a review on the air pollution-monsoon climate interactions in Monsoon Asia.

2. Characteristics of Air Pollution and Monsoon Climate in Asia

Asia is one of the main regions suffering from heavy air pollution. The main characteristics of air pollution in the Monsoon Asia can be summarized as the following two aspects:

Firstly, there are very high emission rate of fossil fuel (FF) combustion sources (see global sulfur dioxide (SO_2) emission as an example shown in Fig. 1) and hence a high concentration of primary and secondary pollutants (such as SO_2, carbon dioxide (CO), NO_x, ozone (O_3), black

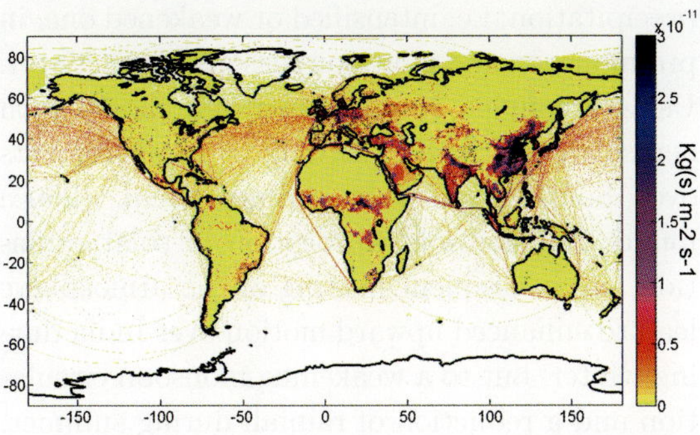

Fig. 1. Global emission rate of sulfur dioxide in 2008 (data from EDGAR V4.2).

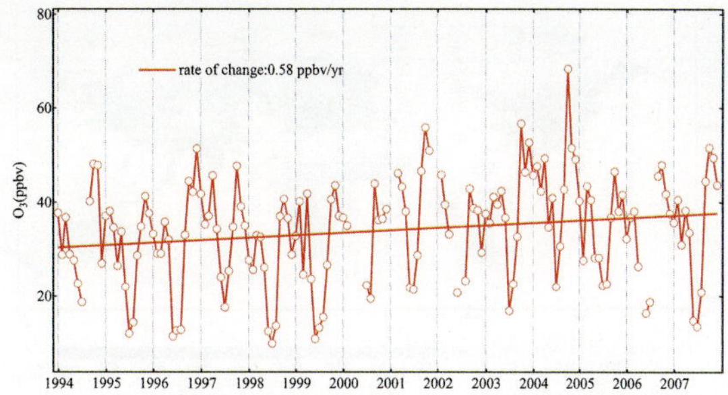

Fig. 2. Increased tropospheric ozone recorded at Hok Tsui background site in South China, 1994–2007. Results reported by Wang *et al.* (2009).

carbon (BC), sulfate, nitrate and organic particles etc.) in Asia, especially in coastal regions of China and Ganges valley of India. These species plays important roles on climate change through their direct and/or indirect impacts on radiative transfer, cloud and precipitations, and they are so called as short-lived climate forcers (SLCF).

Secondly, besides anthropogenic FF combustion pollutants, the Monsoon Asia is also influenced by intensive pollution from biomass burning activities and dust storms in specific seasons. For example, intensive forest burning activities often appear in South Asia in spring and in Siberia in summer (Hao and Liu 1994; Jaffe *et al.* 2004; Zhou *et al.* 2013), and intensive man-made burning activities of agricultural straw occurred in the North and East China Plains in June (Huang *et al.* 2012; Ding *et al.* 2013b). Dust storm frequently occurred from Taklimakan and Gobi deserts in Northwest China and heavy dust aerosols were often transported over the eastern China, southern China and also the Pacific Ocean even the globe (Zhang *et al.* 1997; Nie *et al.* 2012, 2014). These biomass burning or mineral dust aerosols have been found to cause complex interactions in the climate system, after they are mixed with the anthropogenic pollutants.

East Asia not only has the highest concentration of air pollutants, such as SO_2, NO_x and particular matters, but is still experiencing an increasing trend of emission rate of some pollutants (Streets *et al.* 2008). Several works based on satellite retrievals all showed that the NO_2 column has been increasing continuously in the past two decades over the East China (Richter *et al.* 2005; Itahashi *et al.* 2013). At the same time, in this area tropospheric O_3, which is the third-most important greenhouse gas followed CO_2 and CH_4, and also an important pollutant with negative impact on human health and ecology, showed a positive trend with a quite high increasing rate in Pearl River Delta (PRD) in South China (Wang *et al.* 2009), Yangtze River Delta (YRD) in East China (Xu *et al.* 2008), and Beijing in the North China Plain (Ding *et al.* 2008). However, observations did not show a significant trend of tropospheric ozone in other regions of the world (Oltmans *et al.* 2013 and references cited therein).

Influenced by large-scale thermal contrast between Pacific and India Oceans and the Asian continent, Asia is the largest and strongest monsoon area in the globe. Interacted with other monsoon regions (e.g. Australia and West-Africa monsoons), the Asian Monsoon is an important part of the global monsoon system (Fu and Zheng 1998; Fu and Wen 1999; Ding and Chan 2005; Li and Zeng 2005). Previous studies suggested that the Asian monsoon region owns the largest inter-annual and inter-decadal change in rainfall in the world (see Fig. 3) (Fu and Wen 1999). On one hand, owning over 60% of the world population, Asia has the largest impact

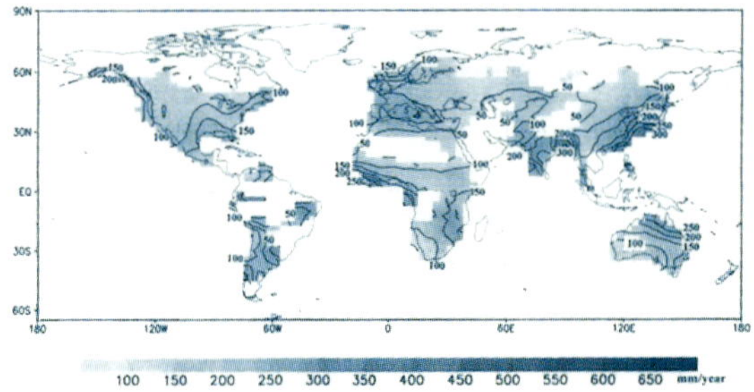

Fig. 3. Global distribution of the interannual variability of precipitation (1951–1990) (unit: mm/year). Results from Fu and Wen (1999).

from the monsoons system by their influence on the water resource, frequency of disasters, air quality, agriculture and human health. On the other hand, here the monsoon system is also easily influenced by the intensive human activities such as fast land-use/land-cover change and heavy air pollution (Fu 2003; Jiang *et al.* 2013).

3. Air Pollution–Monsoon Climate Two-way Interactions in Asia

3.1. *Impacts of air pollution on monsoon climate*

The high concentrations of natural and anthropogenic air pollutants, mainly aerosols, have been known to significantly influence the regional climate in the Monsoon Asia (e.g. Menon *et al.* 2002; Nakajima *et al.* 2007; Li *et al.* 2011). The potential effects of aerosol to Indian monsoon have been relatively well-understood. With model simulations based on INDOEX measurements, Chung *et al.* (2002) found that aerosols could intensify the pre-monsoon rainfall in India. The main mechanism is that aerosols induced surface cooling by reducing solar radiation at the surface and a warming of the lower troposphere by absorption, and then the enhancement of meridional temperature gradient in the atmosphere resulted in a dynamical response which further led to a stronger pre-monsoon rainfall. Some studies also reported distinct effects of aerosols on precipitation, i.e. intensified or weakened one, in pre-monsoon and monsoon seasons, respectively. Using a coupled ocean–atmosphere model with aerosols prescribed according to measurements over South Asia, Ramanathan *et al.* (2005) found that aerosols' absorption of solar radiation and consequent heating of the atmosphere lead to enhanced upward motion over India during winter, but to a weakening monsoon circulation and a reduction of rainfall during summer. These results were further supported by Meehl *et al.* (2008), who found that BC aerosols lead to an increase of pre-monsoon rainfall over India but to a decrease in the monsoon season.

The Tibetan Plateau has been found to influence the monsoon circulation, regional climate and large-scale transport and accumulation of air pollutants in South Asia. Lau *et al.* (2006a; 2006b) proposed an "elevated heat pump" hypothesis and suggested that anomalous accumulation of absorbing aerosols, such as dust aerosols from the nearby deserts and BC from regional sources and the southern slopes of the Himalayas, induces a large-scale upper-level heating anomaly over the Tibetan Plateau in April and May, which reinforces the meridional temperature gradient and intensifies the monsoon over India in June and July. Another co-effect of Tibetan Plateau and aerosols is the BC on glaciers, which has been found to be an impact factor influencing regional climate over the entire Asia region through hydrological cycle (e.g. Qian *et al.* 2011).

Changes in summer precipitation over China have been linked to the change of aerosol loading. Increased summer precipitation has been observed over the southern China but a decreased one in the northern China, presenting a "north drying and south wetting" pattern (Xu *et al.* 2006). Menon *et al.* (2002) attributed the change of rainfall pattern to aerosols. The increased aerosol optical depth (AOD) caused a cool-zone over south-central China, which modified the previous land-sea thermal contract and caused weakened East Asian summer monsoon (EASM) circulations.

The weakened EASM might restrain the rain belt in the southern China longer with less moisture penetrating further north in summer season.

Several numerical simulation studies carried out to further investigate this mechanism. By running the regional climate model RegCM3, Wu *et al.* (2008) found the direct climatic effects of BC could cause large scale change of air temperature and humidity in different regions over the Asian continent. Wu *et al.* (2009) further carried out simulations on the direct climate effects of combined total aerosol on East Asia, including sulfate, dust, BC, and organic carbon, using the NCAR Community Atmospheric Model. They found that the increase in vapor content and precipitation in northern and northwestern China contradicts recent trends of flooding in the south and drought in the north of China. An increase of precipitation (up to 9%) was found in the northern and northwestern China, while a decrease (up to 12%) was obtained in the southern part of the Tibetan Plateau, the Sichuan Basin, and most of the southern, southeastern, and northeastern China.

Jiang *et al.* (2013) used a newer version of NCAR Community Atmospheric Model (version 5) to simulate the impact of anthropogenic aerosols on the summer monsoon clouds and precipitation in East Asia by considering the direct, semi-direct and indirect effects of various aerosol components, such as BC, sulfate, primary organic matter (POM) based on simulations of pre-industry (PI) and present day (PD) scenarios. They found that anthropogenic sulfate and POM reduce the solar flux reaching the surface, and induce a cooling at surface over the land through directly scattering the solar flux and indirectly increasing of CCN concentration and cloud radiative forcing. The cooling at the surface over the land increases the atmospheric stability and thus suppresses the deep convection, and consequently the condensational heating from deep convection is reduced in North China. As a result, the cooled free troposphere and decreased thermal contrast between the land and ocean weaken the EASM and cause reduced precipitations in North China and increased precipitation over the adjacent oceans.

Black carbon shows a different effect from other aerosols (Ding *et al.* 2016). The reduction of surface solar flux by anthropogenic BC is relatively weaker compared to anthropogenic sulfate and POM because BC reduces the cloud liquid water content by its semi-direct effect resulting in a weaker (less negative) shortwave cloud forcing (Jiang *et al.* 2013). The heating of the lower troposphere by anthropogenic BC leads to a stronger deep convection at the latitudes of 25°N–30°N. The combined effect of all aerosols suggests a strong surface dimming leading to the surface and free tropospheric cooling over the land, which suppresses the precipitation in North China and enhances the precipitation in South China and adjacent oceans (Fig. 4).

3.2. *Impacts of monsoon climate on air quality*

Monsoon climate change has been found to play important roles on the long-range transport of air pollutants and consequently to influence the seasonal cycle, interannual and decadal variations, and even long-term trend of air quality in monsoon regions in Asia (Zhou *et al.* 2013; Zhang *et al.* 2016).

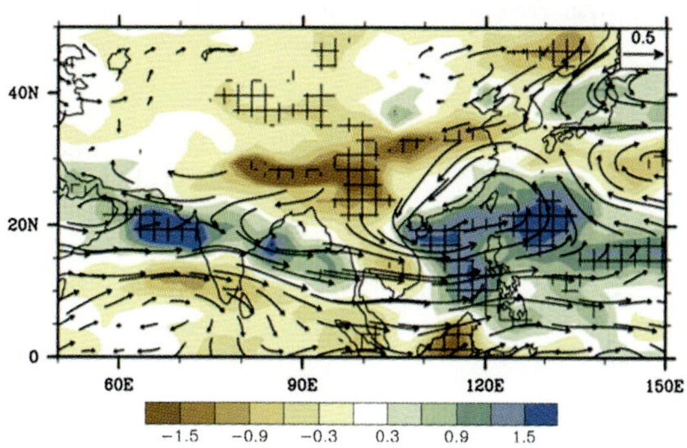

Fig. 4. Summer surface total precipitation (shading, in mmd^{-1}) and 850 hPa wind (vector) changes induced by anthropogenic aerosols. *Note*: the figure shows the differences between simulation of PD and PI scenarios (i.e. PD-PI). Results from Jiang *et al.* (2013).

Observation and numerical studies suggest that the annual cycle of monsoon circulations controls the seasonal patterns of trace gases and aerosols. For example, observational studies showed totally different seasonal patterns of tropospheric O_3 at Hong Kong of South China and Beijing in the North China Plain (Ding *et al.* 2008, Wang *et al.* 2009; see Fig. 5). The O_3 concentration in Hong Kong reaches to a minimum in summer because of transport of marine clean inflows from the South China Sea carried by the summer monsoon, while in Beijing it reaches a maximum in summer because of photochemical production in the long-range transported plumes from the East China and the North China Plain from the south. Observations of O_3 and fine particles ($PM_{2.5}$) at Nanjing also show that both species in the Yangtze River Delta are strongly influenced by the Asian monsoon (Ding *et al.* 2013c).

The influence of monsoon circulations on air pollution can be well addressed by various models. For example, Zhang *et al.* (2010) simulated the seasonal variations of aerosols in the eastern China using a global chemical transport model Geos-Chem, and found that the Asian summer monsoon controls the seasonal patterns of aerosol over the monsoon regions. Using a Lagrangian dispersion model HYSPLIT, Ding *et al.* (2013a) conducted Lagrangian backward dispersion simulations of carbon monoxide and found that monsoon circulations influenced the long-range transport and caused distinct seasonal variations of air pollution from different regions over the Monsoon Asia (Fig. 6).

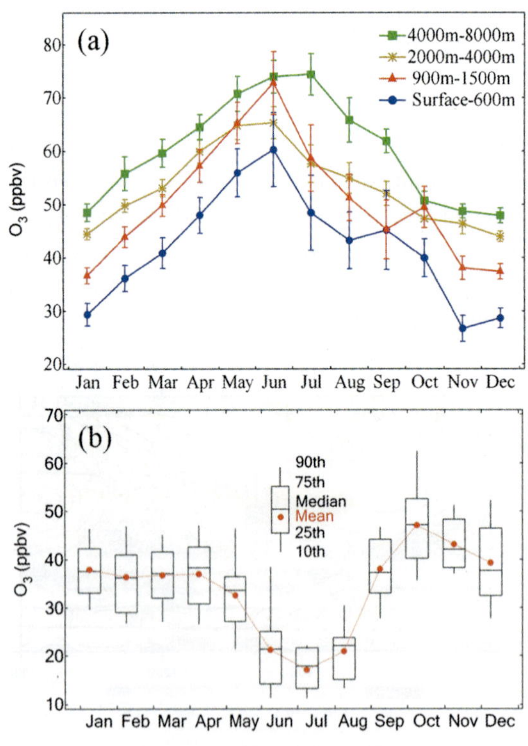

Fig. 5. Seasonal variation of tropospheric ozone (a) over Beijing and (b) in Hong Kong. Results from Ding *et al.* (2008) and Wang *et al.* (2009).

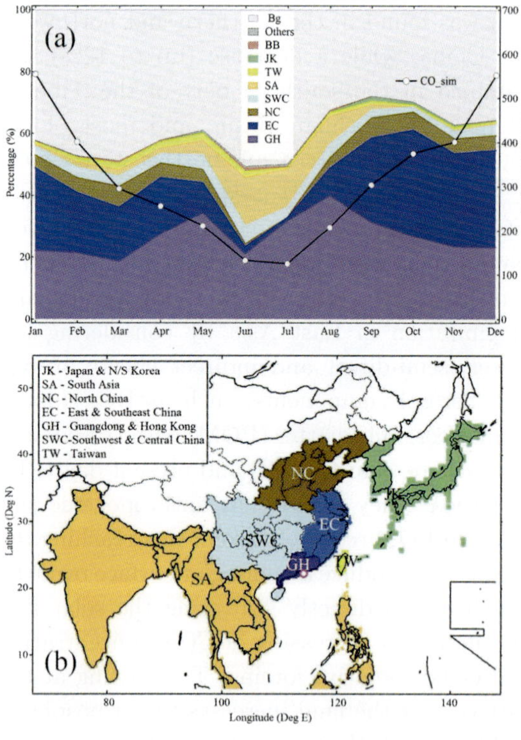

Fig. 6. (a) Seasonal variations of simulated CO and its apportionment with contribution of anthropogenic emission from different regions in Asia, which was defined in (b). Results from Ding *et al.* (2013a).

It has been known that the Asian monsoon has the strongest inter-annual and decadal variations. Several studies showed that the inter-annual/decadal variations of monsoon play important roles on the year-to-year difference of main air pollutants, like O_3 and $PM_{2.5}$ in the monsoon regions. For example, a decadal-scale weakening of the East Asian summer monsoon caused an increase in aerosol concentrations over eastern China, a weak Southeast Asian summer monsoon resulting in an increase of organic carbon concentration in the southern China and the western North Pacific by influencing wet-deposition, deep convection and large scale circulations (Zhang *et al.* 2010, 2010; Zhu *et al.* 2012). Zhu *et al.* (2012) found that monsoon circulations can cause an year-to-year difference in mean $PM_{2.5}$ concentration up to ~20% in the East China (see Fig. 7a), and a calculation of monsoon index suggested that a continuous weaken East Asian summer monsoon contributes to a substantial increase of $PM_{2.5}$ in the past few decades (see Fig. 7b) (Zhu *et al.* 2012; Chin *et al.* 2012).

Tropospheric O_3 in some regions in Asia also shows strong linkage with the intensity of monsoon. For example, Zhou *et al.* (2013) found that the lower tropospheric O_3 in South China showed very good correlation with the monsoon index on an inter-annual scale (Fig. 8a) and the main reasons were attributed to year-to-year difference in long-range transport of continental outflows associated with the anomaly of winter monsoons. Studies (e.g. Zhang *et al.* 2013) also suggested that the frequency and intensity of dominant synoptic weather patterns control inter-annual variability of lower tropospheric O_3

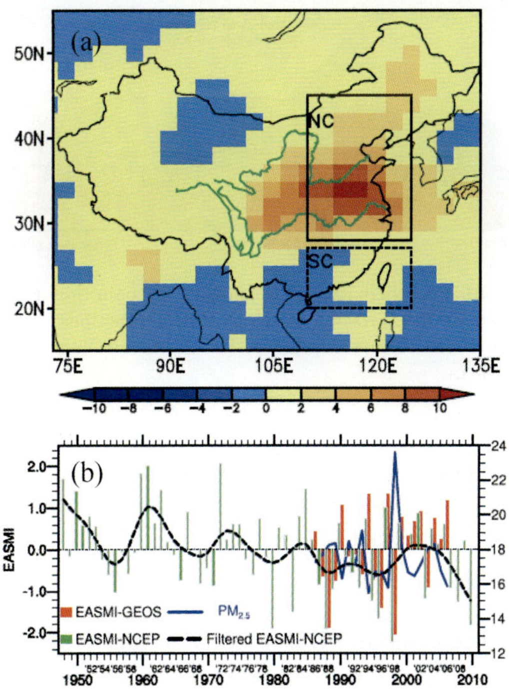

Fig. 7. (a) Differences in simulated JJA $PM_{2.5}$ between the weakest and strongest EASM years; (b) Normalized EASMI and the simulated JJA surface $PM_{2.5}$ concentrations. Results from Zhu *et al.* (2012).

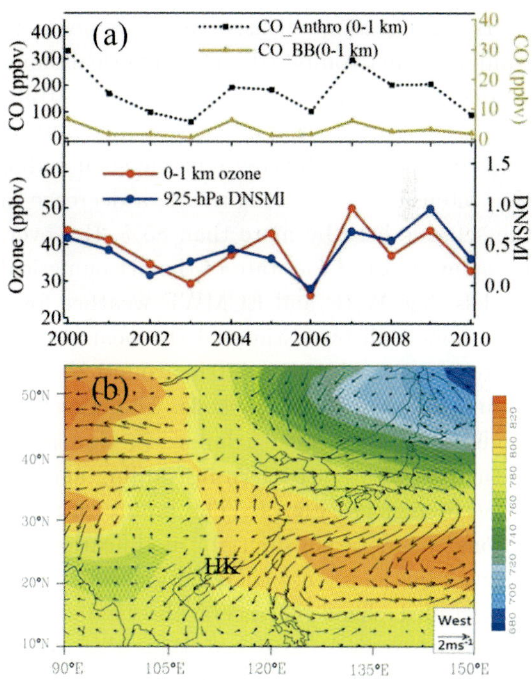

Fig. 8. (a) Averaged PBL O_3, 925-hPa DNSMI, and simulated CO in spring (March–April) during 2000–2010; (b) 925-hPa wind vector difference in spring between years of PBL O_3 peaks and valleys plotted on averaged geopotential height. Results from Zhou *et al.* (2013).

and also influence the O_3 trend in southern China in the latest decade.

3.3. *Needs of considering two-way interactions between monsoon climate and air pollution*

The above two sessions have demonstrated the important relationship between air pollution and monsoon climate. However, most of the numerical modeling works reviewed here were mainly from the perspective of one-way direction, i.e. understanding the impact of air pollution on climate or the impact of monsoon climate on air pollution. Here we will show how important the two-way interactions for both air quality and climate point of view through an observational synoptic evidences reported by Ding *et al.* (2013b) and Huang *et al.* (2016).

By using comprehensive measurements in Nanjing China compared with numerical simulations and reanalysis data, Ding *et al.* (2013b) found that during an intensive air pollution episode there was a decrease in the solar radiation intensity by more than 70%, a decrease in the sensible heat by more than 85%. However, nowadays most of the state-of-the-art numerical models, like WRF and ECMWF weather forecast models all overestimated the near surface air temperature by about 10 K around Nanjing during the air pollution episode. The numerical models also all showed a failure in simulation of the precipitation pattern in Nanjing and the adjacent regions (Fig. 9). The failure of simulations reveals a substantial "modification" of weather by the intense air pollution.

Detailed chemical tracers observed during the study period suggested that this server air pollution episode was caused by intensive agriculture burning plumes mixed the fossil fuel combustion, and it was found that the air pollution-boundary layer dynamics and aerosol-radiation-cloud feedbacks played important roles on the two-way interactions and then

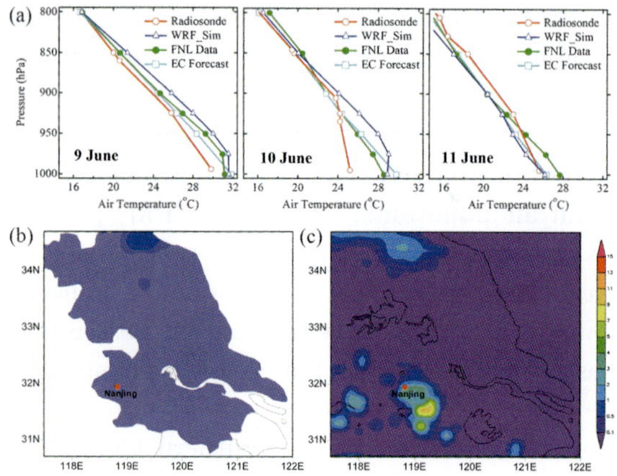

Fig. 9. Comparisons of observed and simulated (a) air temperature and (b) observed and (c) simulated rainfall patterns in Jiangsu province during an intensive air pollution episode. Results from Ding *et al.* (2013b).

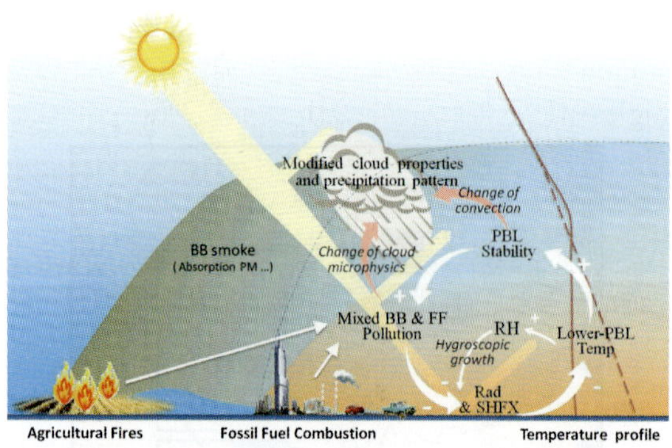

Fig. 10. A schematic figure showing air pollution–PBL dynamics and aerosol–radiation cloud interactions under a condition of mixed plumes. Results from Ding *et al.* (2013b).

substantial modified of the weather (Fig. 10). The main mechanisms of air pollution-PBL-weather interactions in Ding *et al.* (2013b) and Huang *et al.* (2016) can be summarized as:

- An enhanced PBL stability initiated by the pollution suppresses the vertical mixing and dispersion of the pollutants, resulting in more intense pollution in the lower PBL.
- The cooling of the PBL and resulting in increase of the relative humidity, which

amplified the feedback further by increasing the aerosol scattering coefficient through the hydroscopic effect.

- The changed PBL stability and the mixed aerosols modify cloud properties and precipitation patterns.

The above findings from this case are suggestive of strong feedback between a heavy aerosol loading, radiative transfer, air temperature profile/stability and precipitation. To our knowledge, so far most of the models used to assess the impact of air pollution on climate or to understand the impact of monsoon climate on air pollution are off-line. In fact, as that has been shown, these processes have strong two-way interactions. To narrow down the uncertainty, there is a need of using online coupled models, with a horizontal resolution high enough to identify the spatial distribution of different pollution sources, and high vertical resolution and better parameterization to well character these complex air pollution-PBL interactions.

On the other hand, for the aspect of air pollution control measures, the above work also suggests that changes in the PBL stability caused by upwind regional plumes may further enhance the accumulation of local anthropogenic pollutants in lower PBL and cause extreme air pollution around the surface. Recent works done by Wang *et al.* (2014) and Ding *et al.* (2016) showed that a consideration of PBL feedback from radiation effects of aerosols does substantially improve the simulation extreme air pollution in North and East China.

Therefore, from either climate or air quality point of view, these works suggest an urgent need of developing advanced models and comprehensive measurement platforms, which are the two key tools for improving the current understanding the interaction of air pollution and monsoon climate in the Asian monsoon region.

4. Summary

Air pollution and climate change are two important issues related to the social development in the Monsoon Asia. This review gives a summary of the studies related to air pollution-climate change interactions in the monsoon Asian region. The main conclusions are given as follows:

(1) Asia is a region with unique monsoon climate, with strongest (seasonal) monsoon, largest inter-annual and decadal change in precipitation, and many regions suffering from heavy regional pollution. With these feature, Asia is a unique area for studying the two-way coupling process of air-pollution and climate change.

(2) The effects of aerosols on regional climate have been intensively studied by analyzing observational data and numerical modeling in South Asia and East Asia in the past decades. The effects are seasonal dependent and aerosol composition dependent.

(3) Observational and modeling studies showed that the Asian monsoon controls the seasonal, inter-annual and decadal variations even long-term trend in air quality of some regions in Asia, suggesting an important need of developing fully on-line coupled models for assessing aerosol's climate effects and for policy making of decadal-scale air quality measures.

(4) To understand well the impact of interactions between air pollution and monsoon climate, online coupled models and comprehensive measurement platforms are all needed and also should be well co-used.

Acknowledgments

This work was supported by Jiangsu Provincial Collaborative Innovation Center for Climate Change (2011 program). Aijun Ding's work was supported by the National Natural

Science Foundation of China (D0512/ 41422504; D0512/91544231).

References

Chin, M., 2012: Dirtier air from a weaker monsoon. *Nature Geosci.*, **5**, 449–450.

Chung, C. E., V. Ramanathan, and J. T. Kiehl, 2002: Effects of the South Asian absorbing haze on the northeast monsoon and surface–air heat exchange. *J. Climate*, **15**, 2462–2476.

Ding, A. J., T. Wang, M. Zhao, T. J. Wang, and Z. K. Li, 2004: Simulation of sea-land breezes and a discussion of their implications on the transport of air pollution during a multi-day ozone episode in the Pearl River Delta of China, *Atmos. Environ.*, **38**, 39, 6737–6750.

Ding, A. J., T. Wang, V. Thouret, J.-P. Cammas, and P. Nédélec, 2008: Tropospheric ozone climatology over Beijing: Analysis of aircraft data from the MOZAIC program, *Atmos. Chem. Phys.*, **8**, 1–13. doi:10.5194/acp-8-1-2008.

Ding, A. J., T. Wang, and C. B. Fu, 2013a: Transport characteristics and origins of carbon monoxide and ozone in Hong Kong, South China. *J. Geophys. Res.*, **118**, 16, 9475–9488.

Ding, A. J., C. B. Fu, X. Q. Yang, J. N. Sun, T. Petäjä, V.-M. Kerminen, T. Wang, Y. Xie, E. Herrmann, L. F. Zheng, W. Nie, Q. Liu, X. L. Wei, and M. Kulmala, 2013b: Intense atmospheric pollution modifies weather: A case of mixed biomass burning with fossil fuel combustion pollution in eastern China. *Atmos. Chem. Phys.*, **13**, 10545–10554. doi:10.5194/acp-13-10545-2013.

Ding, A. J., C. B. Fu, X. Q. Yang, J. N. Sun, L. F. Zheng, Y. N. Xie, E. Herrmann, W. Nie, T. Petaja, V.-M. Kerminen, and M. Kulmala 2013c: Ozone and fine particle in the western Yangtze River Delta: An overview of 1 yr data at the SORPES station. *Atmos. Chem. Phys.*, **13**, 11, 5813–5830.

Ding, Y. and J. C. L. Chan, 2005: The East Asian summer monsoon: An overview. *Meteorol. Atmos. Phys.*, **89**, 117–142.

Ding, A. J., X. Huang, W. Nie, J. N. Sun, V.-M. Kerminen, T. Petäjä, H. Su, Y. F. Cheng, X. Q. Yang, M. H. Wang, X. G. Chi, J. P. Wang, A. Virkkula, W. D. Guo, J. Yuan, S. Y. Wang, R. J. Zhang, Y. F. Wu, Y. Song, T. Zhu, S. Zilitinkevich, M. Kulmala, and C. B. Fu 2016: Enhanced haze pollution by black carbon in megacities in China, *Geophys. Res. Lett.*, **43**, 2873–2879.

Fu, C. B., 2003: Potential impacts of human-induced land cover change on East Asia monsoon. *Global and Planetary Change*, **37**, 3, 219–229.

Fu, C. B. and G. Wen, 1999: Variation of ecosystems over East Asia in association with seasonal, interannual and decadal monsoon climate variability. *Climatic Change*, **43**, 477–494.

Fu, C. B. and Z. M. Zheng, 1998: Monsoon regions: The highest rate of precipitation changes observed from global data. *Chinese Sci. Bull.* **43**, 8, 662–666.

Gratz, L. E., G. J. Keeler, and E. K. Miller, 2009: Long-term relationships between mercury wet deposition and meteorology. *Atmos. Environ.*, **43**, 39, 6218–6229.

Hao, W. M. and M.-H. Liu, 1994: Spatial and temporal distribution of tropical biomass burning. *Global Biogeochem. Cycle*, **8**, 4, 495–503.

Huang, X., Y. Song, M. M. Li, J. F. Li, Q. Huo, X. H. Cai, T. Zhu, M. Hu, and H. S. Zhang, 2012: A high-resolution ammonia emission inventory in China. *Global Biogeochem. Cycles*, 26, GB1030, doi:10.1029/2011GB004161.

Huang, X., A. Ding, L. Liu, Q. Liu, K. Ding, X. Niu, W. Nie, Z. Xu, X. Chi, M. Wang, J. Sun, W. Guo, and C. Fu 2016: Effects of aerosol-radiation interaction on precipitation during biomass-burning season in East China, *Atmos. Chem. Phys.*, **16**, 10063–10082, doi:10.5194/acp-16-10063-2016.

IPCC, 2013: Climate Change 2013: The Physical Science Basis. Contribution of Working Group I to the Fifth Assessment Report of the Intergovernmental Panel on Climate Change. [Stocker, T. F., D. Qin, G.-K. Plattner, M. Tignor, S. K. Allen, J. Boschung, A. Nauels, Y. Xia, V. Bex and P. M. Midgley (eds.)], Cambridge University Press, Cambridge, UK and New York, NY, USA.

Itahashi, S., I. Uno, H. Irie, J. Kurokawa, and T. Ohara, 2013: Trend analysis of tropospheric NO2 column density over East Asia during 2000–2010: Multi-satellite observations and model simulations with the updated REAS emission inventory. *Atmos. Chem. Phys. Discuss.*, **13**, 11247–11268, doi:10.5194/acpd-13-11247-2013.

Jacob, D. J. and D. A. Winner, 2009: Effect of climate change on air quality. *Atmos. Environ.*, **43**, 51–63.

Jaffe, D., I. Bertschi, L. Jaegle, P. Novelli, J. S. Reid, H. Tanimoto, R. Vingarzan, and D. L. Westphal, 2004: Long-range transport of Siberian biomass burning emissions and impact on surface ozone in western North America, *Geophys. Res. Lett.*, **31**, L16106. doi: 10.1029/2004GL020093.

Jiang, Y. Q., X. H. Liu, X.-Q. Yang, and M. H. Wang, 2013: A numerical study of the effect of different aerosol types on East Asian summer clouds and precipitation. *Atmos. Environ.*, doi: 10.1016/j.atmosenv.2012.12.039.

Lau, K.-M. and K.-M. Kim, 2006a: Observational relationships between aerosol and Asian monsoon rainfall, and circulation. *Geophys. Res. Lett.*, **33**, L21810. doi:10.1029/2006GL027546.

Lau, K., M. Kim, and K. Kim, 2006b: Asian summer monsoon anomalies induced by aerosol direct forcing: the role of the Tibetan Plateau. *J. Clim. Dyn.*, **26**, 855e864.

Lau, K.-M. and K. M. Kim, 2011: Effects of dust and black carbon on varibility of the South Asian Monsoon. The Global Monsoon system — Research and forecast, 2nd Edition (by edited by C.-P. Chang *et al.*), pp. 569–583, World Scientific.

Leung, L. R. and W. I. Gustafson, 2005: Potential regional climate change and implications to US air quality. *Geophys. Res. Lett.*, **32**, 16, L16711.

Li, J. P. and Q. C. Zeng, 2005: A new monsoon index, its interannual variability and relation with monsoon precipitation. *Climatic and Environmental Research* (in Chinese), **10**, 3, 351–365.

Li, Z. Q., C. Li, H. Chen, S.-C. Tsay, B. Holben, J. Huang, B. Li, H. Maring, Y. Qian, G. Shi, X. Xia, Y. Yin, Y. Zheng, and G. Zhuang, 2011: East Asian studies of tropospheric aerosols and their impact on regional climate (EAST-AIRC): An overview. *J. Geophys. Res.*, **116**, D00K34. doi:10.1029/2010JD015257.

Meehl, G. A., J. M. Arblaster, and W. D. Collins, 2008: Effects of black carbon aerosols on the Indian Monsoon. *J. Climate*, **21**, 12, 2869–2882. doi:10.1175/2007JCLI1777.1.

Menon, S., J. Hansen, L. Nazarenko, and Y. Luo, 2002: Climate effects of black carbon aerosols in China and India. *Science*, **297**, 2250–2253.

Nakajima, T., S.-C. Yoon, V. Ramanathan, and G.-Y. Shi *et al.* 2007: Overview of the Atmospheric Brown Cloud East Asian Regional Experiment 2005 and a study of the aerosol direct radiative forcing in East Asia. *J. Geophys. Res.*, **112**, D24S91. doi:10.1029/2007JD009009.

Nie, W., T. Wang, L. K. Xue, A. J. Ding, X. F. Wang, X. M. Gao, Z. Xu, Y. C. Yu, C. Yuan, Z. S. Zhou, R. Gao, X. H. Liu, Y. Wang, S. J. Fan, S. Poon, Q. Z. Zhang, and W. X. Wang, 2012: Asian dust storm observed at a rural mountain site in southern China: Chemical evolution and heterogeneous photochemistry. *Atmos. Chem. Phys.*, **12**, 11985–11995. doi:10.5194/acp-12-11985-2012.

Nie, W., A. Ding, T. Wang, V.-M. Kerminen, C. George, L. Xue, W. Wang, Q. Zhang, T. Petäjä, X. Qi, X. Gao, X. Wang, X. Yang, C. Fu, and M. Kulmala 2014: Polluted dust promotes new particle formation and growth, *Sci. Repts.*, **4**, 6634, doi:10.1038/srep06634.

Oltmans, S. J., A. S. Lefohn, D. Shadwick, J. M. Harris, H. E. Scheel, I. Galbally, D. W. Tarasick, B. J. Johnson, E.-G. Brunke, H. Claude, G. Zeng, S. Nichol, F. Schmidlin, J. Davies, E. Cuevas, A. Redondas, H. Naoe, T. Nakano, and T. Kawasato, 2013: Recent tropospheric ozone changes — A pattern dominated by slow or no growth. *Atmos. Environ.*, **67**, 331–351.

Qian, Y., M. G. Flanner, L. R. Leung, and W. Wang, 2011: Sensitivity studies on the impacts of Tibetan Plateau snowpack pollution on the Asian hydrological cycle and monsoon climate. *Atmos. Chem. Phys.*, **11**, 1929–1948. doi:10.5194/acp-11-1929-2011.

Ramanathan, V., C. Chung, D. Kim, T. Bettge, L. Buja, J. T. Kiehl, W. M. Washington, Q. Fu, D. R. Sikka, and M. Wild, 2005: Atmospheric brown clouds: Impacts on South Asian Climate and hydrological cycle. *PNAS*, **102**, 15, 5326–5333. doi:10.1073/pnas.0500656102.

Ramanathan, V. and Y. Feng, 2009: Air pollution, greenhouse gases and climate change: Global and regional perspectives. *Atmos. Environ.*, **43**, 1, 37–50.

Richter, A., J. P. Burrows, H. Nuss, C. Granier, and U. Niemeier, 2005: Increase in tropospheric nitrogen dioxide over China observed from space. *Nature*, **437**, 129–132.

Squizzato, S., M. Masiol, A. Brunelli, S. Pistollato, E. Tarabotti, G. Rampazzo, and B. Pavoni, 2013: Factors determining the formation of secondary inorganic aerosol: A case study in the Po Valley (Italy). *Atmos. Chem. Phys.*, **13**, 1927–1939.

Streets, D. G., C. Yu, and Y. Wu, 2008: Aerosol trends over China, 1980–2000. *Atmos. Res.*, **88**, 174–182.

Wang, J. D., S. X. Wang, J. K. Jiang, A. J. Ding, M. Zheng, B. Zhao, D. Wong, W. Zhou, G. J. Zheng, L. Wang, J. Pleim, and J. M. Hao, 2014: Impact of aerosol-meteorology interactions on fine particle pollution during China's severe winter haze episode in January 2013. *Environ. Res. Lett*, **9**, 9, 094002.

Wang, T., X. L. Wei, A. J. Ding, C. N. Poon, K. S. Lam, Y. S. Li, L. Y. Chan, and M. Anson, 2009: Increasing surface ozone concentrations in the background atmosphere of Southern China, 1994–2007. *Atmos. Chem. Phys.*, **9**, 6217–6227. doi:10.5194/acp-9-6217-2009.

Wesely, M. L. and B. B. Hicks, 2000: A review of the current status of knowledge on dry deposition. *Atmos. Environ.*, **34**, 12–14, 2261–2282.

Wu, J., C. B. Fu, Y. Xu, W. Wang, and Z. Wang, 2008: Simulation of direct effects of black carbon aerosol on temperature and hydrological cycle in Asia by a regional climate model. *Meterol. Atmos. Phys.*, **100**, 179–193.

Wu, J., C. B. Fu, Y. Y. Xu, J. P. Tang, Z. W. Han, and R. J. Zhang, 2009: Effects of total aerosol on temperature and precipitation in East Asia. *Climate Res.*, **40**, 1, 75–87.

Xiao, H.-Y. and C. Q. Liu, 2002: Sources of nitrogen and sulfur in wet deposition at Guiyang, southwest China. *Atmos. Environ.*, **36**, 33, 5121–5130.

Xu, M., C. P. Chang, C. B. Fu, Y. Qi, A. Robock, D. Robinson, and H. Zhang, 2006: Steady decline of east Asian monsoon winds, 1996–2000: Evidence from direct ground measurements of wind speed. *J. Geophys. Res.*, **111**, D24111. doi:10.1029/2006JD007337.

Xu, X., W. Lin, T. Wang, P. Yan, J. Tang, Z. Meng, and Y. Wang, 2008: Long-term trend of surface ozone at a regional background station in eastern China 1991–2006: Enhanced variability. *Atmos. Chem. Phys.*, **8**, 2595–2607. doi:10.5194/acp-8-2595-2008.

Zhang, L., H. Liao, and J. P. Li, 2010: Impacts of Asian summer monsoon on seasonal and interannual variations of aerosols over eastern China. *J. Geophys. Res.*, **115**, D00K05. doi:10.1029/2009JD012299.

Zhang, Q., J. N. Quan, and X. X. Tie *et al.*, 2015: Effects of meteorology and secondary particle formation on visibility during heavy haze events in Beijing, China. *Sci. of Total Environ.*, **502**, 578–584.

Zhang, X. Y., R. Arimoto, and Z. S. An, 1997: Dust emission from Chinese desert sources linked to variations in atmospheric circulation. *J. Geophys. Res.*, **102**, 28041–28047. doi:10.1029/97jd02300.

Zhang Y., H. T. Mao, A. J. Ding, D. R. Zhou, and C. B. Fu, 2013: Impact of synoptic weather patterns on spatio-temporal variation in surface O_3 levels in Hong Kong during 1990–2011. *Atmos. Environ.*, **73**, 41–50.

Zhang, Y., A. Ding, H. Mao, W. Nie, D. Zhou, L. Liu, X. Huang, and C. Fu 2016: Impact of synoptic weather patterns and inter-decadal climate variability on air quality in the north China plain during 1980–2013, *Atmos. Environ.*, **124**, 119–128.

Zhu, J. L., H. Liao, and J. P. Li, 2012: Increases in aerosol concentrations over eastern China due to the decadal-scale weakening of the East Asian summer monsoon, *Geophys. Res. Lett.*, **39**, L09809. doi:10.1029/2012GL051428.

Chapter 26

Decadal Change of East Asian Summer Monsoon: Contributions of Internal Variability and External Forcing

Tianjun Zhou*,‡, Fengfei Song*, Kyung-Ja Ha† and Xiaolong Chen*

*LASG, Institute of Atmospheric Physics, Chinese Academy of Sciences, Beijing, China

†Division of Earth Environmental System, Department of Atmospheric Sciences, College of Natural Science, Pusan National University, Busan, Korea

‡zhoutj@lasg.iap.ac.cn

The East Asian Summer Monsoon (EASM) has exhibited robust inter decadal changes. In this paper, the authors present a review on our current understanding of the observed changes. The weakening phase of the EASM in the 2nd half of the 20th century is demonstrated to be primarily forced by the positive phase of IPO (Inter-decadal Pacific Oscillation) /PDO (Pacific Decadal Oscillation), and secondarily driven by the increased aerosol emission. The dominance of IPO/PDO to EASM is also evidenced by the recent recovery of EASM since the 1990s, associated with the transition of IPO/PDO from positive to negative phases. Both data diagnosis and numerical model experiments indicate that the decadal change of EASM is dominated by internal variability mode of IPO/PDO and also partly driven by anthropogenic external forcings. The IPO/PDO is traditionally regarded as internal climate variability mode at decadal time scale, but recent studies suggested that external forcings may also trigger the phase transition of IPO/PDO, this has posed a new issue calling for further study. In addition to long-term changes, the EASM also exhibits inter-decadal shift of interannual variability mode. The suggested mechanisms are reviewed, including changes in mean circulation, interannual variability and its ENSO relationship, and the climatological intraseasonal oscillation.

1. Introduction

The East Asian Summer Monsoon (EASM) has a complex structure and multi-timescale variability, which ranges from intraseasonal, interannual to inter-decadal variability. In the community of East Asian monsoon studies, a stronger summer monsoon is defined as a northward penetration of surface monsoon circulation along with northward shift of monsoon rainband (Zhou *et al.* 2009). The interannual variability of EASM, its link to western North Pacific summer monsoon and their associations with El Niño/Southern Oscillation (ENSO) activities have been reviewed by Zhou *et al.* (2011), Ha *et al.* (2012), and recently updated by Hsu *et al.* (2014), and Zhou *et al.* (2014).

In addition to interannual variability, the EASM also exhibits considerable variability on inter-decadal time scales. For example, since the late 1970s, the EASM circulation has exhibited a weakening tendency. Following the weakened monsoon circulation, there was a trend toward increasing drought in North China but excessive rainfall in South China along the Yangtze River valley. This weakening tendency of EASM from the late 1970s to the end of the 20th century has been of great concern to the climate research community (see Zhou *et al.* 2009 for a review). However, recent studies found

The Global Monsoon System: Research and Forecast (3rd Edition)
Edited by C. P. Chang et al.

that the EASM circulation has been recovering since the early 1990s, although the strength of EASM is still weaker than that averaged over the period of 1965–1980. Associated with the recovery of EASM circulation, the monsoon rain belt has a tendency of moving northward to 30°–35°N along eastern China (Liu *et al.* 2012). Thus how to understand the earlier and recent decadal changes of EASM has been of great concern to both greater society and climate research community. Although up to now it is difficult to get a consensus on the mechanisms of the observed monsoon changes, there are increasing evidences supporting that the monsoon changes over East Asia were driven by both internal variability such as the Pacific Decadal Oscillation (PDO)/Inter-decadal Pacific Oscillation (IPO) and external forcing agents such as greenhouse gases and anthropogenic aerosols. In this paper, a review of our current understanding of the issue is given. In addition to inter-decadal change, the interannual variability of East Asian summer monsoon also exhibits inter-decadal variations (Yun *et al.* 2009; Song and Zhou 2015). The progresses in this field are also reviewed.

2. Forcing of IPO/PDO to EASM at Interdecadal Time Scale during the 20th Century

Data diagnosis has revealed that the weakening tendency of the EASM since the late 1970s is not a local phenomenon; rather it is a regional manifestation of global land monsoon changes. The rain gauge data over the land has shown a similar decreasing trend in the total precipitation amount accumulated in the global (mainly northern hemisphere) land monsoon domains (Zhou *et al.* 2008a). To understand the mechanisms, numerical experiments with AGCMs forced by historical sea surface temperature (SST) variation are done. The results show that the observed weakening tendency of both global land monsoon precipitation and EASM circulation during 1950–2000 are reasonably reproduced by the models (Zhou *et al.* 2008b; Li *et al.* 2010). Both changes are driven by inter-decadal changes of tropical ocean SST, which is a tropical lobe of the natural variability mode of PDO/IPO (Zhou *et al.* 2008b; Li *et al.* 2010). The phase transition of PDO/IPO from negative to positive phases around 1976/1977 has led to warming of Tropical Ocean which favors a weakened monsoon circulation through reducing the large-scale land-sea thermal contrast (Li *et al.* 2010).

The forcing of PDO/IPO to EASM is more evident if we extend our analysis from the 2nd half to the end of the 20th century. A weaker monsoon circulation is associated with deficient precipitation and thus more droughts in northern China. Qian and Zhou (2014) analyzed the dry-wet changes in northern China for the period of 1900–2010 on the basis of self-calibrated Palmer drought severity index (PDSI) data. The ensemble empirical mode decomposition method is used to detect multidecadal variability. A transition from significant wetting to significant drying is detected around 1959/60. Approximately 70% of the drying trend during 1960–90 originates from 50–70-yr multidecadal variability related to PDO phase changes (Fig. 1). The PDSI in northern China is significantly negatively correlated with the PDO index, particularly at the 50–70-yr time scale. This kind of significant negative correlation relationship is stable during 1900–2010.

How does PDO/IPO impact the EASM? In addition to the weakened land-sea thermal contrast in East Asia mechanism from a negative to a positive PDO phase proposed by Li *et al.* (2010), the Pacific–Japan/East Asian-Pacific pattern like teleconnection is highlighted by Qian and Zhou (2014). Composite differences between two positive PDO phases (1922–45 and 1977–2002) and one negative PDO phase (1946–76) for summer exhibit an anomalous Pacific–Japan/East Asian-Pacific pattern

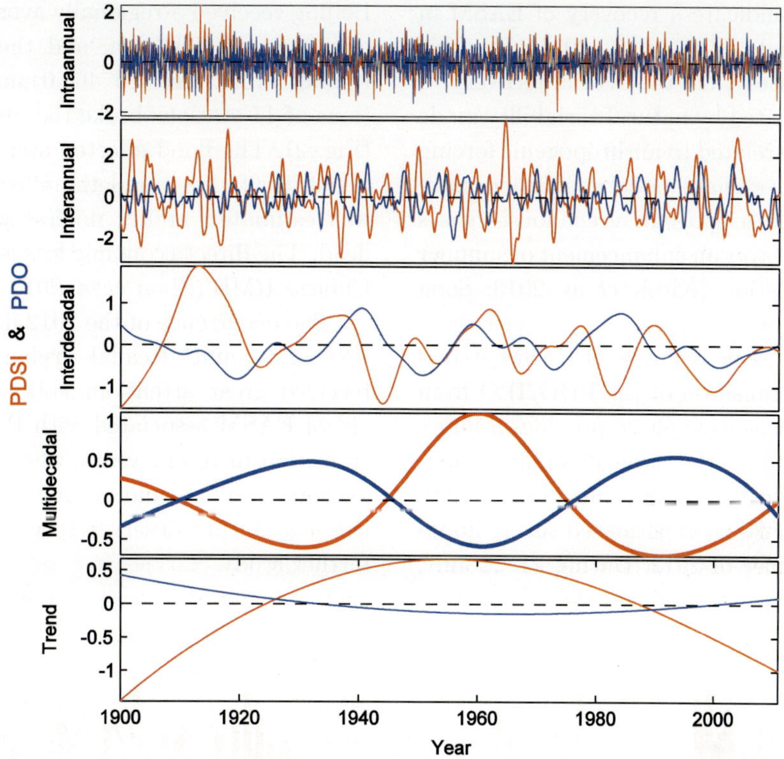

Fig. 1. Decompositions of monthly NC-PDSI (red) and PDO index (blue) during 1900–2010 into five major time scales determined using the EEMD filter. In the last subpanel, the nonlinear trends of NC-PDSI and PDO have been both subtracted by their mean values to facilitate comparison. (After Qian and Zhou 2014.)

like teleconnection, which develops locally in response to the PDO-associated warm SST anomalies in the tropical Indo-Pacific Ocean and meridionally extends from the tropical western Pacific to northern China along the East Asian coast. Northern China is dominated by an anomalous high pressure system at mid-low levels and an anticyclone at 850 hPa, which are favorable for dry conditions (Qian and Zhou 2014).

The recent recovery of EASM circulation is also driven by PDO/IPO related tropical SST anomalies. The phase transition of PDO/IPO from positive to negative phases has led to a SST cooling in the tropical eastern Pacific (Liu *et al.* 2012). This kind of SST cooling is regarded as mega-ENSO signals and has been suggested as one driving mechanisms for the recent enhancement of global monsoon (Wang *et al.* 2012; Lin *et al.* 2014). The strengthening EASM since the early 1990s is linked to the inter-decadal change of land-sea thermal contrast, which is mainly dominated by positive SLP anomalies over the western Pacific and "warmer land-colder ocean" thermal contrast change (Liu *et al.* 2012). Numerical modeling study based on SST-driven AGCM simulation indicates that the shift of PDO/IPO from positive to negative phase induces a warming over the Lake Baikal and a weakened subtropical westerly jet, which further lead to an increase of summer rainfall along the Huang-Huai River valley (32°–36°N, 110°–121°E) while a decrease along the Yangtze River valley (28°–31°N, 110°–121°E) during 2000–2008 in comparison to 1979–1999. These precipitation anomalies also indicate a recovery of EASM in the recent decade (Zhu *et al.* 2011). All this evidence indicates that the recent recovery of EASM is dominated by natural variability mode and may not be related to anthropogenic forcing agents, although climate model simulations indicate that the anthropogenic greenhouses gases emission would favor an enhancement of summer monsoon circulation (Kitoh *et al.* 2013; Song *et al.* 2014).

The decadal scale recovery of EASM associated with the transition of the PDO/IPO from a positive to a negative phase provides a large-scale background for the stronger summer monsoon in 2012. Nothern China including Beijing (~39°52′N, 116°28′E) experienced severe flooding in the summer of 2012. During 21–22 July, Beijing received a regionally averaged total precipitation of 190.3 mm, and the center of the rainfall event received 460.0 mm; the observations of 11 stations broke the historical records (Fig. 2). The flood affected area in Beijing was about 16,000 km^2, and the affected population was estimated at 1.9 million with 77 people dead. The direct economic loss is over 10 billion Chinese RMB (Zhou *et al.* 2013).

The occurrence of the 2012 flood in the context of a multi-decadal drying tendency has received great attention. Following the recovery of EASM associated with PDO/IPO phase transition in recent years, whether this kind of extreme events would occur more frequently in the near future deserves special attention and further study.

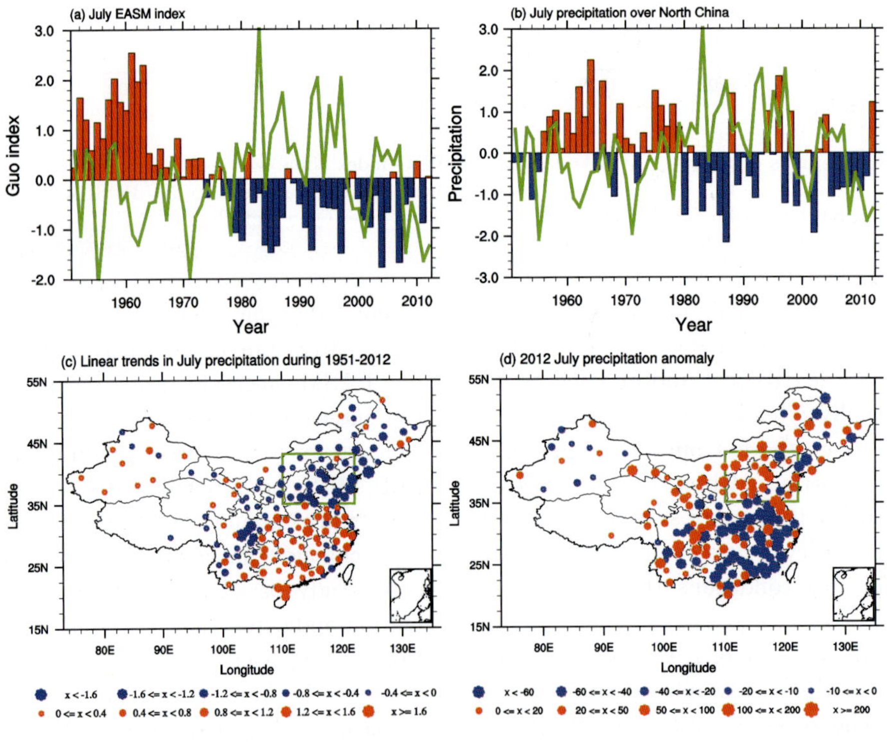

Fig. 2. (a) Normalized Jul EASM index derived from NCEP/NCAR reanalysis based on Guo (1983). The green line indicates the PDO index derived from http://jisao.washington.edu/pdo/PDO.latest. (b) Normalized Jul precipitation amount averaged over North China (35°N–43°N, 110°E–122°E; 23 stations included; dimensionless). The green line indicates the PDO. (c) Linear trends of July precipitation during 1951–2011 (mm yr^{-1}; the absolute values of anomalies larger than 0.4 are statically significant at the 5% level), the green box indicates the North China region (35°N–43°N, 110°E–122°E). (d) Anomalies of July precipitation in 2012 relative to 1981–2010. (After Zhou *et al.* 2013.)

3. Anthropogenic Forcings to EASM at Inter-decadal Time Scale

Through analyzing the long-term changes in Baiu rainfall during 1901–2009, it is found that there is no significant trend in the entire Baiu season over all regions, which indicates that the external forcing's influence on the Baiu is small (Endo 2011). However, the heavy precipitation has increased during the last century, indicative of the external forcing's effect (Fujibe *et al.* 2006). Along the Yangtze River valley, the summer precipitation during the whole 20th century also shows no significant trends (Zhou *et al.* 2009). The decadal/interdecadal change is more evident during the 2nd half of the 20th century (Zhang and Zhou 2011). Previous studies suggested that the anthropogenic changes of atmospheric composition, including greenhouse gases (GHG) and aerosol, may also impact the EASM during the 2nd half of the 20th century (see Zhou *et al.* 2009 for a review). For example, using an atmospheric model, Menon *et al.* (2002) found that the "southern-flood-northern-drought" precipitation pattern, one important feature of the EASM weakening, can be reproduced when the observed black carbon is also considered. The influence of sulphate aerosol on the precipitation structure was investigated by Qian *et al.* (2009). Based on observation and model simulation, they found that decreased light rainfall events in China during 1956–2005 are partly due to the increased anthropogenic aerosols, since they increase the cloud droplet number concentration and reduce droplet sizes.

But the effect of aerosol in inter-decadal EASM change is contested by many modeling studies. For example, there is evidence demonstrating that the specified anthropogenic aerosol forcing recommended by CMIP project failed in reproducing the observed features of monsoon changes (Li *et al.* 2007; Li *et al.* 2010).

The discrepancy of previous studies may partly be related to model limitations. The performances of climate models have generally improved in the past 5 years. In the latest Coupled Model Intercomparison Program phase 5 (CMIP5), there are different external forcing historical simulations, which allows investigation of the different influences of the external forcings in the EASM. Song *et al.* (2014) used 17 CMIP5 models and investigated the response of the EASM to these external forcings during the second half of the 20th century. The different forcing runs from the multi-model ensemble (MME) are analyzed to investigate their contributions to the declining trend of the EASM (Fig. 3). The positive trends of SLP over northern China and negative trends of SLP over the northwestern Pacific (NWP) are seen in both the observation and the all-forcing run (Figs. 3a, b). The anthropogenic-forcing dominates the responses in the all-forcing (Fig. 3c). In contrast, the anticyclone and southerly winds dominate eastern China in the GHG-forcing runs (Fig. 3d). The response of SLP in the aerosol-forcing run features a meridional dipole pattern, with positive anomalies over northern China and negative anomalies over southern China (Fig. 3f). Given that the anthropogenic aerosol loading over East Asia is larger than elsewhere (Wang *et al.* 2013), the surface cooling over the East Asian continent is the strongest. Hence, the surface cooling over eastern China weakens the land-sea thermal contrast and further induces the positive SLP trends over northern China, leading to a weakening trend of the EASM.

However, we should note that the decreasing trend of the EASM under external forcing is far weaker than in the observation. This discrepancy indicates that the internal variability mode of PDO/IPO may play a dominant role in the monsoon weakening as previously suggested (Li *et al.* 2010) and recently re-emphasized (Yu *et al.* 2015), while the aerosol forcing plays a secondary or complementary role. It should also be noted that the PDO/IPO are natural modes, even if climate models are able to simulate

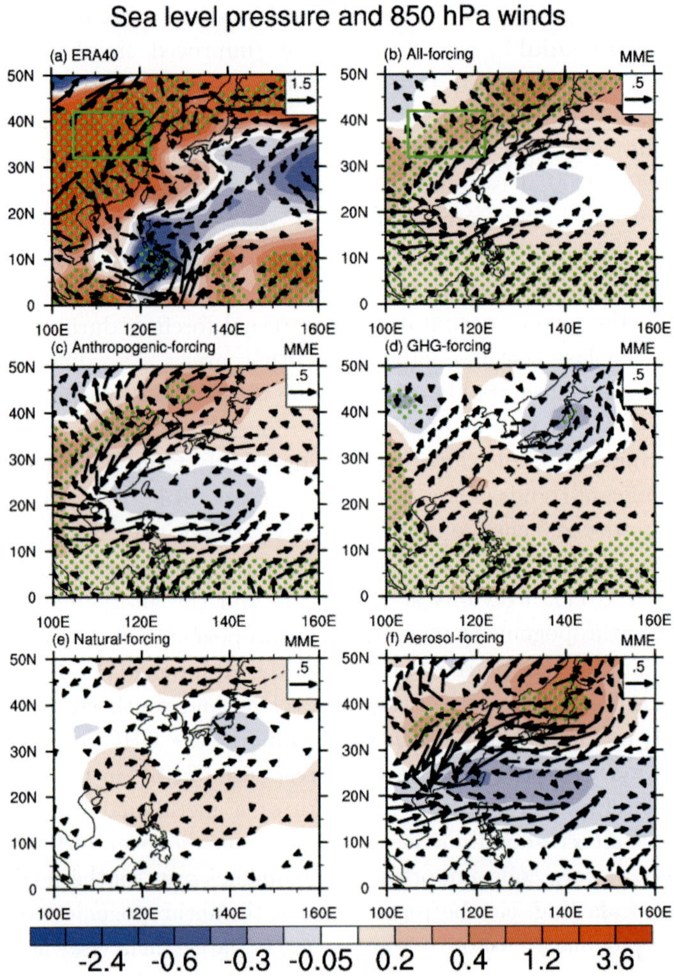

Fig. 3. The linear trends of SLP (shaded; hPa (44 year)$^{-1}$) and 850 hPa winds (vectors; m s^{-1} (44 year)$^{-1}$) in JJA during 1958–2001. (a) Observations (SLP and 850 hPa winds from ERA40); (b) all-forcing run; (c) anthropogenic-forcing run; (d) GHG-forcing run; (e) natural-forcing run; (f) aerosol-forcing run from MME. The green box in (a) and (b) is northern China (32°N–42°N, 105°E–122°E). The dotted areas indicate that the precipitation trends are statistically significant at the 10% level. The MME is constructed by using 35 realizations from 17 CMIP5 models. (After Song *et al.* 2014.)

the PDO/IPO reasonably well, their temporal phases are not reasonably reproduced because models are not initialized. The external forcings may also impact the phase of PDO/IPO (Dong *et al.* 2014a). This poses a challenge for the reproduction of the observed EASM changes.

In summary, the anthropogenic emission of aerosol does contribute to but is not a dominant factor for the weakening tendency of the EASM since the late 1970s to the end of the 20th century.

4. Inter-decadal Shift of EASM Interannual Variability

Interannual variability features of the EASM also exhibit inter-decadal variations. Lee *et al.* (2014) examined the change in principal mode

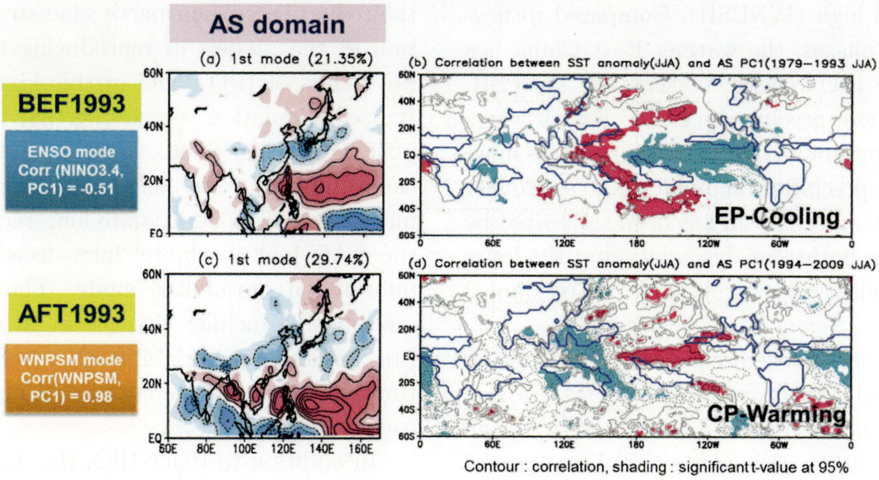

Fig. 4. Spatial patterns of first EOF mode of summer (JJA) precipitation over the Asian summer (AS) domain during (a) 1979–1993 and (b) 1994–2009. Distributions of simultaneous correlation coefficients between the time series of PC1 and JJA SST anomalies during (c) 1979–1993 and (d) 1994–2009. Shading area in (c) and (d) denote regions significant at the 95% confidence level. (After Lee *et al.* 2014.)

of the EASM precipitation (Fig. 4). It is found that during 1979–1993 the first EOF mode is related with ENSO, showing negative correlation to Nino3.4, whereas during 1994–2009 the western North Pacific summer monsoon (WNPSM)-related variability is dominant, with a very significant correlation with WNPSH index. Consequently, the result indicates that this WNPSM-related variability is related to central-Pacific (CP)-type ENSO than the eastern-Pacific (EP) type ENSO. In fact, Kwon *et al.* (2005) have discovered the decadal change over East Asian summer monsoon (EASM) around 1993/1994. The decadal change has been largely explained by three factors: changes in mean circulation (Kwon *et al.* 2007), interannual variability and its ENSO relationship (Kwon *et al.* 2005; Yim *et al.* 2008), and the climatological intraseasonal oscillation (Kim *et al.* 2011). The increasing role of CP-type ENSO on the NH summer monsoon may be caused by the increasing frequency of CP-type ENSO (Yeh *et al.* 2009). Model experiments show that the CP warming after 1993 causes a more active response vertically than the EP cooling before 1993 does, leading to a stronger ascending motion over the warming region. The CP warming induces the enhanced anticyclonic anomaly over the WNP (WNPSH). It confirms that the CP-type ENSO is related to the dominant WNPSM-related variability.

The interdecadal change after the late 1970s is also found in the monsoonal NPISO (northward propagation of intraseasonal oscillation)-ENSO relationship (Yun *et al.* 2010). The preceding winter ENSO affects the early summer NPISO activity before the late 1970s, while a stronger NPISO-ENSO in the preceding winter relationship appears during the later summer after the late 1970s. The altered NPISO-ENSO relationship has been explained by a strengthening of tropical atmospheric bridge process involving the Walker circulation and Rossby wave propagation.

Recently, Song and Zhou (2015) found that EASM-ENSO relationship exhibits significant decadal variation during the entire 20th century, which is modulated by the phase shift of IPO. The IPO modulates the EASM-ENSO relationship through the western North Pacific subtropical high (WNPSH). Compared to negative IPO phases, the warmer East China Sea in positive IPO phases weakens the WNPSH, inducing more precipitation. The Kelvin wave-induced interannual divergence suppresses more mean-state precipitation and leads to a stronger WPAC. This mechanism has been supported by diagnosis of both reanalysis data and the long-term pre-industrial control simulation of coupled models.

In addition, decadal changes of the interannual variability features of the EASM are also witnessed in the Pacific-Japan pattern (Chen and Zhou 2014), the connection between the western North Pacific Subtropical High and tropical SST (He and Zhou 2015), and the tropospheric temperature meridional gradient over East Asia (Zhang and Zhou 2015).

5. Concluding Remarks

The East Asian summer monsoon circulation has exhibited robust inter-decadal variations, with a weakening phase in the 2nd half of the 20th century and a recovery in recent decades. Both data diagnosis and numerical modeling have presented evidences that the weakening tendency of the EASM in the 2nd half of the 20th century and the recovery of the EASM in recent decades are dominated by the phase transitions of IPO/PDO, suggesting that it is primarily a natural variability mode. The driving of IPO/PDO to the EASM is more evident if we examine their relationship in the whole 20th century. Analysis of CMIP5 models that have conducted different external forcing historical climate simulations shows that the specified aerosol forcing resulted in a weakened EASM circulation, while the greenhouse gases (GHGs) favor an enhanced summer monsoon circulation. Thus both the GHG and aerosol forcings have significant impacts on the EASM changes. But the simulated weakening trend of CMIP5 multi-model ensemble is far weaker than the observation partly due to the limitation of the models in reproducing the temporal phases of IPO/PDO in this kind of uninitialized simulation, suggesting that the anthropogenic forcing agents can contribute to but cannot dominate the long-term changes of summer monsoon. In addition to long-term changes, the EASM also exhibits inter-decadal shift of interannual variability mode. The suggested mechanisms include changes in mean circulation, interannual variability and its ENSO relationship, and the climatological intraseasonal oscillation.

In addition to PDO/IPO, the Atlantic multidecadal oscillation (AMO) has been suggested as effective SST signal that influence the East Asian summer monsoon (Lu *et al.* 2006; Gu *et al.* 2009), although up to now there is no consensus on the mechanisms. Based on the experiments of coupled models, Lu *et al.* (2006) suggested a "non-local mechanism" where the positive AMO phase leads to a stronger East Asian summer monsoons through coupled atmosphere-ocean feedbacks in the western Pacific and Indian Oceans and tropospheric temperature changes over Eurasia in response to the imposed AMO-related SST anomaly forcing in the Atlantic. Traditionally, the PDO/IPO and AMO are regarded as two independent natural variability modes, recent studies such as Dong and Zhou (2014) suggested that the AMO may also drive the phase change of PDO/IPO. How the AMO drives the PDO/IPO in phase transition is a new issue calling for further study.

Acknowledgments

The works of Tianjun Zhou, Fengfei Song, and Xiaolong Chen are supported by the National Natural Science Foundation of China under Grant Nos. 41125017, 41330423 and R&D Special Fund for Public Welfare Industry (meteorology) (GYHY201506012).

References

Chen, X. and T. Zhou, 2014: Relative role of tropical SST forcing in the 1990s periodicity change of the Pacific-Japan pattern interannual variability. *J. Geophys. Res. Atmos.*, **119**, 13043–13066.

Dong, L., T. Zhou and X. Chen, 2014: Changes of Pacific Decadal Variability in the twentieth century driven by internal variability, greenhouse gases and aerosols. *Geophys. Res. Lett.*, **41**, doi:10.1002/2014GL062269.

Dong, L. and T. Zhou 2014: The formation of the recent cooling in the eastern tropical Pacific Ocean and the associated climate impacts: A competition of global warming, IPO, and AMO. *J. Geophys. Res. Atmos.*, **119**, 11272–11287. doi:10.1002/2013JD021395.

Endo, H., 2011: Long-term changes of seasonal progress in Baiu rainfall using 109 years (1901–2009) daily station data. *SOLA*, **7**, 5–8.

Fujibe, F., N. Yamazaki, and K. Kobayashi, 2006: Long-term changes of heavy precipitation and dry weather in Japan (1901–2004). *J. Meteor. Soc. Japan*, **84**(6): 1033–1046.

Gu, W., C. Y. Li, X. Wang, and W. Zhou, 2009: Linkage between Mei-Yu precipitation and North Atlantic SST on the decadal timescale. *Adv. Atmos. Sci.*, **26**(1), 101–108. doi: 10.1007/s00376-009-0101-5.

Ha, K.-J., K.-Y. Heo, S.-S. Lee, K.-S. Yun, and J.-G. Jhun, 2012: Variability in the East Asian Monsoon: A review. *Meterol. Appl.*, **19**, 200–215.

He, C. and T. Zhou, 2015: Decadal change of the connection between summer western North Pacific Subtropical High and tropical SST in the early 1990s. *Atmos. Sci. Lett.*, **16**, 253–259.

Hsu, H. H., T. Zhou, and J. Matsumoto, 2014: East Asian, Indochina and Western North Pacific Summer Monsoon — An update. *Asia-Pac. J. Atmos. Sci.*, **50**(1), 45–68.

Kim, W., J.-G. Jhun, K.-J. Ha, and M. Kimoto, 2011: Decadal changes in climatological intraseasonal fluctuation of subseasonal evolution of summer precipitation over the Korean Peninsula in the mid-1990s. *Adv. Atmos. Sci.*, **28**, 591–600.

Kitoh, A., H. Endo, K. Krishna Kumar, I. F. A. Cavalcanti, P. Goswami, and T. Zhou, 2013: Monsoons in a changing world: A regional perspective in a global context. *J. Geophys. Res. Atmos.*, **118**, doi:10.1002/jgrd.50258.

Kwon, M. H., J.-G. Jhun, B. Wang, S.-I. An, and J.-S. Kug, 2005: Decadal change in relationship between East Asian and WNP summer monsoons. *Geophys. Res. Lett.*, **32**, L16709.

Kwon, M., J.-G. Jhun, and K.-J. Ha, 2007: Decadal change in East Asian summer monsoon circulation in the mid-1990s. *Geophys. Res. Lett.*, **34**, L21706. doi:10.1029/2007GL031977.

Lee, E.-J., K.-J. Ha, and J.-G. Jhun, 2014: Interdecadal changes in interannual variability of the global monsoon precipitation and interrelationships among its subcomponents. *Clim. Dyn.*, **42**(9), 2585–2601. doi:10.1007/s00382-013-1762-4.

Li, H., A. Dai, T. Zhou, and J. Lu, 2010: Responses of East Asian summer monsoon to historical SST and atmospheric forcing during 1950–2000. *Clim. Dyn.*, **34**, 501–514. doi:10.1007/s00382-008-0482-7.

Li, L., B. Wang, and T. Zhou, 2007: Contributions of natural and anthropogenic forcings to the summer cooling over eastern China: An AGCM study. *Geophys. Res. Lett.*, **34**, L18807. doi:10.1029/2007GL030541.

Lin, R., T. Zhou, and Y. Qian, 2014: Evaluation of global monsoon precipitation changes based on five reanalysis datasets, *J. Climate*, **27**(3), 1271–1289.

Liu, H., T. Zhou, Y. Zhu, and Y. Lin, 2012: The strengthening East Asia summer monsoon since the early 1990s. *Chinese Sci. Bull.*, **57**, 1553–1558. doi:10.1007/s11434-012-4991-8.

Lu, R., B. Dong, and H. Ding, 2006: Impact of the Atlantic Multidecadal oscillation on the Asian summer monsoon. *Geophys. Res. Lett.*, **33**, L24701. doi:10.1029/2006GL027655.

Menon, S., J. Hansen, L. Nazarenko, and Y. Luo, 2002: Climate effects of black carbon aerosols in China and India. *Science*, **297**, 2250–2253.

Qian, C. and T. Zhou, 2014: Multidecadal variability of North China aridity and its relationship to PDO during 1900–2010, *J. Climate*, **27**(3), 1210–1222. doi: http://dx.doi.org/10. 1175/JCLI-D-13-00235.1.

Qian, Y., D. Gong, J. Fan, L. Leung, R. Bennartz, D. Chen, and W. Wang, 2009: Heavy pollution suppresses light rain in China: observations and modeling. *J. Geophys. Res.*, **114**, D00K02. doi:10.1029/2008JD011575.

Song, F., T. Zhou, and Y. Qian, 2014: Responses of East Asian summer monsoon to natural and anthropogenic forcings in the 17 latest CMIP5 models. *Geophys. Res. Lett.*, **41**, doi: 10.1002/2013GL058705.

Song, F. and T. Zhou, 2015: The crucial role of internal variability in modulating the decadal variation of the East Asian summer monsoon-ENSO relationship during the 20th century, *J. Climate*, **28**, 7093–7107.

Wang, B., J. Liu, H. J. Kim, P. J. Webster, and S. Y. Yim, 2012: Recent Change of the Global Monsoon Precipitation (1979–2008). *Clim Dyn.*, **39**, 1123–1135. doi:10.1007/s00382-011-1266-z.

Wang, T., H. Wang, H. Ottera, Y. Gao, L. Suo, T. Furevik, and L. Yu, 2013: Anthropogenic forcing of precipitation pattern in Eastern China in late 1970s. *Atmos. Chem. Phys.*, **13**, 12,433–12,450. doi:10.5194/acp-13-12433-2013.

Yeh, S.-W., J.-S. Kug, B. Dewitte, M.-H. Kwon, B. P. Kirtman, and F.-F. Jin, 2009: El Niño in a changing climate. *Nature*, **461**, 511–514.

Yim, S.-Y., J.-G. Jhun, and S.-W. Yeh, 2008: Decadal change in the relationship between east Asian–western North Pacific summer monsoons and ENSO in the mid-1990s. *Geophys. Res. Lett.*, **35**, L20711. doi:10.1029/2008GL035751.

Yu, L., T. Furevik, O. H. Otterå, and Y. Gao, 2015: Modulation of the Pacific Decadal Oscillation on the summer precipitation over East China: A comparison of observations to 600-yr control run of Bergen Climate Model. *Clim. Dyn.*, **44**(1): 475–494.

Yun, K.-S., K.-H. Seo, and K.-J. Ha, 2010: Interdecadal change in the relationship between ENSO and the intraseasonal oscillation in East Asia. *J. Climate*, **23**, 3599–3612.

Zhang, L. and T. Zhou, 2011: An assessment of monsoon precipitation changes during 1901–2001. *Clim. Dyn.*, **37**, 279–296.

Zhang, L. and T. Zhou, 2015: Decadal change of East Asian summer tropospheric temperature meridional gradient around the early 1990s. *Science China: Earth Science* **58**(9), 1609–1622.

Zhou, T., L. Zhang, and H. Li, 2008a: Changes in global land monsoon area and total rainfall accumulation over the last half century. *Geophys. Res. Lett.*, doi:10.1029/2008GL034881.

Zhou, T., R. Yu, H. Li, and B. Wang, 2008b: Ocean Forcing to Changes in global monsoon precipitation over the recent half-century. *J. Climate*, **21**(15), 3833–3852.

Zhou, T., D. Gong, J. Li, and B. Li, 2009: Detecting and understanding the multi-decadal variability of the East Asian Summer Monsoon? Recent progress and state of affairs. *Meteorologische Zeitschrift*, **18**(4), 455–467.

Zhou, T., H.-H. Hsu, and J. Matsuno, 2011: Summer monsoons in East Asia, Indochina, and the western North Pacific, in *The Global Monsoon System: Research and Forecast (2nd Edition)*, edited by Chih-Pei Chang *et al.*, World Scientific, pp. 43–72.

Zhou, T., F. Song, R. Lin, X. Chen, and X. Chen, 2013: The 2012 North China floods: Explaining an extreme rainfall event in the context of a long-term drying tendency [in Explaining Extreme Events of 2012 from a Climate Perspective]. *Bull. Amer. Meteor. Soc.*, **94**(9), S49–S51.

Zhou T., B. Wu, and L. Dong, 2014: Advances in Research of ENSO Changes and the associated Impacts on Asia-Pacific Climate, *Asia-Pac. J. Atmos. Sci.*, **50**(4), 405–422.

Zhu, Y. L., H. J. Wang, and W. Zhou, 2011: Recent changes in the summer precipitation pattern in East China. *Clim. Dyn.*, **36**, 1463–1473.

Part V
Field Experiment

Chapter 27

CINDY/DYNAMO Field Campaign: Advancing Our Understanding of MJO Initiation

Chidong Zhang
RSMAS, University of Miami, Miami, FL, USA
czhang@rsmas.miami.edu

Kunio Yoneyama
Japan Agency for Marine-Earth Science and Technology, Yokosuka, Japan

An international field campaign CINDY/DYNAMO took place in the Indian Ocean and its surrounding regions during October 2011–March 2012. The main goal of the field campaign was to collect observations that are needed for advancing our understanding and forecast of MJO initiation over the Indian Ocean. Three scientific hypotheses on interaction between convection and its environmental moisture, evolution in cloud population, and air-sea interaction were proposed to guide the design and operation of the field campaign. Unprecedented data were collected by numerous instruments of advanced technology from ground, airplanes, ships, moorings and other platform. These data cover detailed processes through all stages of convective initiation of three very different MJO events. Field observations were released for public use one year after the completion of the field campaign. Studies using these observations have increasingly emerged. These studies confirmed some existing knowledge, challenge some well-accepted paradigms, and reveal new phenomena that motivate novel thinking and approaches to solve the puzzle of MJO initiation.

1. Nature of the Problem

The Madden-Julian Oscillation (MJO) is the most prominent intraseasonal or subseasonal phenomenon in the tropics. It is an integrated part of the Asian-Australian monsoon (Goswami 2012; Hsu 2012; Wheeler and McBride 2012). It affects weather and climate globally (Zhang 2013). Since it was first documented by Madden and Julian (1971, 1972), tremendous progress has been made in the study of the MJO (see reviews by Madden and Julian 1994; Zhang 2005; Lau and Waliser 2012; Zhang *et al.* 2013). Yet, our understanding of the MJO remains incomplete and our ability of forecasting it unsatisfactory. Among many unsolved problems related to the MJO, its initiation over the Indian Ocean was perhaps the least studied up to recently. Outstanding issues pertaining to MJO initiation include:

(a) *Location.* Without any fancy statistical manipulation, one can easily appreciate the existence of the MJO from a simple time-longitude diagram of precipitation (Figs. 1 and 3). The eastward propagating precipitation centers, almost following straight lines, are signals of the MJO. They are much stronger than other types of equatorial perturbations (e.g. the equatorial Rossby and Kelvin waves). Also clear from such diagrams are the starting points of the most obvious eastward propagating precipitation signals, which are over the Indian Ocean. But why the Indian Ocean? Why not Africa, South America, or the Maritime Continent? If a warm ocean is needed, then why not the western

The Global Monsoon System: Research and Forecast (3rd Edition)
Edited by C. P. Chang et al.

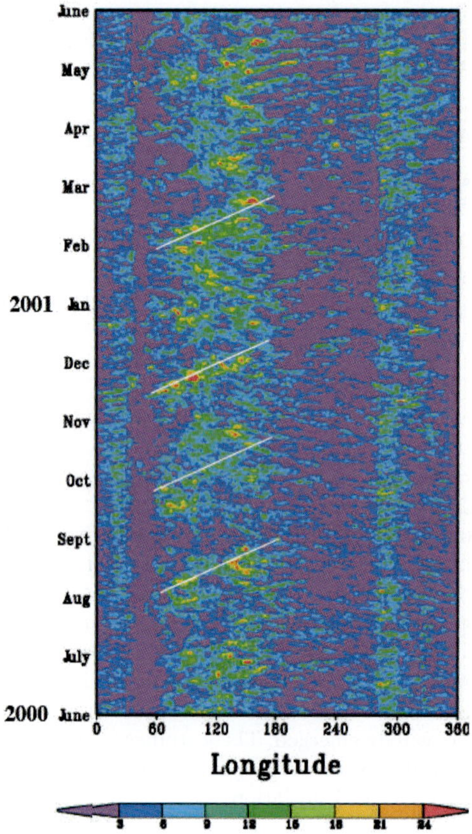

Fig. 1. Time-longitude diagram of TRMM precipitation (mm day^{-1}) illustrating eastward propagating signals of the MJO (white lines).

Pacific and tropical Atlantic? Atmospheric deep convection can be active over all these regions. The uniqueness of the Indian Ocean must play a role. Surrounded by landmasses on three sides, the Indian Ocean hosts the strongest monsoon flows in the world. It is the only tropical ocean where a warm surface coexists with a shallow thermocline below (the Seychelles-Chagos thermocline ridge) and a mean atmospheric subsidence above. It is the only tropical ocean with an equatorial eastward ocean surface current (the Wyrtki jet). It is the only tropical ocean without an equatorial cold tongue. All these unique aspects nevertheless, the precise reason for the Indian Ocean to be the most prominent birthplace of the MJO is unknown.

(b) *Irregularity and timescale.* Another striking feature seen from the time-longitude diagrams of precipitation is that the MJO signals, those eastward propagating precipitation centers, are highly irregular. The interval between two adjacent strong MJO events can be as short as 30 days, and as long as almost a year. This irregularity sets the timescale of the MJO spreading over a wide range of 30–90 days, as shown by the intraseasonal peak of the spectral power of precipitation that is distinct from that of lower-frequency variability (Zhang 2005). The reason for such irregularity is unknown. None of the existing hypotheses and theories of the MJO even attempts to explain this.

(c) *Internal vs. external drives.* All proposed and hypothesized mechanisms for MJO initiation can be grouped into two categories. In one, processes local to the Indian Ocean (internal) are key to MJO initiation. These processes may involve with convective self-regulation (commonly known as convective "discharge-recharge") and/or air-sea interaction. Atmospheric convection, deep and shallow, is at the center of this category, which may be referred to as thermal initiation. In another category, external influences, either from upstream (west) or higher latitudes, or both, are responsible for MJO initiation. Such influences can come as perturbations in the large-scale circulation or stochastic processes related to high-frequency eddies. This category may be referred to as dynamic initiation. In both categories, interaction between convection and the large-scale circulation is crucial. But convective activities come first in thermal initiation, perturbations in the circulation come first in dynamic initiation. While both might be valid, their fundamental causes for MJO initiation are distinct from each other. Both are supported by some evidence from observations and modeling. Neither, in any form of its incarnations, explains the most basic issues of (a) and (b).

(d) *Atmosphere vs. ocean.* There is little doubt that the dynamic core of the MJO is in the atmosphere. But the degree to which the ocean is also important to the MJO has always been debated. Many, although not all, studies that compare MJO simulations and forecast by an atmosphere model with prescribed sea surface temperature (SST) and by the same atmospheric model coupled with an ocean model, show better results from coupled models. The exact mechanism for the ocean to feedback to the atmosphere and thereby to help MJO simulations and forecast is, however, unclear. It is certain that the first step in this feedback is intraseasonal fluctuations in SST induced by the MJO. The second step is perturbations in sea surface sensible and latent heat fluxes caused by intraseasonal fluctuations in SST. From there, we had no clue how the relative small additional perturbations in sea surface sensible and latent heat fluxes caused by intraseasonal fluctuations in SST ($\leq$10% of total perturbations in surface fluxes) would lead to identifiable changes in the atmosphere that matters to the MJO. Part of the problem is that we still do not know the exact role of surface fluxes in the MJO. Any attempt at explaining oceanic feedback to the MJO without considering this unknown third step, no matter how brilliant the idea may sounds, is hand waiving at best. Several studies have recognized that SST perturbations in the Indian Ocean induced by the MJO are larger than those in the western Pacific, mainly due to the shallow mixed layer above the Seychelles-Chagos thermocline ridge. But how they act to help MJO, or its convection, initiates over the Indian Ocean is still unknown.

(e) *Moist vs. dry mode.* When convective activities, such as precipitation and convective cloud, are considered central to the MJO, it is inevitable that moisture must be involved in a significant way. Many observations and modeling studies have documented the evolution in the moist field through the life cycle of the MJO, especially during the transition from local convective inactive to active stages of the MJO. The main signature of this evolution in moisture is an increase in lower tropospheric humidity prior to the peak of local MJO convection. This can also be viewed as an increase in total column water vapor, or a deepening of the moist layer, which is the lowest part of the troposphere where relative humidity is greater than 70%. There are two possible sources for this water vapor increase in the lower troposphere. One is the vertical moisture transport out of the boundary layer by shallow and congestus convection. The other is horizontal moisture transport and convergence by the large-scale circulation. While they are not necessarily mutually exclusive, the nature of MJO dynamics is different depending on which one dominates. The role of moisture in the MJO has recently shaped the theoretical thinking to propose that the MJO is a moist mode. This means if the MJO can be treated as an instability problem, moisture must be included to destabilize the atmosphere. The opposite possibility is the MJO is a dry mode. This means there is a fundamental dynamic mode of the MJO, similar as the equatorial Rossby and Kelvin modes. While the Rossby and Kelvin modes are analytical solutions to the linear shallow-water equations on an equatorial β-plane, the MJO mode is not. It may involve with a certain degree of nonlinearity and its analytical solution has yet to be found. The hypothetical dry MJO mode may share many similarities with the Rossby and Kelvin modes in the troposphere. It is neutral and can sustain only when it is associated with certain energy sources. Its dynamic structure can promotes or demotes deep convection, depending on its phases. Latent heat release from convection is likely to be an efficient energy source for the dry MJO mode. That is why the most prominent signals of the MJO in observations are associated with convection. The idea of a dry MJO mode is attractive because it may dilute the seemingly competition among various proposed MJO mechanisms that emphasize on convective latent heating, moisture, stochastic forcing, upstream and lateral influences, etc. These processes are nothing special from each other but all provide energy to the dry MJO mode in their own way. Finding the dry MJO mode analytically or diagnostically remaining an unmet challenge and it must be done to prove the existence of the dry MJO mode.

2. The Field Campaign

The unsolved issues related to MJO initiation, outlined in the previous section, motivated a field campaign in the Indian Ocean and its surrounding regions during October 2011–March 2012. This was a coordinated project represented internationally by CINDY2011 (*Cooperative Indian Ocean Experiment on Intraseasonal Variability in the Year 2011*) and DYNAMO (*Dynamics of the MJO*), which included US projects AMIE (*ARM MJO Investigation Experiment*) and LASP (*Littoral Air-Sea Processes*). Researchers, including roughly 100 students, from more than 60 institutes of 16 countries (Australia, France, India, Indonesia, Japan, Kenya, Korea, Maldives, Papua New Guinea, Poland, Seychelles, Singapore, Sri Lanka, Taiwan, U.K., and U.S.) participated in this campaign. Yoneyama *et al.* (2013) provided a detailed description of the field campaign. A brief summary is given here.

The main goal of the field campaign was to collect observations that are needed for advancing our understanding and forecast of MJO initiation over the Indian Ocean. Apparently, no field observation can solve all outstanding problems of MJO initiation. The CINDY/DYNAMO field campaign was designed to address three issues that cannot be addressed without field observations. They are expressed in three hypotheses:

Hypothesis I: Deep convection can be organized into an MJO convective envelope only when the lower-tropospheric moist layer has become sufficiently deep over a region of the MJO scale; the pace at which this moistening occurs determines the duration of the pre-onset stage.

Hypothesis II: Specific convective populations at different stages play their unique roles in MJO initiation.

Hypothesis III: The barrier-layer, wind- and shear-driven mixing, shallow thermocline, and mixing-layer entrainment all play essential roles in MJO initiation over the IO by controlling the upper-ocean heat content and sea surface temperature, and thereby surface flux feedback.

The field campaign included the following main platforms (Fig. 2): (a) Two intensive sounding arrays embedded in a sounding network covering a large region from East Africa to the western Pacific; (b) Four research vessels that were part of the intensive sounding arrays at fixed locations and also made measurement along zonal and meridional transects; (c) Two research airplanes; (d) One ARM tropical site at Manus and one ARM mobile facilities at Gan Island, and (e) Several surface and upper-ocean moorings.

Main instruments deployed included, among others, GPS radiosondes and dropsondes,[a] C-, S-, Ka-, W- and X-band ground and airborne radars, 915 MHz wind profilers, lidars, multiple channel microwave radiometers, a suite of surface and boundary layer meteorological instruments on atolls, ships, and moorings, CTDs, XCDTs/XBTs (from both ships and aircraft), and ADCPs, microstructure (turbulent) profilers, a thermistor chain, Wire Walkers, an Argo float, a sea-glider, a sea-soar, and biogeochemical water sampling.

The field campaign observed several intraseasonal convective events, in October, November, December 2011 and March 2012 (Fig. 3), which are described in detail by Gottschalck *et al.* (2013). The November MJO event was the best observed when all deployed platforms were in place. The March MJO event was observed only at Manus after the field operation was terminated in early February because

[a]98% of the sounding data were transmitted to operations center to prepare initial conditions of real time forecast.

Fig. 2. Observation network of the CINDY/DYNAMO field campaign. The intensive sounding arrays are formed by Colombo, Male, Addu Atoll, Diego Garcia, R/V Revelle and Mirai as outlined by red dashed lines. The broader sounding network is shown on the top with stars marking the land based enhanced observing sites, red dots the locations of the ships, and green dots operational sounding sites. Yellow diamonds mark locations of CINDY/DYNAMO moorings, and yellow dots RAMA moorings. Photos of main platforms are given to the left and right.

of the civil unrest at Addu Atoll after the political regime change of the Maldives.

The MJO events observed during the field campaign differ from each other in several respects. The October one can be viewed as a "primary" MJO event (Matthews 2008) without any preceding MJO over the Indian Ocean. The convectively suppressed condition over the Indian Ocean lasted for several months before the convective initiation of this MJO event. Strong 2-day wave signals in precipitation were observed during its convectively active period.

In contrast, the November MJO event was a typical "successive" one (Matthews 2008) because it followed immediately the October MJO event. The suppressed period before its convective initiation lasted less than two weeks. The dominant synoptic-scale perturbations during this MJO event were two Kelvin waves.

The December event was not a conventional MJO event by any standard. It has been disputed as whether it was an MJO event at all. The controversy comes from the discrepancy between eastward slow moving signals in precipitation over the Indian Ocean and a lack of counterclockwise track on the RMM phase diagram, which has been commonly used to identify MJO events (Wheeler and Hendon 2004). The inconsistency suggests decoupling between convection and the large-scale zonal circulation, which challenges our conventional thinking that the MJO is a convection-circulation coupled phenomenon.

All field observations, along with their auxiliary data (e.g. from satellites and numerical models) are archived at three data centers at NCAR EOL, JAMSTEC, and ARM. About 97% of quality controlled data were released for public use one year after the completion of the

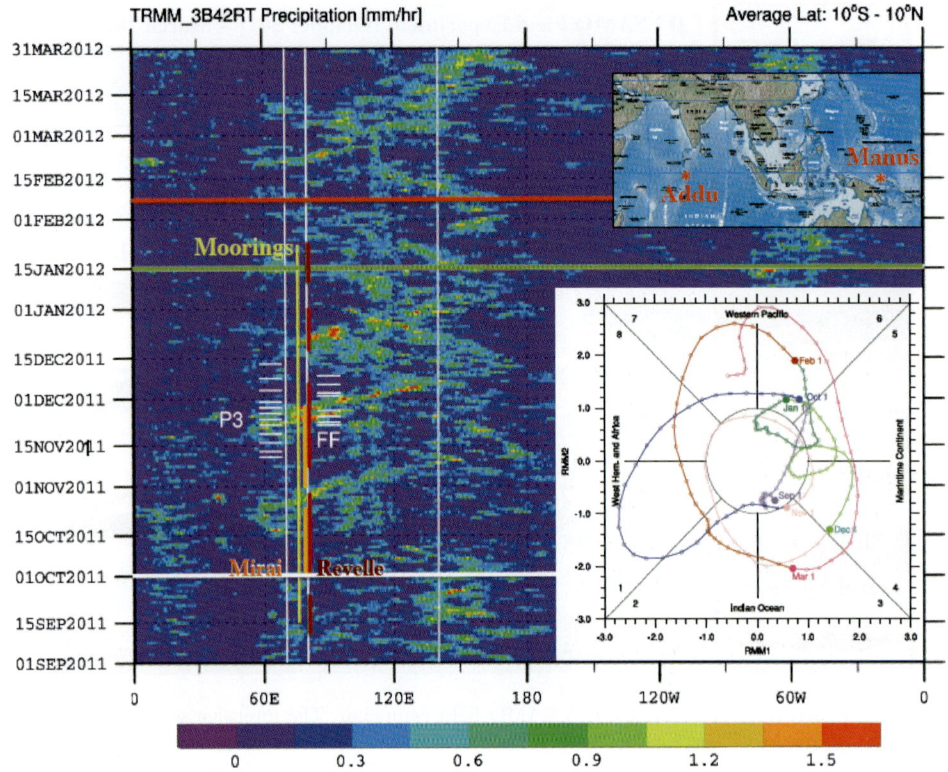

Fig. 3. Time-longitude diagram of TRMM precipitation (mm day^{-1}) during the CINDY/DYNAMO field campaign. Lower inlet is an RMM phase diagram for the same period. Upper inlet shows the location of Addu Atoll and Manus Island. Other symbols are: Vertical white lines for the DYNAMO sounding array (73–81°E), and the longitude of Manus (147°E); Horizontal white line for starting date of the field campaign (October 1, 2011); Horizontal green line for the end of the IOP (January 15, 2012); Horizontal red line for the end of the field campaign (Feb 8, 2012) over the Indian Ocean; Vertical purple lines for the station time of R/V Revell; Vertical orange lines for the station time of R/V Mirai; Vertical yellow line for the deployment time of DYNAMO moorings; Horizontal white sticks for flight missions of French Falcon (FF) and NOAA P3.

field campaign. The speed and quantity with which the field observations were released have no precedent in the history of tropical weather-climate field campaigns.

3. Current Results

The field campaign did not only collect unprecedented observational data that are needed for advancing our understanding of the physical processes deemed key to MJO initiation, it also motivated scientists with different backgrounds in atmospheric and oceanic sciences to focus their attention on MJO initiation, monitor all details at various stages of the MJO events during the field campaign, and view the MJO initiation problems from new angles. All these activities have led to confirmation of some existing knowledge and development of new ideas. Some interesting and intriguing phenomena that have never been documented caught research attention during and after the field campaign. New results from data analysis and modeling of MJO initiation are continuously emerging. Only some of them are briefly discussed here.

(a) The DYNAMO sounding time series have revealed that the top of the moist layer was kept below the 850 hPa level during convectively suppressed phases of the MJO. The increase in low-level moisture leading to the local convective peak of the MJO as expected in Hypothesis I was not smooth and continuous, but in a stepwise manner with two jumps. The first jump occurred when the top of the moist layer suddenly increased to 550 hPa. This was followed by the second jump when the top of the moist layer increased quickly to above the 200 hPa level (Johnson and Ciesielski 2013).

(b) Increases in convective echo top height (radars) and tropospheric moisture (soundings) leading to MJO onset occurred much faster (3–7 days) than commonly perceive (20–30 days) (Powell and Houze 2014). This is consistent to the stepwise behavior of the moist layer. Over the open ocean, lightning peaked when SST reached its maximum before local rainfall peak. Prior to MJO initiation, convection was dominated by isolated cells with their sizes and heights linearly proportional. As active phases approached, convection grew in both height and size but faster in size. The growth in size might depend on mid and upper level humidity and deep vertical wind shear (Xu and Rutledge 2014). These and other radar observations set a stage to test Hypothesis II.

(c) Radar observations also captured the evolution of the cloud population through 2–4 day convective events (Zuluaga and Houze 2013). Shallow convective clouds existed almost all the time but their frequency reached the maximum 12 hours before the rainfall peak and its minimum during the rainfall peak. Deep convective cores tended to reach their maximum 5–10 hours before the rainfall peak. Wide convection reached its frequency maximum at the rainfall peak, while stratiform region maximized several hours after that.

(d) The simultaneous and overlapping sampling of the scanning S-PolKa radar and vertical pointing Ka-band radar at Addu allowed a direct comparison and merge of the two radar observations (Feng *et al.* 2013). The results are time series of radar echoes of a broader range of cloud population than either of the radar can cover, including non-precipitating shallow clouds, precipitating shallow and deep convective clouds, and stratiform precipitation regions.

(e) Aircraft radar observations revealed distinct feature of mesoscale convective systems (MCSs) in comparison to those in the western Pacific. MCSs over the Indian Ocean were linearly organized more parallel to the low level shear without strong cold pools, their updrafts were weaker but deeper, and the descending rear inflow was absent (Guy and Jorgensen 2014).

(f) For the October and November MJO events, the convectively inactive periods near the equator were accompanied by active convection in the ITCZ south of the equator. Equatorial convective onset of these MJO events can be viewed as equatorward shift of the ITCZ (Johnson and Ciesielski 2013; Yoneyama *et al.* 2013). In this view, the transition from pre- to post-onset of MJO convection should be more accurately described as a regime change from ITCZ and MJO convection. Such convective regime change related to the ITCZ occurred in about 50% of MJO initiation over the Indian Ocean during the past 30 years. The possible role of the ITCZ in MJO and its initiation over the Indian Ocean is one of the new discoveries of the field campaign.

(g) One possible mechanism for the equatorward movement of the ITCZ during the initiation of the MJO is intrusion of extratropical dry air (Kerns and Chen 2014b). It tends to destroy an off-equatorial ITCZ. The demise of the ITCZ and its associated equatorial subsidence would allows equatorial convection to develop, which is instrumental to MJO initiation (Zhang and Ling 2012).

(h) South of the equator, sudden decreases in SST and mixed layer temperature in November were found to be caused by an advection of cold water of low salinity associated with a passage of the annual oceanic downwelling Rossby wave.

This cooling may have been another reason to the northward movement of the ITCZ and instigation of the initiation of the November MJO event (Seiki *et al.* 2013).

(i) Multiple instruments onboard of R/V Revelle recorded detailed impacts on the ocean by the MJO and its westerly wind bursts with speed greater than $10\,\mathrm{ms}^{-1}$. In response, the Yoshida-Wyrtki Jet at the equator accelerated from less than $0.5\,\mathrm{ms}^{-1}$ to more than $1.5\,\mathrm{ms}^{-1}$ in 2 days. The jet sustained to energize shear driven entrainment at its base (near 100 m depth) after the MJO wind bursts subsided, thereby further modifying SST for a period of several weeks after the MJO had passed (Moum *et al.* 2014).

(j) During convectively active periods cold pools can induce an increase in sensible heat flux by 15–20 Wm^{-2} due primarily to an increase in the air-sea temperature differences and an increase in the latent heat flux by 30–70 Wm^{-2} due to enhanced surface wind speeds. A succession of cold pools leads to a greater influence on the surface fluxes than those occurring individually. The cold pool intensity, measured by minimum surface air temperature and maximum surface wind speed enhancement, is correlated with the convective activity measured by radar-estimated rainfall and radar echo coverage (Yokoi *et al.* 2014).

(k) Observations from R/V Revelle indicate that before the rainfall peak, aerosol was dominated by fine particles of continental sources. During the convectively active period of the MJO, these particles were washed out. Meanwhile, strong surface wind increased sea spray and sea salt became dominant after the rainfall peak (DeWitt *et al.* 2013). This is the first time such aerosol regime change through the life cycle of the MJO was observed.

(l) Skills of MJO forecast during the field campaign have been assessed using the common RMM index (Nasuno 2013; Fu *et al.* 2014). A new MJO tracking method was developed to quantify the strength, propagation speed, and timing of MJO precipitation. This method was used to assess MJO forecast skill by ECMWF models during the field campaign (Ling *et al.* 2014). It is found that while the RMM index suggest useful skill up to 15 days and beyond, the forecasted MJO amplitude in precipitation decreases to one-third of the observed at day 7 and skill of predicting MJO precipitation propagation speed is even less. The better skill of ECMWF in comparison to other models is likely to come from its ability of capturing both intraseasonal and synoptic variability (Kerns and Chen 2014a).

(m) ECMWF performed three numerical experiments for the period of the field campaign. In the first two, sounding data over the Indian Ocean and Meteosat-7 motion vector wind data covering the entire Indian Ocean were each removed from the initial conditions and the forecasts were rerun (i.e. reforecast). In the third one, tropospheric humidity over the Indian Ocean was relaxed toward that from the reanalysis during forecast integration. The impact of field soundings and Meteosat-7 data over the Indian Ocean on the forecast is limited to the Indian Ocean and lasts less than two days. No significant change in the MJO forecast skill was found in all three experiments. These results suggest that MJO forecast skill depend on global fidelity of the model and its initial conditions (Ling *et al.* 2014).

4. Concluding Remarks

The field campaign was a success thanks to the dedication and professionalism of participating scientists (including nearly 100 students), engineers, technicians, project managers, and staff. Support and cooperation from funding agencies were indispensible. Nature also cooperated.

It is expected that the CINDY/DYNAMO field data will be widely used in research on tropical weather-climate, especially on MJO initiation. Recent studies have pointed out that MJO initiation over the Indian Ocean can be fully understood only when local processes are put in a broader context of the large-scale circulation and dynamics. Specifically, MJO initiation should be investigated together with large-scale convective initiation of non-MJO events, and with the Indian Ocean and Maritime Continent both included (Ling *et al.* 2013; Kim *et al.* 2013).

Field observations have been use to constrain and validate cloud-permitting simulations with domains over the sounding arrays, the Indian Ocean, and the entire globe. Global models with parameterized cumulus convection can also use the field observations to evaluate their simulated cloud statistics, tropospheric profiles of winds, temperature and humidity, surface fluxes, and upper-ocean properties, if only at a single or few grid points.

The CINDY/DYNAMO field campaign helped fill a huge gap in the global coverage of *in situ* atmosphere-ocean observations. But it is all such field observations in combination that is critical to the progress in our understanding of tropical processes that must be represented correctly in numerical models for forecasting and simulating the weather-climate continuum. CINDY/DYNAMO is but a recent addition to the repertoire of tropical field observations. More are needed as the observing technology advances and new ideas emerge.

Acknowledgments

The writing of this chapter was supported by an NSF grant AGS1450582 and a NOAA grant NA13OAR4310161.

References

DeWitt, L. H., D. J. Coffman, K. J. Schulz, W. Alan Brewer, T. S. Bates, and P. K. Quinn, 2013: Atmospheric aerosol properties over the equatorial Indian Ocean and the impact of the Madden-Julian Oscillation. *J. Geophys. Res. Atmos.*, **118**, 5736–5749.

Feng, Z., S. A. McFarlane, C. Schumacher, S. Ellis, and N. Bharadwaj, 2014: Constructing a merged cloud-precipitation radar dataset for tropical convective clouds during the DYNAMO/AMIE experiment at Addu Atoll. *J. Atmos. Ocean Tech.* **31**, 1021–1042.

Fu, X., J.-Y. Lee, P.-C. Hsu, H. Taniguchi, B. Wang, W. Wang, and S. Weaver, 2013: Multi-model MJO forecasting during DYNAMO/CINDY period. *Clim. Dyn.*, **41**, 1067–1081.

Goswami, B. N., 2012: South Asian monsoon. *Intraseasonal Variability of the Atmosphere–Ocean Climate System*, 2nd ed. W. K.-M. Lau and D. E. Waliser (Eds.), Springer, pp. 21–72.

Gottschalck, J., P. E. Roundy, C. J. Schreck III, A. Vintzileos, and C. Zhang, 2013: Large-scale atmospheric and oceanic conditions during the 2011–12 DYNAMO field campaign. *Mon. Wea. Rev.*, **141**, 4173–4196.

Guy, N. and D. P. Jorgensen, 2014: Kinematic and precipitation characteristics of convective systems observed by airborne Doppler radar during the life cycle of a Madden-Julian Oscillation in the Indian Ocean. *Mon. Wea. Rev.*, **142**, 3585–1405.

Hsu, H.-H., 2012: East Asian monsoon. *Intraseasonal Variability of the Atmosphere–Ocean Climate System.* 2nd ed. W. K.-M. Lau and D. E. Waliser (Eds.), Springer, pp. 73–110.

Johnson, R. H. and P. E. Ciesielski, 2013: Structure and properties of Madden–Julian Oscillations deduced from DYNAMO sounding arrays. *J. Atmos. Sci.*, **70**, 3157–3179.

Kerns, B. W. and S. S. Chen, 2014a: ECMWF and GFS model forecast verification during DYNAMO: Multi-scale variability in MJO initiation over the equatorial Indian Ocean. *J. Geophys. Res. Atmos.*, **119**, 3736–3755.

Kerns, B. W. and S. S. Chen, 2014b: Equatorial dry air intrusion and related synoptic variability in MJO initiation during DYNAMO. *Mon. Wea. Rev.*, **142**, 1326–1343.

Kim, D., J.-S. Kug, and A. H. Sobel, 2014: Propagating versus nonpropagating Madden-Julian Oscillation events. *J. Climate*, **27**, 111–125.

Lau, W. K. M. and D. E. Waliser, Eds., 2012: *Intraseasonal Variability of the Atmosphere–Ocean Climate System.* 2nd ed. Springer, 613 pp.

Ling, J., P. Bauer, P. Bechtold, A. Beljaars, R. Forbes, F. Vitart, M. Ulate, and C. Zhang,

2014: Global vs. local MJO forecast skill of the ECMWF model during DYNAMO. *Mon. Wea. Rev.*, **142**, 2228–2247.

Ling, J., C. Zhang, and P. Bechtold, 2013: Large-scale distinctions between MJO and non-MJO convective initiation over the tropical Indian Ocean. *J. Atmos. Sci.*, **70**, 2696–2712.

Madden, R. A. and P. R. Julian, 1971: Detection of a 40–50 day oscillation in the zonal wind in the tropical Pacific. *J. Atmos. Sci.*, **28**, 702–708.

______, and ______, 1972: Description of global-scale circulation cells in the tropics with a 40–50 day period. *J. Atmos. Sci.*, **29**, 1109–1123.

______, and ______, 1994: Observations of the 40–50 day tropical oscillation: A review. *Mon. Wea. Rev.*, **112**, 814–837.

Matthews, A. J., 2008: Primary and successive events in the Madden-Julian Oscillation. *Quart. J. Roy. Meteor. Soc.*, **134**, 439–453.

Moum, J. N., S. P. de Szoeke, W. D. Smyth, J. B. Edson, H. L. DeWitt, A. J. Moulin, E. J. Thompson, C. J. Zappa, S. A. Rutledge, R. H. Johnson, and C. W. Fairall, 2014. Air-sea interactions from westerly wind bursts during the November 2011 MJO in the Indian Ocean. *Bull. Am. Meteor. Soc.*, **95**, 1185–1199.

Nasuno, T., 2013: Forecast skill of Madden-Julian Oscillation events in a global nonhydrostatic model during the CINDY2011/DYNAMO observation period, *SOLA*, **9**, 69–73.

Powell, S. W. and Robert A. Houze, 2014: The cloud population and onset of the Madden-Julian Oscillation over the Indian Ocean during DYNAMO-AMIE. *J. Geophys. Res. Atmos.*, **118**, 11,979–11,995.

Rowe, A. K. and R. A. Houze Jr., 2014: Microphysical characteristics of MJO convection over the Indian Ocean during DYNAMO, *J. Geophys. Res. Atmos.*, **119**, 2543–2554.

Seiki, A., M. Katsumata, T. Horii, T. Hasegawa, K. J. Richards, K. Yoneyama, and R. Shirooka, 2013: Abrupt cooling associated with the oceanic Rossby wave and lateral advection during CINDY2011, *J. Geophys. Res. Oceans*, **118**, 5523–5535.

Suzuki, J., M. Fujiwara, T. Nishizawa, R. Shirooka, K. Yoneyama, M. Katsumata, I. Matsui, and N. Sugimoto, 2014: The occurrence of cirrus clouds associated with eastward propagating equatorial $n = 0$ inertio-gravity and Kelvin waves in November 2011 during the CINDY2011/DYNAMO campaign. *J. Geophys. Res. Atmos.*, **118**, 12,941–12,947.

Wheeler, M. C. and H. H. Hendon, 2004: An all-season real-time multivariate MJO index: Development of an index for monitoring and prediction. *Mon. Wea. Rev.*, **132**, 1917–1932.

Wheeler, M. C. and J. L. McBride, 2012: Australian monsoon. *Intraseasonal Variability of the Atmosphere–Ocean Climate System*, 2nd ed. W. K.-M. Lau and D. E. Waliser (Eds.), Springer, pp. 147–198.

Xu, W. and S. A. Rutledge, 2014: Convective characteristics of the Madden-Julian Oscillation over the central Indian Ocean observed by shipborne radar during DYNAMO. *J. Atmos. Sci.*, **71**, 2859–2877.

Yokoi, S., M. Katsumata, and K. Yoneyama, 2014: Variability in surface meteorology and air-sea fluxes due to cumulus convective systems observed during CINDY/DYNAMO, *J. Geophys. Res. Atmos.*, **119**, 2064–2078.

Yoneyama, K., C. Zhang, and C. N. Long, 2013: Tracking pulses of the Madden-Julian Oscillation. *Bull. Amer. Met. Soc.*, **94**, 1871–1891.

Zhang, C., 2005: Madden-Julian Oscillation. *Rev. Geophys.*, **43**, RG2003. doi:10.1029/2004RG000158.

Zhang, C., 2013: Madden-Julian Oscillation: Bridging weather and climate. *Bull. Amer. Met. Soc.*, **94**, 1849–1870.

Zhang, C., J. Gottschalck, E. D. Maloney, M. W. Moncrieff, F. Vitart, D. E. Waliser, B. Wang, and M. C. Wheeler, 2013: Cracking the MJO nut. *Geophys. Res. Lett.*, DOI: 10.1002/grl.50244.

Zhang, C. and J. Ling, 2012: Potential vorticity of the Madden-Julian Oscillation. *J. Atmos. Sci.*, **69**, 65–78.

Zuluaga, Manuel D., and Robert A. Houze, 2013: Evolution of the population of precipitating convective systems over the equatorial Indian Ocean in active phases of the Madden-Julian Oscillation. *J. Atmos. Sci.*, **70**, 2713–2725.

Chapter 28

Intraseasonal Air-Sea Interaction and Convection Observed in DYNAMO/CINDY/AMIE and TOGA COARE

Simon P. de Szoeke* and James B. Edson†

*College of Earth, Ocean, and Atmospheric Sciences, Oregon State University, Corvallis, OR 97331, USA
sdeszoek@coas.oregonstate.edu

†Department of Marine Sciences, University of Connecticut, Groton, CT 06340, USA
james.edson@uconn.edu

Intraseasonal (30–90 day) variations of surface wind speed and solar radiation associated with the Madden-Julian Oscillation (MJO) induce intraseasonal variations of net surface heat flux of $\pm 30\,\mathrm{W\,m^{-2}}$ in phase with atmospheric convective anomalies, and sea surface temperature (SST) variation of $\pm 0.1^{\circ}$C in quadrature with wind speed and convective anomalies. The DYNAMO (2011) and TOGA COARE (1992) experiments observed local processes and MJO events with stronger than average intraseasonal amplitude. Calm periods between intraseasonal convection permit diurnal warm layers to raise the SST, increasing the minimum value of the sensible and latent heat flux. The number of atmospheric cold pools increases during the convective phase, increasing the sensible heat flux by lowering air temperature, and increasing sensible and latent heat flux by wind gusts.

1. Introduction

The dominant mode of intraseasonal variability in the tropical Indian and western Pacific Oceans is the Madden-Julian Oscillation (MJO, Madden and Julian 1971), a convective anomaly that propagates eastward along the equator with a period of 30–90 days. The MJO is strongest while moving over warm ocean waters from the central Indian Ocean to the western Pacific Ocean. Convective anomalies also propagate northward in northern hemisphere summer to interact with the Asian monsoon circulation. Improved prediction of MJO initiation could improve prediction skill on 10–30 day time scales, but the mechanisms by which the MJO is initiated in the Indian Ocean remain unclear.

Intraseasonal Madden-Julian Oscillations are comprised of a calm convectively suppressed phase, and a windy convectively active phase (McPhaden and Hayes 1991; Anderson *et al.* 1996). MJO deep convection, wind stress, insolation, and SST anomalies propagate eastward from the Indian Ocean, across the Maritime Continent, and over the western Pacific Ocean. Westerly winds are found under and west of the deep convection (Hendon and Glick 1997; Shinoda *et al.* 1998; Woolnough *et al.* 2000). Deep convective clouds shade the sea surface from incident solar radiation. Intraseasonal variations in net surface heat flux and sea surface temperature (SST) in the equatorial central Indian and western Pacific Oceans result from this cloud radiative forcing and evaporation anomalies associated with the large-scale zonal wind anomalies (Stephens *et al.* 2004).

The MJO can be diagnosed from global tropical satellite or reanalysis cloud, rainfall, and wind products. It explains most of the intraseasonal tropical anomalies of outgoing longwave

The Global Monsoon System: Research and Forecast (3rd Edition)
Edited by C. P. Chang et al.

radiation (OLR), zonal wind, and rain. The MJO is a robust empirical signal in longitude-time series whether identified by intraseasonal filtering, spectral analysis, or empirical orthogonal function/principal component analysis. Yet the dynamics of its genesis in the central Indian Ocean and of its eastward migration (at speeds slower than Kelvin waves) elude simple theoretical and complex numerical weather prediction models. General circulation models tend to have more realistic MJO variability when they simulate a range of vertical scales of convection (Maloney and Hartmann 2001; Zhang and Mu 2001; Inness *et al.* 2001), or when they are coupled to an ocean model (Waliser *et al.* 1999; Inness and Slingo 2003; Marshall *et al.* 2008; DeMott *et al.* 2014). It is difficult to attribute success of a simulation to a particular process. The MJO may depend on interactions between processes on a wide range of time- and spatial scales, from the long memory of thermal anomalies in the ocean mixed layer to the mesoscale organization of convective clouds.

Large-scale field experiments conducted over several months have simultaneously measured the evolution of the atmosphere, including the large-scale regional convergence of sensible and latent heat, and the fluxes of momentum, heat, and moisture across the air-sea interface. This chapter reviews two such experiments, Tropical Ocean-Global Atmosphere Coupled Ocean Atmosphere Response Experiment (TOGA COARE, Webster and Lukas 1992; Weller *et al.* 1996; Anderson and Weller 1996; Lin and Johnson 1996) conducted in 1992–1993 and Dynamics of the Madden-Julian Oscillation (DYNAMO, Yoneyama *et al.* 2013) conducted in 2011–2012. These experiments provide detailed observations suitable for diagnosing the response of the upper ocean to intraseasonal anomalies in the surface meteorology, and the role of air-sea interaction in perturbing the overlying atmosphere, throughout the different phases of the MJO.

TOGA COARE, located at about 2°S, 155°W, was designed to investigate coupled ocean-atmosphere processes over the western equatorial Pacific warm pool (Webster and Lukas 1992), including the effects of intraseasonal variability, particularly the MJO, on the ocean (Feng *et al.* 2000). Mean conditions during TOGA COARE were characterized by typically low wind speed and high insolation. However, westerly wind bursts lasting from days to weeks provided significant variability in the wind record over the region (Weller and Anderson 1996). The westerly wind bursts generally accompanied convective anomalies observed in TOGA COARE (Chen *et al.* 1996).

More recently, in 2011 DYNAMO (in international and interagency cooperation with the CINDY, AMIE, and LASP experiments) measured various processes of the MJO in the Indian Ocean. A series of atmospheric soundings from an array in the central Indian Ocean (see Fig. 1 in Yoneyama *et al.* 2013) allowed calculation of large-scale heat, moisture, and mass transports (Johnson and Ciesielski 2014), while precipitation radar observations measured the evolving convective population responsible for releasing latent heat in the atmosphere (Powell and Houze 2013). Cloud, precipitation, and radiation properties were measured with multiple remote sensors at Gan Island. High-resolution surface radiation and turbulence fluxes collocated with ocean currents and turbulent microstructure were measured from the research vessel *Roger Revelle*. Three convectively active MJO events traversed the array during the intensive observation period, allowing a new look at the evolution of temperature, humidity, and momentum profiles in the atmosphere; the evolution of the convective population; the role of westerly wind bursts; the fresh water and salinity stratified barrier layers within the ocean surface layer; and equatorial ocean currents in the central Indian Ocean.

A number of the above studies during TOGA COARE investigated the role of shorter time-scale processes on air-sea coupling and its

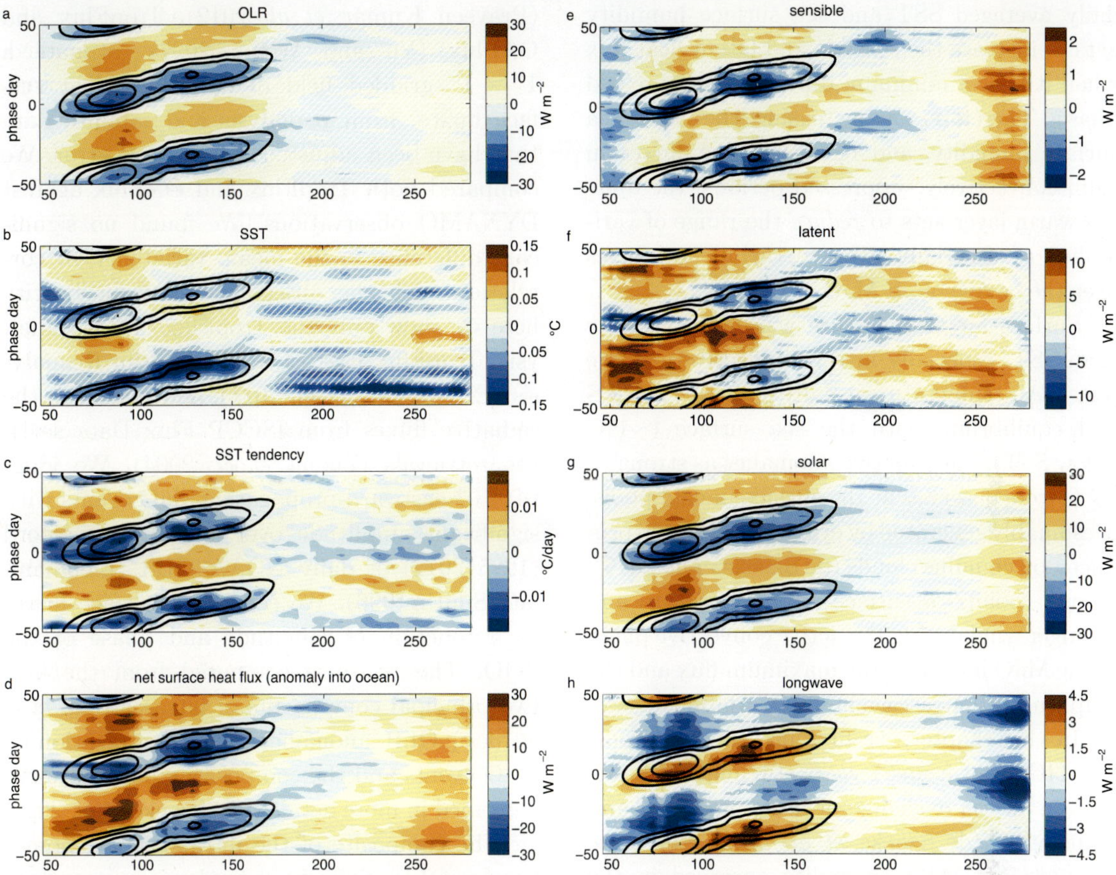

Fig. 1. Longitude-time plots of (a) OLR, (b) SST, (c) SST tendency and (d–h) heat flux anomalies averaged over 15°S–15°N for an MJO composite of 103 strongly eastward-propagating convective events. Positive fluxes warm the ocean. The black contours in each panel show the low-OLR anomaly of the convective phase of the MJO. White hatching covers SST and component flux anomalies that are not statistically significant at 95% confidence (adapted from de Szoeke *et al.* 2015).

feedback on MJO variability. For example, the cumulative effect of the diurnal cycle has been shown to modulate intraseasonal SST in the equatorial Western Pacific warm pool (Shinoda and Hendon 1998; 2001; Shinoda 2004). Slingo *et al.* (2003) and Ruppert and Johnson (2015) investigated the role of the diurnal cycle of SST in cumulus convection, and the subsequent influence on the atmosphere's moisture content, which increases from the inactive to the active phase. Other investigations have shown that intraseasonal fluctuations in SST are driven by variations in the surface radiative and heat fluxes due to the MJO (e.g., Hendon and Glick 1997; Jones *et al.* 1998; Shinoda and Hendon 2001, de Szoeke *et al.* 2015).

Sub-daily variability of wind, air, and sea surface temperature increase the mean surface fluxes. Diurnal warm SST layers were observed in calm, sunny (mean insolation of 250 W/m^2) convectively suppressed periods. The diurnal warm layer increases the SST in the afternoon on days when winds are weak. When winds are weak in the suppressed phase of the MJO, heat fluxes are relatively small compared to the active phase of the MJO. However, the increase in the daily averaged SST and sea surface humidity acts to increase the sensible and latent heat flux when it is at a minimum during the suppressed phase. Therefore, by increasing the SST, sea surface humidity, and resultant heat fluxes in calm convectively suppressed periods, the diurnal warm layer acts to *reduce* the range of variability of the daily average turbulence fluxes over the course of an MJO.

Surface air temperature was more variable than SST, especially in the convectively active phase. Air temperature usually remained in thermal equilibrium with the sea surface ($\sim$1°C below SST), but negative anomalies as strong as –3°C were observed at the surface in cold pools, presumably generated by evaporation of rain produced by cumulus clouds (de Szoeke *et al.* 2016). Strong cooling of surface air temperature and stronger wind during the active convective phase of the MJO increased the maximum flux and the amplitude of intraseasonal flux variability.

In this chapter, we first present a composite MJO event from 27 years of reanalysis-based surface air-sea exchange of heat, moisture, and momentum (de Szoeke *et al.* 2014). Next, we compare the MJO anomalies observed in the TOGA COARE and DYNAMO field experiments with anomalies explained by the global pattern of the MJO variability, and show the observed time series of net fluxes and SST. We diagnose the influence of diurnal warm layers of SST and atmospheric cold pools on the surface turbulent fluxes, and estimate how much of the evaporation anomaly can be attributed to cold pools. Finally, we discuss differences between the MJO in the Central Indian Ocean compared to the Western Pacific Ocean, and its implications for the intraseasonal events observed in DYNAMO and TOGA COARE.

2. Air-sea Interaction in a Typical Intraseasonal Convective Event

This analysis utilizes the 26.5 year time series of surface fluxes from the TropFlux data set (Praveen Kumar *et al.* 2012). TropFlux and OAFlux (Yu and Weller 2007) estimate a 1° × 1° gridded field of daily turbulent surface fluxes, from reanalysis and satellite data that have been adjusted to minimize error. We compared both TropFlux and OAFlux against DYNAMO observations. We found no significant difference in the mean, variability, or correlation with the observations, so for brevity here we show only the TropFlux gridded analysis. TropFlux uses ECMWF Interim Reanalysis (ERA-I, Dee *et al.* 2011), and estimates the radiative fluxes from ISCCP Flux Data satellite retrievals (Zhang *et al.* 2004). We identify 103 eastward-propagating MJO events with significant amplitude over a 26.5 year record (1985–2012) of daily NOAA OLR (Liebmann and Smith 1996). We composite surface fluxes as a function of longitude and phase of the MJO. The phase is computed from the first two principal components of intraseasonal (15–90 day) OLR anomalies. The identification of MJO events and the compositing procedure are described in more detail in de Szoeke *et al.* 2015.

The composite of the intraseasonal anomalies for OLR, SST, SST tendency, net surface flux, and the radiative and turbulent flux components are shown in Fig. 1. Though the anomalies were composited by intraseasonal phase, the phase coordinate is expressed in nominal days for an oscillation with a 50-day average period. Anomalies of all the variables propagate eastward beneath the OLR anomaly. The black contours in each panel show the low-OLR anomaly of the convective phase of the MJO.

The net heat flux (Fig. 1d) averaged equatorward of ±15° latitude varies with a cycle of $\pm 30\,\mathrm{W\,m^{-2}}$ in the Indian and West Pacific Oceans for the 103 MJO events, mostly due to variability in the evaporation and insolation. The net heat flux is in phase with the OLR, with a net cooling during the convective phase. The convective anomaly is accompanied by reduction of solar radiation (Fig. 1g) due to widespread deep, optically thick clouds, and increased evaporation from the ocean surface in sudden bursts that extend over most of the Indian Ocean and Western Pacific, slightly following the minimum of OLR (Fig. 1f).

MJO circulations strongly modulate the surface zonal wind, so that zonal wind variability is responsible for most of the intraseasonal variability in wind speed. This drives intraseasonal variability in the sensible heat flux H, and latent heat flux, E. The reason is illustrated by the bulk heat flux formulae:

$$H = \rho_a c_{pa} C_H U_r [\mathrm{SST} - T_{\mathrm{air}}] \quad (1)$$

$$E = \rho_a L_v C_E U_r [q_{\mathrm{sat}}(\mathrm{SST}) - q_{\mathrm{air}}] \quad (2)$$

where ρ_{a} and c_{pa} are the density and specific heat at constant pressure of air, respectively; L_v is the latent heat of vaporization, C_H and C_E are the transfer coefficients for heat and mass, respectively; U_r is the wind speed relative to the water; T_{air} and q_{air} are the mean air temperature and specific humidity, respectively; and $q_{\mathrm{sat}}(\mathrm{SST})$ is the equilibrium (saturation) specific humidity over saline sea water. Therefore, both wind speed and air-sea temperature and humidity differences drive the heat fluxes. However, the dominant influence of the wind speed in driving variability in the heat fluxes is readily seen in the correlation coefficients between these variables. For example, during the passage of the second MJO event in DYNAMO, the correlation coefficient was 0.92 between the latent heat flux and wind speed, while the correlation was only 0.08 between the latent heat flux and the sea-air humidity difference $q_{\mathrm{sat}}(\mathrm{SST}) - q_{\mathrm{air}}$.

The relationship between wind stress and the heat fluxes that applies to individual events from DYNAMO also extends to the composite results. The MJO composite of zonal wind stress is shown in Fig. 2. The zonal stress anomalies are in phase with the latent heat flux in Fig. 1. The relative amplitude of the zonal wind stress shown in Fig. 2 implies a ±10% modulation of evaporative flux. Specific humidity was remarkably constant over the *Revelle* DYNAMO research cruises, with a value of about 18–19 g/kg at a height of 10 m. The sea-air humidity difference was also nearly constant at 7 g/kg during the entire cruise.

Influence of surface heat fluxes on the upper ocean heat content

Solar radiation is the only net warming term on the ocean surface, balanced by the sum of longwave and turbulent surface flux cooling, internal ocean advection and vertical ocean mixing. A positive value of the net heat flux is therefore expected to increase the surface layer heat content of the ocean. For example, if we assume that the net heat flux is solely responsible for changes in the heat content in the vertically mixed layer,

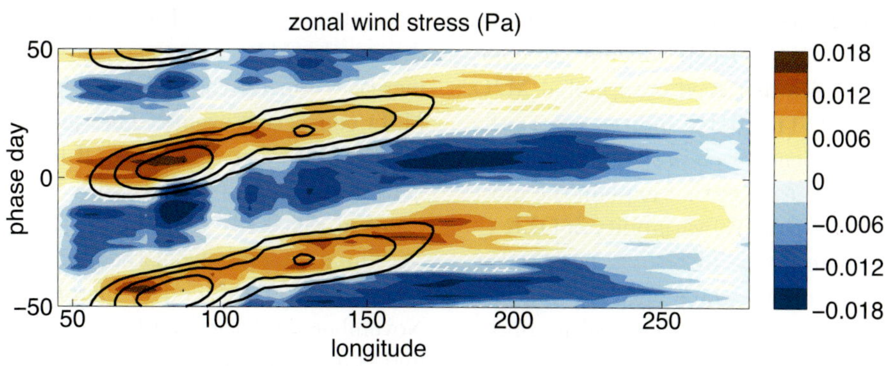

Fig. 2. MJO composite zonal surface wind stress over the Indian and Pacific Oceans, as in Fig. 1 (adapted from de Szoeke *et al.* 2015).

then the simplified heat budget becomes:

$$\rho_s c_{ps} h \frac{d\text{SST}}{dt} = Q_{net} \tag{3}$$

where ρ_s and c_{ps} are the density and specific heat at constant pressure of seawater, respectively; and h is the depth of the mixed layer. The ratio of the net surface heat flux to the SST tendency estimated from their composite intraseasonal amplitudes in Fig. 1 implies a mixed layer with a heat capacity, $\rho_s\, c_{ps} h$, of 80–160 MJ/K/m^2, depending on longitude, equivalent to a mixed layer depth of 20–40 m. Numerous other processes such as horizontal advection, vertical mixing, entrainment at the base of the mixed layer, and penetrating solar radiation also contribute to variability in the mixed layer heat content in the subtropics (e.g. McPhaden and Hayes, 1991). For example, Moum *et al.* (2014) found up to 1/3 of the SST cooling at the onset of the convective phase at 80°E was due to subsurface ocean mixing. This additional subsurface cooling would imply a greater heat capacity and the participation of deeper temperature anomalies in the intraseasonal cycle of SST.

3. TOGA COARE and DYNAMO Intraseasonal Convection and SST Variability

The local anomalies observed over the TOGA COARE and DYNAMO experiments are compared to anomalies regressed onto global MJO indices in Fig. 2. The global indices are 1) the two linearly-independent co-propagating principal component time series that explain the most intraseasonal (15–100 day) variability of OLR — used for the composites in Sec. 1, and 2) the Realtime Multivariate MJO (RMM, Wheeler and Hendon 2004) indices. The regressions of the *global* indices (dashed and dot-dashed) underestimate the *local* amplitude of OLR and SST variations sampled at the DYNAMO and TOGA COARE locations during the field experiments by a factor of 3.

The time series of OLR and SST in Fig. 3 demonstrate that the intraseasonal convective events observed in TOGA COARE and DYNAMO were unusually strong compared to typical MJO variability. The dotted gray lines show the normalized magnitude of the global intraseasonal oscillation (measured by 50-day running averages of the magnitude of the first two PCs). The three MJO events observed over the central Indian Ocean during DYNAMO had an intraseasonal magnitude of 1.4 standard deviations. During TOGA COARE the magnitude was 1 for 1992 October–December, and then briefly rose to 3 times the standard deviation in 1993 January. This shows that the global pattern of anomalies during the two experiments was highly correlated to the canonical patterns that explain the most amplitude of intraseasonal variability. In addition, the local variations sampled by the experiments were larger than projections on the indices of the MJO (RMM and PCs of OLR) based on all longitudes of the tropics. These strong and well-observed MJO convective events are of particular interest to understanding and predicting MJO genesis and propagation (Zhang 2013).

SST measured on the equator, at 80.5°E aboard the research vessel *Revelle* in DYNAMO in 2011 shows three broad intraseasonal maxima, one each in October, November, and December (Fig. 3a, Fig. 4d). SST decreases during the MJO active convection phase, and increases during the quiescent period between MJO convective events. The quadrature phase relation between SST and OLR anomalies in DYNAMO is consistent with the composite event in Fig. 1.

Three intervals from the DYNAMO *Revelle* time series, at the end of October, the end of November, and the middle of December, show high wind speed, decreasing SST, and relatively weak diurnal cycle of SST. These were periods of active convection at 80.5°E, indicated

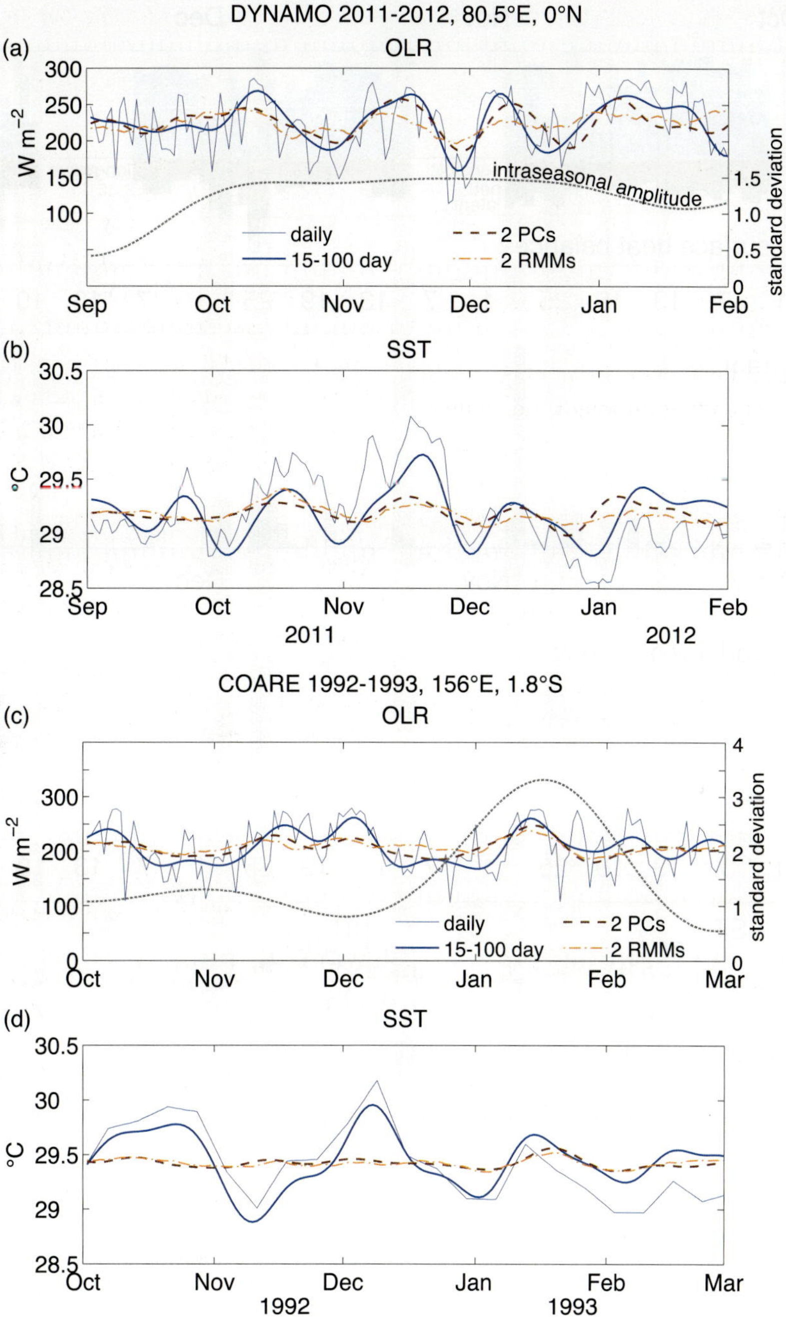

Fig. 3. OLR and SST response at the intensive DYNAMO and TOGA COARE observation arrays during their intensive observation periods. OLR and SST anomalies from linear regression of global indices (the leading two intraseasonal principal components [PCs] of equatorial [±15° latitude] OLR, and the Realtime Multivariate MJO indices [RMMs, Wheeler and Hendon 2004]) underestimate the local intraseasonal variability observed during TOGA COARE and DYNAMO (dashed). Dotted gray lines (a,c) indicate the amplitude of intraseasonal variability, computed as a 50-day running average of the standard deviation of the leading 2 PCs of OLR (adapted from de Szoeke *et al.* 2015).

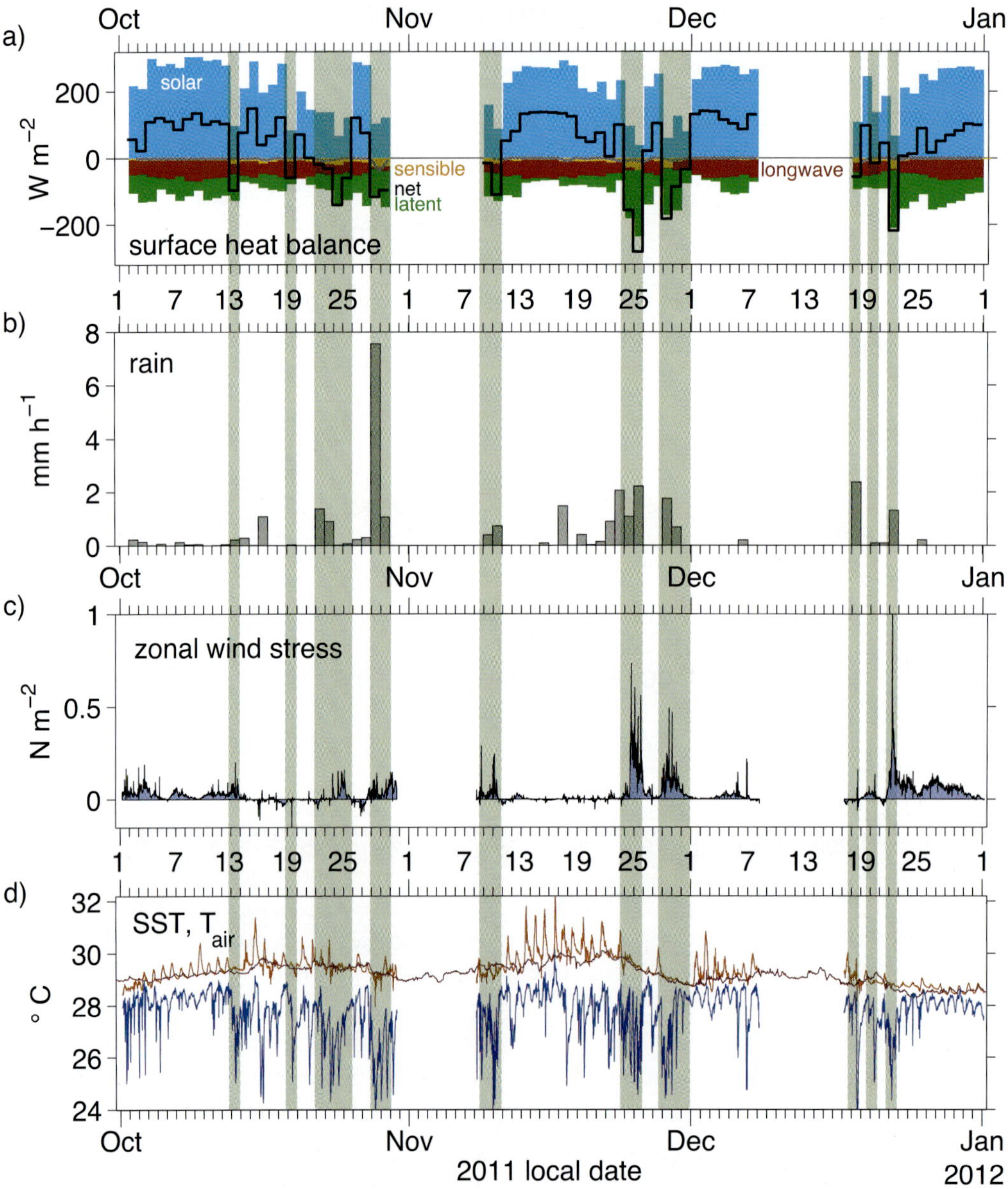

Fig. 4. Daily average surface heat fluxes incident on the ocean (a, positive warms the ocean) indicate intraseasonal variations of solar radiation (blue) and evaporation (green) in DYNAMO. Longwave radiation (red) and sensible heat flux (gold). The net heat flux (black) becomes negative during convectively active westerly phases of the Madden-Julian Oscillation MJO, also indicated by rain (b) and zonal wind stress (c). Sea surface temperature (0.05 m depth SST measured from the *Revelle* at 0°N, 80.5°E; d, orange), 10 m temperature from the APL buoy (0°N, 79°E, dark red), and surface air temperature measured at the *Revelle* (blue) warm and cool similarly in three intraseasonal oscillations measured in the DYNAMO experiment. Gray shading indicates the days of negative net surface heat flux cooling the ocean (adapted from de Szoeke *et al.* 2015).

by low OLR (Fig. 2a). Surface solar radiation measured aboard the *Revelle* was reduced to about half its average value, and evaporation was larger due to the increased surface wind speed (Fig. 4). Increased sensible heat flux nearly compensated reduced longwave radiative cooling of the sea surface. Overall, the net heat flux became negative, cooling the upper ocean under the MJO convection. On a few days the net heat flux exceeded $-100\,\mathrm{W\,m^{-2}}$, where the negative sign indicates a net heat flux out of the ocean.

The intraseasonal cycles of SST sampled in DYNAMO reflect the upper ocean's thermal response to the MJO. Daily net surface heat flux reverses sign from $75\,\mathrm{W\,m^{-2}}$ net warming to $-150\,\mathrm{W\,m^{-2}}$ cooling of the ocean during the convectively active phase of the MJO due mostly to reduced (by 1/3) surface solar radiation (Fig. 4a, blue) and a doubling of the latent heat flux (Fig. 4a, green) driven by enhanced evaporation.

Convective clouds modulate the solar radiation reaching the ocean surface, hence the intraseasonal cloud forcing has a large effect on the sea surface temperature (SST). Wind stress mixes solar heating absorbed near the surface through the depth of the mixed layer. Figure 5d shows SST (0.5 m) and 10 m ocean temperature from a buoy in COARE. The APL buoy ocean temperature at 10 m depth (Chi *et al.* 2014) and the *Revelle* temperature at 0.05 m depth in Fig. 4d are actually separated by 160 km, yet they have very similar intraseasonal variability.

The November convective MJO event was particularly strong and well sampled by the measurements on the *Revelle* (Moum *et al.* 2014). The arrival at the ship of two eastward-propagating Kelvin waves within the November MJO was associated with westerly wind bursts that increased the wind stress by up to 5 times relative to its mean. The surface stress increased suddenly to a maximum of 0.7 $\mathrm{N/m^2}$ at the front of the first wind burst (November 24). Wind stress of at least 0.15 $\mathrm{N/m^2}$ lasted for 36 hours. The wind burst had a deep (0–6 km) westerly zonal wind of more than 10 m/s that started at the surface and deepened with time. The propagation and vertical structure of the November wind bursts resembled convectively coupled Kelvin waves (reviewed in Kiladis *et al.* 2009). Westerly wind bursts associated with Kelvin waves were also observed over the western Pacific in COARE (Lin and Johnson 1996; Smyth *et al.* 1996).

The persistent clouds and intense westerly wind bursts associated with the MJO radically changed the upper ocean momentum and heat balance, changing the sign of the net surface heat flux. SST measured at the *Revelle* decreased by 1.5°C from November 23–30. The surface wind stress accelerated the upper 75 m of the equatorial ocean current (0.1 m/s before November 24) to about 1 m/s after the two wind bursts. Momentum flux convergence measured in the ocean explains the pronounced acceleration of the Yoshida equatorial jet under the November 24 westerly wind burst (Moum *et al.* 2014).

SST and air temperature in Figs. 4 and 5 also show sub-daily variability that indicates faster air-sea interaction processes. The 0.05 m ocean temperature (measured by a thermistor at 0.05 m depth in a floating "sea snake") has nearly the same intraseasonal variability as at 10 m, but with a diurnal cycle of solar warming imposed on top that impacts the surface fluxes. We note that the use of nighttime-only or 10-m buoy temperature data to compute the fluxes would underestimate the effect of the diurnal warm layer on SST, and hence would overestimate the intraseasonal variability of the heat fluxes compared to a more accurate calculation using diurnally-resolved near surface measurements. Surface air temperature (blue line, Fig. 3d) follows the intraseasonal cycle of SST, but has sharp cold events lasting minutes to hours. These "cold pools" are more frequent

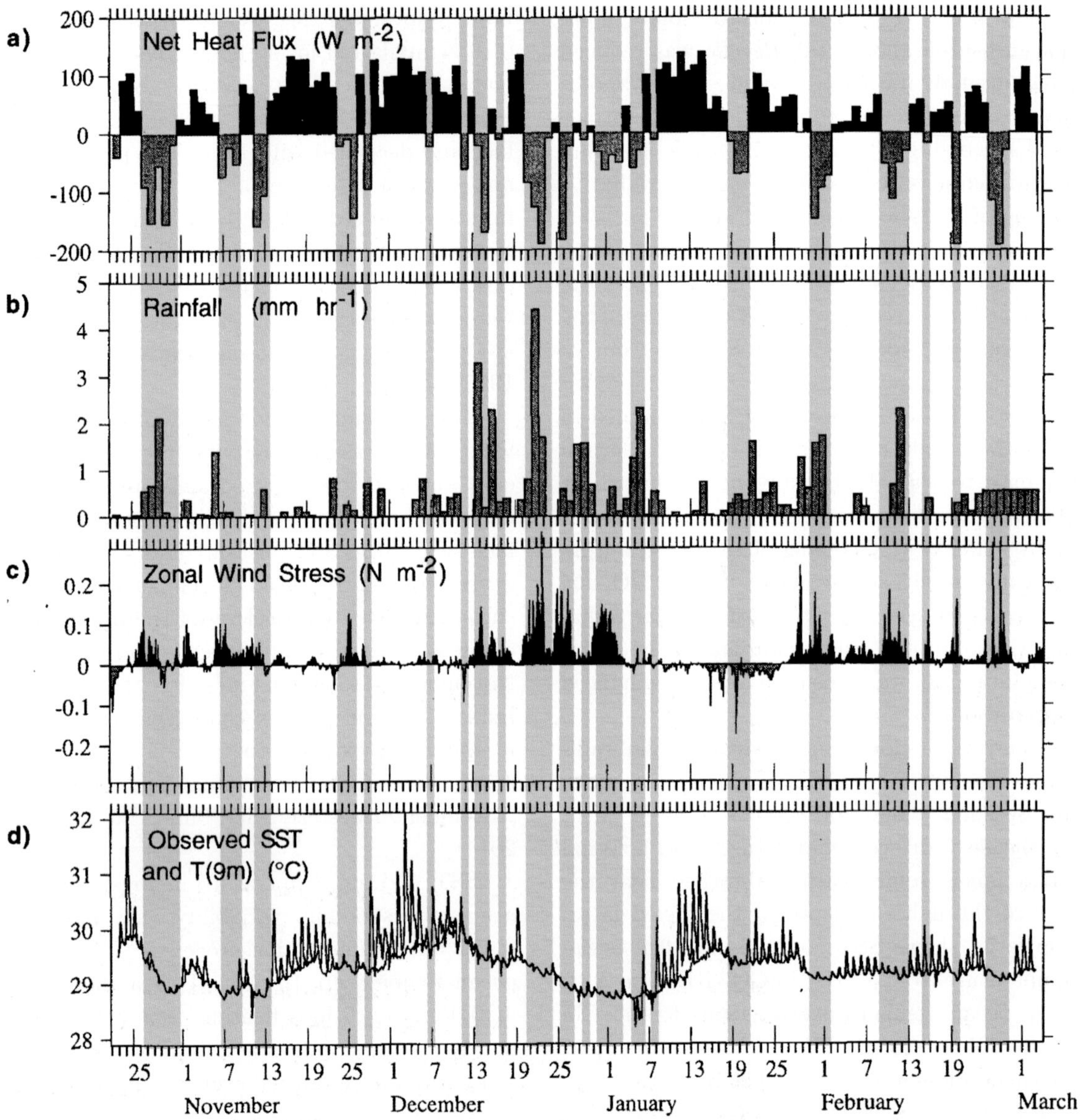

Fig. 5. Daily average net heat flux, rain rate, and 2-hour smoothed zonal wind stress, SST, and 9 m ocean temperature. Shading indicates days of negative net surface heat flux cooling the ocean (from Anderson *et al.* 1996 Fig. 3). (© American Meteorological Society. Used with permission.)

during the convective phase of the MJO. Both the diurnal warm layer and the atmospheric cold pools increase the sea-air temperature difference and the evaporation and sensible heat flux, but they have different effects on the day-to-day and intraseasonal variability of the heat fluxes.

The diurnal warm layer

During the convectively suppressed phase of the MJO, the skies were often clear, allowing strong solar radiation (blue, Fig. 3b) to enter the ocean. At this time the winds were weak, allowing the solar radiation to warm the surface more than the underlying part of the nocturnally mixed layer, stratifying the upper ocean, and creating a diurnal warm layer. Overall, the diurnal warm layer increases the SST as the upper ocean warms during the quiescent phase of the MJO. Conversely, winds are strong and clouds shade the surface from solar radiation in the convective phase of the MJO, which results in weak to absent diurnal warm layers.

Therefore, the strength of the diurnal warm layer depends on the phase of the MJO. For example, the 10-minute resolution time series in Fig. 4d from DYNAMO and Fig. 5 (adapted from Anderson and Weller 1996) from TOGA COARE clearly show the sub daily variations of SST associated with diurnal warm layers. However, both experiments show a preference for strong diurnal warm layers to form during in the afternoons of the convectively suppressed phase when the insolation is strong and the winds are relatively calm. The temperature excursion of the warm layers from the nocturnal temperature minimum reaches a maximum of about 3°C for DYNAMO in mid-November, and 2.5°C for TOGA COARE in December.

Consistent with the model of Price *et al.* (1986) the SST is a minimum at the end of the night, when the upper ocean is thermally mixed and the SST is representative of the mixed layer temperature. The heating by the penetrating solar radiation decreases with depth and the upper ocean becomes stably temperature-stratified from this solar warming. Some of this thin warm surface layer mixes down at night and some is lost to enhanced heat and radiative flux. Overall, the retained heat causes the ocean mixed layer and SST to gradually warm over the days before the most intense convective events.

During the convective westerly phase of the MJO (e.g. 2011 October 22–29 and November 24–30), surface buoyancy flux and wind stress mixed the upper ocean, and diurnal warm layers were not observed in the SST. As discussed above, the reduced solar and enhanced latent heat flux due to enhanced surface wind speed drove a net heat flux out of the ocean during this phase of the MJO. Therefore, observations during both TOGA COARE and DYNAMO clearly show that diurnal SST variability is suppressed during days when the net surface heat flux is negative (cooling the ocean). These days of negative net heat flux are shaded gray in Figs. 4 and 5. The above discussion and Fig. 4d shows that the strength of diurnal warm layers is correlated to the intraseasonal warming and cooling tendency. This is because the diurnal warm layer responds sensitively to the intraseasonal variations of the wind speed and solar radiation associated with MJO deep atmospheric convection. The convection is indicated by rain (Figs. 4b, 5b) and westerly wind bursts (Figs. 4c and 5c). The observations shown in Figs. 4 and 5 indicate that the net daily surface flux absorbed by the ocean correlates well with the convection indicated by rain, wind stress, and by the large-scale pattern of OLR.

Diurnal variation of the latent and sensible heat fluxes

To show the distribution of sea and air temperature over the diurnal cycle, Fig. 6a shows the entire 10-minute SST and air temperature time series from DYNAMO repeated over a 2-day local hour time axis. The distribution of air temperature is negatively skewed because of cold pools. While the days with strongest diurnal warming stand out in Fig. 6 and Fig. 4d, most days experience some diurnal warming. The mean of the diurnal SST distribution rises by about 0.5°C in the early afternoon. One might expect this warm layer to increase the sea-air temperature difference, and hence the surface heat fluxes, in the afternoon. Air temperature in Fig. 6 shows that air temperature follows the diurnal warming of SST, largely mitigating

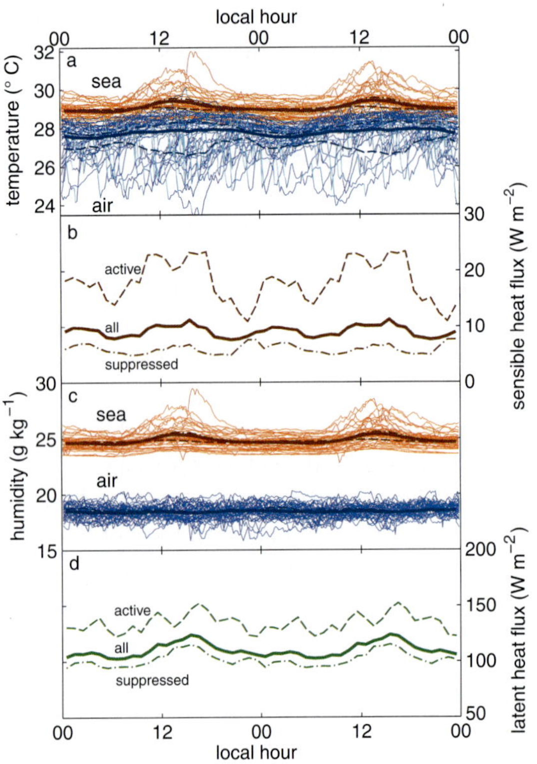

Fig. 6. (a) The entire DYNAMO SST (orange) and T_{air} (blue) time series repeated on a 2 day axis. Solid bold lines are the mean diurnal composite, dashed lines are averages for the 18 convectively active days (identified in Fig. 4 by negative net heat flux cooling the ocean), and dot-dashed lines are averages for the 53 convectively suppressed days (identified by positive net heat flux into the ocean). Temperature lines are cyan when T_{air} decreases $-2.4°\mathrm{C/hour}$ or more. Strong negative T_{air} anomalies or "cold pools" are equally likely any time of day. (b) Mean diurnal cycle of sensible heat fluxes (red) and wind speed (black). (c) As in (a) but for sea surface saturation specific humidity (red) and 10 m air specific humidity (blue). (d) Latent heat flux composited like the sensible heat flux in (b).

the sea-air temperature difference. The mean air temperature decreases in the early morning before sunrise. The sea-air temperature difference in Fig. 6a thus has two daily maxima: one in the afternoon (14LT), and one in the early morning (02LT). The wind speed is nearly constant over the diurnal cycle with a mean of $5.0 \pm 0.3\,\mathrm{m/s}$.

The sensible heat flux also has a semidiurnal cycle (Fig. 6b), with maxima at 01LT in the morning and 16LT in the afternoon. The early morning maximum in sea-air temperature difference and heat flux is due to the air temperature. The diurnal variability of latent heat flux (Fig. 6d) is weak and governed by the sea surface saturation humidity (Fig. 6c). Specific humidity of the air stays relatively constant (18–$19\,\mathrm{g\,kg^{-1}}$) throughout the diurnal cycle, and throughout the different phases of the MJO.

Day-to-day variability observed in DYNAMO explains more flux variability than the diurnal cycle. The dashed lines in Fig. 6 are diurnal averages from the 18 convectively active days and the dash-dotted lines are averages of 53 convectively suppressed days (defined by net heat flux into the ocean, to which sensible heat flux is a minor contributor) sampled aboard the *Revelle*. While SST decreases slightly, sensible heat flux on convectively active days doubles (from 9 to $18\,\mathrm{W\,m^{-2}}$) in response to 26% stronger wind speed and $-0.7°\mathrm{C}$ colder air temperature (46% larger sea-air temperature difference).

Evaporation increased 22% on convectively active days, which can be attributed to the higher wind, as the sea-air humidity difference actually decreased slightly in the active phase. The sea-air humidity difference is reduced under strong winds when net heat flux cools the ocean, because the diurnal warm layer of SST and its saturation specific humidity are reduced by ocean mixing. The $\sim 1\,\mathrm{g/kg}$ diurnal variability of the air-sea humidity difference provides approximately 10% of the total variability in the evaporation. The rest is driven by the much larger intraseasonal variability in the wind speed.

Atmospheric cold pools

The largest temperature differences in Fig. 6 come from distinct negative air temperature excursions of -1 to $-4°\mathrm{C}$, which occur episodically, with little preference for any time of day. We identify events with cold air temperatures below 27°C as atmospheric *cold pools*. Cold pools thus are diagnosed for 16% of the DYNAMO time series. These cold pools tend to depart from the usual temperature distribution with a rapid cooling, followed by a more gradual warming (de Szoeke *et al.* 2016).

During convectively suppressed periods the air temperature gradually follows the warming trend of the sea surface. Most of the surface air temperature variability in Fig. 4a is due to cold pools during convectively active days. The cold pools reduce mean air temperature on convectively active days, especially in the afternoon (Fig. 4a blue dashed). During convectively active periods the air temperature decreased more than the SST, demonstrating the influence of atmospheric thermodynamics. The temperature decreased at least 0.3°C within 6 minutes in more than 200 sharp cold spikes in legs 2 and 3 (October–December 10), with stronger and more frequent cold pools during the convectively active days of the MJO. These cold pools are correlated to wind and rain at the *Revelle*. The source of these rapid drops of the surface temperature is air aloft evaporatively cooled by falling rain. The cooled negatively buoyant air sinks, and spreads out once it reaches the surface. Surface temperature station time series are more likely to sample older cold pools that have spread out over a larger area, than to sample younger stronger cold pools that occupy a smaller area.

The effect of the cold pools is to increase the evaporation and sensible heat flux measured at the equator in DYNAMO. The average sensible heat flux during the cold pools was $27\,\mathrm{W\,m^{-2}}$, triple the all-cruise average sensible heat flux of $9\,\mathrm{W\,m^{-2}}$. The 30% increased evaporation in the cold pools was mostly due to their 27% stronger wind speed. Mean evaporation during cold pools was $140\,\mathrm{W\,m^{-2}}$, compared to the all-cruise mean evaporation of $110\,\mathrm{W\,m^{-2}}$. Cold pools increased the sea-air temperature difference by a factor of 2–3, but their effect on the sea-air specific humidity difference was weak (Fig. 6c).

In the 18 days of the convectively active phase, the temperature falls below the 27°C threshold 43% of the time, while in the 53 days of the suppressed phase the temperature falls below the threshold only 7% of the time. We diagnose the effect of the cold pools on the turbulent heat fluxes in the convective and suppressed periods of the MJO (defined by averaging days of negative and positive net heat flux, respectively) by excluding times when $T_{\mathrm{air}} <$ 27°C. The effect of the cold pools is the difference between the average for each period, and the average computed without the cold pools. Cold pools enhance the total turbulent flux $20\,\mathrm{W\,m^{-2}}$ during the convective phase of the MJO, compared to $1.5\,\mathrm{W\,m^{-2}}$ in the suppressed phase.

4. Summary

Net heat flux varies by $\pm 30\,\mathrm{W\,m^{-2}}$ for a typical MJO event in the Indian and western Pacific Ocean, with maximum flux from the ocean to the atmosphere nearly in phase with low OLR. The largest variations in the surface flux are due to solar radiation and evaporation. Wind speed is the main driver of the turbulent (sensible and latent) flux variability, and clouds are the main driver of the radiative fluxes. SST responds with a cycle of 0.10°C, which would imply a mixed layer of 20–40 m, were there no systematic intraseasonal heat fluxes from below the ocean mixed layer.

The MJO changes the sign of the daily heat flux into or out of the ocean. Anomalous latent heat flux in the convective phase of the MJO contributes up to $12\,\mathrm{W\,m^{-2}}$ more latent heating than the mean in DYNAMO ($\sim 110\,\mathrm{W\,m^{-2}}$, equivalent to 4 mm $\mathrm{day^{-1}}$ precipitation). The mean local latent heat flux accounts for almost half the mean precipitation ($230\,\mathrm{W\,m^{-2}}$, 8 mm $\mathrm{day^{-1}}$) in this part of the central Indian Ocean, yet the maximum evaporation of $240\,\mathrm{W\,m^{-2}}$

accounts for just 17% of the maximum daily precipitation of 50 mm day^{-1} estimated over the DYNAMO northern sounding array (P. Ciesielski, personal communication). Average water budgets from TOGA COARE were similar but with more moderate (20 mm day^{-1}) precipitation maxima (Lin and Johnson 1996). Lateral moisture convergence due to the convectively-coupled atmospheric circulation must import the dominant share of the moisture (de Szoeke *et al.* 2015).

We point out two sorts of sub-daily variability that contribute to the turbulent heat fluxes: diurnal warm layers and atmospheric cold pools. Diurnal warm layers in the upper ocean affect SST, increasing fluxes in the afternoon. Air temperature has a minimum in the early morning. The effect of the sea-air temperature difference is to create two weak maxima in surface sensible heat flux. Atmospheric cold pools 1–3°C cooler than their surroundings are accompanied by wind gusts. Cold pools increase sensible heat fluxes, but are neither systematically moist nor dry, so their latent heat flux is larger chiefly due to their increased wind speed.

Diurnal warm layers and the frequency and strength of cold pools are modulated by the MJO. They increase mean daily SST more strongly during calm, sunny, convectively suppressed phases of the MJO. Compared to a flux computed with predawn minimum SST, the diurnal warm layer increases the net turbulent heat flux 8.3 W m^{-2} in the suppressed phase, but only 1.4 W m^{-2} in the convectively active phase of the MJO. Cold pools, on the other hand, are stronger and more frequent during the convectively active phase of the MJO. Cold pools enhance turbulent fluxes 20 W m^{-2} during the convective phase of the MJO, compared to 1.5 W m^{-2} in the suppressed phase.

The intraseasonal variation of the diurnal warm layer demonstrates the importance of the difference between the *bulk* mixed layer ocean temperature representative of the average heat stored in the upper ocean, and the *interfacial* SST that influences surface fluxes. The effect of the diurnal warm layer on the fluxes is to trap solar heating close to the surface where it quickly flows into the atmosphere by surface fluxes, primarily evaporation and secondarily longwave thermal radiation. The difference between the interfacial and bulk temperature has an influence on intraseasonal air-sea heat exchange. It warms the atmosphere in the quiescent phase at the expense of heat mixed down and stored in the ocean. By increasing SST when winds are weak, the diurnal warm layer maintains more constant evaporation than if the solar absorption were always evenly mixed throughout the upper ocean mixed layer.

References

Anderson, S. P., R. A. Weller, and R. B. Lukas, 1996: Surface buoyancy forcing and the mixed layer of the western Pacific warm pool: Observations and 1D model results. *J. Climate*, **9**, 3056–3085.

Chen, S. S., R. A. Houze, and B. E. Mapes, 1996: Multiscale variability of deep convection in relation to large-scale circulation in TOGA COARE. *J. Atmos. Sci.*, **53**, 1380–1409.

Chi, N.-H., R.-C. Lien, Eric A. D'Asaro, and B. B. Ma, 2014: In-situ observation of surface mixed layer heat budget in the Central Indian Ocean during Madden-Julian Oscillation events, *J. Geophys. Res.* (in revision).

Clayson, C. A. and A. S. Bogdanoff, 2013: The effect of diurnal sea surface temperature warming on climatological air-sea fluxes. *J. Climate*, **26**, 2546–2556. 10.1175/JCLI-D-12-00062.1.

Dee, D. P., *et al.*, 2011: The ERA-Interim reanalysis: Configuration and performance of the data assimilation system. *Quart. J. Roy. Meteor. Soc.*, **137**, 553–597. 10.1002/qj.828.

DeMott, C. A., C. Stan, D. A. Randall, and M. D. Branson, 2014: Intraseasonal variability in coupled GCMs: The roles of ocean feedbacks and model physics. *J. Climate*, 10.1175/JCLI-D-13-00760.1.

de Szoeke, S. P., J. B. Edson, J. R. Marion, and C. W. Fairall, 2015: The MJO and air-sea interaction in TOGA COARE and DYNAMO. *J. Climate*, **28**, 597–622, 10.1175/JCLI-D-14-00477.1.

de Szoeke, S. P., E. D. Skyllingstad, P. Zuidema, and A. S. Chandra, 2016: Cold pools and their influence on the tropical marine boundary layer. *J. Atmos. Sci.*, submitted.

Feng, M., P. Hacker, R. Lukas, R. Weller, and S. P. Anderson, 2000: Upper ocean heat and salt balances in the western equatorial Pacific in response to the intraseasonal oscillation during TOGA COARE. *J. Climate*, **13**, 2409–2427.

Gentemann, C. L., C. J. Donlon, A. Stuart-Menteth, and F. J. Wentz, 2003: Diurnal signals in satellite sea surface temperature measurements. *Geophys. Res. Lett.*, **30**, 1140, 10.1029/2002GL016291.

Hendon, H. H. and J. Glick, 1997: Intraseasonal air-sea interaction in the tropical Indian and Pacific oceans. *J. Climate*, **10**, 647–661.

Inness, P. M., J. M. Slingo, S. J. Woolnough, R. B. Neale, and V. D. Pope, 2001: Organization of tropical convection in a GCM with varying vertical resolution; implications for the simulation of the Madden-Julian Oscillation. *Clim. Dyn.*, **17**, 777–793. 10.1007/s003820000148.

Inness, P. M. and J. M. Slingo, 2003: Simulation of the Madden-Julian Oscillation in a coupled general circulation model. Part I: Comparison with observations and an atmosphere-only GCM. *J. Climate*, **16**, 345–364. 10.1175/1520-0442(2003)016<0345:SOTMJO>2.0.CO;2.

Johnson, R. H., P. E. Ciesielski, J. H. Ruppert, and M. Katsumata, 2014: Sounding-Based Thermodynamic Budgets for DYNAMO. *J. Atmos. Sci.*, 10.1175/JAS-D-14-0202.1.

Jones, C., D. E. Waliser, and C. Gautier, 1998: The influence of the Madden-Julian Oscillation on ocean surface heat fluxes and sea surface temperature. *J. Climate*, **11**, 1057–1072.

Kiladis, G. N., M. C. Wheeler, P. T. Haertel, K. H. Straub, and P. E. Roundy, 2009: Convectively coupled equatorial waves. *Rev. Geophys.*, **47**, RG2003. 10.1029/2008RG000266.

Liebmann, B. and C. A. Smith, 1996: Description of a complete (interpolated) outgoing longwave radiation dataset. *Bull. Amer. Meteorol. Soc.*, **77**, 1275–1277.

Lin, X. and R. H. Johnson, 1996: Kinematic and thermodynamic characteristics of the flow over the western Pacific warm pool during TOGA COARE. *J. Atmos. Sci.*, **53**, 695–715, 10.1175/1520-0469(1996)053<0695:katcot>2.0.co;2.

Liu, P., B. Wang, K. R. Sperber, T. Li, and G. A. Meehl, 2005: MJO in the NCAR CAM2 with the Tiedtke Convective Scheme*. *J. Climate*, **18,** 3007–3020.

Madden, R. A. and P. R. Julian, 1971: Detection of a 40–50 day oscillation in the zonal wind in the tropical Pacific. *J. Atmos. Sci.*, **28**, 702–708.

Maloney, E. D. and D. L. Hartmann, 2001: The sensitivity of intraseasonal variability in the NCAR CCM3 to changes in convective parameterization. *J. Climate*, **14**, 2015–2034, 10.1175/1520-0442(2001)014<2015:tsoivi>2.0.co;2.

Marshall, A. G., O. Alves, and H. H. Hendon, 2008: An enhanced moisture convergence–evaporation feedback mechanism for MJO air-sea interaction. *J. Atmos. Sci.*, **65**, 970–986. 10.1175/2007JAS2313.1.

McPhaden, M. J. and S. P. Hayes, 1991: On the variability of winds, sea surface temperature, and surface layer heat content in the western Equatorial Pacific, *J. Geophys. Res.*, **96**, 3331–3342.

Moum, J. N., *et al.*, 2014: Air-sea interactions from westerly wind bursts during the November 2011 MJO in the Indian Ocean. *Bull. Amer. Meteorol. Soc.*, **95**, 1185–1199. 10.1175/BAMS-D-12-00225.1.

Praveen Kumar, B., J. Vialard, M. Lengaigne, V. S. N. Murty, and M. J. McPhaden, 2012: TropFlux: air-sea fluxes for the global tropical oceans — description and evaluation. *Clim. Dyn.*, **38**, 1521–1543. 10.1007/s00382-011-1115-0.

Price, J. F., R. A. Weller, and R. Pinkel, 1986: Diurnal cycling: Observations and models of the upper ocean response to diurnal heating, cooling, and wind mixing. *J. Geophys. Res. Oceans*, **91**, 8411–8427. 10.1029/ JC091iC07p08411.

Ruppert, J. H., Jr. and R. H. Johnson, 2015: Diurnally Modulated Cumulus Moistening in the Preonset Stage of the MaddenJulian Oscillation during DYNAMO. *J. Atmos. Sci.*, **72**, 1622–1647, doi:10.1175/JAS-D-14-0218.1.

Shinoda, T., H. H. Hendon, and J. Glick, 1998: Intraseasonal variability of surface fluxes and sea surface temperature in the tropical western Pacific and Indian Oceans. *J. Climate*, **11**, 1685–1702. doi:10.1175/1520-0442(1998)011<1685:IVOSFA>2.0.CO;2.

Shinoda, T. and H. H. Hendon, 1998: Mixed layer modeling of intraseasonal variability in the tropical western Pacific and Indian Oceans. *J. Climate*, **11**, 2668–2685.

Shinoda, T. and H. H. Hendon, 2001: Upper ocean heat budget in response to the Madden-Julian oscillation in the western equatorial Pacific. *J. Climate*, **14**, 4147–4165.

Shinoda, T., 2005: Impact of the diurnal cycle of solar radiation on intraseasonal SST variability in the western equatorial Pacific. *J. Climate*, **18**, 2628–2636.

Slingo, J. M., P. Inness, R. Neale, S. Woolnough, and G.-Y. Yang 2003: Scale interactions on diurnal to seasonal timescales and their relevance to model systematic errors. *Ann. Geophys.*, **46**, 139–155.

Smyth, W. D., D. Hebert, and J. N. Moum, 1996: Local ocean response to a multiphase westerly wind burst: 2. Thermal and freshwater responses. *J. Geophys. Res. Oceans*, **101**, 22513–22533, 10.1029/96jc02006.

Stephens, G. L., P. J. Webster, R. H. Johnson, R. Engelen, and T. L'Ecuyer, 2004: Observational evidence for the mutual regulation of the tropical hydrological cycle and tropical sea surface temperatures. *J. Climate*, **17**, 2213–2224. doi:10.1175/1520-0442(2004)017<2213:OEFTMR>2.0.CO;2.

Waliser, D. E., K. M. Lau, and J.-H. Kim, 1999: The influence of coupled sea surface temperatures on the Madden-Julian Oscillation: A model perturbation experiment. *J. Atmos. Sci.*, **56**, 333–358, 10.1175/1520-0469(1999)056<0333:TIOCSS>2.0.CO;2.

Weller, R. A. and S. P. Anderson, 1996: Surface meteorology and air-sea fluxes in the western equatorial Pacific warm pool during the TOGA Coupled Ocean–Atmosphere Response Experiment. *J. Climate*, **9**, 1960–1990.

Wheeler, M. C. and H. H. Hendon, 2004: An all-season real-time multivariate MJO index: Development of an index for monitoring and prediction. *Mon. Wea. Rev.*, **132**, 1917–1932.

Woolnough, S. J., J. M. Slingo, and B. J. Hoskins, 2000: The relationship between convection and sea surface temperature on intraseasonal timescales. *J. Climate*, **13**, 2086–2104, 10.1175/1520-0442(2000)013<2086:trbcas>2.0.co;2.

Yoneyama, K., C. Zhang, and C. N. Long, 2013: Tracking pulses of the Madden-Julian oscillation. *Bull. Amer. Meteor. Soc.*, **94**, 1871–1891. 10.1175/ BAMS-D-12-00157.1

Zhang, C., 2013: Madden–Julian oscillation: Bridging weather and climate. *Bull. Amer. Meteor. Soc.*, **94**, 1849–1870. 10.1175/BAMS-D-12-00026.1

Zhang, G. J. and M. Mu, 2005: Simulation of the Madden-Julian oscillation in the NCAR CCM3 using a revised Zhang-McFarlane convection parameterization scheme. *J. Climate*, **18**, 4046–4064.

Zhang, Y. C., W. B. Rossow, A. A. Lacis, V. Oinas, and M. I. Mishchenko, 2004: Calculation of radiative fluxes from the surface to top of atmosphere based on ISCCP and other global data sets: Refinements of the radiative transfer model and the input data. *J. Geophys. Res.*, **109**.

Chapter 29

An Overview of the Asian Monsoon Years 2007–2012 (AMY) and Multi-Scale Interactions in the Extreme Rainfall Events over the Indonesian Maritime Continent

Jun Matsumoto

Research Center for Climatology and Department of Geography,
Tokyo Metropolitan University,
1-1, Minami-Osawa, Hachioji 192-0397, Japan
Japan Agency for Marine-Earth Science and Technology (JAMSTEC),
2-15, Natsushima-Cho, Yokosuka 237-0061, Japan
jun@tmu.ac.jp

Bin Wang

International Pacific Research Center and Department of Meteorology,
University of Hawaii, Honolulu, Hawaii,
Honolulu, HI 96816, USA

Guoxiong Wu

State Key Lab of Atmospheric Sciences and Geophysical Fluid Dynamics (LASG)
Institute of Atmospheric Physics, Chinese Academy of Sciences,
Beijing 200029, China

Jianping Li

College of Global Change and Earth System Science (GCESS),
Beijing Normal University,
Beijing 100875, China

Peiming Wu, Miki Hattori, Shuichi Mori, Manabu D. Yamanaka and Shin-ya Ogino

Japan Agency for Marine-Earth Science and Technology (JAMSTEC),
2-15, Natsushima-Cho, Yokosuka 237-0061, Japan

Hamada Jun-Ichi

Research Center for Climatology and Department of Geography,
Tokyo Metropolitan University,
1-1, Minami-Osawa, Hachioji 192-0397, Japan

Fadli Syamsudin

Agency for the Assessment and Application of Technology (BPPT),
Jl. M. H. Thamrin 8, Jakarta 10340, Indonesia

The Global Monsoon System: Research and Forecast (3rd Edition)
Edited by C. P. Chang et al.

Toshio Koike, Katsunori Tamagawa, Eiji Ikoma and Hiroko Kinutani
The University of Tokyo,
7-3-1, Bunkyo-ku, Tokyo 113-8656, Japan

Hirotaka Kamahori, Kenji Kamiguchi and Yayoi Harada
Meteorological Research Institute,
1-1, Nagamine, Tsukuba 305-0052, Japan

The Asian Monsoon Years 2007–2012 (AMY) program, is a cross-cutting coordinated observation and modeling initiative that is part of the International Monsoon Study under the World Climate Research Programme (WCRP). From 2008 to 2010, the AMY program coordinated an Intensive Observing Period (IOP). More than twenty international and national projects joined in this IOP. In particular, multi-scale interactions in the extreme rainfall events that characterize the Asian monsoon region were explored. Here, we briefly review multi-scale interactions on intra-seasonal, synoptic, and diurnal time-scales on the Indonesian Maritime Continent, particularly the role of the Asian winter cold-surge and the Madden-Julian Oscillation (MJO) in heavy rainfall events on Jawa Island.

The observational data obtained in the AMY coordinated observations were input for the AMY reanalysis conducted by the Meteorological Research Institute (MRI), Japan. Additionally, in-situ observational data have been archived and made available to the worldwide research community through the Data Integration & Analysis System (DIAS), developed at the University of Tokyo.

1. Introduction

About 60% of the world's population inhabits the region affected by the Asian monsoon. Vegetation, economy, and society across this region are critically influenced by the variability of the monsoon. Many Asian countries are developing countries whose economies are fast growing, yet vulnerable to anomalous climate events and extreme weather. Future changes in the Asian monsoon climate are also of great concern to the world economy and sustainable development. The scientific importance of the Asian monsoon cannot be overemphasized. The giant Asian monsoon system dominates the entire Eastern Hemisphere. It interacts with the El Niño/Southern Oscillation (ENSO) and extratropical circulations, and has a far-reaching impact on the global climate and environment. The Asian monsoon exemplifies the most complex interactions on the Earth's surface with the ocean, atmosphere, hydrosphere, and biosphere.

Monsoon science has been advanced enormously over the last four decades owing to a wealth of new data from satellite observations, operational and re-analysis atmospheric datasets, and field experiments, and due to advances in computing power and mathematical representations of coupled climate systems by global and regional weather and climate models. Historically, for example, the Monsoon Experiment (MONEX) was conducted in 1978–1979, targeting both summer (Summer MONEX; Krishnamurti 1985) and winter (Winter MONEX; Johnson and Chang 2007) monsoons as subprograms of the First Global Atmospheric Research Program (GARP) Global Experiment (FGGE). The Global Energy and Water Experiment (GEWEX) Asian Monsoon Experiment (GAME) was accomplished in 1996–2005, predominantly on the Asian land regions of Siberia (GAME-Siberia), the Huaihe River Basin (HUBEX), the Tibetan Plateau (GAME-Tibet), and Thailand and surrounding Southeast Asia (GAME-T) with the Intensive Observation Period (IOP) set for the summer of 1998 jointly with the South China Sea Monsoon Experiment (SCSMEX). This resulted in a special issue of GAME in the *Journal of the Meteorologivcal Society of Japan* (Yasunari 2001). These advances in Asian monsoon studies are comprehensively reviewed by Chang *et al.* (2002; 2011), Chang (2004) and Wang (2006).

Driven by the needs to better understand and predict monsoon on all time scales from daily weather to long-term climate change, and its connection with aerosol, extensive monsoon researches have been carried out in Asian monsoon region, and Asian Monsoon Years 2007–2012 (AMY) activities were coordinated.

2. AMY

The AMY program is a cross-cutting coordinated observation and modeling initiative that is part of the International Monsoon Study under the World Climate Research Programme (WCRP). Its long-term goal is to improve Asian monsoon predictions for the benefit of society, as well as benefiting each participating project through improved understanding of monsoon variability and predictability. The scientific details and implementation plan for the AMY program were published in Wang *et al.* (2010) and Matsumoto *et al.* (2010), respectively. Further details on the AMY program are available on the AMY web page (http://www.wcrp-amy.org/).

The objectives of the AMY program are to enhance understanding of ocean-land-atmosphere-biosphere interactions, multiple time-scale interactions, from diurnal to intraseasonal, and the aerosol-water cycle interaction in the Asian monsoon system, in order to improve their physical representations in coupled climate models, and to develop data assimilation for the ocean-atmosphere-land system in the Asian monsoon region. These objectives were fulfilled through coordination of participating field experiments and modeling projects in the Asian monsoon region. For implementing AMY, Scientific Steering Committee (SSC) was organized co-chaired by the two leading authors of the present paper, Bin Wang, and Jun Matsumoto. AMY International Program Office (IPO) was set in the Institute of Atmospheric Physics at Beijing China, directed by the fourth author of this paper, Jianping Li. Four working groups were organized in order to manage AMY coordinated studies, WG1: Observation Coordination, WG2: Central data archiving and management, WG3: Medeling and prediction, and WG4: Aerosol-monsoon. Members of these organizations are listed in Matsumoto *et al.* (2010). Here, only part of the activities in WGs 1 and 2 will be presented, and other activities will be introduced elsewhere in future.

The AMY program stems from grassroots scientific and societal imperatives. It was endorsed by the Joint Scientific Committee (JSC) of the World Climate Research Programme (WCRP), as well as the WCRP, the Climate Variability and Predictability (CLIVAR), and the Global Energy and Water Exchanges Project (GEWEX) in 2007. From the Asian-Australian Monsoon Panel (AAMP) played an important role, while from the GEWEX side, the Monsoon Asian Hydro-Atmosphere Scientific Research and Prediction Initiative (MAHASRI) acted as a main player. The MAHASRI is conducted under the Global Hydroclimatology Panel (GHP)/GEWEX/WCRP aiming at establishing hydrometeorological prediction system, particularly up to seasonal time-scales through better scientific understanding of Asian monsoon variability from 2006–2015 (the International Drafting Committee of the Post GAME Planning Working Group 2006). As such some participating projects' period were longer than the main targeting period of AMY 2007–2012, but these results are also included in the AMY research results. Its earlier scientific results of the *MAHASRI* and AMY were published in a special issue of *MAHASRI* in the *Journal of the Meteorologivcal Society of Japan* (Matsumoto *et al.* 2011). It has also been identified as a cross-cutting weather and climate activity by the World Meteorological Organization (WMO) Secretariat in the World Weather Research Programme (WWRP) Strategic Plan.

One of the main targets of the AMY program is the study of the dynamics and predictability of intraseasonal variability (ISV), with a 30–60-day and a biweekly period, through observations and modeling. Land-atmosphere and ocean-atmosphere interactions should be reexamined through intensive observations focusing on their roles in the ISVs. The time-space structures of the ISVs seem to have been changing over the past several decades (e.g. Kajikawa *et al.* 2009), which may, at least partly, be related to the anthropogenic forcing, including the impact of aerosols and other pollutants.

The AMY program coordinated field experiments have been organized by coordinating more than twenty national and international field experiments (Fig. 1). They conducted the majority of their observations during 2008–2010. These observations are classified into three categories: ocean observations, land observations, and special process observations. Special process observations include meso-scale experiments for observing heavy rainfall and tropical cyclones, and aerosol-cloud-radiation experiments. The AMY program seeks to coordinate field campaigns among various individual research projects. The target periods of these observations can be classified as (1) the pre-monsoon period in March–May; (2) the monsoon onset phase in May–June; (3) the monsoon mature phase in July–August; and (4) the winter monsoon from December to February.

After the previously conducted intensive observation period (IOP) of the GAME and SCSMEX in 1998, the Meteorological Research

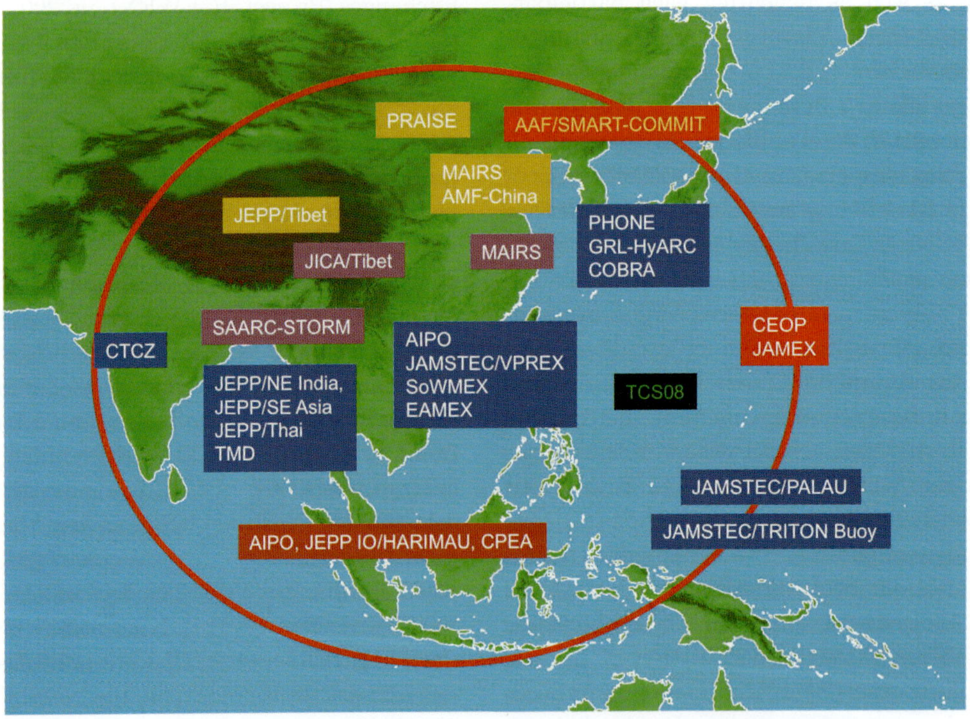

Fig. 1. AMY coordinated observations.

Institute (MRI), Japan Meteorological Agency (JMA) conducted the GAME reanalysis using all the off-line GAME and SCSMEX observations from April to October 1998. Details of the GAME reanalysis are provided in Yamazaki *et al.* (2000) and Yatagai *et al.* (2000), and comparisons with JRA-25 (Japanese 25-year Reanalysis; Onogi *et al.* 2007), as well as those between versions 1.1 and 1.5 of the GAME reanalysis, are presented in Yatagai *et al.* (2007).

Likewise, the AMY reanalysis was conducted by the MRI, JMA, to attain a high quality atmospheric dataset using AMY off-line observational data. They utilized the same model scheme as the JRA-55 reanalysis by the JMA (Ebita *et al.* 2011). Table 1 summarizes the comparisons between the GAME reanalysis and the AMY reanalysis. In Tables 2 and 3, the number of additional stations utilized in the AMY reanalysis is shown. Also participating project information is summarized in Table 4. Although the number of additional stations is not very large, except in the pre-monsoon to monsoon period in the summer of 2008, the model performance nevertheless improved after the GAME reanalysis period as shown in Table 1. We are expecting higher quality datasets, in particular for the Asian monsoon region, and the AMY reanalysis product is under final calculation in 2016, and will be released for the AMY community in 2017, and made available to all the research community in 2019.

AMY *in-situ* observational data have been archived and opened to the worldwide research community through the Data Integration & Analysis System (DIAS http://dias-dss.tkl.iis.u-tokyo.ac.jp/ddc/finder) hosted by the University of Tokyo, Japan. As of October 2016, ten datasets are available: TRITON buoy, HARIMAU sonde/AWS, CTCZ-Pilot, MAHASRI-AMY, GRL-HyARC radiosonde, HARIMAU-WPR, COBRA radiosonde, Palau-AMY, JICA-Tibet and VPREX-AMY_Mactan. They are also available from the AMY (http://www.wcrp-amy.org/two.asp?lmid=6&id=31) and/or MAHASRI (http://hydro.iis.u-tokyo.ac.jp/mahasriwiki/wiki/data/index.html). The input data for the AMY reanalysis were also archived in the DIAS. These activities were conducted in part of the WG2 activities in AMY.

As of 2015, eight AMY workshops were organized to plan and implement activities. These workshops were held as follows: The first in Beijing, China, in April 2009; the second in Bali, Indonesia in September 2007; the third in Yokohama, Japan in January 2008; the fourth in Busan, Korea in June 2008; the fifth in Beijing, China in October 2008; the sixth in Kunming China in November 2009; the seventh in Pune, India in July 2010; and the eighth in Beijing, China in May 2011. Additionally,

Table 1. Comparison of the GAME reanalysis and the AMY reanalysis.

	GAME reanalysis	AMY reanalysis
Input data	GTS Satellite GAME-offline	GTS Satellite AMY-offline
Horizontal resolution	T213(60 km)	TL959(20 km)
Vertical levels	40	100
Model top	0.4 hPa	0.01 hPa
Assimilation algorithms	3-dimensional optimum interpolation	4-dimensional variational method
Target period	Apr. 1998–Oct. 1998 Only in summer	Jan. 2008–Dec. 2010 All season
Other		Various improvement in physical processes after the GAME reanalysis

Table 2. Surface meteorological data input for the AMY reanalysis.

ID	Project name	Data periods	Number of stations
01	TRITON buoy (North Pacific Ocean)	2008/01–2010/12	18
02	MAHASRI-Vietnam (Vietnam)	2009/03–2010/12	3
03	PRAISE (Mongolia)	2003/01–2010/08	1
05	EAMEX (Taiwan)	2008/01–2010/12	25
06	JEPP-Thai (Thailand)	2008/01–2010/12	13
07	HARIMAU (Indonesia)	2008/01–2010/12	5
09	JEPP-NE India-Bangladesh (Bangladesh)	2008/01–2010/12	23
19	JICA_Tibet (China)	2008/01–2008/07	77

Table 3. Upper air meteorological data input for the AMY reanalysis.

ID	Project name (region or nation)	Data periods	Number of stations
04	AIPO (Northern South China Sea)	2007/08, 2008/08, 2009/09, 2010/09	1
05	EAMEX_UpperAir (Taiwan)	2008/01–2010/12	10
07	HARIMAU-UpperAir (Indonesia)	2010/01–2010/02	4
07	HARIMAU-WP (Indonesia)	2008/01–2010/12	4
08	MAHASRI-AMY (Indonesia)	2008/08, 2008/12, 2009/01, 2009/08, 2010/12	2
10	Palau-AMY (Palau)	2008/06–2008/07, 2010/05, 2010/06	1
11	VPREX-AMY_DaNang (Vietnam)	2010/10–2010/12	1
12	VPREX-AMY_Mactan (Philippines)	2010/09–2010/12	1
13	CTCZ-Pilot (Bay of Bengal)	2009/07–2009/08	1
14	TMD_Upper Air (Thailand)	2008/01–2010/12	5
15	COBRA Project (Japan)	2008/05–2009/06	1
16	SoWMEX (Taiwan area and northern boundary of South China)	2008/05–2008/06	12
17	SAARC_STORM_UpperAir (Bangladesh, India)	2010/04–2010/05	12
18	AMF-China (China)	2008/05–2008/12	1
19	JICA_Tibet_Upper Air (China)	2008/01–2008/09	39
20	GRL-HyARC (Japan)	2010/06	1
21	PHONE (East China Sea area)	2008/06–2008/07, 2009/06–2009/07, 2010/07	2

an AMY data workshop was held in Tokyo, Japan, in June 2010. Besides these workshops, science sessions have been organized at the Asia Oceania Geosciences Society (AOGS) Annual Meetings, the American Geophysical Union (AGU), the European Geosciences Union (EGU) General Assemblies, the International Association of Meteorology and Atmospheric Sciences (IAMAS), the International Union of Geodesy and Geophysics (IUGG), and other relevant meetings. In order to summarize AMY activities, the Open Science Conference (AMY-OSC) was organized in October 26–27, 2013, at Zhuhai, China (http://amy2013.csp.escience.cn/dct/page/1).

3. Multi-scale Interactions over the Indonesian Maritime Continent

The Indonesian Maritime Continent (IMC) is well known as an area that sees heavy rainfall. For this region, Robertson *et al.* (2011) reviewed the spatio-temporal climatic variability, and the interactions between the strong diurnal cycle, sub-seasonal weather types, and ENSO-related inter-annual variability. Researches, particularly satellite observations, have revealed that a diurnal cycle is predominant in this area (e.g. Mori *et al.* 2004; Sakurai *et al.* 2005; Araki *et al.* 2006). The Madden-Julian Oscillation (MJO; Madden and Julian 1971; 1972) is one of the

Table 4. List of project names and their information.

Project abbreviation	Project name	URL
TRITON	Triangle Trans Ocean Buoy Network	http://www.jamstec.go.jp/jamstec/TRITON/real_time/
MAHASRI	Monsoon Asian Hydro-Atmosphere Scientific Research and Prediction Initiative	http://mahasri.cr.chiba-u.ac.jp/wiki/index.php/Main_Page
PRAISE	Post RAISE Research Activities	http://raise.suiri.tsukuba.ac.jp/new/index.html
EAMEX	East Asian Monsoon Experiment	http://eamex.iastate.edu/index.html/
HARIMAU	Hydrometeorological ARray for ISV-Monsoon AUtomonitoring	http://www.jamstec.go.jp/iorgc/harimau/HARIMAU.html
JICA-Tibet	Japan International Cooperation Agency/Japan-China Meteorological Disaster Reduction Corporation Research Center Project (JICA/Tibet Project)	https://www.jstage.jst.go.jp/browse/jmsj/90C/0/_contents
AIPO	Ocean-Atmosphere Interaction over the Joining Area of Asia and Indian-Pacific Ocean and Its Impact on the Short-Term Climate Variation in China	http://973aipo.lasg.ac.cn/ (Only in Chinese)
Palau	Pacific Area Long-term Atmospheric Observation for Understanding Climate Change	http://www.jamstec.go.jp/dcop/e/observations/palau.html
VPREX	Vietnam Philippine Rainfall Experiment	http://www.jamstec.go.jp/rigc/j/tcvrp/mhcrt/vprex.html
CTCZ	Continental Tropical Convergence Zone Programme	http://www.incois.gov.in/portal/datainfo/ctczsp.jsp
TMD	Thai Meteorological Department	http://www.tmd.go.th/en/
COBRA	CRL Okinawa Bisastic Polarimetric Radar	http://okinawa.nict.go.jp/EN/COBRA/index.html
SoWMEX	Southwest Monsoon Experiment	http://p3rd.as.ntu.edu.tw/
STORM	Severe Thunderstorms — Observations & Regional Modeling Programme	http://www.imd.gov.in/SciencePlanofFDPs/STORM%20Science%20Plan.pdf
AMF	Deployment of the Atmospheric Radiation Measurement Mobile Facility	https://www.arm.gov/publications/programdocs/doe-sc-arm-0802.pdf
GRL	Global Research Laboratory, Pukyong National University, Korea	http://grl.pknu.ac.kr/eng/grl/grl_intro.jsp
HyARC	Hydroshperic Atmospheric Research Center, Nagoya University, Japan	http://www.hyarc.nagoya-u.ac.jp/Eng/index-en.html
PHONE	Baiu Frontal Rainfall Experiment in Miyako and Tarama	

major modulators interacting with the diurnal cycle over the Maritime Continent (e.g. Ichikawa and Yasunari 2006; 2007; 2008; Fujita *et al.* 2011; Rauniyar and Walsh 2011; Oh *et al.* 2012). Very recently, Yamanaka (2016) comprehensively reviewed climatological features of the IMC in various time-scales.

In order to reveal the detailed processes of the diurnal cycle in convective activities, and its interaction with the MJO over the Maritime Continent, the first observation campaign, the HARIMAU (Yamanaka *et al.* 2008), called "HARIMAU2006", was conducted over the southwestern coastline of Sumatra

Island. This campaign mainly used two X-band Doppler radars (XDRs) and intensive atmospheric soundings, and was conducted in 2006 (Mori *et al.* 2011; Sakurai *et al.* 2011). One year's worth of XDR observation results from this campaign on the modulation of the diurnal rainfall cycle by the MJO are investigated in Kamimera *et al.* (2012). A distinct diurnal cycle in the water level of the Ciliwung River, which flows through Jakarta on Jawa Island, was also found using the HARIMAU2010 IOP data (Sulistyowati *et al.* 2014). Another campaign observation (HARIMAU2011) was collaborated with CINDY2011 (Cooperative Indian Ocean Experiment on Intraseasonal Variability in the Year 2011; Yoneyama *et al.*, 2013) and DYNAMO (Dynamics of the MJO; Zhang 2013; Johnson *et al.*, 2013) projects in the Indian Ocean.

Although precisely speaking, HARIMAU-2006 year was out of the research target of AMY 2007–2012, it was included in the MAHASRI project period (2006–2015) participating in AMY. In addition HARIMAU related surface meteorological and upper wind profiler data, and HARIMAU2010 upper air sonde data were put into AMY reanalysis as shown in Tables 2 and 3.

Owing to the strong influence of local land-sea and mountain-valley circulations on the convective rainfalls, coarse resolution global circulation models (GCMs) substantially underestimated the precipitation amount over the IMC (Neale and Slingo 2003). In CMIP3, precipitation underestimation is still observed over the IMC (Randall *et al.* 2007), on the other hand, it changed into overestimation in CMIP5 models (Flato *et al.* 2013). Even using fine-scale (20 km grid) GCMs, the diurnal cycle of large islands for this region was incorrectly simulated (Hara *et al.* 2009). In contrast, finer-scale regional climate model (RCM) (Qian 2008; Wu *et al.* 2008; Hara *et al.* 2009; Oh *et al.* 2013) and/or a very fine-scale global cloud-resolving model (GCRM), known as the Nonhydrostatic Icosahedral Atmospheric Model (NICAM) (Sato *et al.* 2009), can more accurately model such local-circulation-related rainfall. Severe local thunderstorms and rainstorms are prevalent hazardous extreme weather events that have high socio-economic significance.

Improving our understanding of these severe local storms is of immense scientific importance and must be addressed. Multiple time-scale interaction, from diurnal to intra-seasonal was one of the main research target of AMY as introduced in the previous section. Here, as an example of the multi-scale interactions in monsoon rainfall, heavy rainfall events over the IMC mainly in Jakarta in northern Jawa Island, will be discussed.

Over the IMC, severe rainfall events also occur frequently, and can cause severe damages, for example, in the Indonesian capital city of Jakarta. The interactions between the cold-surge from Siberia and the diurnally varying local circulation were important to the continuous torrential rains that occurred in Jakarta in late January to early February, 2007 (Wu *et al.* 2007, Fig. 2). During that winter, heavy rainfall also occurred in Malaysia (Tangang *et al.* 2008)

As shown in Fig. 2, during this period of heavy rainfall, there was almost no cloud cover over Jawa Island in the morning, while strong convective activity occurred in the evening over the central mountainous area. This activity propagates to the northern coastal area at night, interacting with a strong cold-surge flow. It should be noted that the MJO was not in an active phase over the Maritime Continent. In this case, interactions between diurnal convection and synoptic-scale cold-surges are important. Upper air soundings at Soekarno-Hatta International Airport, Jakarta, for 1200 UTC (1900 local time) on January 31 and February 1, 2007, are shown in Fig. 3. Both dew-point profiles indicated the existence of a distinct dry layer at the mid-level of 600–300 hPa, which induced a strong convective instability over the

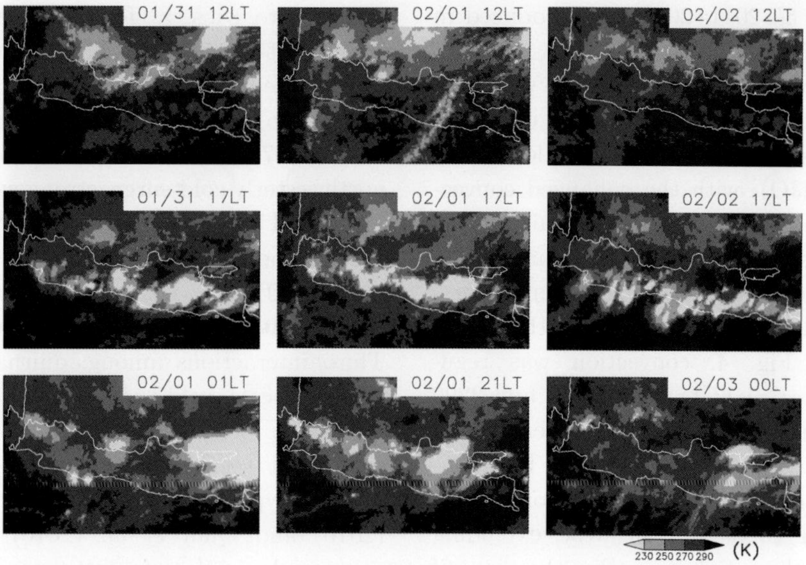

Fig. 2. Geostationary Meteorological Satellite (GMS) infrared (IR) images over Jawa Island from January 31 to February 3, 2007 (Wu *et al.* 2007).

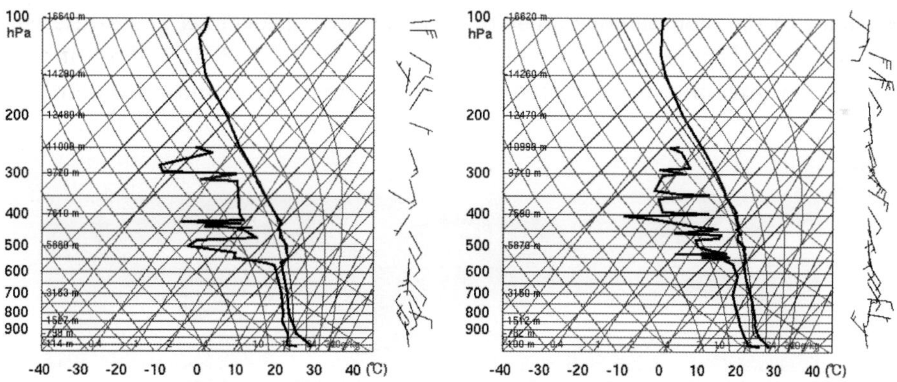

Fig. 3. The Skew-T Log-P diagrams obtained at Jakarta for 1200 UTC (1900 local time) on (left) January 31 and (right) February 1, 2007. The right-hand line in each of the plots is the temperature profile, and the left-hand line is the dew-point profile. Winds are plotted as wind barbs at the relevant height on the right-hand edge of the plots (Wu *et al.* 2007).

island. Meanwhile, the wind profiles indicated that while a southeasterly wind prevailed above 700 hPa, there was a layer of northwesterly winds in the levels below about 800 hPa. The wind directions at 900 and 700 hPa were almost opposite, with a large vertical wind shear of 15 m s^{-1} through the 900–700 hPa layers. It was also notable that the upper prevailing winds over Jawa Island were weak in this case, providing favorable conditions for the development of thermally induced valley-mountain circulation. This scenario differs from the heavy rainfall events in 2013 described below. In contrast, the low-level northwesterly winds due to the strong trans-equatorial monsoon flow from the Northern Hemisphere were common to both events. This heavy rainfall case was investigated in detail using a numerical non-hydrostatic model

(Trilaksono *et al.* 2011; 2012) as introduced in Yoden *et al.* (2016).

As a contrasting example, a heavy rainfall event in Jakarta in mid-January, 2013, coincided with an active phase of the Madden-Julian Oscillation (MJO), with the enhanced convective phase centered on the western Pacific, by the all-season, real-time multivariate MJO index (http://www.bom.gov.au/climate/mjo/) (Wheeler and Hendon 2004). In this case, as shown in Fig. 4, convection was regularly initiated at night between 1900 and 2200 (local time), from January 14–17, over the sea to the northwest of Jawa Island. The convection then solidified into a line orientated north-south or northeast-southwest, extending for about 100 km. Subsequently, the line of storms migrated southeastward over western Jawa Island overnight. It took 4–5 hr for the rain systems to move from the northwestern edge of the island to the mountainous areas to the south of Jakarta. The propagation speed of the precipitation system was estimated to be close to 8 ms^{-1}. The initiation and propagation of convection were markedly different from those during the heavy rain event of late January to early February, 2007, reported by Wu *et al.* (2007). In this case, above the near-surface northwesterly cold-surge, strong westerly winds, associated with strong MJO activity in the east of the Maritime Continent, were present up to 400 hPa, with a maximum wind speed in the 900–600 hPa layer, as shown in Fig. 5. Thus, interactions among, diurnal convection, synoptic-scale cold-surges, and the MJO are important for this case. It is also notable that this event seems to correspond to the daily weather type (WT5) discussed by Moron *et al.* (2010) and Qian *et al.* (2010), when cross-equatorial air flow is distinct and rainfall is abundant both over Jawa Island and the adjacent sea.

Numerical simulations of the extreme rain event by Trilaksono *et al.* (2012), which are also introduced in Yoden *et al.* (2016). They reproduced two types of initiation and propagation of

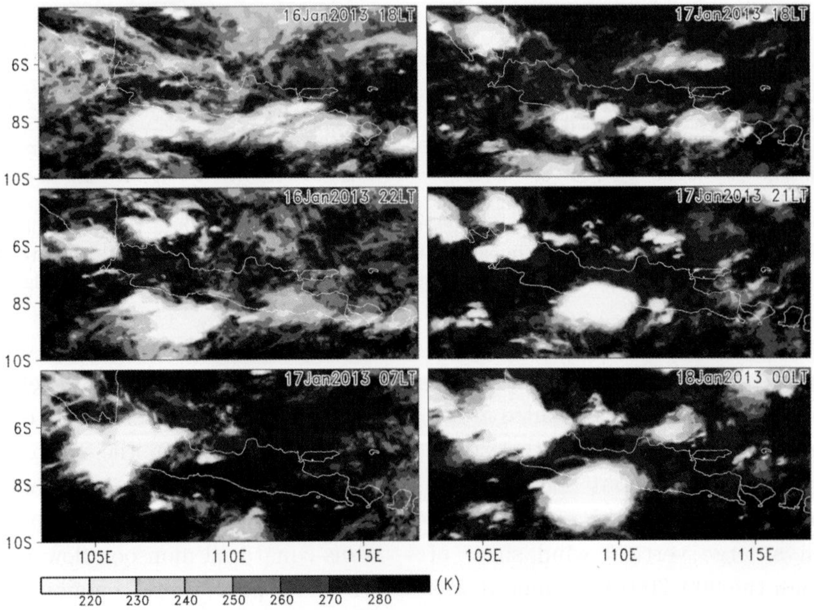

Fig. 4. GMS satellite infrared (IR) images over Jawa Island from January 16–18, 2013 (Wu *et al.* 2013).

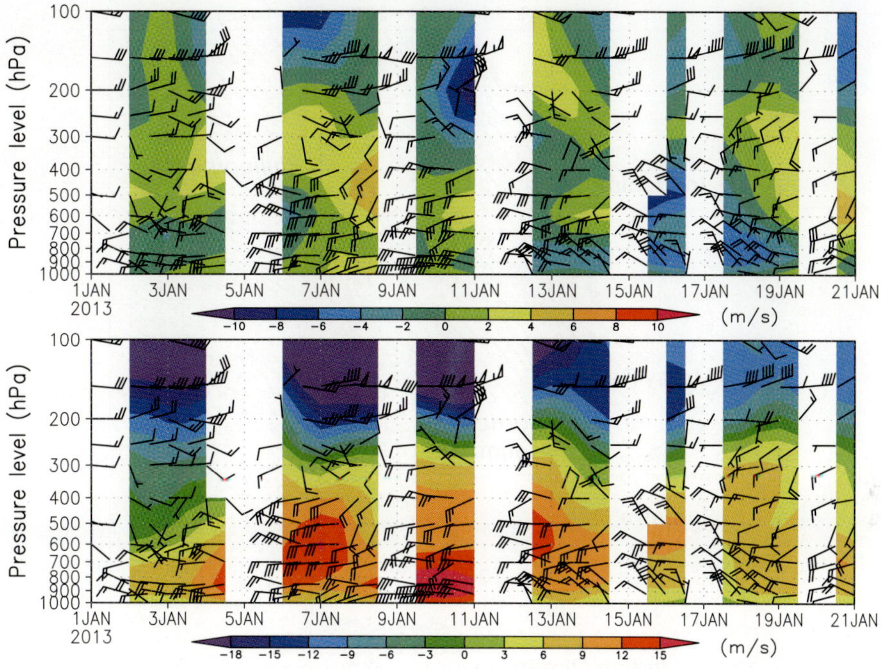

Fig. 5. Profile of horizontal winds obtained from upper air soundings at Jakarta as a function of time, January 2013. The shades in the upper and lower panels denote the meridional wind component and zonal wind component, respectively. The tick marks on the x-axes indicate 0000 UTC (0700 local time) (Wu *et al.* 2013).

convection, that is, convection that initiated in the evening near the foot of the mountain propagates northward over the northern plains, and the convection that initiated in the early morning over the Jawa Sea propagates southward to western Jawa Island. Maybe, these two patterns interacting cold surge and/or MJO will be an important factors for generating heavy rainfall in northern Jawa Island.

It should be noted that cross-equatorial northerly flow beyond the equator reaches Jawa Island whose north shore is located at approximately 6°S from mid-December to late March even in a climatological condition (Figs. 6 and 7). On the other hand, in each year situation, for example, in 2008–2009 boreal winter, strong cross-equatorial northerly surge (CENS) events, northerly wind exceeding $5\,\mathrm{m\,s^{-1}}$ over the oceanic region in 105°E–115°E, 5°S–EQ based on the Quick Scatterometer (QuikSCAT) sea surface wind data (Freilich and Dunbar 1999; Chelton and Freilich 2005) are observed only occasionally as indicated by white stars in Fig. 5 in Yoden *et al.* (2016). Such events occurred around 21 December, 13 January, 7 February and 1 March. Among them in the second and third cases they are accompanied with high pressure and lower temperature accompanied cold surge from the Asian Continent, while in the last case, it occurred with lower pressure and no temperature decrease (Fig. 8). Therefore, CENSs are not always accompanied by the so-called cold surges from the Asian Continent.

Considering the above findings, Hattori *et al.* (2011) distinguished synoptic-scale CENSs as cold-surges (CSs), MJO-surges, and a combination of both and made a statistical composite analysis for the period 2000–2009. If we see the Tropical Rainfall Measurement Mission (TRMM) rainfall composite based on TRMM 3B42 data (Huffman *et al.* 2007), in case of CS-pattern, convective activities were enhanced

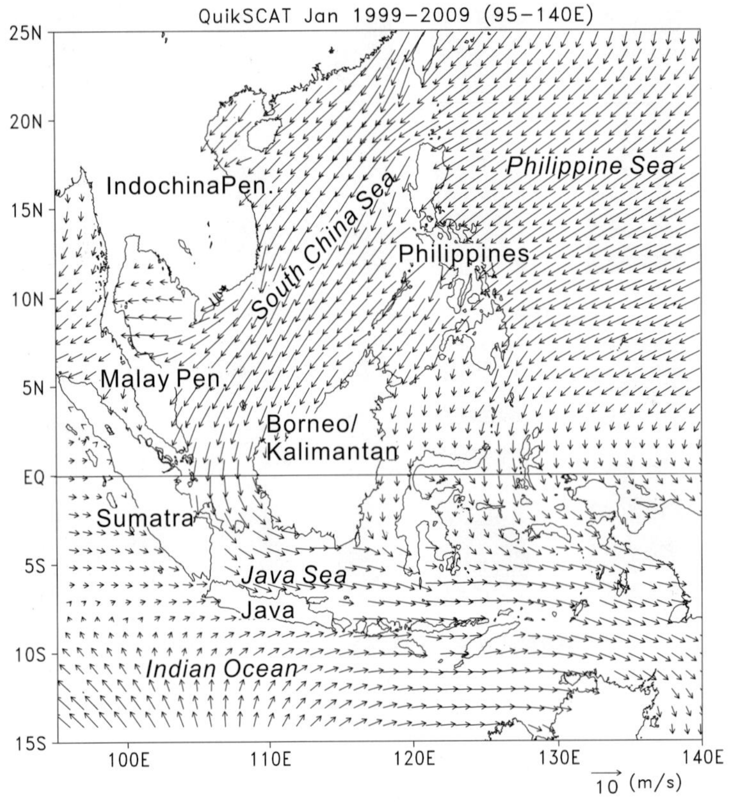

Fig. 6. The 10-year averaged sea surface wind over the South China Sea and the Maritime Continent for January 2000 to 2009 based on the QuikSCAT data. (Adapted from Hattori *et al.*, 2011.)

over the Jawa Sea and the northern part of Jawa Island (Fig. 9a). This was also seen in the results of Chang *et al.* (2005). The heavy rainfall event in 2007, discussed above, seems to correspond to this pattern. In case of MJO-pattern, convective activities were activated mainly over the sea west of Sumatra and south of Jawa, but not over Jawa Island (Fig. 9b). For combined CS–MJO pattern activities, the increase in precipitation over the Maritime Continent was much larger than for the other two patterns (Fig. 9c). The CS–MJO pattern had characteristics of both the CS and the MJO patterns in that the increase in precipitation was widely induced over the Maritime Continent both over the sea and islands, in particular, in northern Jawa Island. Another heavy rainfall event in 2013 seems to correspond to this pattern.

As described in this section, multi-scale interactions among diurnal, synoptic and intraseasonal variations were shown to be important for the occurrence of heavy rainfall events in Jakarta, northern Jawa Island, both in case studies and climatological studies. However, only synoptic-scale conditions have been analyzed so far, due partly to the shortage of local-scale observation data which have not been obtained under AMY project. Here we presented two heavy rainfall events in Jakarta and climatological composite study also implies that there seem to be some common synoptic conditions in the heavier rainfall in northwestern Jawa Island (Fig. 9). However, what kind of differences are existent between heavy rainfall events in other regions in northwestern Jawa have not been analyzed yet.

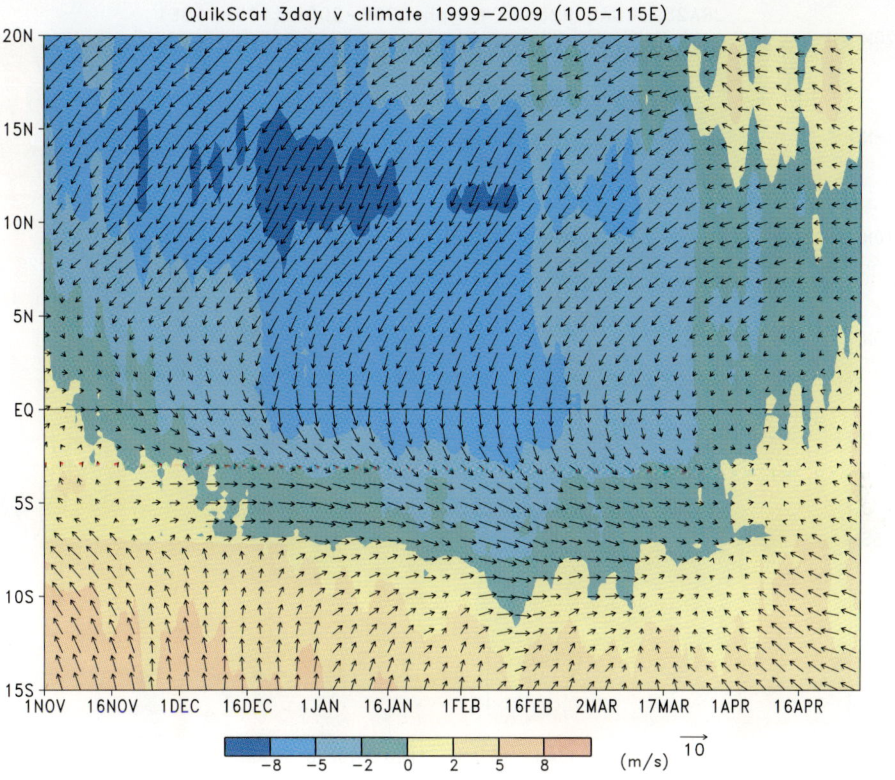

Fig. 7. Time-latitude cross-section of the 10-winter averaged meridional sea surface wind component indicated by shading and the wind vectors from November to April over the South China Sea between 105°E and 115°E. (Adapted from Hattori *et al.* 2011.)

As for the interannual variations, Jakarta rainy season rainfall is found to be more abundant in neutral years both in El Niño Southern Oscillation (ENSO) and Indian Ocean Dipole (IOD) (Hamada *et al.* 2012). Hamada *et al.* (2012) showed that northerly wind over the South China Sea, probably related with northern hemisphere winter monsoon, is closely related with rainfall variations in Northwestern Jawa Island around Jakarta in its peak rainy season in November–January (Fig. 10). Fill the gap between such seasonal-scale climate variations and synoptic heavy rainfall events is also a future problem.

Furthermore, Villafuerte and Matsumoto (2015) presented the long-term trend of extreme daily heavy rainfall (RX1day) over the whole Southeast Asian region for the period 1951–2007 by utilizing gauge-based gridded dataset, the Asian Precipitation-Highly Resolved Observational Data Integration Towards Evaluation of Water Resources (APHRODITE; Yatagai *et al.* 2009; 2012). In general, daily extreme rainfall tends to decrease over the IMC contrasting with increasing trend over the Indochina Peninsular and central Philippines. No significant trend is detected in annual basis in northern Java Island (Fig. 11), while decreasing trend is observed in some small regions there during dry season April to September (Fig. 12). However, since the number of data utiziled in APHRODITE gridded dataset is not uniform throughout the period, the results should be verified by the homogeneous station-based data.

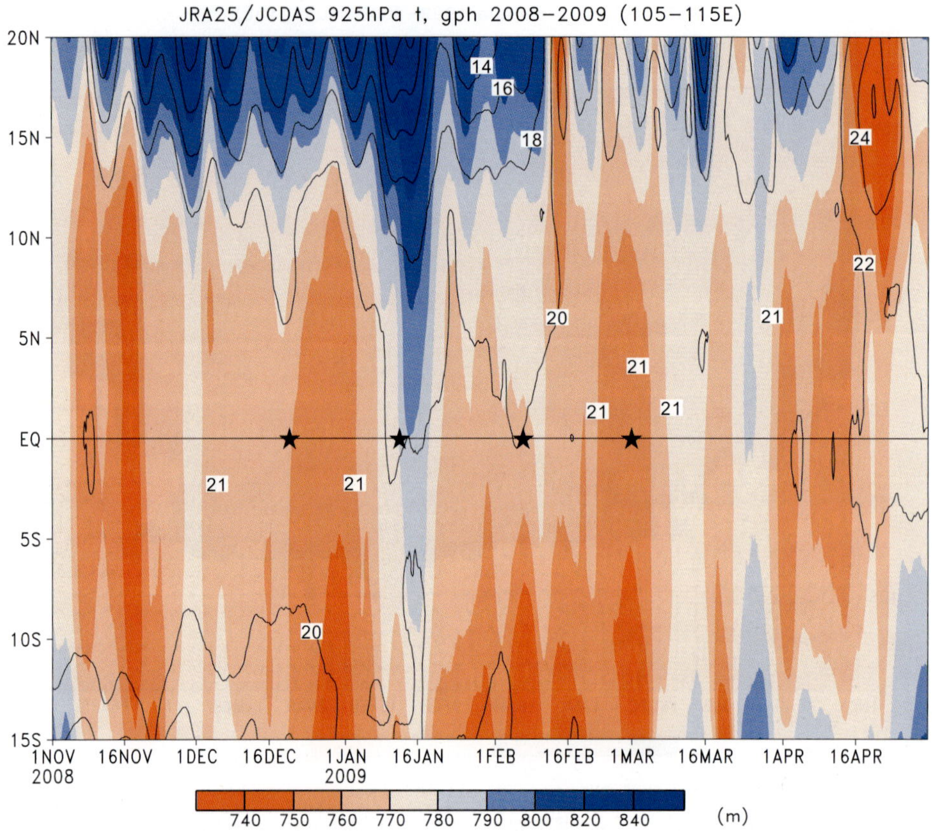

Fig. 8. As in Fig. 6 but for the temperature indicated by contour lines and geopotential height at 925 hPa indicated by shading for the year 2008–2009. Contours are at 12, 14, 16, 18, 20, and 21°C. Starts represent cases of southward extension of the strong northerly wind indicated in the text. (Adapted from Hattori *et al.* 2011.)

4. Remarks

AMY was the first grassroots-driven international research program under the WCRP that was composed mostly of Asian research projects. It provides high-accuracy in-situ observations and a reanalysis dataset for 2008–2010 over a wide Asian monsoon region, allowing GCMs or RCMs to be used for studying Asian monsoon variability, seasonal predictions, and the forecasting of extreme weather. Coordination of the various national and international projects will facilitate mutual data exchange and new scientific findings, leading to a deeper understanding of the Asian monsoon, particularly the link between regional components (for example, monsoon variability over India, the Tibetan Plateau, East Asia, and the Western North Pacific, as well as connections with mid-latitudes). As a result of these research activities, research collaboration in the Asian region has been enhanced, and understanding of atmosphere-ocean-land-biosphere interactions, multiple time-scale interactions, and aerosol-cloud-monsoon interactions has progressed.

Here, we have introduced AMY related results for the IMC, particularly the important role of multi-scale interactions among diurnal cycles, the MJO, and the Asian winter monsoon in heavy rainfall events in Jakarta. A number of studies noted a distinct influence of the

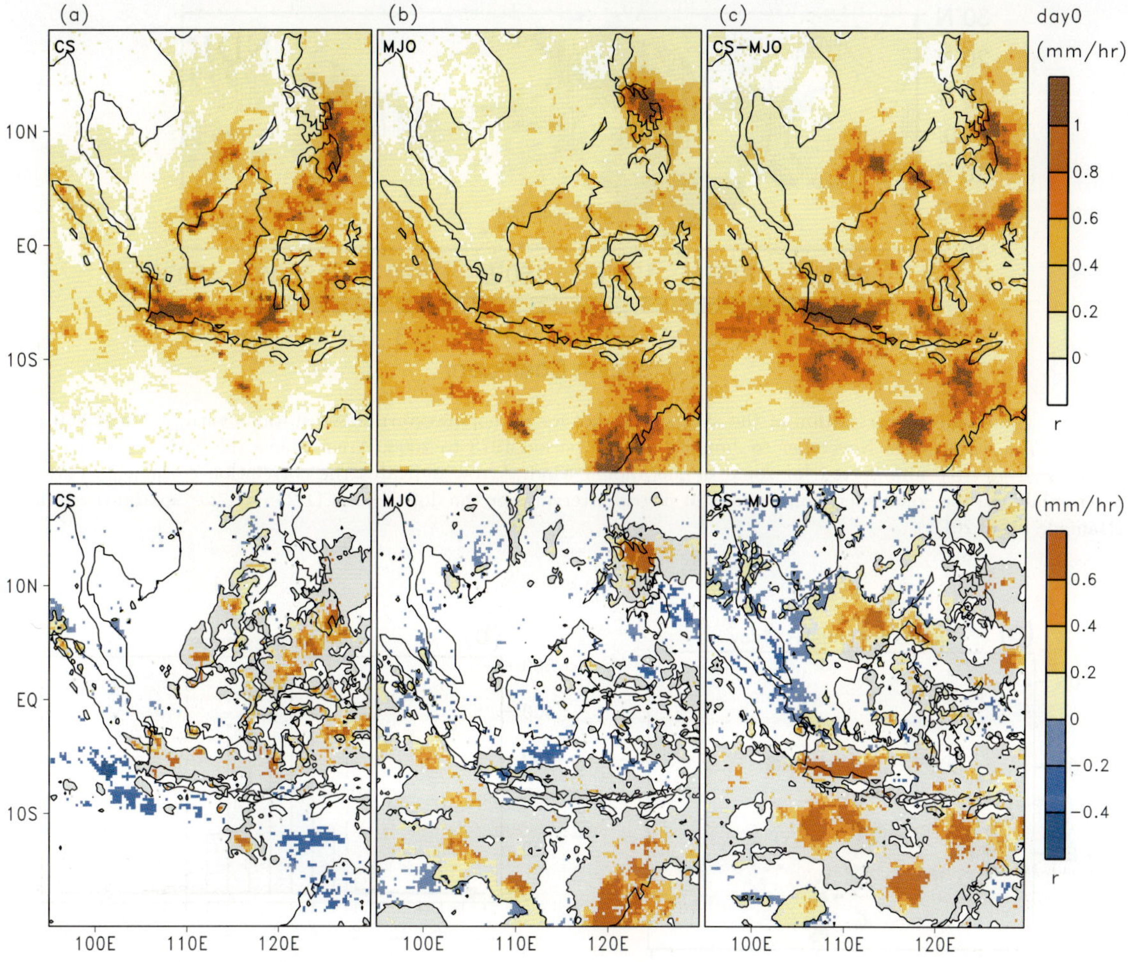

Fig. 9. Composite precipitations (upper panels) and the anomalies from the 11-winter mean (lower panels) for DJFM (a) cold-surges (CS), (b) Madden-Julian Oscillation (MJO), and (c) CS–MJO patterns. The solid contours in the lower three panels represent an anomaly of 0 mm hr^{-1}. Positive anomalies are indicated by light gray shading, and only anomalies above the 90% significance level are represented by color shading. (Adapted from Hattori *et al.* 2011.)

MJO on the diurnal cycle over the Indonesian Maritime Continent. Climatological studies showed stronger rainfall over the islands in the MJO inactive phase, but over the adjacent ocean areas in the active phase (e.g. Rauniyar and Walsh 2011; Oh *et al.* 2012). Relationships between the monsoon winds and diurnal cycles were also discussed (Qian *et al.* 2010). However, our studies show that both MJO and the monsoon affect rainfall on Jawa Island, and the influence is not simple, as demonstrated by the heavy rainfall events discussed above. Although heavy rainfall events are rare, they cause severe damage in this region, and therefore, more observational and modeling studies are needed. Additionally, rapid economic development within this region requires improved disaster prevention and mitigation. However, we still do not specify in what region heavy rainfall occurs under specific synoptic-scale conditions and need improvement of both observation and modeling.

As introduced in Yoden *et al.* (2016) numerical prediction of severe rainfall events over

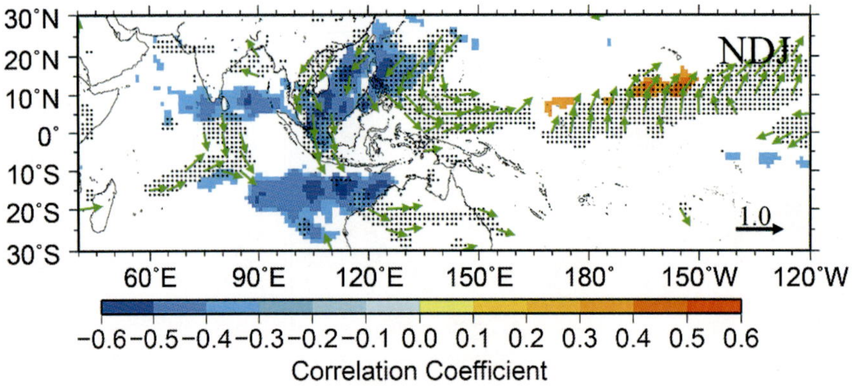

Fig. 10. Simultaneous correlation of the average rainfall for all northwestern Jawa stations with SST and 850 hPa wind for the season from November to January for the period 1973 to 2008. Areas above the 95% confidence level are colored for SST and dotted for zonal or meridional wind. Arrow indicates the correlation coefficients between the rainfall and the zonal/meridional winds at 5°grid interval over the dotted area (1.25° × 1.25° spatial resolution) (Hamada *et al.* 2012.)

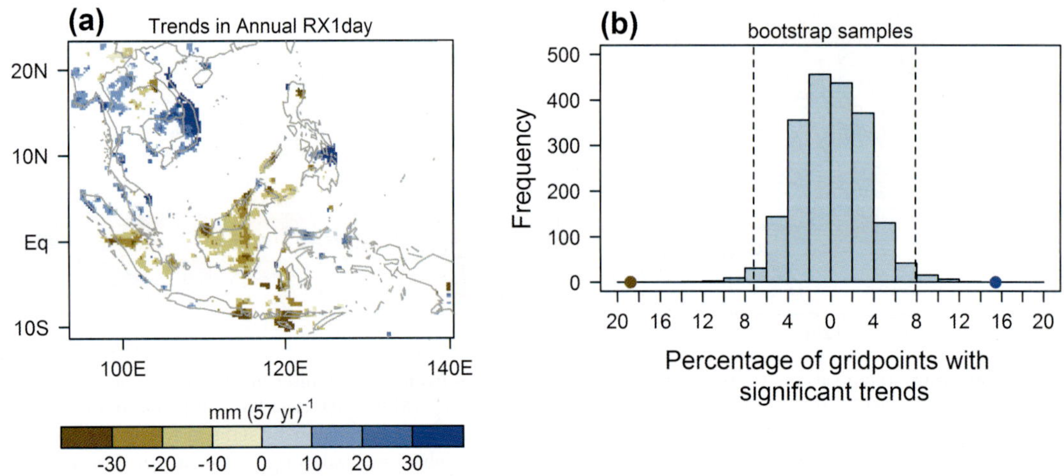

Fig. 11. (a) Spatial distribution of the long-term trends (1951–2007) in annual maximum daily rainfall (RX1day) that are significant at the 5% level based on the Mann–Kendall trend test, and (b) the percentage of grid boxes that showed significant decreasing (left side of zero in the abscissa) and increasing (right side of zero in the abscissa) trends in annual RX1day. The histogram in (b) is based on the 1000 bootstrap samples of annual RX1day in Southeast Asia, while the brown and the blue dots are based on the original data. The two dashed-lines in (b) mark the 95% of the distribution that occurred by random chance (Villafuerte and Matsumoto 2015).

the IMC is still not satisfactory, although finer model resolution, in general, gives better performance. Complex island/sea distribution along with mountainous topography in islands provide difficulties both in observation and modeling over this region. For example, although local meteorological service (Badan Meteorologi, Klimatologi dan Geophysica: BMKG) have recently developed nation-wide automatic weather station and radar networks, number of automatic stations are still not sufficient considering their complex topographic features. In addition, radar

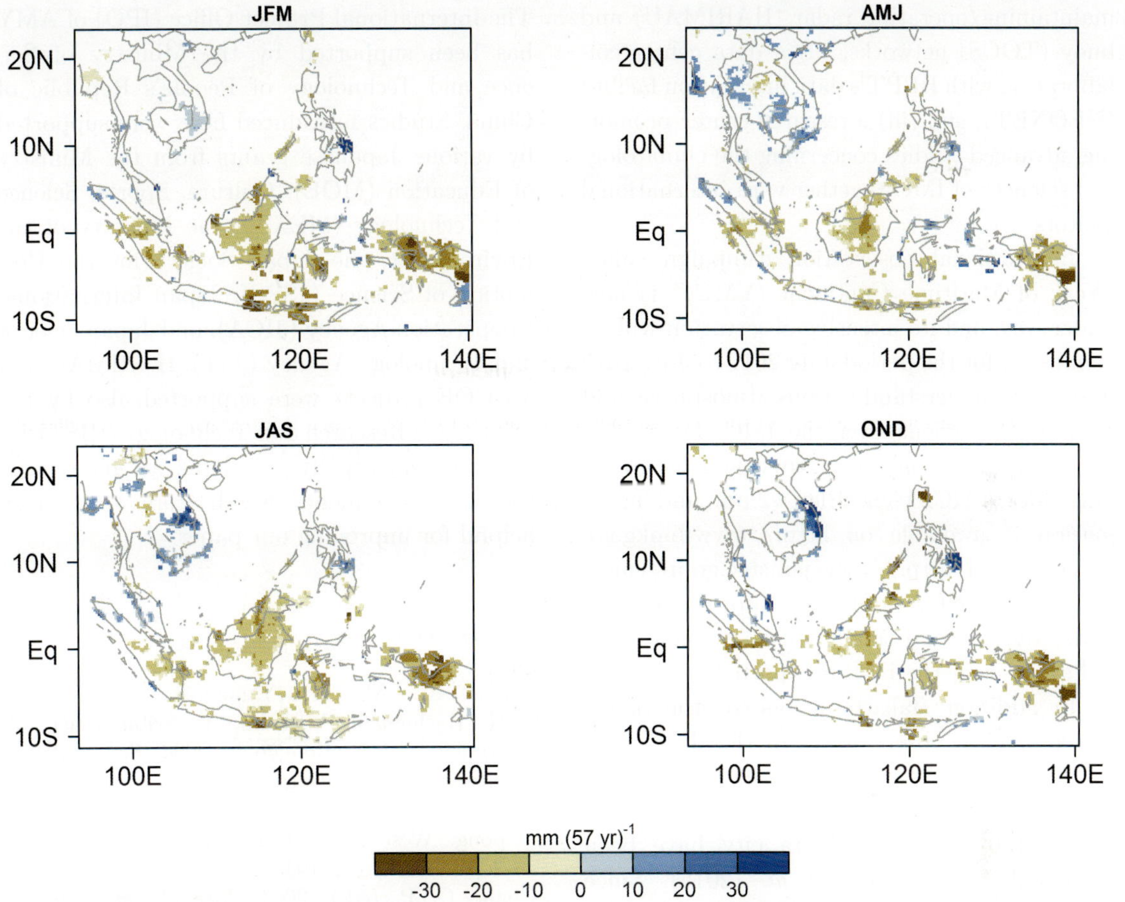

Fig. 12. Long-term trends (1951–2007) in seasonal RX1day [the seasons are defined as January–March (JFM), April–June (AMJ), July–September (JAS), and October–December (OND)]; only those that are significant at 5% level based on the Mann–Kendall trend test are shown (Villafuerte and Matsumoto 2015).

network composed by different radar companies hinders development of radar composite products and/or radar/raingauge composite products. They set limitations for early warning system for heavy rainfall events. Furthermore, Indonesian archipelago is composed of complicated land/sea distribution. The oceanic observations along the island are also important for improving our understanding of air-sea interactions near the coastline. Recently, Ogino *et al.* (2016) found that the tropical rainfall is concentrated within approximately 300 km of the coast line based on TRMM 2A25 data. Large islands in the IMC are typical cases where such abundant coastal rainfalls occur, and minute processes generating such abundant rainfall will be revealed by studies in the IMC.

A number of activities have been conducted in order to develop observational facilities and atmospheric science over the IMC which are comprehensively reviewed by Yamanaka (2016). All the instruments have been taken over by an international center of tropical climatology, meteorology and oceanography, called "Maritime Continent Center of Excellence" (MCCOE), which was established under BPPT in November 2013. The MCCOE has three functions: (i) an observation center maintaining/operating radar (HARIMAU) and buoy (TOCS) networks, (ii) a data center collaborating with BPPT's data integration facility (NEONET), and (iii) a research center promoting advanced studies concerning the climatological sciences of IMC together with international visitors.

International observation campaign called "Year of Maritime Continent (YMC)" is now under way, and its intensive observation will be conducted for the period July 2017 to July 2019 in order to understand various atmospheric and oceanic phenomena over the IMC. Its science plan is available on http://www.jamstec.go.jp/ymc/docs/YMC_SciencePlan_v2.pdf and information is available on http://www.bmkg.go.id/ymc/ and http://www.jamstec.go.jp/ymc/. With such efforts further improvement of understanding and prediction of extreme rainfall events over the IMC is expected.

In addition, Jakarta is one of the oldest meteorological observatories in the deep tropics. Observation started in the early 19th centrury. Very recently, changes in diurnal cycles during 130 years in Jakarta city have been presented by Siswanto *et al.* (2015). Since Jakarta experienced very rapid urban expansion in recent years, such long-term data are precious for detecting the effect of human activities on local climate. However, huge volumes of data are still not digitized in the pre-World War period in Indonesia. Tremendous efforts are still needed for the data rescue activities.

Under the AMY program, a number of studies have been conducted over the whole Asian Monsoon Region besides the IMC on various topics. The scientific results of these studies will be reported elsewhere.

Acknowledgments

The AMY program has been an international coordinated activities and each participating project was funded by various national funding. The International Project Office (IPO) of AMY has been supported by the Ministry of Science and Technology of People's Republic of China. Studies introduced here were supported by various Japanese grants from the Ministry of Education (MOE), Culture, Sports, Science and Technology (MEXT), the Ministry of the Environment, the Japan Society for the Promotion of Science (JSPS), Japan International Cooperation Agency (JICA), and Japan Science and Technology Agency (JST). HARIMAU and MCCOE projects were supported also by the Ministry of Research and Technology (RISTEK) of Indonesia. Finally, we thank an anonymous reviewer's comments which were very much helpful for improving our paper.

References

Araki, R., M. D. Yamanaka, F. Murata, H. Hashiguchi, Y. Oku, T. Sribimawati, M. Kudsy, and F. Renggono, 2006: Seasonal and interannual variations of diurnal cycles of local circulation and cloud activity observed at Serpong, West Jawa, Indonesia. *J. Meteor. Soc. Japan*, **84A**, 171–194.

Chang C. P. (ed.), 2004: *East Asian Monsoon.* World Scientific, 564 pp.

Chang C. P., G. Wu, B. Jou and C. Y. Lam (eds.), 2002: *East Asia and Western Pacific Meteorology and Climate.* World Scientific, 440 pp.

Chang, C. P., P. A. Harr, and H. J. Chen, 2005: Synoptic disturbances over the equatorial South China Sea and western maritime continent during boreal winter. *Mon. Wea. Rev.*, **133**, 489–503.

Chang C. P., Y. Ding, N. C. Lau, R. H. Johnson, B. Wang and T. Yasunari (eds.), 2011: *The Global Monsoon System: Research and Forecast (2nd ed.).* World Scientific, 594 pp.

Chelton, D. B. and M. H. Freilich, 2005: Scatterometer-based assessment of 10-m wind analyses from the operational ECMWF and NCEP numerical weather prediction models. *Mon. Wea. Rev.*, **133**, 409–429.

Ebita, A., S. Kobayashi, Y. Ota, M. Moriya, R. Kumabe, K. Onogi, Y. Harada, S. Yasui, K. Miyaoka, K. Takahashi, H. Kamahori, C. Kobayashi, H. Endo, M. Soma, Y. Oikawa, and T. Ishimizu, 2011: The Japanese 55-year reanalysis "JRA-55": An interim report. *SOLA*, **7**, 149–152.

Flato, G., J. Marotzke, B. Abiodun, P. Braconnot, S. C. Chou, W. Collins, P. Cox, F. Driouech, S. Emori, V. Eyring, C. Forest, P. Gleckler, E. Guilyardi, C. Jakob, V. Kattsov, C. Reason and M. Rummukainen, 2013: Evaluation of climate models. In: *Climate Change 2013: The Physical Science Basis. Contribution of Working Group I to the Fifth Assessment Report of the Intergovernmental Panel on Climate Change.* (Eds.) Stocker, T. F., D. Qin, G.-K. Plattner, M. Tignor, S. K. Allen, J. Boschung, A. Nauels, Y. Xia, V. Bex and P. M. Midgley. Cambridge University Press.

Freilich, M. H. and R. S. Dunbar, 1999: The accuracy of the NSCAT 1 vector winds: Comparisons with National Data Buoy Center buoys. *J. Geophys. Res.*, **104**, 11231–11246.

Fujita, M., K. Yoneyama, S. Mori, T. Nasuno, and M. Sato, 2011: Diurnal convection peaks over the eastern Indian Ocean off Sumatra during different MJO phases. *J. Meteor. Soc. Japan*, **89A**, 317–330.

Hamada, J.-I., S. Mori, H. Kubota, M. D. Yamanak, U. Haryoko, S. Lestari, R. Sulistyowati, and F. Syamsudin, 2012: Interannual rainfall variability over Northwestern Jawa and its relation to the Indian Ocean Dipole and El Niño-Southern Oscillation events. *SOLA*, **8**, 069–072.

Hara, M., T. Yoshikane, H. G. Takahashi, F. Kimura, A. Noda, and T. Tokioka, 2009: Assessment of the diurnal cycle of precipitation over the Maritime Continent simulated by a 20 km mesh GCM using TRMM PR data. *J. Meteor. Soc. Japan*, **87A**, 413–424.

Hattori, M., S. Mori, and J. Matsumoto, 2011: The cross-equatorial northerly surge over the maritime continent and its relationship to precipitation patterns. *J. Meteor. Soc. Japan*, **89A**, 27–47.

Huffman, G. J., R. F. Adler, D. T. Bolvin, G. J. Gu, E. J. Nelkin, K. P. Bowman, Y. Hong, E. F. Stocker, and D. B. Wolff, 2007: The TRMM multisatellite precipitation analysis (TMPA): Quasi-global, multiyear, combined-sensor precipitation estimates at fine scales. *J. Hydrometeor.*, **8**, 38–55.

Ichikawa, H. and T. Yasunari, 2006: Time-space characteristics of diurnal rainfall over Borneo and surrounding oceans as observed by TRMM-PR, *J. Climate*, **19**, 1238–1260.

Ichikawa, H. and T. Yasunari, 2007: Propagating diurnal disturbances embedded in the Madden-Julian Oscillation. *Geophys. Res. Lett.*, **34**, L18811. doi:10.1029/2007GL030480.

Ichikawa, H. and T. Yasunari, 2008: Intraseasonal variability in diurnal rainfall over New Guinea and the surrounding oceans during Austral Summer, *J. Climate*, **21**, 12, 2852–2868.

Johnson, R. H. and C. P. Chang, 2007: Winter MONEX: A quarter century and beyond. *Bull. Amer. Meteor. Soc.*, **88**, 385–388.

Johnson, R. H. and P. E. Ciesielski, 2013: Structure and properties of Madden-Julian Oscillations deduced from DYNAMO sounding arrays. *J. Atmos. Sci.*, **70**, 3157–3179.

International Drafting Committee of the Post GAME Planning Working Group, 2006: *Monsoon Asian Hydro-Atmosphere Scientific Research and Prediction Initiative (MAHASRI)*, 54 pp. Available on line at: http://hydro.iis.u-tokyo.ac.jp/mahasriwiki/before_wiki/documents/MAHASRI_SciencePlan_v4.1.pdf

Kajikawa, Y., T. Yasunari, and B. Wang 2009: Decadal change in intraseasonal variability over the South China Sea. *Geophys. Res. Lett*, **36**, L06810. doi:10.1029/2009GL037174.

Kamimera, H., S. Mori, M. D. Yamanaka, and F. Syamsudin, 2012: Modulation of diurnal rainfall cycle by the Madden-Julian Oscillation based on one-year continuous observations with a meteorological radar in West Sumatera. *SOLA*, **8**, 111–114.

Krishnamurti, T. N., 1985: Summer Monsoon Experiment — A review. *Mon. Wea. Rev.*, **113**, 1590–1626.

Madden, R. A. and P. R. Julian, 1971: Detection of a 40–50 day oscillation in the zonal wind in the tropical Pacific. *J. Atmos. Sci.*, **28**, 702–708.

Madden, R. A. and P. R. Julian, 1972: Description of global-scale circulation cells in the tropics with a 40–50 day periods. *J. Atmos. Sci.*, **29**, 1109–1123.

Matsumoto, J., T. Satomura, J. Asanuma, T.-C. Chen, Y. Fujiyoshi, A. Higuchi, H. Ishikawa, M. N. Islam, S. Kanae, R. Kawamura, Y. Kodama, J.-P. Li, T. Nakazawa, F. Syamsudin, I. Tamagawa, H. Vathananukij, M. D. Yamanaka, B. Wang, and M.-C. Yen (eds.), 2011: Special Issue on MAHASRI–Monsoon Asian Hydro-Atmosphere Scientific Research and Prediction Initiative–. *J. Meteorol. Soc. Japan*, **89A**, 464 pp.

Matsumoto, J., B. Wang, J. Li, and G. Wu, 2010: *The Implementation Plan for Asian Monsoon Years (AMY 2007–2012), A cross-cutting WCRP initiative.* China Meteorological Press, 38 pp. Available on line at http://www.wcrp-amy.org /about.asp?lmid=2&id=18.

Mori, S., J.-I. Hamada, Y. I. Tauhid, M. D. Yamanaka, N. Okamoto, F. Murata, N. Sakurai, H. Hashiguchi, and T. Sribimawati, 2004: Diurnal land-sea rainfall peak migration over Sumatra Island, Indonesian maritime continent, observed by TRMM satellite and intensive rawinsonde soundings. *Mon. Wea. Rev.*, **132**, 2021–2039.

Mori, S., J. I. Hamada, N. Sakurai, H. Fudeyasu, M. Kawashima, H. Hashiguchi, F. Syamsudin, A. A. Arbain, R. Sulistyowati, J. Matsumoto, and M. D. Yamanaka, 2011: Convective systems developed along the coastline of Sumatra Island, Indonesia observed with an X-band Doppler radar during the HARIMAU2006 campaign. *J. Meteor. Soc. Japan*, **89A**, 61–81.

Moron, V., A. W. Robertson, and J.-H. Qian 2010: Local versus regional-scale characteristics of monsoon onset and post-onset rainfall over Indonesia. *Clim. Dyn.*, **34**, 281–299.

Neale, R. and J. Slingo, 2003: The Maritime Continent and its role in the global climate: A GCM study. *J. Climate*, **16**, 834–848.

Ogino, S.-Y., M. D. Yamanaka, S. Mori, and Matsumoto, 2016: How much is the precipitation amount over the tropical coastal region? *J. Climate*, **29**, 12131–1236.

Oh, J.-H., K.-Y. Kim, and G. H. Lim, 2012: Impact of MJO on the diurnal cycle of rainfall over the western Maritime Continent in the austral summer. *Clim. Dyn.*, **38**, 1167–1180.

Oh, J.-H., B.-M. Kim, K.-Y. Kim, H.-J. Song, and G. H. Lim, 2013: The Impact of the diurnal cycle on the MJO over Maritime Continent: A modeling study assimilating TRMM rain rate into global analysis. *Clim. Dyn.*, **40**, 893–911.

Onogi, K., J. Tsutsui, H. Koide, M. Sakamoto, S. Kobayashi, H. Hatsushika, T. Matsumoto, N. Yamazaki, H. Kamahori, K. Takahashi, S. Kadokura, K. Wada, K. Kato, R. Oyama, T. Ose, N. Mannoji, and R. Taira, 2007: The JRA-25 reanalysis. *J. Meteorol. Soc. Japan*, **85**, 369–432.

Qian, J.-H. 2008: Why precipitation is mostly concentrated over islands in the Maritime Continent. *J. Atmos. Sci.*, **65**, 1428–1441.

Qian, J.-H., A. W. Robertson, and V. Moron, 2010: Interactions among ENSO, the monsoon, and diurnal cycle in rainfall variability over Java, Indonesia. *J. Atmos. Sci.*, **67**, 3509–3524.

Randall, D. A., R. A. Wood, S. Bony, R. Colman, T. Fichefet, J. Fyfe, V. Kattsov, A. Pitman, J. Shukla, J. Srinivasan, R. J. Stouffer, A. Sumi, and K. E. Taylor, 2007: Climate Models and Their Evaluation. In *Climate Change 2007: The Physical Science Basis. Contribution of Working Group I to the Fourth Assessment Report of the Intergovernmental Panel on Climate Change.* Eds. S. Solomon, D. Qin, M. Manning, Z. Chen, M. Marquis, K. B. Averyt, M. Tignor and H. L. Miller, Cambridge Univ. Press, pp. 589–662.

Rauniyar, Surendra P., and Kevin J. E. Walsh, 2011: Scale interaction of the diurnal cycle of rainfall over the Maritime Continent and Australia: Influence of the MJO. *J. Climate*, **24**, 325–348.

Robertson, A. W., V. Moron, J.-H. Qian, C.-P. Chang, F. Tangang, E. Aldrian, T. Y. Koh, and L. Juneng, 2011: The Maritime Continent monsoon, In *The Global Monsoon System: Research and Forecast*, Chap. 6, Eds. C.-P. Chang *et al.*, World Scientific, pp. 85–98.

Sakurai, N., F. Murata, M. D. Yamanaka, S. Mori, J.-I. Hamada, H. Hashiguchi, Y. I. Tauhid, T. Sribimawati, and B. Suhardi, 2005: Diurnal cycle of cloud system migration over Sumatera Island. *J. Meteorol. Soc. Japan*, **83**, 835–850.

Sakurai, N., S. Mori, M. Kawashima, Y. Fujiyoshi, J. Hamada, S. Shimizu, H. Fudeyasu, Y. Tabata, W. Harjupa, H. Hashiguchi, M. D. Yamanak, J. Matsumoto, Emrizal, and F. Syamsudin, 2011: Migration process and 3D wind field of precipitation systems associated with a diurnal cycle in west Sumatera: Dual Doppler radar analysis during the HARIMAU2006 campaign. *J. Meteorol. Soc. Japan*, **89**, 341–361.

Sato, T., H. Miura, M. Satoh, Y. N. Takayabu, and Y. Wang, 2009: Diurnal cycle of precipitation in the tropics simulated in a global cloud-resolving model. *J. Climate*, **22**, 4809–4826.

Siswanto, S., G. J., van Oldenborgh, G., van der Schrier, R. Jilderda, and B. van den Hurk, 2015: Temperature, extreme precipitation, and diurnal rainfall changes in the urbanized Jakarta city during the past 130 years. *Int. J. Climatol.*, **36**, 3207–3225. doi: 10.1002/joc.4548.

Sulistyowati, R., R. I. Hapsari, F. Syamsudin, S. Mori, S. T. Oishi, and M. D. Yamanaka, 2014: Rainfall-driven diurnal variations of water level in the Ciliwung River, West Jawa, Indonesia. *SOLA*, **10**, 141–144.

Tangang, F. T., L. Juneng, E. Salimun, P. N. Vinayachandran, Y. K. Seng, C. J. C. Reason, S. K. Behera, and T. Yasunari, 2008: On the roles of the northeast cold surge, the Borneo vortex, the Madden-Julian Oscillation, and the Indian Ocean Dipole during the extreme 2006/2007 flood in southern Peninsular Malaysia. *Geophys. Res. Lett.*, **35**, L14S07. doi:10.1029/2008GL033429.

Trilaksono, N. J., S. Otsuka, and S. Yoden, 2011: Dependence of model-simulated heavy rainfall on the horizontal resolution during the Jakarta flood event in January–February 2007. *SOLA*, **7**, 193–196.

Trilaksono, N. J., S. Otsuka, and S. Yoden, 2012: A time-lagged ensemble simulation on the modulation of precipitation over west Java in January–February 2007. *Mon. Wea. Rev.*, **140**, 601–616.

Villafuerte II, M.Q. and J. Matsumoto, 2015: Significant influences of global mean temperature and ENSO on extreme rainfall in Southeast Asia. *J. Climate*, **28**, 1905–1919.

Wang, B. (ed.), 2006: *Asian Monsoon.* Springer/Praxis Publishing Ltd., 679 pp.

Wang, B., J. Matsumoto, G. Wu, and J. Li, 2010: *The Science Plan for Asian Monsoon Years (AMY 2007–2012), A cross-cutting WCRP initiative.* China Meteorological Press, 67 pp. Available on line at http://www.wcrp-amy.org/about.asp? lmid=2&id=17.

Wheeler, M. and H. Hendon, 2004: An all-season real-time multivariate MJO index: Development of an index for monitoring and prediction. *Mon. Wea. Rev.*, **132**, 1917–1932.

Wu, P., M. Hara, H. Fudeyasu, M. D. Yamanaka, J. Matsumoto, F. Syamsudin, R. Sulistyowati, and Y. S. Djajadihardja, 2007: The impact of trans-equatorial monsoon flow on the formation of repeated torrential rains over Java Island. *SOLA*, **3**, 93–96.

Wu, P., M. D. Yamanaka, and J. Matsumoto, 2008: The formation of nocturnal rainfall offshore from convection over western Kalimantan (Borneo) Island. *J. Meteor. Soc. Japan*, **86A**, 187–203.

Wu, P., A. A. Arbain, S. Mori, J. Hamada, M. Hattori, F. Syamsudin, and M. D. Yamanaka, 2013: The effects of an active phase of the Madden-Julian Oscillation on the extreme heavy precipitation event over western Java Island in January 2013. *SOLA*, **9**, 79–83.

Yamanaka, M. D., 2016: Physical climatology of Indonesian maritime continent: An outline to comprehend observational studies. *Atmos. Res.*, **178–179**, 231–259.

Yamanaka, M. D., S. Mori, P.-M. Wu, J.-I. Hamada, N. Sakurai, H. Hashiguchi, M. K. Yamamoto, Y. Shibagaki, M. Kawashima, Y. Fujiyoshi, T. Shimomai, T. Manik, Erlansyah, W. Setiawan, B. Tejasukmana, F. Syamsudin, Y. S. Djajadihardia, and J. T. Anggadiredja, 2008: HARIMAU radar profiler network over Indonesian maritime continent: A GEOSS early achievement for hydrological cycle and disaster prevention. *J. Disaster Res.*, **3**, 78–88.

Yamazaki N., H. Kamahori, A. Yatagai, K. Takahashi, H. Ueda, K. Aonashi, K. Kuma, Y. Takeuchi, H. Tada, Y. Fukutomi, H. Igarashi, H. Fujinami, and Y. Kajikawa, 2000: On the release of GAME reanalysis products. *Tenki*, **47**, 659–663 (in Japanese).

Yasunari, T. (ed.), 2001: Special Issue, GEWEX Asian Monsoon Experiment (GAME). *J. Meteor. Soc. Japan*, **79–1B**, 605 pp.

Yatagai A., N. Yamazaki, H. Kamahori, K. Takahashi, H. Ueda, K. Aonashi, K. Kuma, Y. Takeuchi, and H. Tada, 2000: On the GAME reanalysis. *Journal of Japan Society of Hydrology and Water Resources* **13**, 486–495 (in Japanese).

Yatagai, A., N. Yamazaki, and T. Kurino, 2007: The products and validation of GAME reanalysis and JRA-25, part 1: Surface fluxes. *Hydrol. Process.* **21**, 2061–2071.

Yatagai, A., O. Arakawa, K. Kamiguchi, H. Kawamoto, M. I. Nodzu, and A. Hamada, 2009: A 44-year daily gridded precipitation dataset for Asia based on a dense network of rain gauges. *SOLA*, **5**, 137–140.

Yatagai, A., K. Kamiguchi, O. Arakawa, A. Hamada, N. Yasutomi, and A. Kitoh, 2012: APHRODITE: constructing a long-term daily gridded precipitation dataset for Asia based on a dense network of rain gauges. *Bull. Amer. Meteor. Soc.*, **93**, 1401–1415.

Yoden, S., S. Otsuka, N. J. Trilaksono, and T. W. Hadi, 2016: Recent progress and achievements in research on the Maritime Continent monsoon, *The Global Monsoon System Research and Forecast III, Chap. 6*, Eds. C.-P. Chang *et al.*, World Scientific.

Yoneyama, K., C. Zhang, and C. N. Long, 2013: Tracking pulses of the Madden-Julian oscillation. *Bull. Amer. Meteor. Soc.*, **94**, 1871–1891.

Zhang, C., 2013: Madden-Julian oscillation bridging weather and climate. *Bull. Amer. Meteor. Soc.*, **43**, 1849–1870.